MW00354415

Probability

Lawrence M. Leemis
Department of Mathematics
The College of William & Mary
Williamsburg, Virginia

Library of Congress Cataloging-in-Publication Data

Leemis, Lawrence M.
      Probability / Lawrence M. Leemis.
      Includes bibliographic references and index.
      ISBN 978-0-9829174-0-4
      1. Probability
      QA 273.L44 2011

The author and publisher of this book have used their best efforts in preparing this book. These efforts include the development, research, and testing of the mathematics and programs to determine their effectiveness. The author and publisher make no warranty of any kind, expressed or implied, with regard to the mathematics or programs or the documentation contained in this book. The author and publisher shall not be liable in any event for incidental or consequential damages in connection with, or arising out of, the furnishing, performance, or use of the mathematics or programs.

Printed in the United States of America

10 9 8 7 6 5 4 3 2

ISBN 978-0-9829174-0-4

For Jill, Lindsey, Mark, and Logan

Contents

Preface

This text provides a calculus-based introduction to probability for a student who has completed courses in multivariate calculus and linear algebra. The material is also appropriate as the first course in a mathematical statistics sequence of classes, and covers all of the topics associated with Exam P given by the Society of Actuaries and Exam 1 given by the Casualty Actuarial Society. There is a secondary emphasis on Monte Carlo simulation and symbolic algebra as tools to check the solutions to probability problems that are achieved by analytic methods.

The ordering of the material is consistent with most probability texts. I have chosen to delay the introduction of multivariate distributions as long as possible to allow students to get used to working with univariate random variables. When multivariate distributions are introduced in Chapter 6, significant time is spent with just two random variables at a time before moving to the n-variable case.

The Maple-based symbolic probability language APPL (A Probability Programming Language) has been used throughout the text. My thanks goes to Andy Glen, Diane Evans, Billy Kaczynski, and Jeff Yang for writing, maintaining, and extending the language. APPL can be downloaded for non-commercial use at `applsoftware.com`.

The R language has also been used throughout the text for graphics, computation, and Monte Carlo simulation. R can be downloaded at `r-project.org`. The vast majority of the R code also works in S-Plus.

The text is organized into chapters and sections. When there are several topics within a section, they are set off by boldface headings. Definitions and theorems are boxed; examples are indented; proofs are terminated with a box. Proofs are only included when they are instructive to the material being presented. Exercises are numbered sequentially at the end of each chapter. Computer code is set in monospace font, and is not punctuated. Indentation is used to indicate nesting in code and pseudocode. A solutions manual and teaching slides are available from the author.

Some care has been taken to differentiate between exact probabilities, often written as fractions, and their floating-point approximations to a few digits. A floating-point approximation can easily be compared to an intuitive initial guess at the solution to the problem.

The support from The College of William & Mary in the form of a QEP/Mellon grant to help with the preparation of the material that is contained in the text is gratefully acknowledged. My thanks also goes to Mr. Joseph J. Plumeri II for his support of the preparation of this text.

There have been many people who have caught errors along the way: Kevin Cummiskey removed dozens of typos in the first draft; Jim Hartman from The College of Wooster made hundreds helpful comments and edits in the second draft; Bruce Schmeiser from Purdue made dozens of suggestions for the first two chapters. The students in my Math 401 class, particularly typo czars Troy Bupp, Mike Joseph, Jasper Lu, Katie Moretti, Kim Mount, Allison Oldham, Benjamin O'Neil, Maria Pawlosky, Mike Schilling, and Amy Then, spotted hundreds of typos. Ross Iaci and Tanujit Dey from The College of William & Mary were helpful in guiding which material to include in the text. Tanujit Dey, Jason Owen from the University of Richmond, and Andy Glen from West Point were the first to class-test the completed text. My thanks goes to Lindsey Leemis for the cover design.

There are no references cited in the text for readability. The sources of materials in the various chapters are cited in the paragraphs below.

Chapter 1 notes: The exact statement of the "Monty Hall problem" was taken from the associated Wikipedia website. The ball bearing data set was first published in J. Lieblein and M. Zelen (1956), "Statistical Investigation of the Fatigue Life of Deep-Groove Ball Bearings," *Journal of Research of the National Bureau of Standards*, Volume 57, Number 5, pages 273–316. The data on the eruptions of the Old Faithful geyser came from A. Azzalini and A.W. Bowman (1990), "A Look at Some Data on Old Faithful Geyser," *Journal of the Royal Statistical Society, Series C*, Volume 39, pages 357–366. The cork deposit data can be obtained from D.J. Hand, F. Daly, A.D. Lunn, K.J. McConway and E. Ostrowski (1994), *Small Data Sets*, Chapman and Hall. The original data was published in Rao (1948), "Tests of Significance in Multivariate Analysis," *Biometrika*, Volume 35, pages 58–79. The U.S. House of representatives data was taken from page 73 of G. Will (1992), *Restoration: Congress, Term Limits and the Recovery of Deliberate Democracy*, Free Press, A Division of Macmillan, New York. The problem involving pulling billiard balls from a bag was inspired by a more general example on page 14 of G. Casella and R.L. Berger (2002), *Statistical Inference*, 2nd edition, Duxbury.

Chapter 2 notes: The Monte Carlo simulation program that simulates the game of craps was adapted from a C program written by Steve Park from The College of William & Mary.

Chapter 3 notes: The last Chebyshev's inequality example was adapted from an example in R.E. Hogg, J.W. McKean, and A.T. Craig (2005), *Introduction to Mathematical Statistics*, 6th edition, Prentice–Hall. The counter-intuitive result in Exercise 3.48 concerns what are known as "nontransitive" or "Efron" dice, named after their inventor, Brad Efron.

Chapter 4 notes: The horse kick data was taken from L. Bortkiewicz (1898), *Das Gesetz der Kleinen Zahlen*, Leipzig: Teubner. The Benford distribution was first recognized by what Simon Newcomb, but popularized as "Benford's Law" by F. Benford (1938), "The Law of Anomalous Numbers," *Proceedings of the American Philosophical Society*, Volume 78, pages 551–572.

Chapter 5 notes: The 3×3 matrix of plots associated with the beta distribution was suggested to me by Lee Schruben from the University of California at Berkeley. A similar 4×4 matrix appears on page 220 of N.L. Johnson, S. Kotz, and N. Balakrishnan (1995), *Continuous Univariate Distributions*, Volume 2, 2nd edition, John Wiley & Sons.

Chapter 6 notes: The theorem about factoring the joint probability density function is proved in R.E. Hogg, J.W. McKean, and A.T. Craig (2005), *Introduction to Mathematical Statistics*, 6th edition, Prentice–Hall. Don Campbell from the Economics Department at The College of William & Mary helped me determine economic measures for the five variable example. The bivariate distribution with the L-shaped support introduced in Exercise 6.10 was devised by Nick Loehr from Virginia Tech. The example illustrating the fact that that mutual independence does not imply pairwise independence is attributed to S. Bernstein and is given on page 120 in R.E. Hogg, J.W. McKean, and A.T. Craig (2005), *Introduction to Mathematical Statistics*, 6th edition, Prentice–Hall.

Chapter 7 notes: Example 7.19 was inspired by a similar example on page 298 in Wackerly, Mendenhall, and Scheaffer (2008), *Mathematical Statistics with Applications*, 7th edition, Thompson Brooks/Cole. The Box–Muller algorithm was introduced by G.E.P. Box and M.J. Muller (1958), "A Note on the Generation of Random Normal Deviates," *Annals of Mathematical Statistics*, Volume 29, pages 610–611.

Chapter 8 notes: The relationship between convergence in probability and convergence in distribution is proved in R.E. Hogg, J.W. McKean, and A.T. Craig (2005), *Introduction to Mathematical Statistics*, 6th edition, Prentice–Hall.

Williamsburg, VA

Larry Leemis
August 2011

Chapter 1

Introduction

I was standing under a large tree on a pleasant and still day several years ago. The stillness was interrupted by a large branch, about six inches in diameter, crashing to the ground about ten feet from where I was standing. I walked away knowing that I was just feet away from a major, random catastrophe. We live in a world of events that occur at random, and probability is a way of measuring and predicting this randomness. Consider the following examples:

- a reliability engineer contemplates the number of spare parts needed to support a fleet of mining trucks;

- an actuary analyzes data compiled in life tables to determine appropriate premiums for term life insurance policies;

- a medical doctor decides what action to take based on two tests—one with a positive result and one with a negative result;

- a college student scans the sky to help determine whether to take an umbrella to class;

- a toddler slowly reaches up toward the counter to grab a forbidden cookie, contemplating likely outcomes;

- a young mother sends her child off to kindergarten for the first time, wondering if the school has adequate evacuation plans if it is hit by a meteor.

Each of these people, in one fashion or another, at one level of sophistication or another, and at one level of rationality or another, is assessing probabilities. The notion of probability is very intuitive, and all of us make dozens of decisions every day based on probability assessments.

The purpose of this book is to hone your already-intuitive probability notions into a mathematical framework. In this way, when you are confronted with a complex problem involving probability, you will be able to confidently use this framework to craft a solution.

Although intuition typically works well when it comes to probability, it occasionally breaks down. The following two examples are probability questions whose solutions defy intuition.

Example 1.1 The "birthday problem" is usually stated along the following lines:

> If 40 people are gathered in a room, what is the probability that two or more people have the same birthday?

The years in which the 40 people were born are not considered in this problem. People typically guess too low when asked to estimate this probability. One of the more common guesses that I encounter is 40/365. The probability is actually about 0.89, so it is *very* likely that one or more birthdays will match in a room of 40 people. This problem will be solved using complementary probabilities and the multiplication rule in Chapter 2.

Example 1.2 The second problem is alternatively called the "car and goats" problem, the "Monty Hall" problem, or the "Let's Make a Deal" problem after the popular television game show.

> Suppose you're on a game show, and you're given the choice of three doors. A car is placed behind one door; goats are placed behind the other two doors. The car and the goats were placed randomly behind the doors before the show. The rules of the game show are as follows: After you have chosen a door, the door remains closed for the time being. The game show host, Monty Hall, who knows what is behind the doors, now has to open one of the two remaining doors, and the door he opens must have a goat behind it. If both remaining doors have goats behind them, he chooses one randomly. After Monty Hall opens a door with a goat, he will ask you to decide whether you want to stay with your first choice or to switch to the last remaining door. Imagine that you chose Door 1 and the host opens Door 3, which has a goat. He then asks you: Do you want to switch to Door Number 2? Is it to your advantage to change your choice?

The intuitive answer to this question is that there is no advantage to switching doors. The car is behind one door and the second goat is behind the other door, so the two results are equally likely. This problem was stated in a slightly different form in a letter to Marilyn vos Savant's *Ask Marilyn* column in *Parade* magazine in 1990. The solution, which states that changing doors doubles the probability of getting the car from $1/3$ to $2/3$, created a barrage of about 10,000 letters, nearly 1000 of which came from PhD's, stating that the solution was wrong. We will use the rule of Bayes in Chapter 2 to show that her solution was indeed correct.

The next section gives a sampling of a few more probability questions that will appear subsequently in the book, along with some pointers toward one of the most common applications of probability: the analysis of data using statistical techniques.

1.1 Applications

Probability is a branch of mathematics that describes experiments whose outcome can't be predicted with certainty prior to performing the experiment. It was first studied by Blaise Pascal and Pierre de Fermat in the 17th century when they applied probability to gambling games. Here is an example of such a gambling game.

Example 1.3 Toss a pair of dice 24 times. You win if you roll double sixes at least once. Find the probability of winning.

This problem can be easily solved using the tools provided in this book. There are certain assumptions that can be made, for example, the dice are fair and the rolls are independent. Once these assumptions are made, the axioms and results provided in Chapter 2 will yield a solution to this problem of

$$
\begin{aligned}
P(\text{winning}) &= 1 - P(\text{losing}) \\
&= 1 - P(\text{tossing no double sixes}) \\
&= 1 - \left(\frac{35}{36}\right)^{24} \\
&\cong 0.4914.
\end{aligned}
$$

The fact that the probability is close to 0.5 will draw gamblers to the game; the fact that the probability is slightly less than 0.5 assures that the house will draw revenue from the game in the long run.

The questions that can be addressed via probability techniques apply to many important applied fields beyond gambling games. Application areas include genetics, actuarial science, casualty insurance, meteorology, stock market analysis, economics, quality control, reliability, medicine, biostatistics, marketing, strength of materials, human factors, and sociology, just to name a few.

To illustrate the variety of problems that can be addressed using the tools of probability, two more simple examples are presented below.

Example 1.4 Find the probability of dealing a five-card poker hand containing a full house from a well-shuffled deck of playing cards.

Questions of this nature also require assumptions. For example, assume that we are playing with a full deck and that all of the possible shufflings are equally likely. Again, using the techniques from Chapter 2, the probability of dealing a full house (three cards having one rank and two cards having another rank) is

$$P(\text{full house}) = \frac{\binom{13}{1}\binom{12}{1}\binom{4}{3}\binom{4}{2}}{\binom{52}{5}} = \frac{13 \cdot 12 \cdot 4 \cdot 6}{2,598,960} = \frac{3744}{2,598,960} = \frac{6}{4165} \cong 0.00144.$$

The binomial coefficient $\binom{n}{r}$ will be defined in the next section. Since the probability of dealing a full house is just a bit over 1 in 1000, one can conclude that this will not occur often.

Leaving the realm of gambling games, we now switch from working with problems involving discrete outcomes to a problem involving a continuous outcome.

Example 1.5 Probability problems involving sums of random quantities often arise. In this example, let $X_1, X_2, \ldots, X_{10}$ be independent random variables that are uniformly distributed between 0 and 1. That is, each of the random variables assumes a continuous value between 0 and 1 with equal likelihood. Most calculators and computer programs have a *random number generator* capable of producing such numbers. Find the probability that their sum lies between 4 and 6, that is,

$$P\left(4 < \sum_{i=1}^{10} X_i < 6\right).$$

A well-known approximation, known as the central limit theorem (introduced in Chapter 8), yields only one digit of accuracy for this particular problem. Another approximation technique, known as Monte Carlo simulation, requires custom computer programming, and the result is typically stated as an interval around the true value. This problem can also be solved exactly using some of the techniques and software provided in this text, yielding

$$P\left(4 < \sum_{i=1}^{10} X_i < 6\right) = \frac{655177}{907200} \cong 0.7222.$$

Notwithstanding the obvious benefit of probability calculations to a gambler, a more significant application of probability theory lies in the field of *inferential statistics*, which has the goal of drawing inferences (conclusions) about the population from which a data set was drawn. The field of statistics was first studied as numerical data was collected on political units (for example, a census). This has eventually evolved into what is now known as "political science." The following examples illustrate the graphical and numerical analysis of a data set using standard statistical techniques. The letter n is used nearly universally to denote the *sample size*, which is the number of data values collected.

Example 1.6 (Ball bearing failure times) Consider the data set of $n = 23$ ball bearing failure times (measured in 10^6 revolutions):

$$
\begin{array}{cccccccc}
17.88 & 28.92 & 33.00 & 41.52 & 42.12 & 45.60 & 48.48 & 51.84 \\
51.96 & 54.12 & 55.56 & 67.80 & 68.64 & 68.64 & 68.88 & 84.12 \\
93.12 & 98.64 & 105.12 & 105.84 & 127.92 & 128.04 & 173.40.
\end{array}
$$

There are several things that one can do to analyze such a data set. Computing certain numerical measures that summarize a data set is common, particularly with large data sets. The two most commonly used sample statistics are the *sample mean* and the *sample variance*. Using the notation $x_1, x_2, \ldots, x_n$ to denote the data values, the formulas for the sample mean, $\bar{x}$, and sample variance, s^2, are

$$
\bar{x} = \frac{1}{n} \sum_{i=1}^{n} x_i = 72.22 \qquad \text{and} \qquad s^2 = \frac{1}{n-1} \sum_{i=1}^{n} (x_i - \bar{x})^2 = 1405.4.
$$

The sample mean is a measure of the central tendency of a data set; the sample variance is a measure of the dispersion of a data set. The sample mean has the same units as the data values. The positive square root of the sample variance, known as the *sample standard deviation*, $s = 37.49$ for this data set, also has the same units as the data values. These two quantities are *random* in the sense that a data set of $n = 23$ other ball bearings would produce different values for $\bar{x}$ and s^2. These quantities are to be distinguished from the *population mean*, μ, and the *population variance*, σ^2, which would be obtained if we sampled the entire population of ball bearings.

In addition to summarizing the data set with numerical values such as $\bar{x}$ and s, there are also some graphical procedures that can be applied to a data set. A *histogram* is useful for determining the shape of a probability distribution; it is the statistical analog of a function to be introduced in Chapter 3 known as a probability density function. A histogram for the ball bearing lifetimes is shown in Figure 1.1. The horizontal axis is the failure time and the vertical axis is the number of ball bearing failure times that fall in each of the cells of width 20. The histogram reveals a clumping of the data around 50 million revolutions and also reveals that the largest of the ball bearing failure times, 173.40 million revolutions, lies significantly to the right of the others. The data set appears to come from a population with a single mode (peak) near 50 million revolutions. Issues associated with a histogram include choosing the number of cells and cell boundaries, which are arbitrary decisions made by the data analyst. Unfortunately, histograms are not good graphical instruments for comparing two or more distributions.

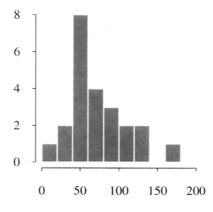

Figure 1.1: Histogram of ball bearing failure times.

Statistical packages are useful time-saving tools that can quickly perform numerical calculations and produce graphical displays associated with a data set. This text will use the statistical programming language R for such calculations and displays. R is available for free download on the web and is a powerful package that provides useful graphics, programming capability, and numerical calculations. It is becoming a standard that is used by statisticians. The R code to compute the sample mean and sample variance, and display the histogram for the ball bearing failure times is given below.

```
bearings = c(17.88, 28.92, 33.00, 41.52, 42.12, 45.60, 48.48, 51.84,
             51.96, 54.12, 55.56, 67.80, 68.64, 68.64, 68.88, 84.12,
             93.12, 98.64, 105.12, 105.84, 127.92, 128.04, 173.40)
mean(bearings)
var(bearings)
hist(bearings)
```

The ball bearing data illustrates what statisticians refer to as a *univariate* data set, since only a single variable has been collected on each ball bearing sampled. Many data sets involve collected *pairs* of data values, resulting in a *bivariate* data set, as illustrated by the following two examples.

Example 1.7 (Old Faithful Geyser eruptions) A data set of $n = 299$ data pairs, (x_i, y_i), for $i = 1, 2, \ldots, 299$, was collected on the waiting time x_i and the eruption duration y_i at the Old Faithful geyser in Yellowstone National Park in Wyoming. All observations are recorded in minutes. The data pairs are plotted in Figure 1.2. The data set exhibits some rather unique characteristics. First, there is an unusual clumping of the eruption duration around 4 minutes, which could be a natural phenomenon or could be due to rounding by those who collected the data. Second, there appears to be a tri-modal joint distribution of the (x_i, y_i) pairs. Look carefully at the scatterplot in Figure 1.2 to see if you can spot the three modes.

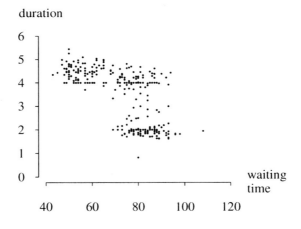

Figure 1.2: Geyser data.

Scatterplots like this are easily generated in R. The R command

```
plot(faithful$waiting, faithful$eruptions)
```

generates a scatterplot for a similar data set that is pre-loaded into R. The first argument to the `plot` function, the vector `faithful$waiting`, contains the waiting times for plotting on the horizontal axis; the second argument, the vector `faithful$eruptions`, contains the associated eruption times for plotting on the vertical axis.

Example 1.8 (Automobile warranty claims) As a second example of a bivariate data set, consider one particular make and model of an automobile that has a warranty that expires after 3 years or 36,000 miles, whichever occurs first. Warranty claim times (measured in both mileage and age in years) for a bivariate sample of size $n = 260$ are plotted in Figure 1.3.

age (years)

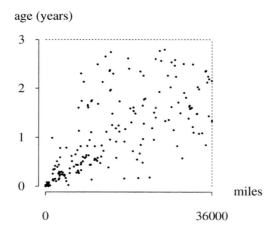

Figure 1.3: Scatterplot of warranty claim times.

The scatterplot of mileage and age reveals a significantly different pattern from the geyser data. First, we know the boundaries of the support, or the allowable values for the data pairs: they must fall in the rectangle with opposite corners at $(0, 0)$ and $(36000, 3)$. Second, there is a clustering of the data near the origin that corresponds to cars with problems that appear soon after being purchased. Third, there is what statisticians refer to as a *positive sample correlation* between the mileage and the age of an automobile that is taken to the dealer for a warranty claim. The two measures that reflect the aging of the automobile tend to increase together, resulting in (x_i, y_i) pairs that tend to be on the same sides of their sample means $\bar{x}$ and $\bar{y}$ more often than not. Fourth, if the histograms of the x_i and y_i values are plotted separately, as in Figure 1.4, one can clearly

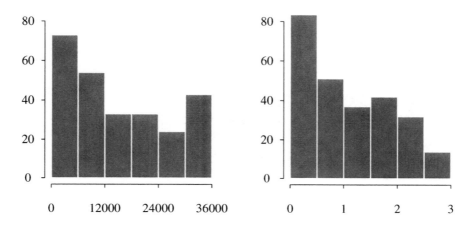

Figure 1.4: Histograms of warranty claim times (mileage on left; years on right).

see the early failures on both histograms. The histogram on the left (mileage), however, also has a mode near 36,000 miles that corresponds to drivers being aware of a warranty expiring after 36,000 miles. The histogram on the right (age in years) does not have a second mode near 3 years because drivers are less aware of the 3-year anniversary of their purchase. In this case, the two histograms confirm our intuition about automobile owners and expiring warranties. The notion of the distribution of one of two variables (ignoring the other variable) will be developed in Chapter 6 as a *marginal* distribution.

The previous two examples have presented *bivariate* data sets for the Old Faithful geyser data and the car warranty claim data. When there are more than two data values collected, histograms and scatterplots can become problematic. A *boxplot* is a more compact way of looking at the shape of the distribution, as illustrated in the following example, where four data values are collected on each observational unit.

Example 1.9 (Cork deposits) The weights of cork deposits (in centigrams) of $n = 28$ trees is collected in the four directions: north, east, south, and west. The data is given in Table 1.1.

N	E	S	W
72	66	76	77
60	53	66	63
56	57	64	58
41	29	36	38
32	32	35	36
30	35	34	26
39	39	31	27
42	43	31	25
37	40	31	25
33	29	27	36
32	30	34	28
63	45	74	63
54	46	60	52
47	51	52	43
91	79	100	75
56	68	47	50
79	65	70	61
81	80	68	58
78	55	67	60
46	38	37	38
39	35	34	37
32	30	30	32
60	50	67	54
35	37	48	39
39	36	39	31
50	34	37	40
43	37	39	50
48	54	57	43

Table 1.1: Cork deposit weights.

A casual inspection of this data set reveals a positive correlation among the cork deposits in the four directions. Check out the tree with the spectacular 100 centigrams of cork deposits on its south side. All four of its cork deposit weights, that is $(91, 79, 100, 75)$, are either the heaviest or second heaviest in their respective directions. In other words, when one of the four directions tends to have heavier cork deposits, the other directions are more likely to also have heavier cork deposits. The notion of positive correlation between the deposits captured in the four directions can be captured in a 4×4 *correlation matrix*, which for this data set is

$$\begin{bmatrix} 1.00 & 0.89 & 0.90 & 0.88 \\ 0.89 & 1.00 & 0.83 & 0.77 \\ 0.90 & 0.83 & 1.00 & 0.92 \\ 0.88 & 0.77 & 0.92 & 1.00 \end{bmatrix}.$$

The rows and columns of this matrix can be thought of as N, E, S, and W, rather than the usual 1, 2, 3, and 4. This is a *symmetric* matrix, so values on the opposite sides of the diagonal are equal. The diagonal elements are all 1.0 and imply that there is perfect positive correlation between the data in each direction and itself. The off-diagonal elements are the correlations between the different directions. Consider the $(1, 2)$, or more exactly, the (N, E) element of the matrix, which has the value 0.89. This value indicates that there is a high positive correlation (all correlations must fall between -1 and 1) between the weights of the cork deposits in the north and east directions. In other words, the deposits in the north and east directions tend to be on the same sides of their means together.

Another question that comes to mind is how the cork deposits in each direction vary individually. A histogram is not a good graphical device for comparing four distributions, but a *boxplot* can be used to compare several distributions simultaneously. The four boxplots displayed side-by-side in Figure 1.5 capture the essence of the four distributions for comparison. For each of the four directions, the middle half of the distribution is displayed vertically in a box. In other words, the top and the bottom of the box are the estimates of what is known as the *25th percentile* and the *75th percentile* of the probability distribution. Thus, the *quartiles* of a distribution are apparent from a boxplot. The horizontal line in the middle of the distribution shows the sample *median* or the middle value of the data set. This is an estimate of the *50th percentile* of the distribution. For a data set with an odd sample size n, this is just the middle value of the sorted data values. For an even sample size n, the two middle values are averaged. The fact that the median tends to fall consistently in the lower half of the box in all four directions indicates that the distribution of cork deposits is a *non-symmetric* distribution. The shortest box is associated with the weights of the deposits taken from the eastern side of the tree and the tallest box is associated with weights of the deposits taken from the southern side of

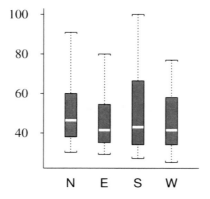

Figure 1.5: Boxplot of cork deposits in the four directions.

the tree. This implies that the variability of the weights is greater on the southern side of the tree, even though the medians appear to be nearly identical. Finally, the *whiskers* in a boxplot extend to the smallest and largest data values, although this convention is not universal. The upward whisker in the southern direction extends to 100.

The R code to read the data from an external data set into an array named `cork`, calculate the correlation matrix, and generate the boxplot is given below.

```
cork = read.table("cork.dat")
cor(cork)
boxplot(cork, names = c("N", "E", "S", "W"))
```

Statisticians often encounter problems where a function needs to be fitted to a data set. The simplest function to fit is a line. The next example illustrates a line being fit to data.

Example 1.10 (U.S. House of Representatives turnover) The average turnover percentages for the 12 decades following the end of the Civil War (these percentages are the averages of the five turnover percentages for elections held during the decade) are given in Table 1.2.

Decade	Mean turnover percentage
1870	50.3
1880	40.6
1890	40.8
1900	23.9
1910	28.5
1920	21.1
1930	26.3
1940	22.2
1950	14.6
1960	15.2
1970	16.6
1980	12.9

Table 1.2: U.S. House of Representatives turnover percentages.

It is clear that there is a downward trend in the data over time. It appears to be increasingly difficult to vote incumbent politicians out of office. It is impossible to fit a single line that will pass through all of these data values simultaneously, so we attempt to find the best line possible. One criterion for determining this best line is to find the *least squares* line that minimizes the sum of the squared vertical deviations between the line and the data points. Thus, the model for what is known as a *simple linear regression* is

$$Y = a + bX,$$

where X is the decade and Y is the turnover percentage in this particular setting. The slope and intercept of the regression line can be calculated by hand or by using any standard statistical package. Using the integers $1, 2, \ldots, 12$ to denote the $n = 12$ decades, the slope and intercept of the regression line are $b = -3.03$ and $a = 45.78$. The interpretation of slope is that the turnover percentage is decreasing by about 3% per decade on average. Two possible explanations for this decrease are the power of incumbency and the rise of the "career politician." A scatterplot of the data and the associated regression line is shown in Figure 1.6. The lengths of the vertical distances between the data points and the regression line are known as the *residuals*.

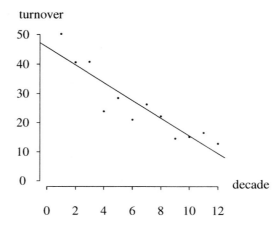

Figure 1.6: U.S. House of Representatives turnover percentages.

The R code to generate this figure is

```
decade   = 1:12
turnover = c(50.3, 40.6, 40.8, 23.9, 28.5, 21.1,
             26.3, 22.2, 14.6, 15.2, 16.6, 12.9)
plot(decade, turnover)
reg = glm(decade, turnover)
abline(reg$coef)
```

The first statement sets the R vector decade to the first 12 positive integers. The second statement assigns the 12 turnover percentages to the vector turnover. The 12 data pairs are then plotted using the plot statement. The glm function (for general linear model) performs the linear regression, and the least squares line is plotted using abline after extracting the regression coefficients (that is, a and b).

Before introducing probability in the next chapter, two important mathematical tools will be introduced that are often used in solving probability problems. These tools are *counting techniques* and *set theory*.

1.2 Counting

In many problems that arise in probability and statistics, it is useful to list (enumerate) or count the number of outcomes of an experiment. Counting is easy when there are only a handful of outcomes to count; when there are thousands or millions of outcomes, a more systematic approach is required.

Enumeration involves listing all of the possible outcomes to an experiment. Tree diagrams can be helpful, as will be seen in the following example.

> **Example 1.11** Imagine the unimaginable: The Chicago Cubs and the Chicago White Sox are playing in the World Series. The best-of-seven series is tied at two games apiece. What are the possible outcomes to the series?

The question is not asking for the *number* of possible outcomes, but rather a list of the possible outcomes. This will be accomplished using a *tree diagram* illustrated in Figure 1.7. The tree diagram reveals six possible endings to this particular World Series,

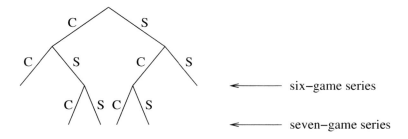

Figure 1.7: World Series outcomes.

two of which correspond to a six-game series and four of which correspond to a seven-game series, where C denotes a Cub victory and S denotes a Sox victory.

We now turn to *counting techniques* (or combinatorics or combinatorial methods), which are used when enumerating is cumbersome or infeasible. The field originated with a thirteenth-century Catalan missionary named Ramon Llull. We'll consider the following three techniques and some variations.

1. The multiplication rule.

2. Permutations (which is a special case of the multiplication rule).

3. Combinations.

Multiplication rule

We begin the discussion of counting techniques with the *multiplication rule*. The multiplication rule is also known as the fundamental theorem of counting, the basic principle of counting, the counting rule for compound events, and the rule for the multiplication of choices.

Theorem 1.1 (Multiplication rule) Assume that there are r decisions to be made. If there are n_1 ways to make decision 1, n_2 ways to make decision 2, ..., n_r ways to make decision r, then there are $n_1 n_2 \ldots n_r$ ways to make all decisions.

$$\underbrace{n_1} \ \underbrace{n_2} \ \underbrace{n_3} \ \ldots \ \underbrace{n_r}$$

Proof To show why the multiplication rule holds, consider the case of $r = 2$ decisions. In this case, all of the potential outcomes can be displayed in the $n_1 \times n_2$ matrix given below. The rows represent the choices associated with decision 1; the columns represent the choices associated with decision 2.

	Choice 1	Choice 2	$\cdots$	Choice n_2
Choice 1				
Choice 2				
$\vdots$				
Choice n_1				

Thus, for $r = 2$, there are $n_1 \times n_2$ ways to make both decisions. To proceed from $r = 2$ to $r = 3$ decisions results in a rectangular solid consisting of $n_1 \times n_2 \times n_3$ cubes associated with the various ways of making the $r = 3$ decisions as illustrated in Figure 1.8.

The more general result for r decisions continues to follow this pattern, and is proved by induction. Assume that the theorem holds for r decisions. To show that this implies that

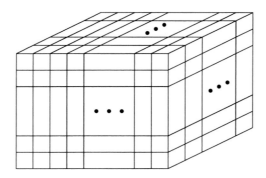

Figure 1.8: Multiplication rule justification for $r = 3$.

the theorem holds for $r + 1$ decisions, each of the $n_1 n_2 \ldots n_r$ ways to make the r decisions is matched with one of the n_{r+1} ways to make decision $r + 1$, resulting in $n_1 n_2 \ldots n_{r+1}$ total ways to make the $r + 1$ decisions, thus proving the theorem by induction.   □

Theorem 1.1 is applied to two simple counting problems involving coins and dogs. In each example, the number of choices for each decision is constant, that is, $n_1 = n_2 = \cdots = n_r$.

Example 1.12 How many different sequences of heads and tails are possible in 16 tosses of a fair coin?

Each of the 16 tosses can be considered a "decision" in terms of Theorem 1.1 with two possible outcomes: heads and tails. Since there are $r = 16$ tosses, there are

$$\underbrace{2 \cdot 2 \cdot 2 \cdot \ldots \cdot 2}_{16} = 2^{16} = 65,536$$

n_i are all identical. b/c you have same amount of decisions per each position.

different sequences.

This problem is easily generalized to similar settings. Two such settings are

- the number of ways to answer a 16-question T/F test,
- the number of integers that can be stored on a 16-bit computer.

Example 1.13 How many ways can a mother give away 8 dogs to her 3 children?

In this example, Mom has $r = 8$ decisions on her hands, one for each dog. Furthermore, each decision can be made in three ways. Thus, there are

$$\underbrace{\overbrace{3}^{Fido} \cdot \overbrace{3}^{Inky} \cdot \overbrace{3}^{Suzy} \cdot \ldots \cdot \overbrace{3}^{Spot}}_{8} = 3^8 = 6,561$$

different ways for her to give away her 8 dogs to her 3 children.

As illustrated in the next two examples, occasions arise when the n_i values in the multiplication rule are not all identical.

Example 1.14 How many ways can a family of 5 line up for a photograph?

The photographer has $r = 5$ decisions to make. The first decision is whom to place on the left (which can be done five ways); the second decision is whom to place next to the person on the left (which can be done four ways), etc. Thus, there are

$$5 \cdot 4 \cdot 3 \cdot 2 \cdot 1 = 5! = 120$$

different ways to line up the family for the photograph.

Example 1.15 How many ways can a family of 5 that consists of 3 men and 2 women line up for a photograph so that men and women alternate?

There are again $r = 5$ decisions for the photographer, but the decisions are restricted by gender, so we expect a lower count of outcomes than in the previous example. There is a choice of one of the 3 men to place on the left, a choice of 2 women to place next to him, etc. Continuing in this fashion yields a total of

$$\overset{M}{3} \cdot \overset{W}{2} \cdot \overset{M}{2} \cdot \overset{W}{1} \cdot \overset{M}{1} = \boxed{12}$$

men and women alternate indicates men start on the left.

different ways to line up the family members in the restricted fashion.

Foreshadowing the introduction of probability in Chapter 2, the previous two examples will be used to determine the probability that men and women alternate when the family of five sits for the photograph in a random order. Using the previous two examples, the desired probability is $P(\text{men and women alternate}) = 12/120 = 1/10$ assuming that all of the 120 possible orderings are equally likely. More details concerning the calculation of probabilities are given in Chapter 2.

Example 1.16 How many ways are there to arrange the letters in "dynamite"?

There are $r = 8$ decisions to be made, so there are

$$8 \cdot 7 \cdot 6 \cdot \ldots \cdot 2 \cdot 1 = 8! = 40,320$$

↳ all different letters so don't need non-discreet permutation.

different ways to arrange the letters.

Example 1.17 How many ways can 3 men, 4 women, and 2 children arrange themselves in a row of nine chairs if

(a) the children insist on sitting together?

(b) the children insist on sitting on the leftmost and rightmost chairs?

(c) the men, women, and children must sit next to one another?

If there were no restrictions, there would be $9! = 362,880$ different ways to line up the $r = 9$ people. Since each part of this question places a restriction on the ordering, the number arrangements must be less than $362,880$.

(a) When the children insist on sitting together, classify the people as 2 children and 7 adults. There are 2! ways to arrange the children. Since the leftmost child can occupy any one of 8 positions and there are 7! ways to arrange the adults, there are a total of

$$2! \cdot 8 \cdot 7! = 2 \cdot 8 \cdot 5040 = 80,640$$

different arrangements with the children sitting together.

(b) When the children insist on sitting on the leftmost and rightmost chairs, there are 2! ways to arrange the children at the extremes and 7! ways to arrange the adults in between them, so there are a total of

$$2! \cdot 7! = 2 \cdot 5040 = 10,080$$

different arrangements with the children sitting at the extremes.

(c) If the men sit in the three leftmost chairs, the women sit in the next four chairs, and the children sit in the two rightmost chairs, there are $3! \cdot 4! \cdot 2!$ arrangements. Since there are 3! arrangements of the men, women, and children as groups, there are

$$3! \cdot 3! \cdot 4! \cdot 2! = 6 \cdot 6 \cdot 24 \cdot 2 = 1728$$

different arrangements with the men, women, and children sitting next to one another.

Example 1.18 Cindy shuffles a deck of playing cards. Is it likely that she is the first person in history to achieve this particular ordering of the cards?

This is another of those problems that defies intuition. Of all of the people in history, almost surely *someone* must have attained the same shuffle as Cindy. By the multiplication rule, there are

$$52! = 80658175170943878571660636856403766975289505440883277824000000000000$$

different shufflings. Yikes! Perhaps Cindy's shuffle is likely unique after all. To address the likelihood of her shuffle being unique, some back-of-the-envelope calculations are required. The world population is about seven billion people. Approximately half of the people that have ever lived are currently alive, so assume that 14 billion people have lived through the ages. Now assume that everyone lives 100 years on average (dubious), and shuffles a deck of cards ten times a day on average (even more dubious), then there have been a total of a mere

$$14000000000 \cdot 100 \cdot 365 \cdot 10 = 5110000000000000$$

total shuffles. Hence Cindy's shuffle is almost certainly unique. Every shuffle of a deck of cards is almost always making playing-card history.

Although simple to state and use, the multiplication rule is a surprisingly versatile tool for addressing counting (combinatorics) problems. There is a special case of the multiplication rule that arises so often that it gets special treatment here. The object of interest is known as a *permutation*.

Permutations

The notion of whether a sample is taken with or without replacement is a critical notion in combinatorics and probability. When a sample of size r, for example, is selected at random and *with replacement* from a set of n distinct objects, there are n^r different ordered samples that can be taken. On the other hand, when the items are selected *without replacement*, the ordered items that are selected are a *permutation*.

> **Definition 1.1** A *permutation* is an ordered arrangement of r objects selected from a set of n objects without replacement.

order is important implies $(a,b) \neq (b,a)$ Thus they are

One key question to be addressed in a counting problem is whether the ordering of the objects is *both* relevant. If the ordering is relevant, then using permutations might be appropriate. *likely events.*

Example 1.19 List the permutations from the set $\{a, b, c\}$ selected 2 at a time.

Applying Definition 1.1 with $n = 3$ and $r = 2$ yields the 6 ordered pairs:

$$(a, b) \quad (b, a)$$
$$(a, c) \quad (c, a)$$
$$(b, c) \quad (c, b).$$

The second column of permutations is the same as the first column in reverse order.

> **Theorem 1.2** The number of permutations of n distinct objects selected r at a time without replacement is
> $$n \cdot (n-1) \cdot (n-2) \cdot \ldots \cdot (n-r+1) = \frac{n!}{(n-r)!}$$
> for $r = 0, 1, 2, \ldots, n$ and n is a positive integer, and $0! = 1$.

Proof Consider the following two cases based on the value of r. Case I: When $r = 0$, there is only one way to choose the sample (don't select any items), and

$$\frac{n!}{(n-0)!} = 1.$$

Case II: When $r = 1, 2, \ldots, n$, the multiplication rule with r decisions yields

$$\underbrace{n \cdot (n-1) \cdot (n-2) \cdot \ldots \cdot (n-r+1)}_{r \text{ factors}} = \frac{n!}{(n-r)!}$$

which establishes the result. $\qquad\square$

The proof shows that finding the number of permutations is just a special case of the multiplication rule.

Example 1.20 How many ways are there to pick a president, vice-president, and treasurer from 7 people?

The "objects" from the previous definition are the $n = 7$ candidates, and there are $r = 3$ of them being selected. The fact that sampling is performed without replacement is implicit in the problem statement in that one person could not occupy all three positions. Furthermore, the fact that order is relevant in their selection is implicit in the problem statement in that they are given distinct titles (president, vice-president, and treasurer). Thus, there are

$$\frac{7!}{(7-3)!} = \frac{7!}{4!} = 7 \cdot 6 \cdot 5 = 210$$

permutations associated with filling the three positions. The $7 \cdot 6 \cdot 5$ part of the equation serves as a reminder that this question could have been addressed directly by the multiplication rule.

Example 1.21 A ship has 3 stands and 12 flags to send signals. How many 3-flag signals can be sent?

Again, implicit in the statement of the problem is the fact that the 12 flags are distinct. Furthermore, to send a 3-flag signal requires sampling without replacement from the 12 flags. Proceeding with $n = 12$ and $r = 3$, there are

$$\frac{12!}{(12-3)!} = \frac{12!}{9!} = 12 \cdot 11 \cdot 10 = 1320$$

different signals.

Example 1.22 In the previous example, what if one or two flags also constitute a signal?

In this setting, the number of signals should simply be summed. Proceeding with $n = 12$ and $r = 1, 2, 3$, there are

$$\frac{12!}{11!} + \frac{12!}{10!} + \frac{12!}{9!} = 12 + 132 + 1320 = 1464$$

different signals that can be sent.

Example 1.23 How many ways are there to drive, in sequence, to four cities from a starting location?

Assuming that the starting location differs from the four cities, there are $n = 4$ objects (the cities) and all $r = 4$ of them must be selected. Thus, there are

$$\frac{4!}{(4-4)!} = \frac{24}{1} = 4 \cdot 3 \cdot 2 \cdot 1 = 24$$

different ways to drive to the four cities.

There are two minor tweaks that can be performed on permutations that are often useful in solving combinatorics problems: circular permutations and nondistinct permutations. Circular permutations consider the placement of n objects in a circle.

Theorem 1.3 The number of permutations of n distinct objects arranged in a circle is $(n-1)!$

Proof Fix one object's position and use the multiplication rule with $n - 1$ decisions to conclude that there are

$$(n-1) \cdot (n-2) \cdot \ \ldots \ \cdot 1 = (n-1)!$$

different ways to arrange the objects in a circle.   □

Example 1.24 How many ways are there to seat 6 people around a round table for dinner?

There are $n = 6$ objects (the diners!) to place around the table. There are

$$5! = 120$$

ways to order them around the table.

Care should be taken when interpreting the solution to the dinner table question. The only thing that matters in the particular seating is who is on your left and who is on your right when considering a circular permutation. For example,

- What if all diners shift one chair clockwise? This would *not* be a new circular permutation.
- What if one seat is a blue throne and it matters who is sitting in the blue throne? In this case a clockwise shift does result in a new circular permutation, so there are $6! = 720$ circular permutations in the blue throne setting.
- What if the order of seating is reversed (clockwise vs. counterclockwise)? This is indeed a new ordering. The diners on the left and right of each person have been interchanged.

Example 1.25 How many circuits can a traveling salesman make of n cities? A reverse route is not considered a unique path.

Assuming that the traveling salesman is beginning at one of the cities, there are

$$\frac{(n-1)!}{2} = \frac{n!}{2n}$$

different routes. Dividing by two prevents double counting reverse circuits. The spectacular factorial growth in this quantity is shown in Table 1.3

n	3	4	8	10	15	50
$n!/(2n)$	1	3	2520	181,440	43,589,145,600	$\sim 10^{62}$

Table 1.3: The number of traveling salesman routes.

Even if you are going around town to run just 8 errands, you have plenty of options. If you need to run 15 errands, for example, there are over 43 billion routes. Finding the shortest of all of these routes is known as the *traveling salesman problem*, which is a classic optimization problem in a field known as *operations research*. The problem is faced daily by package delivery companies. It is particularly difficult to solve because of the factorial growth in the number of routes.

The second tweak of a permutation is known as a nondistinct permutation. In a typical permutation counting problem, all of the n objects are distinct. We now consider the possibility of just r distinct types of objects.

Theorem 1.4 The number of nondistinct permutations of n objects of which n_1 are of the first type, n_2 are of the second type, ..., n_r are of the rth type, is

$$\frac{n!}{n_1! n_2! \dots n_r!}$$

where $n_1 + n_2 + \cdots + n_r = n$.

Proof Let A be the number of nondistinct permutations. We want to show that

$$A = \frac{n!}{n_1! n_2! \dots n_r!}.$$

If all n objects were distinct, the number of permutations is $n!$ or, by the multiplication rule, the number of permutations is $n_1! n_2! \dots n_r! A$. Equating and solving for A yields the desired result. $\square$

Example 1.26 Consider the case of

$$n = 9; r = 3, n_1 = 4, n_2 = 2, n_3 = 3.$$

The number of ways to order the objects

$$a_1 a_2 a_3 a_4 b_1 b_2 c_1 c_2 c_3$$

when the a, b, and c objects can't be distinguished from one another by their subscripts is

$$\frac{9!}{4!3!2!} = \frac{362,880}{24 \cdot 6 \cdot 2} = 1260.$$

Example 1.27 How many ways are there to arrange the letters in the word "door"?

There are $n = 4$ objects (the letters) and $r = 3$ of them are distinct. Thus, there are

$$\frac{4!}{1!2!1!} = \frac{24}{2} = 12$$

different arrangements. The 2! in the denominator accounts for swapping the two indistinguishable o letters.

Example 1.28 How many ways are there to arrange the letters in "puppet"?

Proceeding with the $n = 6$ objects with $r = 4$ distinct letters (p, u, e, t), there are

$$\frac{6!}{3!1!1!1!} = \frac{720}{6} = 120$$

different arrangements.

Example 1.29 How many ways are there to arrange the letters in "wholesome"?

In the word wholesome, there are $n = 9$ letters, $r = 7$ of which are distinct, leading to

$$\frac{9!}{2!2!} = \frac{362,880}{4} = 90,720$$

different arrangements.

Example 1.30 How many ways are there to line up identical twins and identical triplets for a photo if identical-looking people are nondistinct?

There are $n = 5$ people to line up with just $r = 2$ distinct looks. Since there are $n_1 = 2$ twins and $n_2 = 3$ triplets, there are

$$\frac{5!}{2!3!} = \frac{120}{12} = 10$$

different ways to line up the five.

This ends the discussion of permutations and two spinoffs (circular permutations and nondistinct permutations). We now switch to a discussion of *combinations*, which are closely related to permutations.

Combinations

In some situations, we are interested in the number of ways of selecting r objects without considering the *order* that they are selected (for example, a poker hand). These are called *combinations* and are a special case of nondistinct permutations when there are two types of objects.

Definition 1.2 A set of r objects taken from a set of n objects without replacement is a *combination*.

Example 1.31 List the combinations of 2 elements taken from $\{a, b, c, d\}$.

Since the order is not relevant, there will be fewer combinations than permutations:

$$\{a, b\}, \{a, c\}, \{a, d\}, \{b, c\}, \{b, d\}, \{c, d\}. \qquad \binom{4}{2} = \frac{24}{4} = 6$$

The number of combinations of r items selected from n items arises so often in combinatorics and probability that it gets its own symbol, as will be seen in the following theorem. The expression

$$\binom{n}{r}$$

is read "n choose r."

Theorem 1.5 The number of combinations of r objects taken without replacement from n distinct objects is

$$\binom{n}{r} = \frac{n!}{(n-r)!r!}.$$

Proof This result can be proved in two different fashions. First,

$$\binom{n}{r} = \frac{n!}{(n-r)!r!},$$

which is just the formula for the number of permutations divided by $r!$ to account for the number of ways to order the r objects selected.

A second way to prove the theorem is to think of the two groups of objects (those selected and those not selected) as two different types of indistinguishable items. Using the result concerning the number of nondistinct permutations,

$$\binom{n}{r} = \frac{n!}{(n-r)!r!}. \qquad \square$$

We now illustrate the application of Theorem 1.5.

Example 1.32 How many ways are there to pick a *committee* of three people from seven "volunteers"?

This question differs fundamentally from the earlier example involving the selection of a president, vice-president, and treasurer from a group of seven people. The fact that titles were being assigned to the three people selected meant that order was important. In this case, the committee selection process makes no implication with respect to the order that the members are selected, so using combinations is appropriate. Since $r = 3$ people are being selected from the larger group of $n = 7$ people, there are

$$\binom{7}{3} = \frac{7!}{4!3!} = \frac{7 \cdot 6 \cdot 5}{3 \cdot 2 \cdot 1} = 35$$

different committees that can be formed.

Example 1.33 How many ways can a five-card hand be dealt from a standard deck of playing cards?

There are $r = 5$ cards to be selected from the $n = 52$ cards in the deck. Since the order in which the cards are dealt is not relevant, combinations should be used to solve the problem. Using Theorem 1.5, there are

$$\binom{52}{5} = \frac{52!}{47!5!} = \frac{52 \cdot 51 \cdot 50 \cdot 49 \cdot 48}{5 \cdot 4 \cdot 3 \cdot 2 \cdot 1} = 2,598,960$$

different five-card hands that can be dealt.

Example 1.34 How many ways are there to answer a 10-question true/false test with exactly two true answers? *order not important*

One way of thinking about this problem is to consider all of the $2^{10} = 1024$ different sequences of 10 answers: T (for true) and F (for false). Consider dealing out the two *positions* for the T responses. Table 1.4 lists the responses associated with exactly two T responses, along with the positions of the T responses. The *order* that the two positions are selected is not relevant (for example, choosing positions 1 and 2 for the T responses is identical to choosing positions 2 and 1 for the T responses).

Proceeding using the selection of $r = 2$ positions for the T responses of the $n = 10$ positions means that there are

$$\binom{10}{2} = \frac{10!}{8!2!} = \frac{10 \cdot 9}{2 \cdot 1} = 45$$

different ways of answering the true/false exam with exactly two true responses.

Answers	Positions of the T responses
T T F F F F F F F F	1, 2
T F T F F F F F F F	1, 3
T F F T F F F F F F	1, 4
$\vdots$	$\vdots$
F F F F F F F F T T	9, 10

Table 1.4: Ten question true/false test responses.

Example 1.35 A ship has 3 stands and 12 flags to send signals. How many signals can be sent if one, two, or three flags constitute a signal and the stand(s) selected are relevant?

The problem implies that a red flag in stand 1 and a blue flag in stand 2 constitute a different signal than a red flag in stand 2 and a blue flag in stand 3. The solution to this problem requires the use of both combinations and permutations. Combinations are used to pick the stands, then permutations are used to place the flags in order in those stands. There are   *Order not important for the stands. Order important for the flags.*

$$\binom{3}{1}\frac{12!}{11!} + \binom{3}{2}\frac{12!}{10!} + \binom{3}{3}\frac{12!}{9!} = 36 + 396 + 1320 = 1752$$

different signals that can be sent. The three quantities being added in the solution correspond to one-flag, two-flag, and three-flag signals.

Example 1.36 How many ways can 14 people split into two teams of seven for a game of ultimate frisbee?

Since the question concerns *teams*, the order of selection is not relevant, which implies that combinations should be used here. There are

$$\frac{\binom{14}{7}}{2} = 1716$$

different ways to split the 14 people into two teams of seven. Division by two avoids double counting identical teams.

These examples illustrate the wide variety of problems that can be addressed using combinations. Combinations also have a number of interesting mathematical properties which will be given in an outline format below.

1. The well-known *binomial theorem* can be used to expand quantities such as

$$(x+y)^4 = 1x^4 + 4x^3y + 6x^2y^2 + 4xy^3 + 1y^4.$$

   The coefficients in the expansion (namely $1, 4, 6, 4, 1$ in this case) happen to correspond to the number of combinations. For this reason $\binom{n}{r}$ is often referred to as a "binomial coefficient." The general statement of the binomial theorem is

$$(x+y)^n = \sum_{r=0}^{n} \binom{n}{r} x^{n-r} y^r.$$

2. There are several miscellaneous results that are associated with the binomial coefficients. Here are a few such results, stated without proof.

(a) Symmetry: $\binom{n}{r} = \binom{n}{n-r}$, for $r = 0, 1, \ldots, n$; and n is a positive integer

(b) $\binom{n}{r} = \binom{n-1}{r} + \binom{n-1}{r-1}$

(c) $\sum_{r=0}^{k} \binom{m}{r}\binom{n}{k-r} = \binom{m+n}{k}$

3. The binomial coefficient $\binom{n}{r}$ is defined to be 0 when $r < 0$ or $r > n$.

4. Pascal's triangle, which is given by

$$
\begin{array}{ccccccccc}
& & & & 1 & & & & \\
& & & 1 & & 1 & & & \\
& & 1 & & 2 & & 1 & & \\
& 1 & & 3 & & 3 & & 1 & \\
1 & & 4 & & 6 & & 4 & & 1 \\
& & & & \vdots & & & &
\end{array}
$$

consists entirely of binomial coefficients. (Notice the 1, 4, 6, 4, 1 in the fifth row corresponding to the coefficients in the expansion of $(x+y)^4$.) Some other interesting tidbits about Pascal's triangle are listed below.

- The row number is determined by n and the position in the row is determined by r.

- Each row determines the subsequent row. Each entry that is not on the boundary of the triangle is the sum of the two closest entries in the previous row. This is equivalent to the result in 2(b) above.

- The row sums are powers of 2, that is, $\sum_{r=0}^{n} \binom{n}{r} = 2^n$.

- The sums of the first n diagonal elements are n, $n(n+1)/2$, $n(n+1)(2n+1)/6$, ... due to constant, linear, quadratic, ... growth of the diagonal elements.

- The Fibonacci sequence 1, 1, 2, 3, 5, 8, 13, ... can be found in the triangle. See if you can find a way to determine these values.

- Replacing odd numbers in the triangle by 1 and even numbers by 0 yields the "Sierpinski gasket."

- Try replacing each element modulo 3.

- If the digits of the first five rows are concatenated, they yield the powers of 11.

5. The binomial theorem can be extended to the "multinomial theorem" to handle the expansion of expressions like $(x+y+z)^8$. The coefficient for $x_1^{m_1} x_2^{m_2} \ldots x_k^{m_k}$ when expanding $(x_1 + x_2 + \cdots + x_k)^n$ is

$$
\binom{n}{m_1, m_2, \ldots, m_k} = \frac{n!}{m_1! m_2! \ldots m_k!}.
$$

6. Combinations are a special case of partitioning. Consider the following two examples.

Example 1.37 How many ways are there to deal a five-card poker hand?

This problem was encountered earlier and solved using combinations. The problem can also be considered as a partitioning problem. Dealing five cards from a 52-card deck is equivalent to partitioning the deck into five cards (those selected for the hand) and 47 other cards (those not selected for the hand), as shown below. The bar is used to denote the partitioning position.

$$\underbrace{1\ 2\ 3\ 4\ 5}_{r\ \text{here}}\ |\ \underbrace{6\ 7\ \ldots\ 51\ 52}_{n-r\ \text{here}}$$

The next example moves from partitioning a set of objects into two groups to partitioning a set of objects into three groups.

Example 1.38 You have a one, five, twenty, and hundred dollar bill to invest in three stocks: AT&T, Boeing, and Coke. How many ways are there to invest 2 bills in AT&T, 1 bill in Boeing, and 1 bill in Coke?

Let

- O denote the one dollar bill,
- F denote the five dollar bill,
- T denote the ten dollar bill,
- H denote the hundred dollar bill,

and let

- a bill to the left of the bars corresponds to an investment in AT&T,
- a bill between the bars corresponds to an investment in Boeing,
- a bill to the right of the bars corresponds to an investment in Coke.

The 12 possible investment strategies are enumerated below. The bars again represent the partition.

$OF|T|H$	$OF|H|T$	$OT|F|H$	$OT|H|F$
$OH|F|T$	$OH|T|F$	$FT|O|H$	$FT|H|O$
$FH|O|T$	$FH|T|O$	$TH|O|F$	$TH|F|O$

These two examples lead to a more general result which is stated without proof.

Theorem 1.6 The number of ways of partitioning a set of n distinct objects into k subsets with n_1 in the first subset, n_2 in the second subset, ..., n_k in the kth subset, is

$$\binom{n}{n_1, n_2, \ldots, n_k} = \frac{n!}{n_1! n_2! \ldots n_k!},$$

Same as nondistinct permutations or similar.

where $n_1 + n_2 + \cdots + n_k = n$.

The previous example concerning the number of investment strategies is solved using Theorem 1.6 with $n = 4$, $k = 3$, $n_1 = 2$, $n_2 = 1$, and $n_3 = 1$ yielding

$$\frac{4!}{2!1!1!} = 12$$

different investment strategies.

Example 1.39 The Glen family consists of 9 people. How many arrangements are there for them to watch the nightly news seated on four sofas: one that seats three and the others seat two?

Implicit in the problem statement is that the *position* (for example, left, right, middle on the big sofa) occupied by one of the Glens on a particular sofa is not relevant in terms of TV viewing arrangements. Applying Theorem 1.6 with $n = 9$ family members being partitioned onto $k = 4$ sofas with $n_1 = 3$, $n_2 = 2$, $n_3 = 2$, and $n_4 = 2$ family members occupying each sofa, there are

$$\binom{9}{3, 2, 2, 2} = \frac{9!}{3!2!2!2!} = 7560$$

different TV viewing arrangements. Assuming that the Glens have no better form of amusement, they could go over 20 years swapping different TV viewing arrangements each night.

The alert reader will have noticed that nondistinct permutations and partitioning problems both use

$$\frac{n!}{n_1!n_2!\dots n_k!}.$$

The following two examples illustrate how these two approaches are actually solving fundamentally identical problems.

Example 1.40 (Nondistinct permutations) How many ways are there to arrange the letters in the word "bib"?

The b's in "bib" are considered nondistinct so that swapping the b's does not correspond to a new ordering. Enumerating the outcomes yields

$$bbi$$
$$bib$$
$$ibb$$

and the formula from Theorem 1.4 for nondistinct permutations yields

$$\frac{3!}{2!1!} = 3$$

different orderings. The "indistinguishable objects" here are the b's.

Example 1.41 (Partitioning) Preston, Jill and Gretchen are sisters. How many ways are there to sleep the three girls in a double and single bed?

Let P, J, and G denote the three girls. Also, place the two girls in the double bed on the left of the bar and the girl in the single bed on the right of the bar. Enumerating the outcomes yields

$$PJ\,|\,G$$
$$PG\,|\,J$$
$$GJ\,|\,P$$

Using Theorem 1.6 for partitioning problems, there are

$$\frac{3!}{2!1!} = 3$$

different orderings. The "indistinguishable objects" here are the two girls in the double bed. The problem of sleeping the sisters in the beds is fundamentally the same as the ordering of the letters in the word "bib."

We close this section with one final unifying example that stresses the importance of the following two questions associated with a counting problem. (*a*) Is the sampling performed *with replacement* or *without replacement*? (*b*) Is the sample considered *ordered* or *unordered*?

Example 1.42 How many ways are there to select 4 billiard balls from a bag containing the 15 balls numbered $1, 2, \ldots, 15$?

The question as stated is (deliberately) vague. It has not been specified whether

- the billiard balls are replaced (that is, returned to the bag) after being sampled, and
- the order that the balls are being drawn from the bag is important.

So there are really $2 \times 2 = 4$ different questions being asked here. The answers to these questions are given in the 2×2 matrix below.

	Without replacement	With replacement
Ordered sample	$15 \cdot 14 \cdot 13 \cdot 12$	$15 \cdot 15 \cdot 15 \cdot 15$
Unordered sample	$\binom{15}{4}$	$\binom{18}{4}$

These simplify to

$\hookrightarrow$ *important*!

	Without replacement	With replacement
Ordered sample	32,760	50,625
Unordered sample	1365	3060

There are several observations that can be made on the numbers in this 2×2 matrix. First of all, the entries in column 2 are always greater than the corresponding entries in column 1. This is because sampling with replacement allows for more possible draws due to the fact that the size of the population from which a draw is made remains constant rather than diminishing. Secondly, the entries in row 1 are always greater than the corresponding entries in row 2. This is because the count of ordered draws (permutations) will always exceed the corresponding number of unordered draws (combinations).

A further explanation of the lower-right entry of the matrix might be needed. Consider 15 bins and 4 balls, where $\bigcirc$ denotes a billiard ball. One draw of 4 balls is depicted below.

$$| \quad \bigcirc \bigcirc \quad | \quad , \quad | \quad \bigcirc \quad | \quad , \quad | \quad \cdots \quad | \quad , \quad | \quad \bigcirc$$
$$1 \qquad 2 \qquad 3 \qquad 4 \qquad 5 \qquad \cdots \qquad 14 \qquad 15$$

This arrangement of bins and markers corresponds to the unordered draw 2, 2, 4, 15 taken with replacement from the bag. We need to count the number of arrangements of 14 dividers plus 4 balls, or a total of 18 objects. Since the $\bigcirc$'s are indistinguishable, there are

$$\binom{18}{4}$$

different orderings (the outer walls are ignored).

The previous example has highlighted two important issues that arise in combinatorial problems: order and replacement. These concerns lead to a generic class of problems known as "urn models" in which objects are drawn sequentially from an urn.

This has been an unusually long section, so it ends with an outline of the topics considered, and their associated formulas.

1. Multiplication rule: $n_1 n_2 \ldots n_r$

2. Permutations: $\dfrac{n!}{(n-r)!}$

    (a) Circular permutations: $(n-1)!$

    (b) Nondistinct permutations: $\dfrac{n!}{n_1! n_2! \ldots n_r!}$

3. Combinations (binomial coefficients): $\dbinom{n}{r} = \dfrac{n!}{(n-r)!r!}$

1.3 Sets

Sets are often used in solving probability problems. The German mathematician Georg Cantor is generally credited with creating set theory. We provide a brief review of set theory here, and begin with basic definitions.

> **Definition 1.3** A *set* is a collection of objects (elements).

Upper-case letters are typically used to denote sets, for example, A, B. The notion of a set is a very general one, as will be seen in the example below.

Example 1.43 Consider the following four sets.

$$
\begin{aligned}
A &= \{1, 2, \ldots, 100\} \\
B &= \{x \mid x \text{ is a positive integer less than 101}\} \\
C &= \{\text{Bulls, Trailblazers}\} \\
D &= \{(x, y) \mid 0 < x < 1, 0 < y < 2\}
\end{aligned}
$$

The elements of sets A and B are integers; the elements of set C are the names of basketball teams; the elements of set D are points in the interior of a rectangle in the Cartesian coordinate system. Sets B and D are defined by what is known as the *set-builder* notation, and the bar is read as "such that." Thus, the definition of B is read as "the set of all values x such that x is a positive integer less than 101." The sets A and B have identical elements, and their equality is written as $A = B$.

> **Definition 1.4** If an object belongs to a set, it is said to be an *element* of the set. The notation $\in$ is used to denote membership in a set.

Example 1.44 Using the sets defined in Example 1.43,

$$
17 \in A \qquad 99 \in B \qquad \left(\frac{2}{3}, 1\right) \in D \qquad \text{Cubs} \notin C.
$$

The notation $\in$ is read as "is a member of." The notation $\notin$ is read as "is not a member of."

> **Definition 1.5** If every element of the set A_1 is also an element of the set A_2, then A_1 is a *subset* of A_2. The notation $\subset$ is used to denote the subset relationship.

The subset symbol $\subset$ from Definition 1.5 allows for the two sets A_1 and A_2 to be equal. For any set A, for example, $A \subset A$. If A_1 is a subset of A_2, but A_1 is not allowed to equal A_2, then the relationship between A_1 and A_2 is known as a *proper subset*.

Example 1.45 The natural numbers **N**, also known as the positive integers, are a subset of the integers **Z**, which are a subset of the rational numbers **Q**, which are a subset of the real numbers **R**, which are a subset of the complex numbers **C**. These relationships are compactly stated as

$$\mathbf{N} \subset \mathbf{Z} \subset \mathbf{Q} \subset \mathbf{R} \subset \mathbf{C}.$$

Venn diagrams are a useful tool in set theory and in probability for sorting out the relationships between various sets. An example of a Venn diagram containing the sets A and B is shown in Figure 1.9.

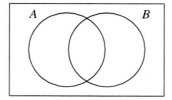

Figure 1.9: Venn diagram.

The external rectangle that is drawn outside of the two sets A and B is often called the *universal set*, and it contains all possible elements under consideration. If it is assumed that $A \subset B$, then the Venn diagram can be modified as in Figure 1.10.

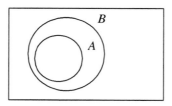

Figure 1.10: Venn diagram for $A \subset B$.

Example 1.46 When there are several subsets involved in a particular application, we often use subscripts, rather than individual letters to denote the sets. Thus, the relationship between

$$A_1 = \{x \mid 0 < x < 1\}$$

and

$$A_2 = \{x \mid 0 < x < 5\}$$

can be described by

$$A_1 \subset A_2.$$

Definition 1.6 A set containing no elements is called the *null set* (also known as the *empty set*). The notation $\emptyset$ is used to denote the null set.

Example 1.47 List the subsets of $\{a, b, c\}$.

By the multiplication rule, there are $2^3 = 8$ such subsets because there are three decisions to be made (whether to include or not include each element in the subset), and each decision can be made in two ways. The subsets are

$$\emptyset, \{a\}, \{b\}, \{c\}, \{a, b\}, \{a, c\}, \{b, c\}, \{a, b, c\}.$$

To generalize, by the multiplication rule, there are always 2^n subsets of any set containing n elements.

This completes the statement of some basic definitions in set theory. We now define the operations that can be applied to a set. We consider just three: union, intersection, and complement.

Definition 1.7 (Union) $A \cup B = \{x \mid x \in A \text{ or } x \in B\}$. Not XOR.

The meaning of "or" in Definition 1.7 is not exclusive: the elements in $A \cup B$ are in A alone, B alone, or in both A and B simultaneously. A Venn diagram with the union of A and B shaded is given in Figure 1.11

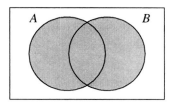

Figure 1.11: Venn diagram for $A \cup B$.

The notion of the union of two sets generalizes naturally to more than two sets.

Definition 1.8 $A_1 \cup A_2 \cup \ldots = \{x \mid x \in A_1 \text{ or } x \in A_2 \text{ or } \ldots\}$, which applies to a finite or infinite number of sets.

logic symbol V.

$\hookrightarrow \bigcup_{i=1}^{n} A_i$ same thing as the definition.

Example 1.48 For the sets

$$A_1 = \{x \mid 0 < x < 1\}$$
$$A_2 = \{x \mid 0 < x < 5\}$$

find the union of A_1 and A_2.

Since $A_1 \subset A_2$, the union of A_1 and A_2 is just A_2,

$$A_1 \cup A_2 = A_2.$$

Example 1.49 Let

$$A_k = \{k, k+1, k+2, \ldots, k^2\}$$

for $k = 1, 2, \ldots$. Find the union of A_3, A_4, and A_5.

Since the three-way union is all of the elements in A_3, A_4, or A_5,

$$A_3 \cup A_4 \cup A_5 = \{3, 4, \ldots, 25\}.$$

The next set operator to be introduced considers elements that belong to two sets simultaneously.

Definition 1.9 (Intersection) $A \cap B = \{x \mid x \in A \text{ and } x \in B\}$

Logic symbol $\wedge$

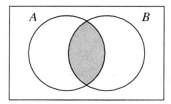

Figure 1.12: Venn diagram for $A \cap B$.

A Venn diagram with the intersection of A and B shaded is given in Figure 1.12. The notion of the intersection of two sets also generalizes naturally to more than two sets.

Definition 1.10 $A_1 \cap A_2 \cap \ldots = (x \mid x \in A_1, x \in A_2, \ldots)$, which applies to a finite or infinite number of sets.

Both union and intersection are symmetric operators, for example, $A \cap B = B \cap A$ and $A \cup B = B \cup A$.

Example 1.50 For the sets

$$\bigcap_{i=1}^{\frown} A_i$$

$$A = \{(x, y) \mid x^2 + y^2 \le 16\}$$
$$B = \{(0, 0), (1, 1), (2, 4), (3, 9), (4, 16), \ldots\}$$

find the intersection of A and B.

Since only the first two ordered pairs fall in the circle with radius 4 centered at the origin,

$$A \cap B = \{(0, 0), (1, 1)\}.$$

Definition 1.11 If $A \cap B = \emptyset$, then A and B are *disjoint* or *mutually exclusive*.

A Venn diagram for the disjoint sets A and B is given in Figure 1.13

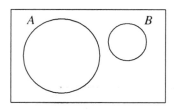

Figure 1.13: Venn diagram for disjoint sets A and B.

The final operation on a set that we present here is *complement*, which corresponds to the elements outside of a particular set.

Definition 1.12 (Complement) $A' = \{x \mid x \notin A\}$.

A Venn diagram with the complement of A shaded is given in Figure 1.14

Example 1.51 Let the set A be the set of all real numbers on the open interval $(0, 1)$, that is

$$A = \{x \mid 0 < x < 1\}.$$

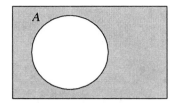

Figure 1.14: Venn diagram for A'.

Find A'.

Assuming that the universal set is the set of real numbers, the complement of A must be all of the real numbers not in the open interval $(0, 1)$, that is

$$A' = \{x \,|\, x \leq 0 \text{ or } x \geq 1\}.$$

There are a number of comments on the small portion of set theory presented here that are given below in outline form.

1. Many authors use A^*, A^c, or $\bar{A}$ for complement, so the choice of A' used in this book is not universal. The symbols $\cup$ and $\cap$ are fairly universal.

2. Many authors use the shorthand $\displaystyle\bigcup_{i=1}^{n}$ and $\displaystyle\bigcap_{i=1}^{n}$ which parallels the use of $\displaystyle\sum_{i=1}^{n}$ for sums and $\displaystyle\prod_{i=1}^{n}$ for products. So, for example,

$$A_1 \cup A_2 \cup A_3 \cup A_4 = \bigcup_{i=1}^{4} A_i.$$

3. There are more operations on sets than the three presented here. One such operator which has applications in computer science is the *exclusive or* operator. The exclusive or of the sets A and B is denoted by $A \oplus B$ and includes all elements that are in set A or in set B, but not both. Figure 1.15 contains a Venn diagram for $A \oplus B$.

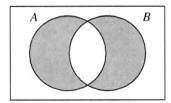

Figure 1.15: Venn diagram for $A \oplus B$.

4. DeMorgan's laws are given by

$$\left(\bigcup_{i=1}^{n} A_i \right)' = \bigcap_{i=1}^{n} A_i'$$

and

$$\left(\bigcap_{i=1}^{n} A_i \right)' = \bigcup_{i=1}^{n} A_i'.$$

5. The distributive laws are given by

$$A_1 \cap (A_2 \cup A_3) = (A_1 \cap A_2) \cup (A_1 \cap A_3)$$

and

$$A_1 \cup (A_2 \cap A_3) = (A_1 \cup A_2) \cap (A_1 \cup A_3).$$

Venn diagrams associated with sets can be useful for unscrambling befuddling counting problems, as illustrated in the following two examples.

Example 1.52 Of 100 boomers polled, 85 said they like Elvis, 62 said they like Zappa, and 5 said that they don't like either. How many of them like both Elvis and Zappa?

Let E be the set of boomers who like Elvis; let Z be the set of boomers who like Zappa. As shown in Figure 1.16, the 5 who said that they didn't like Elvis or Zappa are placed outside the sets E and Z. This leaves 95 for the remaining unfilled slots. Letting the cardinality function $N(\cdot)$ be a function that counts the number of elements in a set, the relationship between the three remaining blank spots on the Venn diagram is

$$N(E) + N(Z) - N(E \cap Z) = 95.$$

Since $N(E) = 85$ and $N(Z) = 62$, there are

$$N(E \cap Z) = 52$$

of the 100 that like both Elvis and Zappa. The remaining numbers have been placed into Figure 1.16.

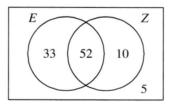

Figure 1.16: Venn diagram for counting fans.

Example 1.53 How many of the first 1000 positive integers are multiples of neither 6 nor 9?

There are 166 multiples of 6, which are

$$6, 12, 18, \ldots, 996$$

because $166 \cdot 6 = 996$. Likewise, there are 111 multiples of 9, which are

$$9, 18, 27, \ldots, 999$$

because $111 \cdot 9 = 999$. An integer is a multiple of both 6 and 9 if it is a multiple of the least common multiple of 6 and 9, which is $\mathrm{lcm}(6,9) = 18$. The 55 integers between 1 and 1000 that are multiples of both 6 and 9 are

$$18, 36, 54, \ldots, 990$$

because $55 \cdot 18 = 990$. Letting the set A denote the multiples of 6 and the set B denote the multiples of 9, the Venn diagram in Figure 1.17 shows the counts of the various numbers of integers in the four regions partitioned by the sets A and B.

To answer the original question, there are 778 integers between 1 and 1000 that are multiples of neither 6 nor 9.

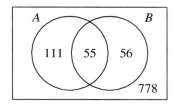

Figure 1.17: Venn diagram counting multiples.

1.4 Exercises

1.1 Each of the following questions has the same answer: 2^{32}. Write two more questions, (d) and (e), that also have 2^{32} as an answer.

    (a) How many ways are there to answer a 32-question true/false test?

    (b) How many integers can be represented on a 32-bit computer (ignoring the sign bit)?

    (c) How many sequences of successes and failures can be recorded when Michael shoots 32 free throws?

1.2 Compute $\binom{4}{2}$.

1.3 A die is rolled four times consecutively. Find the number of possible outcomes if

    (a) the order of the four outcomes is important,

    (b) the order of the four outcomes is not important.

1.4 How many initial possible pairings are there for a single-elimination ping-pong tournament involving n players, $n = 2, 4, 8$?

1.5 How many triangles can be formed by connecting any three of ten distinct points that lie on an ellipse?

1.6 Chip has 4 pennies, 3 dimes, and 5 quarters in his pocket. How many different positive monetary values can he make with these coins? (*Hint*: Not all coins need to be used, but 0 cents is not a monetary value. He could, for example, make 30 cents with the three dimes.)

1.7 A license plate consists of two letters followed by three numbers.

    (a) How many different license plates can be made?

    (b) How many different license plates can be made if no two-character U.S. state abbreviations are allowed for the two letters?

1.8 Skip likes to use clichés. Here are some of his favorites.

    That's the greatest thing since sliced bread.
    This fog is as thick as pea soup.
    An apple a day keeps the doctor away.

Irritated by his "habit," several of Skip's friends and relatives decide to limit him to only three of his clichés per day. In an effort to keep his routine fresh, Skip vows to never use the same set of three clichés from his repertoire for the rest of his life. If Skip plans on living another 60 years (ignore leap years), what is the minimum size of his repertoire in order to achieve his goal and use exactly three clichés each day?

1.9 How many arrangements of six people in a row of six chairs are possible if

  (a) there are no restrictions on the ordering?

  (b) Bill and Sarah must sit together?

  (c) Bill and Sarah must sit apart?

  (d) there are three men and three women and no two people of the same gender can sit next to one another?

  (e) there are four men and two women and the men must sit together?

  (f) there are four men and two women and both the men and women must sit together?

  (g) there are three married couples and the couples must sit next to one another?

1.10 Mr. Oliver North, Mr. Ray Southworth, Mrs. Mary Easterling, and Mr. Paul Westfield are playing cards.

  (a) They decide to play bridge. The game of bridge begins by dealing 13 cards to each of the 4 players. How many different bridge *deals* are possible? Consider the players to be distinct (for example, if Mr. North and Mrs. Easterling swap their cards, it is a different deal).

  (b) They decide to play poker. The game begins by dealing five cards to each person. How many *deals* are possible if the players are considered to be distinct?

1.11 How many ways are there to line up 8 people in a line for a photograph in a row of chairs if Rex and Laurie must sit next to one another and Bonnie and Clyde refuse to sit by one another?

1.12 Logan has 100 indistinguishable one dollar bills that he would like to invest in 4 banks. How many ways can he invest in these banks? (*Hint*: one way of investing is $98 in the first bank, $0 in the second bank, $2 in the third bank, and $0 in the fourth bank).

1.13 A committee must be chosen from 10 Republicans, 12 Democrats, and 5 Independents.

  (a) How many committees of size three are possible if each member of the committee must have the same political affiliation?

  (b) How many committees of size two are possible if the committee members must have a different political affiliation?

  (c) How many committees of size five are possible that consist of two Republicans, two Democrats, and one Independent?

1.14 Of the $\binom{52}{5} = 2,598,960$ different five-card poker hands, how many contain

  (a) the two of clubs?

  (b) four of a kind?

  (c) two pair (for example, KK772 counts as a two pair hand, but KK777 does not, since it is a full house)?

1.15 A laboratory has seven female and six male rabbits. Three females and three males will be selected, then paired for mating. How many pairings are possible?

1.16 How many 6-digit numbers of the form $d_1 d_2 d_3 d_4 d_5 d_6$, which range from 000000 to 999999, have the sum of the digits equal to 12? Use a combinatorics argument, then check your solution by enumeration.

1.17 How many ways can 8 people be seated around a round table?

1.18   Consider a sequence of n binary digits.

(a) How many sequences are possible?

(b) In how many of the possible sequences does the sum of the digits exceed j, where $j = 0, 1, 2, \ldots, n-1$?

1.19   Use the binomial theorem to show that for any positive integer n,

$$\sum_{i=0}^{n} (-1)^i \binom{n}{i} = 0.$$

1.20   Expand:

(a) $(x + 2y^3)^4$ using the binomial theorem,

(b) $(x + y + 3z)^3$ using the multinomial theorem.

Check the results with the Maple expand function.

1.21   Use the binomial theorem to show that for any positive integer n,

$$\sum_{i=0}^{n} \binom{n}{i} = 2^n.$$

1.22   Find the number of trailing 0's at the end of 100000! using a combinatorics argument.

1.23   A 12-digit number of the form $d_1 d_2 d_3 d_4 d_5 d_6 d_7 d_8 d_9 d_{10} d_{11} d_{12}$ (these numbers range from 000000000000 to 999999999999) is said to be *wonderful* if

$$d_1 + d_2 + d_3 + d_4 + d_5 + d_6 = d_7 + d_8 + d_9 + d_{10} + d_{11} + d_{12}.$$

Prove that the number of wonderful numbers is even.

1.24   The game of *FreeCell* begins by dealing all of the cards from a standard deck into eight columns of cards. Four of these columns contain seven cards and four of these columns contain six cards. The order that the cards fall *within* a column is significant, but the order of the columns is not significant. How many different deals are possible?

1.25   A father has nine identical coins to give to his three children.

(a) How many allocations are possible?

(b) How many allocations are possible if each child must receive at least one coin?

(c) How many allocations are possible if each child must receive at least two coins?

1.26   There are n women who try out for a high school basketball team. Of the n, there are m that make the team. Of the m, there are five that start in the first game. Assuming that $n \geq m \geq 5$, use two different combinatorial approaches to find the number of possible ways to select a team and a starting lineup for the first game.

1.27   How many ways can a mother give n_1 identical coins to her n_2 children? Assume that n_1 and n_2 are positive integers satisfying $n_1 > n_2$. She must give all of the coins away. (*Hint*: if $n_1 = 5$ and $n_2 = 3$, then giving three coins to the firstborn, two coins to the middle child and no coins to the youngest is one possibility).

1.28   A family consists of n members. How many (pairwise) relationships are there between members?

1.29 A restaurant offers 3 appetizers, 4 entrees, and 5 desserts. How many ways are there to place an order for one appetizer, one entree, and one dessert?

1.30 How many ways can a seven-card hand be dealt from a standard 52-card deck?

1.31 Arthur, Ivah, Richard, Cindy, Larry, and Nancy need to cross a bridge at night. Exactly two may cross at a time, and they must carry a flashlight. There is only one flashlight. Assume that two people always cross the bridge and that one always returns with the flashlight. How many ways are there to get everyone across? *Illustration:* Here is *one sample* sequence: Arthur and Ivah cross, Arthur returns, Richard and Cindy cross, Richard returns, Larry and Nancy cross, Larry returns, Arthur and Richard cross, Cindy returns, Cindy and Larry cross.

1.32 How many ways can a committee of 11 people be subdivided into four subcommittees containing 4, 3, 2, and 2 people each?

1.33 A bag contains billiard balls numbered $1, 2, \ldots, 15$. How many ways can three balls be selected from the bag when the order of the selection is important?

1.34 Flip can climb a staircase at one, two, or three steps at a time. How many ways can he climb a 12-step staircase? (Note: The order that Flip climbs the steps is relevant.)

1.35 Find the number of solutions to the equation

$$a + b + c = 9,$$

where a, b, and c are positive integers.

1.36 How many of the first 5000 positive integers are neither perfect squares nor perfect cubes?

1.37 The set A consists of all positive integers x from 1 to 15 inclusive such that $\gcd(x, 15) = 1$. List the elements of A.

1.38 A 6-letter "word" is formed by selecting 6 of the 26 letters without replacement. Two examples of such words are

FRISBE and XEALRY.

Let A_1 be the set of all words beginning with X and A_2 be the set of all words ending with Y. Find the numbers of distinct words in

(a) $A_1 \cap A_2$,

(b) $A_1 \cup A_2$.

1.39 A prime number is a positive integer that has exactly two distinct divisors: 1 and itself. Let A be the set of all prime numbers. Let B be the set of all even integers. Let C be the set of all negative integers. Draw a Venn diagram that describes the relationship among the sets A, B, and C.

1.40 Draw a Venn diagram with events A and B and shade $A' \cap B'$.

1.41 Draw a Venn diagram with events A_1, A_2, and A_3 and shade $A_1 \cap (A_2 \cup A_3)$.

1.42 Draw a Venn diagram with events A_1, A_2, and A_3 and shade $(A_1 \cap A_2') \cup A_3$.

1.43 If A and B are two events, use any of the set operations (for example, union, intersection, complement) to describe the event that neither A nor B occurs.

1.44 Let the set A be the perfect squares in the first 30 positive integers. Let the set B be the prime numbers in the first 30 positive integers. Find $N(A' \cap B')$, where N gives the cardinality (number of elements) in a set.

Chapter 2

Probability

The main topic of this textbook, probability, is introduced in this chapter. Probability quantifies uncertainty. Probability theory is traditionally based on a set of three assumptions known as the Kolmogorov axioms. Once these axioms have been established, we can use the counting techniques and set theory from the previous chapter to solve problems that arise in probability. A secondary emphasis is placed on *Monte Carlo simulation* as an alternative way of calculating probabilities. The topics covered in this chapter are experiments, sample spaces, events, Kolmogorov axioms, computing probabilities, Monte Carlo simulation, conditional probability, the rule of elimination, the rule of Bayes, and independence.

2.1 Experiments, Sample Spaces & Events

Before presenting a formal definition of the probability function P, the notion of a random experiment is defined.

> **Definition 2.1** A *random experiment* is one where every possible outcome can be described prior to its execution.

Associated with a random experiment is the set of all possible outcomes to that experiment. For example

- when a quarterback throws a pass in football, there are three possible outcomes: a complete pass, an incomplete pass, and an interception;

- when a gambler bets on "red" in roulette, there are two possible outcomes: winning and losing;

- when a backgammon player rolls a pair of dice, there are 11 different sums that are possible: $2, 3, \ldots, 12$.

This set of all possible outcomes is defined formally below.

> **Definition 2.2** The set of all possible outcomes to a random experiment is called the *sample space* and is denoted by S.

> **Example 2.1** Table 2.1 contains a list of six random experiments and the associated sample spaces. All of the random experiments involve either the rolling of dice or the tossing of a coin. The sample spaces in Table 2.1 have one common attribute: they all correspond to sets that are known as *finite* sets. Each sample space has a finite number of elements, namely, 6, 36, 11, 11, 4, and 3.

Random experiment	Sample space
Roll a die and observe the up face	$S = \{1, 2, 3, 4, 5, 6\}$
Roll two dice and observe the ordered pair	$S = \{(1, 1), (1, 2), \ldots, (6, 6)\}$
Roll two dice and observe the sum of the up faces	$S = \{2, 3, \ldots, 12\}$
Roll a red die and a green die and observe the difference between the red up face and the green up face	$S = \{-5, -4, \ldots, 5\}$
Toss a coin twice and observe the sequence of H's and T's	$S = \{HH, HT, TH, TT\}$
Toss a coin twice and observe the number of H's	$S = \{0, 1, 2\}$

Table 2.1: Random experiments and sample spaces.

A set that is not finite is known as an *infinite* set. Furthermore, a set is *denumerable* if its elements can be placed in a one-to-one correspondence with the natural numbers, as illustrated in the following example. *Countably infinite.*

Example 2.2 Toss a coin repeatedly until a head appears and observe the number of tosses required. Find the sample space S.

In this application, it is impossible to place an upper bound on the number of tosses that are required, so the set of all possible outcomes to the experiment is the sample space

$$S = \{1, 2, 3, \ldots\}.$$

This is not a finite set like the sample spaces in Example 2.1. It is a special case of an infinite set known as a denumerable set.

There is still another way to classify sets. A set is *countable* if it is either finite or denumerable. A set is *uncountable* if it is not countable, as illustrated in the following two examples.

Example 2.3 Observe the lifetime of a light bulb. Find the sample space S.

In this application, the lifetime of the light bulb can assume any nonnegative value, allowing for the (annoying) possibility of a purchase followed by immediate failure. So the sample space is

$$S = \{t \mid t \geq 0\}. \quad \text{No 1-1 correspondence w/ } \mathbb{N}.$$

The infinite set S can be classified as uncountable.

All of the previous examples have illustrated sample spaces that have been written in just one fashion. We end our discussion of sample spaces with an example where there are at least two ways to describe a sample space.

Example 2.4 Observe the position of a dart on a dart board with a 15" radius. Find the sample space S.

If one chooses to use rectangular coordinates, the position of the dart can be described by the sample space

$$S = \{(x, y) \mid x^2 + y^2 < 225\}.$$

On the other hand, if one chooses to use polar coordinates, the position of the dart can be described by the sample space

$$S = \{(\rho, \theta) \mid 0 \leq \rho < 15, 0 \leq \theta < 2\pi\}.$$

> **Definition 2.3** Let the *event A* be a subset of the sample space *S*. If an outcome to a random experiment is in *A* then event *A* has occurred.

Events are a cornerstone in probability theory. Venn diagrams can be helpful in classifying multiple events, as seen in the following example. Like sets, the capital letters $A, B, C, \ldots$ are used to denote events.

Example 2.5 Observe the top card drawn on a 52-card deck. Define the events:

$$A:\quad 4, 5 \text{ or } 6 \text{ of } \heartsuit,$$
$$B:\quad \text{any red card,}$$
$$C:\quad \text{any queen,}$$
$$D:\quad \text{jack of } \clubsuit.$$

Draw a Venn diagram that relates these events to one another.

The sample space is the finite set of 52 elements:

$$S = \{2\clubsuit, 3\clubsuit, \ldots, K\spadesuit, A\spadesuit\}.$$

In Figure 2.1, the sample space *S* is implicitly the outer rectangle in the Venn diagram. The events $A, B, C,$ and D are placed on the Venn diagram in a manner that is consistent with their definitions. There is no relationship between the size of the circles and the number of elements in a set when drawing a Venn diagram.

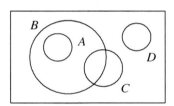

Figure 2.1: Venn diagram for a card drawn from a deck.

2.2 Probability Axioms

This section formally introduces the notion of probability. In the paragraphs that follow, we introduce four techniques for determining probabilities.

The first way to *estimate* probability is the *empirical* or *relative frequency* approach. For someone without mathematical training, this is perhaps the most natural of the four ways to determine the probability of an event occurring.

> **Definition 2.4** Perform a random experiment *n* times. Let *x* be the number of times that the event *A* occurs. The ratio x/n is the *relative frequency* of the event *A* in the *n* experiments.

Definition 2.4 implies that the probability of a particular event *A* occurring can be estimated by simply performing the random experiment *n* times, and calculating the ratio of the number of times *A* occurs to *n*. This ratio might change if the random experiment were performed another *n* times. The relative frequency is a random quantity that provides an *estimate* of the true probability that event *A* occurs, which is a constant denoted by $P(A)$. Preferably, *n* is large, which results in a more accurate estimate of the probability of event *A* occurring. There are situations (for example, crashing test cars into walls), where costs or time prevent using a large *n*. Here is an example of a cheap, quick random experiment.

Example 2.6 Estimate the probability that a coin comes up "heads" on a single toss of a fair coin.

Using the relative frequency approach, we toss a coin $n = 1000$ times and observe $x = 517$ heads. In this case, we can use the relative frequency

$$\frac{517}{1000} = 0.517$$

as an estimate of the probability of tossing a head on a single toss.

Example 2.6 illustrates three of the shortcomings of the relative frequency approach to estimating probability. First, if we know that the coin is fair, the probability of interest is 0.5. Second, the estimate for the probability of tossing a head, 0.517 in the example, will change from one experiment to the next, which is not desirable. Our experiment gave us a value close, but not equal to 0.5. Third, the coin-tossing experiment described above is tiring—both on the thumb of the flipper and the mind of the poor soul who tallied the results. Can computers make this easier?

This is an appropriate moment to introduce a technique for more efficiently generating random experiments, known as *Monte Carlo simulation*. The next example uses a computer to simulate the coin flips.

Example 2.7 Computer languages typically have what is known as a *random number generator* that is used to generate random numbers that are equally likely between 0 and 1. A loop that generates $n = 1000$ random numbers and counts the number of times that a random number is less than 0.5 (simulating a coin toss that comes up heads) can be used to generate a relative frequency estimate of the probability of event A. The following R code simulates 1000 coin flips and prints the fraction of the flips that come up heads.

```
count = 0
for (i in 1:1000)
   if (runif(1) < 0.5) count = count + 1
print(count / 1000)
```

The R function `runif` with the parameter 1 generates a single random number. The variable `count` counts the number of heads that are tossed. The fraction `count / 1000` is the relative frequency. This code is executed after the `set.seed(3)` command that initializes the random number stream to 3. Executing this code yields the estimate 0.489. The code can also be written more compactly as

```
sum(runif(1000) > 0.5) / 1000
```

which returns identical results. R strongly prefers the use of vectors, as in the single-line implementation, over loops in terms of execution speed. The Monte Carlo method has the additional advantage that the experiment can be repeated with very little additional effort. It is good practice to execute a Monte Carlo simulation several times. When the algorithm and associated code is executed seven times, the following relative frequencies are returned:

0.489   0.515   0.501   0.501   0.492   0.492   0.473.

Notice that the seven values hover around the true value of 0.5 for a fair coin.

So far, we have only encountered an *estimate* for the probability of some event A occurring. Although impossible to implement in practice, if we could implement an infinite number of Monte Carlo simulation replications, the result would be the following definition.

Definition 2.5 The *limiting relative frequency* of the event A

$$P(A) = \lim_{n \to \infty} \frac{x}{n}$$

limit as you perform the experiment an ∞ amount of times.

is called the

- probability that the outcome of the random experiment is in A,

- probability of event A.

Definition 2.5 states that the probability of an event is the fraction of time that the event occurs in the long run. The probability that a fair coin comes up heads, defined as event A, for example, is

$$P(A) = \lim_{n \to \infty} \frac{x}{n} = \frac{1}{2}.$$

The second way to calculate probability is the *subjective* approach. The notion here is to consult an "expert" in the field of interest (for example, the stock market or horse racing) and elicit probabilities directly from them. The expertise and experience of the expert influence the precision of the probability estimate. This approach is riddled with problems because the estimate for the probability is nothing more than an educated guess. Some experts are more capable than others; some events can have more accurate probability assessments than others. In some sense *all* probabilities are subjective in that there are always assumptions made in framing an opinion and then calculating a probability.

The third way to calculate probability is the *classical* or *equally likely* approach. This approach can be used when all outcomes are equally likely, as illustrated in the following example.

Example 2.8 Toss a fair coin three times and observe the sequence of heads and tails. Find the probability that all three outcomes are heads.

By the multiplication rule, there are $2^3 = 8$ outcomes to the random experiment, resulting in the sample space

$$S = \{HHH, HHT, HTH, THH, TTH, THT, HTT, TTT\},$$

where H denotes tossing a head and T denotes tossing a tail. The eight possible outcomes listed in the sample space are all equally likely, so if the event A corresponds to tossing three heads

$$P(A) = \frac{1}{8}.$$

The equally likely approach is superior to the relative frequency approach and the subjective approach in that it yields an *exact* probability, rather than an estimate.

The set function $P(\cdot)$ is rather unlike the real-valued functions $y = f(x)$ that typically occur in most courses in mathematics. It is referred to as a set function because its domain is a collection of sets. In the previous example the set A was $A = \{HHH\}$. The range of the probability set function is a real number between 0 and 1. An event with probability 0 corresponds to an *impossible* event; an event with probability 1 corresponds to a *certain* event. In the more general setting where the sample space S consists of equally-likely outcomes, the probability of event A is the ratio of the number of elements in A to the number of elements in S:

$$P(A) = \frac{N(A)}{N(S)}.$$

Probability problems that are addressed using the classical or equally likely approach occur frequently. Unfortunately, it is not always the case that all outcomes are equally likely. Therefore, we introduce a fourth way to compute probabilities.

The fourth way to determine probabilities is the *axiomatic* approach. Probability theory can be built on the three *Kolmogorov axioms*, which were developed by Russian mathematician Andrei Kolmogorov (1903–1987). Let S denote the sample space for a random experiment and let $A \subset S$ be an event of interest. We define a set function $P(A)$ that satisfies certain axioms and yields the probability that event A occurs. Informally, the three Kolmogorov axioms are

- nonnegativity,

- additivity (of disjoint events), and

- unit measure,

which are defined more carefully below. *important in using probability.* *proofs*

Definition 2.6 (Kolmogorov axioms) Consider a random experiment with sample space S and an event $A \subset S$ of interest. If $P(A)$ is defined and

Axiom 1. $P(A) \geq 0$,   *The prob of an event is always ≥ 0*

Axiom 2. $P(A_1 \cup A_2 \cup \cdots) = P(A_1) + P(A_2) + \cdots$, where $A_1, A_2, \ldots$ are disjoint events, and

Axiom 3. $P(S) = 1$,   *Probability of the Sample Space is 1.*

then $P(A)$ is the probability of event A occurring.

Using Definition 2.6 for various different events, $P(A)$ indicates how the probability is "distributed" over the various events. This is the source of the phrase "probability distribution." Axiom 2 extends to finite disjoint unions. This tiny set of three axioms leads to a number of useful results when it comes to calculating probabilities. We now prove a sequence of theorems that will be helpful in working with probabilities.

Theorem 2.1 (Complementary probability) For each $A \subset S$, $P(A) = 1 - P(A')$.

---

partition definition.

Proof Since $S = A \cup A'$ and $A \cap A' = \emptyset$ (that is, A and A' form a *partition* of S), by using Axioms 2 and 3,

$$P(S) = 1 = P(A \cup A') = P(A) + P(A').$$

Rearranging, $P(A) = 1 - P(A')$.   □

The notion of complementary probability will be used throughout the text. For some probability problems, it is easier to calculate $P(A')$ than it is to calculate $P(A)$. Referring to Example 2.8, the probability of tossing two or fewer heads in three tosses of a fair coin can be calculated immediately as

$1 - P(3 \text{ heads})$. $\qquad P(\text{two or fewer heads}) = 1 - P(\text{three heads}) = 1 - \dfrac{1}{8} = \dfrac{7}{8}.$

Theorem 2.2 If $A_1, A_2 \subset S$ such that $A_1 \subset A_2$ then $P(A_1' \cap A_2) = P(A_2) - P(A_1)$, and therefore

$$P(A_1) \leq P(A_2).$$

Proof The events A_1 and $A_1' \cap A_2$ form a partition of A_2 (that is, $A_2 = A_1 \cup (A_1' \cap A_2)$ and $A_1 \cap (A_1' \cap A_2) = \emptyset$). From Axiom 2,

$$P(A_2) = P(A_1) + P(A_1' \cap A_2).$$

From Axiom 1, $P(A_1' \cap A_2) \geq 0$, so $P(A_1) \leq P(A_2)$.   □

This result is consistent with intuition. If one event is a subset of a second event, then the probability of the first event should be smaller. Consider the problem of selecting a single card at random from a 52-card deck. If the event A_1 corresponds to selecting a red face card, and A_2 corresponds to selecting a red card, then $P(A_1) = 6/52 = 3/26$ and $P(A_2) = 26/52 = 1/2$, because there are 26 red cards, 6 of which are face cards ($J\diamondsuit, Q\diamondsuit, K\diamondsuit, J\heartsuit, Q\heartsuit, K\heartsuit$). Thus,

$$P(A_1) \le P(A_2),$$

as expected.

Theorem 2.3 $P(\emptyset) = 0.$

Proof Using Axiom 3 and complementary probability with $A = \emptyset$ and $A' = S$, the probability of the null set is

$$P(\emptyset) = 1 - P(S) = 1 - 1 = 0. \qquad \square$$

This is certainly not a shocking result, since it is obvious that the probability of a set that corresponds to no outcomes is zero, but it is needed to prove the next result. It was stated earlier that the range of a probability set function is a real number between 0 and 1. This notion is formalized by the next theorem.

Theorem 2.4 For every $A \subset S, 0 \le P(A) \le 1.$

Proof Since $\emptyset \subset A \subset S$, by Theorem 2.2

$$P(\emptyset) \le P(A) \le P(S) \qquad \Longrightarrow \qquad 0 \le P(A) \le 1$$

by Axiom 3 and Theorem 2.3. $\qquad \square$

Theorem 2.5 (Addition rule) If $A_1, A_2 \subset S$ then

$$P(A_1 \cup A_2) = P(A_1) + P(A_2) - P(A_1 \cap A_2).$$

Draw venn Diagrams!

Proof Write $A_1 \cup A_2$ and A_2 as the unions of disjoint sets:

$$A_1 \cup A_2 = A_1 \cup (A'_1 \cap A_2) \qquad \text{and} \qquad A_2 = (A_1 \cap A_2) \cup (A'_1 \cap A_2).$$

From Axiom 2,

$$P(A_1 \cup A_2) = P(A_1) + P(A'_1 \cap A_2)$$

and

$$P(A_2) = P(A_1 \cap A_2) + P(A'_1 \cap A_2).$$

$A_1 \cup A_2 = S \cap (A_1 \cup A_2)$
$= (A_1 \cup A'_1) \cap (A_1 \cup A_2)$
$= A_1 \cup (A'_1 \cap A_2)$

Solving the bottom equation for $P(A'_1 \cap A_2)$ and substituting into the top equation yields

$$P(A_1 \cup A_2) = P(A_1) + P(A_2) - P(A_1 \cap A_2). \qquad \square$$

In addition to the proof by analytic methods, the addition rule can also be understood with the aid of the Venn diagram in Figure 2.2. If the probability of interest, $P(A_1 \cup A_2)$, were to be calculated by adding just $P(A_1)$ and $P(A_2)$, that would be too much probability because $P(A_1 \cap A_2)$ has been added twice. Hence it must be subtracted out once.

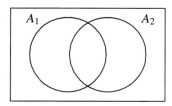

Figure 2.2: Determining $P(A_1 \cup A_2)$.

By the same kind of thinking, the addition rule can be extended to more than two events. For example, with three events, the result is

$$P(A_1 \cup A_2 \cup A_3) = P(A_1) + P(A_2) + P(A_3) - P(A_1 \cap A_2) - P(A_1 \cap A_3) - P(A_2 \cap A_3) + P(A_1 \cap A_2 \cap A_3).$$

Consider the Venn diagram in Figure 2.3. To calculate $P(A_1 \cup A_2 \cup A_3)$, the probability of single events are first added together. But this results in double-counting the two-way intersections, so each of them is subtracted out once. As for the three-way intersection, it has now been added in three times and subtracted out three times, so it must now be added in one final time.

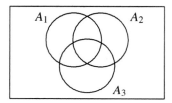

Figure 2.3: Determining $P(A_1 \cup A_2 \cup A_3)$.

Solving for $P(A_1 \cap A_2)$ in Theorem 2.5 yields a special case of <u>Bonferroni's inequality</u>:

$$P(A_1 \cap A_2) \geq P(A_1) + P(A_2) - 1,$$

which can be useful for bounding probabilities.

The next result requires the introduction of a concept known as a partition. If events $A_1, A_2, \ldots, A_n$ are disjoint and their union is the sample space S, then they form a partition of S. These two conditions can be stated in terms of set theory notation as

$$S = \bigcup_{i=1}^{n} A_i \qquad \text{and} \qquad A_i \cap A_j = \emptyset \text{ for } i \neq j.$$

Theorem 2.6 For events $A_1, A_2, \ldots, A_n$ that form a partition of S and another event $A \subset S$,

$$P(A) = \sum_{i=1}^{n} P(A \cap A_i).$$

Proof Since A can be written as the union of k disjoint events as

$$A = (A \cap A_1) \cup (A \cap A_2) \cup \ldots \cup (A \cap A_k),$$

by Axiom 2,

$$P(A) = \sum_{i=1}^{k} P(A \cap A_i). \qquad \square$$

Figure 2.4 is a Venn diagram that illustrates Theorem 2.6 with $n = 4$ events that partition the sample space S. This result extends to the case of a partition of S that consists of an infinite number of events.

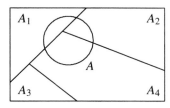

Figure 2.4: Events A_1, A_2, A_3, and A_4 partition the sample space S.

This ends the statement of the results that are direct consequences of the Kolmogorov axioms that define the probability set function $P(A)$. This section ends with three examples that can be solved using the results.

Example 2.9 Let $P(A_1) = 0.3$, $P(A_2) = 0.5$, and $P(A_1 \cup A_2) = 0.6$. Find $P(A_1 \cap A_2)$ and $P(A_1 \cap A_2')$.

The problem will be solved analytically using the axioms, then it will be solved using a Venn diagram. A modification of the addition rule can be used to find $P(A_1 \cap A_2)$:

$$
\begin{aligned}
P(A_1 \cap A_2) &= P(A_1) + P(A_2) - P(A_1 \cup A_2) \\
&= 0.3 + 0.5 - 0.6 \\
&= 0.2.
\end{aligned}
$$

To find $P(A_1 \cap A_2')$, use the fact that A_1 can be expressed as the union of disjoint sets as

$$ A_1 = (A_1 \cap A_2) \cup (A_1 \cap A_2'), $$

so

$$
\begin{aligned}
P(A_1 \cap A_2') &= P(A_1) - P(A_1 \cap A_2) \\
&= 0.3 - 0.2 \\
&= 0.1
\end{aligned}
$$

by Theorem 2.6. The four probabilities that arise in this problem are illustrated in four regions in the Venn diagram in Figure 2.5. Since the problem states that $P(A_1 \cup A_2) = 0.6$, the area outside of the events A_1 and A_2 is labeled as 0.4 because of complementary probabilities. Also, since $P(A_2) = 0.5$, this leaves just 0.1 for $P(A_1 \cap A_2')$, which solves the second question. Since $P(A_1) = 0.3$, this leaves 0.2 for $P(A_1 \cap A_2)$, which solves the first question. Finally, the remaining probability that has not been labeled is filled in with $P(A_1' \cap A_2) = 0.3$ because the four probabilities must sum to one, which completes all probabilities associated with the two events.

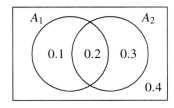

Figure 2.5: Venn diagram with probabilities.

Example 2.10 Toss a fair coin 3 times. Let the event A be observing exactly 2 heads. Find $P(A)$.

Using the multiplication rule, there are $2^3 = 8$ equally-likely outcomes in the sample space:

$$S = \{HHH, HHT, HTH, THH, TTH, THT, HTT, TTT\}.$$

Thus the probability of getting exactly 2 heads is

$$P(A) = \frac{3}{8}.$$

Example 2.11 Roll a pair of fair dice. Define the events

A_1: rolling a total of 7,
A_2: rolling a total of 12,
A_3: rolling doubles,
A_4: rolling 5, either individually or as a total,
A_5: rolling numbers that differ by 3.

Find $P(A_1), P(A_2), \ldots, P(A_5)$.

All five of these problems are addressed by using the 6×6 matrix shown in Figure 2.6. If the dice are red and green, the rows can be thought of as the number of spots on the up face on the red die and the columns can be thought of as the number of spots on the up face on the green die.

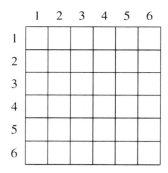

Figure 2.6: Sample space for the rolling of a pair of dice.

Since the dice are fair, the 36 outcomes in the matrix are equally likely. Hence solving this problem is just a matter of counting the number of outcomes associated with the various events $A_1, A_2, \ldots, A_5$. The results are:

$$P(A_1) = \frac{6}{36} \qquad P(A_2) = \frac{1}{36} \qquad P(A_3) = \frac{6}{36} \qquad P(A_4) = \frac{15}{36} \qquad P(A_5) = \frac{6}{36}.$$

Knowing these probabilities can be helpful in playing games of chance involving a pair of dice, such as backgammon, craps, or Monopoly^{TM}.

2.3 Computing Probabilities

→ root of
probability

A variety of examples are presented here illustrating various applications of the Kolmogorov axioms and subsequent results from the previous section. Many of these examples include the flipping of coins, the rolling of dice, and the random sampling of items from an urn or bag. These settings are used to allow for exact results in the case of the analytic solution of the simple problems presented here.

Example 2.12 Three men and two women sit in a row of chairs in a random order. Let the event A be that men and women alternate (that is, MWMWM). Find $P(A)$.

Using the multiplication rule, there are $5! = 120$ equally likely outcomes to the random ordering, and, again by the multiplication rule, there are 12 orderings that correspond to men and women alternating (see Examples 1.14 and 1.15). Therefore,

$$P(A) = \frac{3 \cdot 2 \cdot 2 \cdot 1 \cdot 1}{5 \cdot 4 \cdot 3 \cdot 2 \cdot 1} = \frac{12}{120} = \frac{1}{10}.$$

This analytic solution is exact and correct. It can be checked by Monte Carlo simulation. The R function `sample` can be used to generate a random ordering of the five people, who will be numbered 1, 2, 3, 4, 5 (the women are even and the men are odd, an agreeable convention for the women). The event A associated with men and women alternating is equivalent to chairs 1, 3, and 5 being occupied by the men, so the product of their indices will be 15. Increasing the number of replications from 1000 (in the coin flipping experiment in Example 2.7) to 100,000, the R code for the Monte Carlo simulation experiment is shown below.

```
nrep = 100000
count = 0
for (i in 1:nrep) {
  x = sample(5)
  if (x[1] * x[3] * x[5] == 15) count = count + 1
}
print(count / nrep)
```

Using indices 1, 3, and 5 to denote both the chairs and the men is coincidental. We could have used 1, 2, and 3, for example, to denote the men. After a call to `set.seed(3)` to initialize the random number seed, the code segment is run five times yielding

0.09948	0.10076	0.10100	0.10105	0.09859.

The fact that the five probability estimates are closer to the analytic value than in the previous Monte Carlo simulation experiment is due to the larger number of replications. Two of these estimates of the probability that men and women alternate are less than the true value $(1/10)$ and three of these estimates are greater than the true value, so the analytic solution is considered "verified." Although we will use the terms like *verified* and *confirmed* when Monte Carlo simulation results hover around the analytic value, these terms are a bit misleading. The Monte Carlo simulation results provide supporting evidence, but they do not provide a mathematical verification of an analytic solution. Monte Carlo can be helpful to see when an analytic solution is incorrect, however, because the estimates will not hover around the analytic solution.

Example 2.13 A *hatcheck girl* collects n hats and returns them at random. Let the event A be the proper return of the hats to their owners. Find $P(A)$.

By the multiplication rule, the hats can be returned in $n!$ different orders. Of these orders, only one of the orders is correct. Thus the probability of returning all hats to the correct owners is

$$P(A) = \frac{1 \cdot 1 \cdot \ldots \cdot 1}{n \cdot (n-1) \cdot \ldots \cdot 1} = \frac{1}{n!}.$$

Example 2.14 Roll a pair of fair dice 24 times. Let the event A be rolling double sixes at least once. Find $P(A)$.

Using complementary probabilities, the probability of rolling double sixes is one minus the probability of not rolling double sixes. Considering order to be important, there are 36^{24} different possible outcomes for 24 rolls of the dice by the multiplication rule. Of these possible outcomes, 35^{24} correspond to not rolling double sixes. Therefore, the probability of rolling a double six somewhere in the 24 rolls is

$$P(A) = 1 - P(A') = 1 - \frac{35 \cdot 35 \cdot \ldots \cdot 35}{36 \cdot 36 \cdot \ldots \cdot 36} = 1 - \frac{35^{24}}{36^{24}} \cong 0.4914.$$

Example 2.15 A five-card poker hand is dealt. Let the event A be that there are exactly 2 kings in the hand. Find $P(A)$.

There are $\binom{52}{5} = 2,598,960$ equally-likely five-card poker hands, which goes in the denominator of the expression for $P(A)$. There is the assumption that the *order* in which the cards are dealt into the poker hand is not relevant, and that fact will be reflected in the numerator. Determining how many of these hands contain exactly two kings requires the use of both the multiplication rule and combinations. First, choose the two kings (for example, the king of hearts and the king of diamonds) out of the four kings, which can be done in $\binom{4}{2} = 6$ different ways, where order is not relevant. Likewise, choose the three non-kings out of the 48 non-kings in $\binom{48}{3} = 17,296$ different ways, where order is again not relevant. Finally, the total number of hands with exactly two kings can be found by the multiplication rule by taking the product of the two binomial coefficients, yielding

(margin handwritten: choose 2 kings from 4 then choose other 3 cards in hand.)

$$P(A) = \frac{\binom{4}{2}\binom{48}{3}}{\binom{52}{5}} = \frac{103776}{2598960} = \frac{2162}{54145} \cong 0.0399.$$

This example illustrates that it is helpful to express the solution as an exact fraction, but also to express the solution as a decimal. Approximately 4% of the random deals of a five-card poker hand will contain exactly 2 kings.

Example 2.16 A five-card poker hand is dealt. Let the event A be a full house. Find $P(A)$.

A full house consists of three cards having one rank and two cards having another rank, for example $7\heartsuit, 7\clubsuit, 7\spadesuit, J\diamondsuit, J\clubsuit$. As in the previous exercise, we assume that there are $\binom{52}{5}$ different equally-likely five-card poker hands. To determine how many of these correspond to a full house, there are 13 ways to select the rank for the threesome, leaving 12 ways to select the rank for the pair. Once the two ranks have been selected, there are $\binom{4}{3} = 4$ ways to select the suits associated with the threesome, and $\binom{4}{2} = 6$ ways to select the suits associated with the pair. So the probability of dealing a full house is

(margin handwritten: 13 Choose 1 for the first card, 12 choose 1 for the second card then 4 choose 3 for the suit of the 3-some and $\binom{4}{2}$ for the 2-some)

(handwritten arrow: # of full house hands.)

$$P(A) = \frac{\binom{13}{1}\binom{12}{1}\binom{4}{3}\binom{4}{2}}{\binom{52}{5}} = \frac{\boxed{3,744}}{2,598,960} = \frac{6}{4165} \cong 0.00144.$$

This is an exceedingly rare event; there are only 3744 full-house hands out of the $2,598,960$ possible five-card poker hands.

Example 2.17 A five-card poker hand is dealt. Let the event A be two pair. Find $P(A)$.

A five-card poker hand containing two pair has one pair having one rank, a second pair of a different rank, and a single card of a third rank, for example, $8\heartsuit, 8\clubsuit, 2\spadesuit, 2\diamondsuit, A\clubsuit$.

The denominator is the same as the previous two examples. This time, the order of the choice of the two ranks associated with the pairs is not relevant, so there are $\binom{13}{2} = 78$ ways to select the ranks of the pairs, then $\binom{4}{2} = 6$ ways to select the suits of the pairs, then $11 \cdot 4 = 44$ ways to choose the single card. Thus, the probability of dealing two pair is

[handwritten annotations: "ranks of pairs ←", "suits of pairs →", "rank of single card →", "suit of single."]

$$P(A) = \frac{\binom{13}{2}\binom{4}{2}\binom{4}{2}\binom{11}{1}\binom{4}{1}}{\binom{52}{5}} = \frac{123,552}{2,598,960} = \frac{198}{4165} \cong 0.0475.$$

Not surprisingly, this probability is higher than the probability of dealing a full house.

Example 2.18 A five-card poker hand is dealt from a well-shuffled deck. Let the event A be a straight. Find $P(A)$.

A straight consists of five cards, not all of the same suit, which have consecutive ranks, for example, $8\heartsuit, 9\heartsuit, 10\spadesuit, J\diamondsuit, Q\spadesuit$. Without loss of generality, assume that the ace can be played low or high. We again put the usual binomial coefficient in the denominator that counts the number of different five-card poker hands, which ignores the order that the cards are dealt. To determine the number of these hands that correspond to a straight, use the multiplication rule on the following two quantities: (a) there are 10 ways to select the lowest rank in the straight $(A, 2, 3, \ldots, 10)$, (b) there are 4^5 ways to select the 5 cards with consecutive ranks in the straight via the multiplication rule (the 4 that is subtracted from this quantity corresponds to all of the cards having the same suit). Thus, the probability of a straight is

$$P(A) = \frac{10\left(4^5 - 4\right)}{\binom{52}{5}} = \frac{10,200}{2,598,960} = \frac{5}{1274} \cong 0.0039.$$

Example 2.19 A dozen eggs contains 3 defectives. If a sample of 5 is selected from the dozen at random,

(a) find the probability that the sample contains exactly 2 defectives,

(b) find the probability that the sample contains 2 or fewer defectives.

Since the order that the sample is selected is not important in both of the questions, there are $\binom{12}{5} = 792$ ways to sample 5 eggs from the dozen.

(a) To find the probability that exactly 2 are defective eggs in the sample, we combine the multiplication rule and combinations to determine the number of samples with exactly two defectives. There are $\binom{3}{2} = 3$ ways to select the two defective eggs and $\binom{9}{3} = 84$ ways to select the nondefective eggs. Therefore

$$P(\text{exactly two defectives}) = \frac{\binom{3}{2}\binom{9}{3}}{\binom{12}{5}} = \frac{252}{792} = \frac{7}{22} \cong 0.3182.$$

(b) To find the probability that 2 or fewer are defective follows the same pattern, but a summation is required to add the probabilities of the disjoint events:

*[handwritten annotation at left:
$1 - P(3\,\text{defective})$
$= 1 - \dfrac{\binom{3}{3}\binom{9}{2}}{\binom{12}{5}}$
$= 1 - \dfrac{1}{22} = \dfrac{21}{22}$]*

$$P(\text{two or fewer defectives}) = \sum_{i=0}^{2} \frac{\binom{3}{i}\binom{9}{5-i}}{\binom{12}{5}} = \frac{21}{22} \cong 0.9545.$$

The exact fractional value can be calculated with the Maple statement

```
sum(binomial(3, i) * binomial(9, 5 - i) / binomial(12, 5),
    i = 0 .. 2);
```

Example 2.20 Let A be the event that two or more people have the same birthday in a room filled with n people. Ignore leap years and assume that each day is equally likely as a birthday. Find $P(A)$.

This is the famous "birthday problem" whose solution is surprisingly high. To find the probability that two or more people have the same birthday, begin by using complementary probability to alter the problem to finding the probability that all n of the people have different birthdays. Think of assigning the birthdays to the n people as drawing n balls, numbered $1, 2, \ldots, 365$, from an urn with replacement. Using this model, there are 365^n ways to sample n balls from the urn via the multiplication rule, which is the denominator. In this denominator, order has been assumed to be important, which must be reflected in the numerator. To determine how many of these samples correspond to all n of the balls having different numbers, the multiplication rule again yields $365 \cdot 364 \cdot \ldots \cdot (365 - n + 1)$. So the probability that two or more people have the same birthday is

$$P(A) = 1 - P(A') = 1 - \frac{365 \cdot 364 \cdot \ldots \cdot (365 - n + 1)}{365 \cdot 365 \cdot \ldots \cdot 365} = 1 - \frac{365!}{(365 - n)! 365^n}.$$

The probabilities are computed for several values of n and listed below.

n	10	20	30	40	50	60
$P(A)$	0.117	0.411	0.706	0.891	0.970	0.994

These values can be plotted using the R code below.

```
n = 1:60
p = rep(1, 60)
for (i in 2:60)
  p[i] = p[i - 1] * (365 - i + 1) / 365
plot(n, 1 - p)
```

The plot appears in Figure 2.7. Surprisingly, it only takes $n = 23$ people in a room for this probability to exceed $1/2$. This may provide an opportunity for you to extract some funds from your non-probability friends on the consideration of a wager.

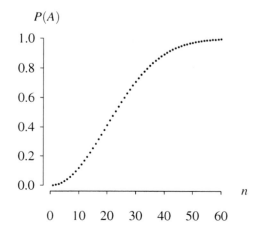

Figure 2.7: Birthday problem probabilities.

Example 2.21 A die is loaded in such a fashion that the probability of a particular face coming up is proportional to the number of spots on the face. Find the probability that exactly x spots appear on the up face and the probability that a one or a six is tossed.

The problem statement indicates that the probability of x spots on the up face is proportional to x, so

$$P(x \text{ spots appear on the up face}) = cx \qquad x = 1, 2, \ldots, 6,$$

where c is a constant. The sum of the six probabilities must be one, so

$$\sum_{x=1}^{6} P(x \text{ spots appear on the up face}) = \sum_{x=1}^{6} cx = c \sum_{x=1}^{6} x = c \frac{6(6+1)}{2} = 21c = 1.$$

Thus

$$c = \frac{1}{21}$$

$$\sum_{x=1}^{n} x = \frac{n(n+1)}{2}$$

and the probability that exactly x spots appear on the up face is

$$P(\text{exactly } x \text{ spots appear on the up face}) = \frac{x}{21} \qquad x = 1, 2, \ldots, 6$$

or

$$P(\text{exactly } x \text{ spots appear on the up face}) = \begin{cases} 1/21 & x = 1 \\ 2/21 & x = 2 \\ 3/21 & x = 3 \\ 4/21 & x = 4 \\ 5/21 & x = 5 \\ 6/21 & x = 6. \end{cases}$$

The whitespace between the probabilities and the x values in this equation can be interpreted as "if" or "when," which is a convention that is used throughout this text. Finally, the probability that a one or six is tossed is

$$P(1 \text{ or } 6 \text{ is tossed}) = \frac{1}{21} + \frac{6}{21} = \frac{7}{21} = \frac{1}{3}.$$

Example 2.22 Experience has shown that the number of arrivals in one hour to a certain ice cream stand during peak hours has a discrete probability function

Poisson Distribution

$$P(\text{exactly } x \text{ customers arrive in one hour}) = \frac{7^x e^{-7}}{x!} \qquad x = 0, 1, 2, \ldots.$$

What is the probability that there will be fewer than 3 arrivals in a particular hour?

Since 0, 1, or 2 arrivals are each disjoint events, the probability that there will be fewer than 3 arrivals in a particular hour is calculated by summing:

$$
\begin{aligned}
P(0, 1, \text{ or } 2 \text{ customers arrive in one hour}) &= P(0 \text{ customers arrive in one hour}) + \\
&\quad P(1 \text{ customer arrives in one hour}) + \\
&\quad P(2 \text{ customers arrive in one hour}) \\
&= \frac{7^0 e^{-7}}{0!} + \frac{7^1 e^{-7}}{1!} + \frac{7^2 e^{-7}}{2!} \\
&= e^{-7}\left(1 + 7 + \frac{49}{2}\right) \\
&= e^{-7}\left(\frac{65}{2}\right) \\
&\cong 0.02964.
\end{aligned}
$$

Example 2.23 A bag contains 15 billiard balls numbered 1 through 15. Five balls are randomly drawn from the bag without replacement. Let the event A be exactly two odd-numbered balls are drawn from the bag and they occur on odd-numbered draws. Find $P(A)$.

Divide all possible outcomes into $15 \cdot 14 \cdot 13 \cdot 12 \cdot 11$ equally-likely outcomes because the order that the balls are drawn from the bag matters. To determine the number of outcomes corresponding to exactly two odd-numbered balls drawn from the urn and which occur on odd-numbered draws from the bag, consider that

- there are $\binom{8}{2}$ different ways to choose the odd-numbered balls, from the urn (order is not relevant),

- there are $\binom{7}{3}$ different ways to choose the even-numbered balls from the urn (order is not relevant),

- there are three different positions (namely 1, 3, and 5) for the draw number where the one even draw assumes an odd draw number,

- there are 3! ways to order the even numbers,

- there are 2! ways to order the odd numbers.

Using the equally likely approach, the probability that exactly two odd-numbered balls are drawn from the bag and they occur on odd-numbered draws from the bag is

$$P(A) = \frac{\binom{8}{2} \cdot \binom{7}{3} \cdot 3 \cdot 3! \cdot 2!}{15 \cdot 14 \cdot 13 \cdot 12 \cdot 11} = \frac{28 \cdot 35 \cdot 3 \cdot 6 \cdot 2}{360,360} = \frac{35,280}{360,360} = \frac{14}{143} \cong 0.09790.$$

This result can be checked by Monte Carlo simulation. The R code below counts the number of times the event of interest occurs over 100,000 replications.

```
nrep = 100000
count = 0
for (i in 1:nrep) {
  x = sample(15, 5)
  numeven  = x[2]%%2 + x[4]%%2
  numodd   = x[1]%%2 + x[3]%%2 + x[5]%%2
  if ((numeven == 0) && (numodd == 2)) count = count + 1
}
print(count / nrep)
```

The `sample` function returns a random sample of 5 observations selected from the integers 1, 2, ..., 15 without replacement. The *modulo* operator `%%` returns the remainder when the second operator is divided into the first operator. The variable `numeven` counts the number of odd numbered balls selected on even numbered draws; the variable `numodd` counts the number of odd numbered balls selected on odd numbered draws. After calling `set.seed(3)`, this code is executed five times, yielding the probability estimates

0.09751   0.09867   0.09934   0.09666   0.09854.

Since two of these values fall below the calculated exact probability and three of these values fall above the calculated exact probability, the Monte Carlo simulation supports the analytic result.

Example 2.24 Consider an $n \times n$ matrix with all zeroes as entries. If n elements of the matrix are selected at random and set equal to one, find the probability that each row sum and each column sum is equal to one.

All of the examples thus far in this section have considered only a single solution method, typically one where the order of the outcomes is considered important or not important as dictated by the problem statement. In this particular question, the problem can be solved both ways. The key to getting the correct solution is to be consistent on the decision of whether order is important in the numerator and denominator of the solution. The first solution considers order to be important.

Solution 1 (assume that the order of the placement of the ones is relevant). Start with the case of $n = 3$, which corresponds to a 3×3 matrix. Considering the order of the placement of the ones to matter, there are $9 \cdot 8 \cdot 7 = 504$ different ways to place the ones. Now consider the ways to place the ones so that all three row sums and column sums equal 1. The first 1 can be placed in any of the nine positions. Once it is placed, the second 1 can be placed in any of four positions (it can't be in the same row or column as the first 1). Finally, the third 1 can be placed in only one position. Thus for $n = 3$, the probability that the row sums and column sums equal 1 is

$$\frac{9 \cdot 4 \cdot 1}{9 \cdot 8 \cdot 7} = \frac{36}{504} = \frac{1}{14}.$$

For $n = 4$, this becomes

$$\frac{16 \cdot 9 \cdot 4 \cdot 1}{16 \cdot 15 \cdot 14 \cdot 13} = \frac{576}{43,680} = \frac{6}{455}.$$

Proceeding to the general case, the probability that the row sums and column sums all equal 1 is

$$\frac{n^2 \cdot (n-1)^2 \cdot \ldots \cdot 1^2}{n^2 \cdot (n^2 - 1) \cdot \ldots \cdot (n^2 - n + 1)} = \frac{\prod_{i=1}^{n} i^2}{(n^2)!/(n^2 - n)!} = \frac{(n!)^2}{(n^2)!/(n^2 - n)!} = \frac{n!}{\binom{n^2}{n}}.$$

Solution 2 (assume that the order of the placement of the ones is not relevant). Start again with the case of $n = 3$. Considering the order of the placement of the ones to not be relevant, there are $\binom{9}{3} = 84$ different ways to place the ones. Now consider the ways to place the ones so that all three row sums and column sums equal 1. The 1 that occupies the first row can be placed in any of the three positions. Once it is placed, the 1 that occupies the second row can be placed in any of two positions (it can't be in the same column as the first 1). Finally, the 1 in the third row can be placed in only one position. Thus for $n = 3$, the probability that the row sums and column sums equals 1 is

$$\frac{3 \cdot 2 \cdot 1}{\binom{9}{3}} = \frac{6}{84} = \frac{1}{14}.$$

For $n = 4$, this becomes

$$\frac{4 \cdot 3 \cdot 2 \cdot 1}{\binom{16}{4}} = \frac{24}{1820} = \frac{6}{455}.$$

Proceeding to the general case, the probability that the row sums and column sums all equal 1 is

$$\frac{n!}{\binom{n^2}{n}}.$$

2.4   Conditional Probability

The previous section solved a series of probability problems using the probability axioms, the equally likely approach, and Monte Carlo simulation. This section addresses probability questions in light of other additional information, that is, calculating probabilities conditioned on the fact that another event has occurred. Oftentimes, the probability of an event goes up (or down) when you know whether another event has occurred. In other instances, we are interested in only the outcomes in a subset of S. Applications of conditional probability include:

- *Meteorology:* What is the probability that it rains tomorrow given that it is raining today?

- *Stock market:* What is the probability that a stock market index rises today given that it dropped yesterday?

- *Genetics:* What is the probability that a child will have blue eyes given that one parent has blue eyes?

- *Economics:* What is the probability that government revenue will increase next month given that there is a specified small increase in unemployment this month?

In all cases, we seek the probability of one event given that another event has occurred. Consider the following specific example.

> **Example 2.25** Toss a fair die and observe the number of spots on the up face. Let the event A correspond to tossing a 1, 2, or 3. Let the event B correspond to tossing an odd number. What is the probability of A given that B has occurred?
>
> The sample space and two events of interest are
>
> $$S = \{1, 2, 3, 4, 5, 6\},$$
> $$A = \{1, 2, 3\},$$
> $$B = \{1, 3, 5\}.$$
>
> If a single toss of the die has occurred and you do not know of the result, but you are informed that event B has occurred (the result is an odd number), what is the probability that the result is a 1, 2, or 3? Once you are informed that B has occurred, the sample space is restricted to just the three odd numbers. Since two of the odd numbers are in A, the conditional probability is $2/3$.

   This notion of restricting the sample space in order to compute the conditional probability is captured in the following formal definition of conditional probability.

Definition 2.7 If A and B are two events in the sample space S associated with a random experiment, then the probability of A given B is

$$P(A \mid B) = \frac{P(A \cap B)}{P(B)}$$

Probability of Event A given that Event B has occured.

provided $P(B) \neq 0$.

> **Example 2.26** Find $P(A \mid B)$ for the events defined in the previous example.
>
> In this case,
>
> $$A \cap B = \{1, 3\},$$
> $$P(A \cap B) = 2/6,$$

$A \cap B$ → intersect of "AND"

$$P(B) = 3/6.$$

Applying Definition 2.7 to compute the conditional probability,

$$P(A\,|\,B) = \frac{P(A \cap B)}{P(B)} = \frac{2/6}{3/6} = \frac{2}{3}.$$

Example 2.27 The results of a random sample of 100 subjects classified by their gender and eye color is given in Table 2.2. If one of the subjects is selected at random,

(a) find the probability that they have blue eyes given that they are male,

(b) find the probability that they are female given that they have green eyes.

	Blue	Green	Other
Male	26	23	24
Female	13	12	2

Table 2.2: Gender and eye color for 100 subjects.

For an individual subject, define the events as follows:

B: blue eyes,
G: green eyes,
O: other colored eyes,
M: male,
F: female.

Using Definition 2.7, the conditional probabilities of interest are

$$P(B\,|\,M) = \frac{P(B \cap M)}{P(M)} = \frac{26/100}{73/100} = \frac{26}{73}$$

and

$$P(F\,|\,G) = \frac{P(F \cap G)}{P(G)} = \frac{12/100}{35/100} = \frac{12}{35}.$$

Example 2.28 Consider the events A_1 and A_2 with associated probabilities $P(A_1) = 0.3$, $P(A_2) = 0.5$, and $P(A_1 \cap A_2) = 0.2$. Find $P(A_1\,|\,A_2)$ and $P(A_2\,|\,A_1)$.

These are the same probabilities encountered in Example 2.9; the associated Venn diagram is replicated in Figure 2.8.

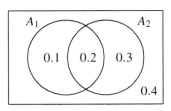

Figure 2.8: Venn diagram with probabilities.

Using Definition 2.7, the conditional probabilities requested are

$$P(A_1 | A_2) = \frac{P(A_1 \cap A_2)}{P(A_2)} = \frac{0.2}{0.5} = \frac{2}{5}$$

and

$$P(A_2 | A_1) = \frac{P(A_1 \cap A_2)}{P(A_1)} = \frac{0.2}{0.3} = \frac{2}{3}.$$

Both of the probabilities were altered by the fact that we have been given the additional information. For example,

$$P(A_1 | A_2) > P(A_1)$$

means that the knowledge that A_2 has occurred increases the probability that A_1 occurs. In this sense, the two events are dependent.

There are five loose ends concerning conditional probability that will be addressed in this paragraph. First, conditional probability, like any other probability, will have a range between 0 and 1, that is,

$$0 \leq P(A | B) \leq 1.$$ Always!   Kolmogorov axiom

Second, it should not be surprising that $P(A_1 | A_1) = 1$. Third, the definition of conditional probability generalizes to having more complicated events for A and B, for example,

$$P(A_1 \cap A_2' | A_3) = \frac{P((A_1 \cap A_2') \cap A_3)}{P(A_3)}$$

or

$$P(A_1 | A_2' \cup A_3) = \frac{P(A_1 \cap (A_2' \cup A_3))}{P(A_2' \cup A_3)}.$$

Fourth, if $A_2, A_3, \ldots$ are disjoint, then $P(A_2 \cup A_3 \cup \ldots | A_1) = P(A_2 | A_1) + P(A_3 | A_1) + \cdots$ provided $P(A_1) \neq 0$. Finally, a simple manipulation of Definition 2.7 yields

$$P(A \cap B) = P(A | B)P(B),$$ returns in Rule of Bayes and rule of

which is often referred to as the multiplication rule. Likewise, by symmetry,   elimination.

$$P(A \cap B) = P(B | A)P(A).$$

Using this operation will be referred to simply as *conditioning* in this text. This simple relationship, which is applied in the next two examples, also leads to the rule of elimination and the rule of Bayes.

Example 2.29 There are 10,000 doctors placed on a low dose of aspirin and 10,000 other doctors placed on a placebo for a year. During this year 107 of those on aspirin have a heart attack, while 187 of those on the placebo have a heart attack. If a doctor from the study is selected at random, what is the probability that the doctor will have been on aspirin and had a heart attack?

Define the events:

$$A: \text{ being on aspirin,}$$
$$H: \text{ having a heart attack.}$$

The information given in the problem statement is

$$P(H | A) = 0.0107$$

and

$$P(A) = \frac{10,000}{20,000} = 0.5.$$

So the probability that a doctor selected at random has been on aspirin and had a heart attack is

$$P(A \cap H) = P(H \mid A)P(A) = (0.0107)(0.5) = 0.00535.$$

Example 2.30 Ray is a professional blackjack player. He looks out over the nine cards that are face up on the table and sees that there are no jacks showing and only one ace showing. What is the probability that Ray will get an ace and a black jack (either order is fine) on the next two cards dealt?

Begin by defining the events

$$A_1: \text{ drawing an ace first,}$$
$$B_2: \text{ drawing a black jack second.}$$

Using conditional probability,

$$
\begin{aligned}
P(A_1 \cap B_2) &= P(B_2 \mid A_1)P(A_1) \\
&= \frac{2}{42} \cdot \frac{3}{43} \\
&= \frac{1}{301}.
\end{aligned}
$$

To determine the probability of the other ordering, define the events

$$B_1: \text{ drawing a black jack first,}$$
$$A_2: \text{ drawing an ace second.}$$

Again using conditional probability,

$$
\begin{aligned}
P(B_1 \cap A_2) &= P(A_2 \mid B_1)P(B_1) \\
&= \frac{3}{42} \cdot \frac{2}{43} \\
&= \frac{1}{301}.
\end{aligned}
$$

Since the two events associated with drawing the ace and black jack are disjoint,

$$P\big((A_1 \cap B_2) \cup (B_1 \cap A_2)\big) = \frac{1}{301} + \frac{1}{301} = \frac{2}{301}.$$

Conditioning can be used repeatedly, for example

$$
\begin{aligned}
P(A_1 \cap A_2 \cap A_3) &= P(A_1 \cap A_2)P(A_3 \mid A_1 \cap A_2) \\
&= P(A_1)P(A_2 \mid A_1)P(A_3 \mid A_1 \cap A_2).
\end{aligned}
$$

Using conditioning repeatedly in this fashion results in the more formally stated *rule of elimination*, also called the *law of total probability*, which is given next. Recall from Section 2.2 that a partition of a sample space S is formed by a set of disjoint events whose union is S.

Theorem 2.7 (Rule of elimination) Let $A_1, A_2, \ldots, A_n$ be a set of events that partition the sample space S, and $P(A_i) > 0$ for $i = 1, 2, \ldots, n$. For any event B,

$$P(B) = \sum_{i=1}^{n} P(B \mid A_i)P(A_i).$$

Proof The probability of B can be written as

[handwritten annotation: →definition using set theory]

$$P(B) = P(B \cap S)$$

$$= P\left(B \cap \left(\bigcup_{i=1}^{n} A_i\right)\right)$$

[handwritten annotation: →create S thaughs partion of sets A_i.]

$$= P\left(\bigcup_{i=1}^{n}(B \cap A_i)\right)$$

[handwritten annotation: →Move $\bigcup$ outside]

$$= \sum_{i=1}^{n} P(B \cap A_i)$$

[handwritten annotation: → definition of disjoint probabilities Axiom 2.]

$$= \sum_{i=1}^{n} P(B|A_i)P(A_i)$$

[handwritten annotation: ↳ conditional prob. definition]

because $A_1, A_2, \ldots, A_n$ form a partition of S and the events $B \cap A_1, B \cap A_2, \ldots, B \cap A_n$ are disjoint. □

Example 2.31 Urn 1 contains three white balls and two black balls. Urn 2 contains one white ball and three black balls. The initial state of the balls is illustrated in Figure 2.9. A ball is selected at random from Urn 1 and transferred to Urn 2. Next, a ball is selected at random from Urn 2 and transferred to Urn 1. Finally, a ball is selected at random from Urn 1. Find the probability that the ball selected on this third draw is black.

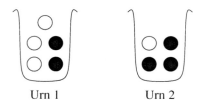

Urn 1   Urn 2

Figure 2.9: The initial state of the two urns.

Define the events

W_i: white ball selected on draw i, for $i = 1, 2, 3$,
B_i: black ball selected on draw i, for $i = 1, 2, 3$.

We want to find $P(B_3)$. Conditioning on the results of the first two draws,

$$
\begin{aligned}
P(B_3) &= P(B_3 \,|\, W_1 \cap W_2)P(W_1 \cap W_2) + P(B_3 \,|\, W_1 \cap B_2)P(W_1 \cap B_2) + \\
&\quad\; P(B_3 \,|\, B_1 \cap W_2)P(B_1 \cap W_2) + P(B_3 \,|\, B_1 \cap B_2)P(B_1 \cap B_2) \\
&= P(B_3 \,|\, W_1 \cap W_2)P(W_2 \,|\, W_1)P(W_1) + P(B_3 \,|\, W_1 \cap B_2)P(B_2 \,|\, W_1)P(W_1) + \\
&\quad\; P(B_3 \,|\, B_1 \cap W_2)P(W_2 \,|\, B_1)P(B_1) + P(B_3 \,|\, B_1 \cap B_2)P(B_2 \,|\, B_1)P(B_1) \\
&= \left(\frac{2}{5}\right)\left(\frac{2}{5}\right)\left(\frac{3}{5}\right) + \left(\frac{3}{5}\right)\left(\frac{3}{5}\right)\left(\frac{3}{5}\right) + \left(\frac{1}{5}\right)\left(\frac{1}{5}\right)\left(\frac{2}{5}\right) + \left(\frac{2}{5}\right)\left(\frac{4}{5}\right)\left(\frac{2}{5}\right) \\
&= \frac{12}{125} + \frac{27}{125} + \frac{2}{125} + \frac{16}{125} \\
&= \frac{57}{125} \\
&= 0.456.
\end{aligned}
$$

The R code for the Monte Carlo simulation check of the analytic solution simulates the movement of the balls from urn to urn. The code given below performs the three random samplings of balls from the urns 100,000 times, keeping a count of the number of times a black ball is selected on the third draw. Within the `for` loop, the variables `w1`, `b1`, `w2`, and `b2`, which represent the number of white and black balls in the two urns, are initialized. The `runif` function generates a random number that is equally likely between 0 and 1 that is used for simulating the randomness of the draws of the balls from the urns. The first `if-else` block simulates the movement of the ball from Urn 1 to Urn 2; the second `if-else` block simulates the movement of the ball from Urn 2 to Urn 1. The number of white balls and black balls in the two urns are incremented or decremented based on the type of ball selected. Finally, if the third draw from Urn 1 results in a black ball, then the `count` variable is incremented. After `nrep` replications of the experiment have been conducted, the program prints the fraction of times that a black ball is selected on the third draw.

```
nrep = 100000
count = 0
for (i in 1:nrep) {
   w1 = 3
   b1 = 2
   w2 = 1
   b2 = 3
   if (runif(1) < b1 / (w1 + b1)) {
      b1 = b1 - 1
      b2 = b2 + 1
   }
   else {
      w1 = w1 - 1
      w2 = w2 + 1
   }
   if (runif(1) < b2 / (w2 + b2)) {
      b2 = b2 - 1
      b1 = b1 + 1
   }
   else {
      w2 = w2 - 1
      w1 = w1 + 1
   }
   if (runif(1) < b1 / (w1 + b1)) count = count + 1
}
print(count / nrep)
```

After a call to `set.seed(3)` to initialize the random number seed, five runs of this code yield the following five estimated probabilities for selecting a black ball on the third draw:

0.45856	0.45667	0.45333	0.45621	0.45520.

Since three of these estimated probabilities fall above the exact value (0.456) and two of these estimated probabilities fall below the exact value, the Monte Carlo simulation lends support to the analytic solution.

Bayes

Bayes established the following result, which has evolved into an entire branch of
Bayesian statistics.

ule of Bayes) Let $A_1, A_2, \ldots, A_n$ be a set of events that partition the sample
space $_i) > 0$ for $i = 1, 2, \ldots, n$. For any event B with $P(B) > 0$,

$$P(A_j|B) = \frac{P(B|A_j)P(A_j)}{\sum_{i=1}^{n} P(B|A_i)P(A_i)}$$

for $j = 1, 2, \ldots, n$.

Proof Using the definition of conditional probability and the rule of elimination,

$$P(A_j|B) = \frac{P(A_j \cap B)}{P(B)} = \frac{P(B|A_j)P(A_j)}{\sum_{i=1}^{n} P(B|A_i)P(A_i)}$$

for $j = 1, 2, \ldots, n$. $P(A|B) = \boxed{P(A \cap B)/P(B)}$ Always remember $\square$

Example 2.32 Moe, Curly and Larry are gas station attendants. Moe handles 30% of
the customers, Curly handles 50% of the customers, and Larry handles the remaining 20% of the customers. They are always supposed to wash the customer's windshield. Moe forgets 1 time in 20, Curly forgets 1 time in 10, and Larry forgets 1 time
in 2.

(a) What is the probability that a windshield has *not* been washed?

(b) Given that a windshield was not washed, what is the probability that it was Curly
who didn't wash it?

Define the events

M: Moe handles the car,
C: Curly handles the car,
L: Larry handles the car,
W: the car's windshield is not washed.

The information that is given in the problem can be written in terms of these events as
follows.

$$P(M) = 0.3 \qquad\qquad P(W|M) = 0.05$$
$$P(C) = 0.5 \qquad\qquad P(W|C) = 0.10$$
$$P(L) = 0.2 \qquad\qquad P(W|L) = 0.50$$

(a) To find the probability of a windshield not being washed, use the rule of elimination:

$$
\begin{aligned}
P(W) &= P(W|M)P(M) + P(W|C)P(C) + P(W|L)P(L) \\
&= (0.05)(0.3) + (0.10)(0.5) + (0.50)(0.2) \\
&= 0.015 + 0.05 + 0.1 \\
&= 0.165.
\end{aligned}
$$

This probability is effectively a weighted average of the three probabilities of not
washing the windshield, where the weights 0.3, 0.5, and 0.2 correspond to the
probabilities of the Three Stooges handling the car.

(b) To find the probability that it was Curly who missed washing a windshield requires the rule of Bayes:

$$P(C|W) = \frac{P(W \cap C)}{P(W)}$$

$$= \frac{P(W|C)P(C)}{P(W)}$$

$$= \frac{(0.10)(0.5)}{(0.05)(0.3) + (0.10)(0.5) + (0.50)(0.2)}$$

$$= \frac{0.05}{0.165}$$

$$\cong 0.3030.$$

∴ P(W) is not a given and is found using rule of elimination, use rule of Bayes.

Example 2.33 Select a number at random from $1, 2, \ldots, n$. Call your number m. Now select a second number at random from $1, 2, \ldots, m$.

(a) Give an expression for the probability that the second number selected is 1.

(b) Give an expression for the probability that the first number was n given that the second number selected was $n - 1$.

Let the events $F_1, F_2, \ldots, F_n$ correspond to the first number selected being $1, 2, \ldots, n$. Let the events $S_1, S_2, \ldots, S_n$ correspond to the second number selected being $1, 2, \ldots, n$.

(a) The desired probability is $P(S_1)$. Using the rule of elimination,

$$P(S_1) = \sum_{i=1}^{n} P(S_1 | F_i)P(F_i)$$

$$= \sum_{i=1}^{n} \frac{1}{i} \cdot \frac{1}{n}$$

$$= \frac{1}{n}\left[1 + \frac{1}{2} + \frac{1}{3} + \cdots + \frac{1}{n}\right].$$

(b) The desired probability is $P(F_n | S_{n-1})$. Using the rule of Bayes,

$$P(F_n | S_{n-1}) = \frac{P(S_{n-1}|F_n)P(F_n)}{\sum_{j=n-1}^{n} P(S_{n-1}|F_j)P(F_j)}$$

$$= \frac{P(S_{n-1}|F_n)P(F_n)}{P(S_{n-1}|F_{n-1})P(F_{n-1}) + P(S_{n-1}|F_n)P(F_n)}$$

$$= \frac{\frac{1}{n} \cdot \frac{1}{n}}{\frac{1}{n-1} \cdot \frac{1}{n} + \frac{1}{n} \cdot \frac{1}{n}}$$

$$= \frac{n-1}{2n-1}.$$

This solution makes sense in the limiting case where

$$\lim_{n \to \infty} P(F_n | S_{n-1}) = \lim_{n \to \infty} \frac{n-1}{2n-1} = \frac{1}{2}$$

because for large n, the probability that the first number is n given that the second number is $n - 1$ should converge to $1/2$ because the probabilities $P(F_n | S_{n-1})$ and $P(F_{n-1} | S_{n-1})$ are equal in the limit as $n \to \infty$.

Example 2.34 The "car and goats" problem, also known as the "Monty Hall" paradox or the "Let's Make a Deal" problem, can be solved using the rule of Bayes. The game show host, Monty Hall, shows you three closed doors. There is a car behind one of the doors and goats behind the other two. If you open the door with the car behind it, you keep the car. You select a door, but before the door is opened, Monty Hall opens one of the other doors to reveal a goat, then gives you the option of switching doors. Is there any advantage to switching?

This is considered a paradox because most people believe that switching doors does not improve your chances of getting the car. The rule of Bayes will show otherwise. Begin by defining the following events.

$$C_1: \quad \text{car is behind door 1,}$$
$$C_2: \quad \text{car is behind door 2,}$$
$$C_3: \quad \text{car is behind door 3.}$$

The information that is given in the problem statement can be written in terms of these events as follows. Since all three of these events are equally likely, $P(C_1) = P(C_2) = P(C_3) = 1/3$. Furthermore define the following events associated with the host's action following your initial door selection.

$$H_1: \quad \text{host opens door 1,}$$
$$H_2: \quad \text{host opens door 2,}$$
$$H_3: \quad \text{host opens door 3.}$$

Without loss of generality, assume that you select door 1 and the host opens door 3. In this case,

$$P(H_3 \,|\, C_1) = 1/2$$

$$P(H_3 \,|\, C_2) = 1$$

$$P(H_3 \,|\, C_3) = 0$$

assuming that the host is equally likely to open either door 2 or door 3 if the car is behind door 1. So using the rule of Bayes,

$$P(C_1 \,|\, H_3) = \frac{P(H_3 \,|\, C_1)P(C_1)}{\sum_{i=1}^{3} P(H_3 \,|\, C_i)P(C_i)} = \frac{\frac{1}{2} \cdot \frac{1}{3}}{\frac{1}{2} \cdot \frac{1}{3} + 1 \cdot \frac{1}{3} + 0 \cdot \frac{1}{3}} = \frac{1}{3}$$

and

$$P(C_2 \,|\, H_3) = \frac{P(H_3 \,|\, C_2)P(C_2)}{\sum_{i=1}^{3} P(H_3 \,|\, C_i)P(C_i)} = \frac{1 \cdot \frac{1}{3}}{\frac{1}{2} \cdot \frac{1}{3} + 1 \cdot \frac{1}{3} + 0 \cdot \frac{1}{3}} = \frac{2}{3}.$$

The optimal strategy is to switch doors. Your probability of winning the car doubles by switching. This counter-intuitive strategy might best be explained by *New York Times* columnist John Tierney:

> ... when you stick with door 1, you'll win only if your original choice was correct, which happens only 1 in 3 times on average. If you switch, you'll win whenever your original choice was wrong, which happens 2 out of 3 times on average.

2.6 Independent Events

When $P(A|B) = P(A)$, one can conclude that the occurrence or nonoccurence of event B has no effect on the probability that event A occurs. In this case, the events A and B are called *independent events*.

By the definition of conditional probability,

$$P(A|B) = \frac{P(A \cap B)}{P(B)}.$$

Since $P(A|B) = P(A)$ for independent events, the equation

$$P(A \cap B) = P(A)P(B)$$

is typically used to define independent events.

Definition 2.8 Events A and B are *independent* if and only if
$$P(A \cap B) = P(A)P(B).$$

$$P(A|B) = \frac{P(A \cap B)}{P(B)}$$
$$\underset{P(A)}{\downarrow}$$

If the occurrence (or nonoccurrence) of one event doesn't affect the probability of another event occurring, then the two events are said to be *independent*. Events that are not independent are said to be *dependent*.

> **Example 2.35** A single card is drawn at random from a 52-card deck. Let the event H be that the suit of the card is hearts. Let the event Q be that the rank of the card is a queen. Are the events H and Q independent?
>
> Since there are 13 hearts, the probability that the card is a heart is
>
> $$P(H) = \frac{13}{52} = \frac{1}{4}.$$
>
> Since there are 4 queens, the probability that the card is a queen is
>
> $$P(Q) = \frac{4}{52} = \frac{1}{13}.$$
>
> Since there is only one queen of hearts, the probability of $H \cap Q$ is
>
> $$P(H \cap Q) = \frac{1}{52}.$$
>
> The events H and Q are independent because
>
> $$P(H \cap Q) = \frac{1}{52} = \frac{1}{4} \cdot \frac{1}{13} = P(H)P(Q).$$
>
> The independence of H and Q is consistent with common sense: the probability that the card is a heart is not altered based on whether the card is a queen, and vice-versa.

When independence is generalized to more than two events, some difficulties arise, as illustrated by the following example.

> **Example 2.36** A fair coin is tossed twice. Show that the events
>
> A: the first toss yields heads,
> B: the second toss yields heads,
> C: the two tosses yield different results,

are pairwise independent, but $P(A \cap B \cap C) \neq P(A)P(B)P(C)$.

The probabilities for the individual events are all equal: $P(A) = P(B) = P(C) = 1/2$. The pairwise probabilities are also all equal: $P(A \cap B) = P(A \cap C) = P(B \cap C) = 1/4$. Thus the three events are pairwise independent. Finally, $P(A \cap B \cap C) = 0$, which differs from $P(A)P(B)P(C) = 1/8$.

Example 2.36 has shown that pairwise independence of events does not imply that three events will satisfy a definition of independence similar to Definition 2.8. So with more than two events, the definition of independence must be a bit more complicated.

Definition 2.9 Events $A_1, A_2, \ldots, A_n$ are *mutually independent* if and only if the probability of occurrence of the intersection of any 2, 3, $\ldots$, or n of these events is equal to the product of their associated probabilities of occurrence.

In the case of three events, the following equations must be satisfied for three events to be mutually independent:

$$P(A_1 \cap A_2) = P(A_1)P(A_2)$$

$$P(A_1 \cap A_3) = P(A_1)P(A_3)$$

$$P(A_2 \cap A_3) = P(A_2)P(A_3)$$

$$P(A_1 \cap A_2 \cap A_3) = P(A_1)P(A_2)P(A_3).$$

Example 2.37 Consider a *series* system of n independent components with probabilities of functioning $p_1, p_2, \ldots, p_n$. If all components must function for the system to function, find the probability that the system functions.

Design engineers often base their product designs on reliability considerations. A block diagram of a series system with $n = 3$ components is shown in Figure 2.10. The reliability (the probability that the system functions) of an n-component series system can be found by multiplying the probabilities associated with independent components:

$$
\begin{aligned}
P(\text{system functions}) &= P(\text{all components function}) \\
&= P(\text{component 1 functions}) \ldots P(\text{component } n \text{ functions}) \\
&= p_1 p_2 \cdots p_n \\
&= \prod_{i=1}^{n} p_i.
\end{aligned}
$$

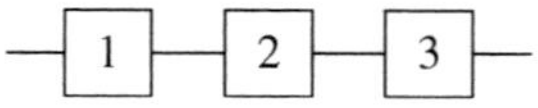

Figure 2.10: Block diagram for a three-component series system.

The series system arrangement is the worst possible way to arrange components in a system because the first component failure causes the system to fail. On the other end of the spectrum is the parallel system.

Example 2.38 Consider a *parallel* system of n independent components with probabilities of functioning $p_1, p_2, \ldots, p_n$. If one or more components must function for the system to function, find the probability that the system functions.

A parallel system has a higher probability of operating than the associated series system because components essentially back one another up in the case of a component failure. A block diagram of a parallel system with $n = 3$ components is shown in Figure 2.11. Using complementary probabilities and the independence of the components,

$$
\begin{aligned}
P(\text{system functions}) &= 1 - P(\text{system fails}) \\
&= 1 - P(\text{all components fail}) \\
&= 1 - P(\text{component 1 fails})\ldots P(\text{component } n \text{ fails}) \\
&= 1 - (1 - p_1)(1 - p_2)\ldots(1 - p_n) \\
&= 1 - \prod_{i=1}^{n}(1 - p_i).
\end{aligned}
$$

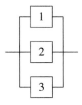

Figure 2.11: Block diagram for a three-component parallel system.

In many applications, there are a finite number of repeated independent *trials*, or sub-experiments, that we are interested in. In the next two examples, the repeated independent trials are playing in a major league baseball game and the rolling a pair of dice.

Example 2.39 Cal Ripken broke Lou Gehrig's record for consecutive major league baseball games played on September 6, 1995 when he played his 2131st consecutive game. If the probability that a player is healthy for any particular game is 0.99 and a player's ability to play is independent from game to game, find the probability that a player could play in 2131 consecutive games.

Because of independence,

$$P(\text{playing 2131 games in a row}) = (0.99)^{2131} = 0.0000000004996.$$

This small probability makes Gehrig's and Ripken's consecutive game streaks all the more impressive.

The chapter closes with a slightly more complicated example—an analysis of the dice game known as "craps."

Example 2.40 The game of craps consists of repeated tosses of a pair of fair dice. You win at craps by throwing a 7 or 11 on the first toss, or by throwing a 4, 5, 6, 8, 9, or 10 on the first toss (this number tossed is referred to as a "point") and subsequently throwing your point before you throw a 7. Find the probability of winning at craps.

Begin by defining the events

$$
\begin{aligned}
W &: \text{ winning the game,} \\
A_k &: \text{ initial roll is } k, \text{ for } k = 2, 3, \ldots, 12.
\end{aligned}
$$

The question seeks the probability of winning $P(W)$. Using the rule of elimination, the probability of winning can be expressed as

$$P(W) = P(W \mid A_2)P(A_2) + P(W \mid A_3)P(A_3) + \cdots + P(W \mid A_{12})P(A_{12}).$$

The probabilities $P(A_2), P(A_3), \ldots, P(A_{12})$ can be determined by the methods illustrated in Example 2.11. Calculating one of the conditional probabilities is illustrated below for $P(W \mid A_4)$, the probability of winning given that a 4 was tossed initially. Since the probability of throwing a 4 is $3/36$ and the probability of throwing neither a 4 nor a 7 is $27/36$,

$$
\begin{aligned}
P(W \mid A_4) &= \left(\frac{3}{36}\right) + \left(\frac{27}{36}\right)\left(\frac{3}{36}\right) + \left(\frac{27}{36}\right)\left(\frac{27}{36}\right)\left(\frac{3}{36}\right) + \cdots \\
&= \left(\frac{3}{36}\right)\left[1 + \frac{27}{36} + \left(\frac{27}{36}\right)^2 + \cdots\right] \\
&= \left(\frac{3}{36}\right)\left[\frac{1}{1 - 27/36}\right] \\
&= \left(\frac{3}{36}\right)\left(\frac{36}{9}\right) \\
&= \frac{1}{3}.
\end{aligned}
$$

The sum in the brackets is a geometric series with common ratio $27/36$. The second column of Table 2.3 shows the results of similar calculations for $k = 2, 3, \ldots, 12$. The rightmost column sum is the desired probability

$$P(W) = \frac{244}{495} \cong 0.4929.$$

The fact that the winning probability is high draws gamblers to the game; the fact that the winning probability is less than 0.5 assures the casino of positive revenue from the game in the long run. Not surprisingly, casinos are very particular about the fairness of the dice that are used in craps and typically have an employee who inspects the dice between every roll.

k	$P(W \mid A_k)$	$P(A_k)$	$P(W \mid A_k) \cdot P(A_k)$
2	0	1/36	0
3	0	2/36	0
4	1/3	3/36	1/36
5	2/5	4/36	2/45
6	5/11	5/36	25/396
7	1	6/36	1/6
8	5/11	5/36	25/396
9	2/5	4/36	2/45
10	1/3	3/36	1/36
11	1	2/36	1/18
12	0	1/36	0
		1	244/495

Table 2.3: Probabilities associated with craps.

This solution can also be checked by Monte Carlo simulation, although the coding is more complicated than in previous examples because the game is more complicated. A recursive algorithm for simulating 100,000 games of craps follows. Two functions are written prior to the main program. The function `roll` simulates the outcome $(2, 3, \ldots, 12)$ associated with one throw of a pair of dice. The function `play` simulates the throws required if a 4, 5, 6, 8, 9, or 10 is thrown on the first toss. All text following the # character is treated as a comment.

```
#
#  simulate the roll of a pair of fair dice; return sum of spots
#
roll = function()
{
  floor(1 + 6 * runif(1)) + floor(1 + 6 * runif(1))
}
#
#  roll until a 7 is thrown (return 0 for loss)
#  or until the point is made (return 1 for win)
#
play = function(point)
{
  sum = roll()
  while ((sum != point) && (sum != 7)) sum = roll()
  if (sum == point) return(1) else return(0)
}
#
#  main program
#
nrep  = 100000                        # number of replications
count = 0                             # number of wins
for (i in 1:nrep) {                   # perform nrep replications
  point = roll()                      # the initial roll
  result = switch(point,              # 0 = lose, 1 = win
                          0,          # 1 is impossible
                          0,          # 2 is a loss
                          0,          # 3 is a loss
                          play(point),# 4 is the point
                          play(point),# 5 is the point
                          play(point),# 6 is the point
                          1,          # 7 is a win
                          play(point),# 8 is the point
                          play(point),# 9 is the point
                          play(point),# 10 is the point
                          1,          # 11 is a win
                          0)          # 12 is a loss
  count = count + result              # count the number of wins
}                                     # end the for loop
print(count / nrep)                   # print the estimate of winning
```

Five runs of this simulation yield

0.49251	0.49500	0.49461	0.49057	0.49124

as estimates for the probability of winning at a game of craps. These estimates hover around the exact analytic solution of $P(W) = 244/495 \cong 0.4929$, so our confidence in the analytic solution is enhanced. As has been emphasized throughout this chapter,

Monte Carlo simulation does not verify or confirm an analytic solution; rather it can be used to identify an error in the analytic solution if the Monte Carlo estimates of a particular probability do not hover around the analytic value. When there is a significant difference between the two, there is either a bug in the Monte Carlo simulation code or there is a mathematical or logical error in the analytic solution.

In summary, this chapter began by defining three important concepts:

- a random experiment (an experiment with an uncertain outcome prior to its performance),

- a sample space S (the set of all possible outcomes to a random experiment),

- an event A (any subset of S).

With these definitions in place, the focus shifted to calculating the probability that event A will occur, which is denoted by $P(A)$. The argument to the probability set function P is an event A, and the range of the probability set function is a real number between 0 and 1 inclusive, where 0 corresponds to an impossible event and 1 corresponds to a certain event. Four ways of determining $P(A)$ are listed below.

- In the *relative frequency approach*, the random experiment is performed n times, keeping track of x, the number of times that event A occurs. The ratio x/n is known as the relative frequency. This approach is the basis for Monte Carlo simulation, which is useful for checking analytic solutions and estimating probabilities when analytic solutions are not possible. The relative frequency approach provides an estimate for $P(A)$.

- In the *subjective approach*, an expert is consulted to provide an estimate of $P(A)$.

- In the *equally likely approach*, the sample space consists of outcomes that occur with equal probabilities, so $P(A)$ is computed by taking the ratio

$$P(A) = \frac{N(A)}{N(S)}.$$

- In the *axiomatic approach*, probability theory is developed based on a set of three assumptions known as the Kolmogorov axioms.

The conditional probability of event A given that event B has occurred is

$$P(A \mid B) = \frac{P(A \cap B)}{P(B)},$$

when $P(B) \neq 0$. The rule of elimination and the rule of Bayes can be used when the sample space is partitioned by events A_1, A_2, ..., A_n. Pairwise independent and mutually independent events arise when the occurrence of an event is not influenced by the outcome of other events. We have emphasized the exact computation of probabilities in an analytic fashion in this chapter. There has been a secondary emphasis placed on Monte Carlo simulation.

2.7   Exercises

2.1   Four people play one round of the rock-scissors-paper (R-S-P) game.

(a) How many outcomes are in the sample space if players are considered distinct?

(b) If players are considered distinct, how many elements of the sample space correspond to:

- all four players getting the same symbol (for example, PPPP)?
- three players getting one symbol and the other player getting a different symbol (for example, SRSS)?
- two players getting one symbol and the two other players getting a different symbol (for example, PRPR)?
- two players getting one symbol and the two other players getting the other two symbols (for example, PPSR)?

2.2 Consider two equally-likely events A and B. If the probability that both occur is 0.2 and the probability that neither occurs is 0.3, find $P(A)$.

2.3 If A and B are disjoint events satisfying $P(A) = 0.3$ and $P(B') = 0.4$, what is $P(A' \cup B)$?

2.4 Let A_1, A_2, and A_3 be three events that partition the sample space S. Find $P(A_1 \cap (A_2 \cup A_3))$.

2.5 Prove Bonferroni's inequality: for any two events A_1 and A_2,

$$P(A_1 \cap A_2) \geq P(A_1) + P(A_2) - 1.$$

2.6 Let A_1, A_2, and A_3 form a partition of the sample space S. Find $P(A_1' \cup A_3')$.

2.7 Let E, F, and G be three events satisfying: $E \cap F = \emptyset$, $E \cap G \neq \emptyset$, and $F \cap G \neq \emptyset$, where $\emptyset$ is the null set. Draw a Venn diagram to illustrate that $P(E \cup F \cup G) = P(E) + P(F) + P(G) - P(E \cap G) - P(F \cap G)$.

2.8 For any two events A and B in S, if $P(A \cap B) = 0.4$, what is $P(A' \cup B')$?

2.9 For $P(A) = 2/3$ and $P(B) = 3/5$, what are the allowable values for $P(A \cap B)$?

2.10 Lucy, Edmond, Susan, and Peter are having an argument over who will get the first ride on the lion's back. They devise the following scheme to determine who will go first. They fill a bag with 15 billiard balls numbered $1, 2, \ldots, 15$, then draw out a ball at random in the following order: Lucy, Edmond, Susan, Peter, Lucy, Edmond, Susan, Peter, etc. The first person to draw out the ball with an 8 on it gets the first ride. Find the probability that Edmond will get the first ride if

(a) sampling from the bag is performed with replacement,

(b) sampling from the bag is performed without replacement.

2.11 If A and B are events such that the probability that at least one of them occurs is $2/3$ and the probability that A occurs but B does not occur is $1/4$, then what is $P(B)$?

2.12 What is the probability that a five-card hand dealt from a well-shuffled deck contains no clubs?

2.13 There are 10 married *couples* in a room. A random sample of 9 *people* is drawn from the 20 people. Find the probability that the sample contains exactly 3 married couples. Check your solution by Monte Carlo simulation.

2.14 Two points are chosen at random on the *perimeter* of a unit square. Find the probability that the two points are on opposite sides of the square.

2.15 Find the probability that a 9-digit social security number code is a palindrome. (A palindrome reads identically from left to right and from right to left, so, for example, 555555555 and 123454321 are palindromes and 123456789 is not a palindrome.) Social security numbers can have leading zeros.

2.16   Five balls are sampled without replacement from a bag containing 15 billiard balls numbered $1, 2, \ldots, 15$. Find the probability that the 1 ball and the 15 ball are in the sample. Write a Monte Carlo simulation in R to support the analytic solution.

2.17   A bowl contains 6 red chips, 7 white chips, and 8 blue chips. If four chips are selected from the bowl at random and without replacement, find the probability that each color is represented in the sample.

2.18   What is the probability that all six faces appear exactly once in six tosses of a fair die?

2.19   Lindsey has ten *pairs* of running shoes in her closet, each pair being a different brand. If Lindsey randomly selects five *shoes* from the closet without replacement, find the probability that she will have one or more matching pairs.

2.20   A survey of the schedules of 100 college freshmen reveals that

- 46 of the schedules contain a calculus class,
- 45 of the schedules contain an English class,
- 23 of the schedules contain a psychology class,
- 18 of the schedules contain both a calculus class and an English class,
- 7 of the schedules contain both a calculus class and a psychology class,
- 9 of the schedules contain both an English class and a psychology class,
- 2 of the schedules contain all three classes.

(a) How many people are enrolled in exactly two of the three types of classes?

(b) If one of the 100 schedules is chosen at random, find the probability that the schedule will contain no calculus class, no English class, and no psychology class.

(c) If ten different schedules are selected at random from the 100 schedules, find the probability that exactly six of the schedules include a calculus class.

2.21   Five fair die are rolled simultaneously. Find the probability that all five show different results on the up faces.

2.22   Abagail, Bert, Chuck, Debbie, Ed, and Francine line up shoulder-to-shoulder for a portrait in random order. Find the probability that Abagail is on one end or the other.

2.23   You roll a fair die five times. What is the probability that you will see a string of three or more consecutive ones?

2.24   Consider 10 tosses of a fair coin.

(a) What is the probability that all tosses are heads?

(b) What is the probability that there are exactly two heads?

(c) What is the probability that the first two tosses are heads?

2.25   What is the probability that all four suits (for example, $\spadesuit, \heartsuit, \diamondsuit, \clubsuit$) are represented in a five-card poker hand dealt from a well-shuffled deck?

2.26   Consider a bag of 15 billiard balls numbered $1, 2, \ldots, 15$. If n balls are drawn without replacement, $n = 1, 2, \ldots, 15$, find the probability that the product of the numbers on the balls selected is even.

2.27   Five fair dice are tossed. Find the probability of getting two pairs (for example, 6, 6, 2, 2, 5 or 1, 1, 4, 4, 6; but neither 4, 4, 4, 4, 2 nor 3, 3, 3, 3, 3, nor 5, 5, 4, 4, 4).

2.28 An urn contains 7 red balls, 8 white balls, and 9 blue balls. What is the probability that there are exactly x red balls and y white balls in a sample of 6 balls selected at random and without replacement from the urn? (*Hint:* make sure to include appropriate restrictions on x and y.)

2.29 A bag contains n balls numbered $1, 2, \ldots, n$, where n is an integer that is greater than or equal to 3.

    (a) If two balls are drawn from the bag at random with replacement, find the probability that the sum of the numbers drawn is less than or equal to n.

    (b) If two balls are drawn from the bag at random without replacement, find the probability that the sum of the numbers drawn is less than or equal to n.

    (c) Find the limit as $n \to \infty$ of your answers to parts (a) and (b).

2.30 Reckless rectilinear Russell is taking a rectilinear walk from point H (home) to point W (work) along the grid lines shown in Figure 2.12. Assuming all of the paths of length 21 from point H to point W (that is, he doesn't go out of his way and each square on the grid is 1×1) are equally likely, find the probability that he will go past the mailbox at point M.

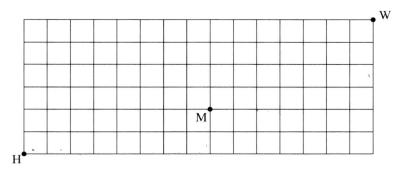

Figure 2.12: Reckless Russell's rectilinear neighborhood.

2.31 Risk-averse rectilinear Russell is taking a rectilinear walk from point H (home) to point W (work) along the grid lines shown in Figure 2.13. Relative to the previous exercise he has decided to avoid some of the neighborhoods in the northwest and southeast. Assuming that all of the paths of length 21 from point H to point W (that is, he doesn't go out of his way and each square on the grid is 1×1) are equally likely, find the probability that he will go past the mailbox at point M.

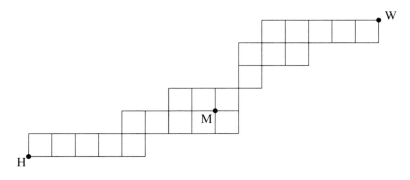

Figure 2.13: Risk-averse rectilinear Russell's neighborhood.

2.32 Six women and four men wait in line to board a commuter airplane. If their order in line is random, find the probability that all of the women are adjacent to one another.

2.33 Shweta rolls n fair dice, where n is a positive integer. Find the probability that all of the up faces are different.

2.34 Elisa rolls n fair dice, where n is a positive integer. Find the probability that the product of the numbers on the up faces is even.

2.35 Twelve people randomly divide into two teams of six each for a game of ultimate frisbee. What is the probability that Nancy and Kevin, two of the twelve, are on the same team?

2.36 Steve, Kay, Bethany, and Rebecca are playing the "rock, scissors, paper" game. If each chooses rock, scissors, and paper with equal probability, find the probability that all four choose paper.

2.37 An urn contains 7 red balls, 8 white balls, and 9 blue balls. What is the probability that all colors are represented in a sample of 4 balls selected at random and without replacement from the urn?

2.38 A bag contains 100 bills, 80 of which are authentic bills and 20 of which are counterfeit bills. Bills are drawn sequentially and without replacement from the bag. What is the probability that exactly two counterfeit bills are drawn before the third authentic bill is drawn?

2.39 A fair die is rolled repeatedly. Find the probability that the fourth 2 occurs on the xth roll.

2.40 A bag contains 6 balls numbered 1, 2, 3, 4, 5, and 6. Three balls are drawn without replacement from the bag. What is the probability that the difference between the largest numbered ball selected and the smallest numbered ball selected (often known as the "range") is equal to 4?

2.41 There are n people in a room. Each of them takes off both of their shoes and places them in a large (smelly) urn. If each person then selects two individual shoes at random and without replacement from the urn, find the probability that everyone has the correct pair of shoes.

2.42 Consider a random experiment consisting of tossing a fair coin repeatedly until the fourth head has appeared. What is the probability that the experiment ends after exactly eight tosses of the coin with a head appearing on both the seventh and eighth tosses?

2.43 If 8 people are chosen at random from a group of 18 married couples, what is the probability that there are exactly x (for $x = 0, 1, 2, 3, 4$) married couples in the group of 8?

2.44 Consider repeated rolls of five fair dice. On each roll, the sum of the spots on the five up faces is totaled. What is the probability of getting ten 28's before two 30's?

2.45 Find the probability that a five-card poker hand dealt from a well-shuffled deck contains:

(a) one pair (for example, QQ72J),

(b) two pair (for example, QQ77J, but not QQ777),

(c) a full house (for example, QQ777),

(d) three of a kind,

(e) four of a kind.

2.46 If m married couples sit in a random order at a round table, find the probability that no wife sits next to her husband in the limit as $m \to \infty$.

2.47   A round in the game of *Yahtzee*™ begins by rolling five fair dice. Find the probability of rolling

   (a) one pair (for example, 33421, but not 33441),

   (b) two pair (for example, 33441, but not 33444),

   (c) three of a kind (for example, 24252, but not 24242).

2.48   An urn contains 6 red balls, 7 white balls, and 8 blue balls.

   (a) If three balls are sampled without replacement, find the probability that all are different colors.

   (b) If three balls are sampled without replacement, find the probability that all are the same color.

   (c) If three balls are sampled with replacement, find the probability that all are different colors.

   (d) If three balls are sampled with replacement, find the probability that all are the same color.

   (e) If n balls are sampled with replacement, find the probability that all are red.

   (f) If n balls are sampled with replacement, find the probability that all are the same color.

   The last two parts to this question should yield sensible results in the limit as $n \to \infty$.

2.49   Three of the 250 children at Edison Elementary School have lice. The school nurse inspects n of the children for lice. Find the smallest value of n such that the probability that the nurse finds one or more infected children exceeds $1/2$.

2.50   Twenty panda bears are tagged and released in a particular area in China that is believed to contain 100 pandas. One year later, four of the pandas are captured. Find the probability that exactly i of the pandas captured have tags.

2.51   Consider a 10×10 matrix that consists of all zeros. Ten elements of the matrix are selected at random and their value is changed from a zero to a one. Find the probability that the ones fall in a line (row-wise, column-wise, or diagonally).

2.52   A waiting line consists of 40 men and 10 women arranged in a random order. What is the probability that no two women in the line are adjacent to one another?

2.53   A five-card hand is dealt from a well-shuffled deck of cards. This hand is a four of a kind (four cards of one rank and one card of another rank). A second five-card hand is dealt from the remaining cards in the deck. Find the probability that the second hand contains a four of a kind.

2.54   Five fair dice are rolled simultaneously. Find the probability that the total number of spots on the up faces totals 28 or more.

2.55   Six men and six women attend an Independence Day dance. Two of the men wear red shirts; two of the men wear white shirts; two of the men wear blue shirts. Likewise, two of the women wear red shirts; two of the women wear white shirts; two of the women wear blue shirts. The men and women are paired off into couples at random.

   (a) What is the probability that all couples are wearing the same colored shirts?

   (b) What is the probability that all couples are wearing different colored shirts?

   Check your solutions by Monte Carlo simulation.

2.56 Consider a single-elimination tournament with 16 teams. Each team is given a unique "seed" value in the range 1 to 16. The probability that a seed i team beats a seed j team is $j/(i+j)$. Given the pairings shown in Figure 2.14, use Monte Carlo simulation to estimate the probability that a 1 seed team wins the tournament. Run the simulation with enough replications so that you can report your estimate to two digits of accuracy. *Note:* In terms of the NCAA basketball tournament (March madness), this question is equivalent to estimating the probability that a number 1 seed team makes it to the final four.

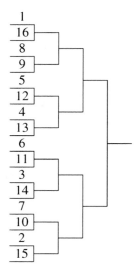

Figure 2.14: Single elimination tournament.

2.57 A bag contains 15 balls numbered $1, 2, \ldots, 15$. If 15 balls are drawn from the bag with replacement, what is the probability that all 15 are unique?

2.58 Consider the events E, F, and G with the following attributes:

- $P(E) = 0.37, P(F) = 0.23, P(G) = 0.18$,
- $P(E \cap F) = 0.10, P(E \cap G) = 0.07, P(F \cap G) = 0.05$,
- $P(E \cap F \cap G) = 0.04$.

Find

(a) $P(E \mid F')$,

(b) $P(E \cap F \mid G)$,

(c) $P(E \cap F' \mid F \cap G)$.

2.59 Urn I contains 7 white balls and 3 black balls. Urn II contains 2 white balls and x black balls. One ball is drawn from Urn I and two balls are drawn without replacement from Urn II. The probability that the three balls drawn are all the same color is exactly $37/210$. Set up and solve an equation in x that can be used to determine the number of black balls in Urn II.

2.60 What can be concluded about the events A and B if $P(A \mid B) = P(B \mid A)$?

2.61 If $P(A \cup B) = \frac{11}{12}$, $P(A) = \frac{5}{12}$, and $P(B') = \frac{1}{4}$, find $P(A \mid B)$.

2.62 Eight cards are drawn without replacement from a well-shuffled deck. Find the probability that at least one king is drawn.

2.63 There are n boys and n girls sitting around the perimeter of a middle-school gymnasium, where n is a positive integer. The dance instructor selects boys one at a time at random. When chosen, the boy walks over to the girl of his choice and asks her if she will dance with him. Out of etiquette, girls always say "yes" to the boy's request. Jordan, one of the boys, has his eye on Greta. If every boy other than Jordan is equally likely to pick any of the remaining girls when his turn is taken, find the probability that Jordan dances with Greta.

2.64 A bag contains fifteen billiard balls, numbered $1, 2, \ldots, 15$. Two balls are drawn without replacement. Find the probability that the largest number drawn is equal to five.

2.65 Urn I contains three white balls and two black balls. Urn II contains one white ball and three black balls. An urn is selected at random, then a ball is drawn at random from the urn selected. Find the probability that the ball selected is black.

2.66 A fair coin is tossed repeatedly until a head appears. Find the conditional probability that the head appeared on the third toss given that the head appeared on an odd-numbered toss.

2.67 An urn contains 11 amber balls and 17 black balls. Two balls are selected at random, without replacement, and removed from the urn. The colors of the two balls removed are not observed. A third ball is then drawn from the urn at random. Find the probability that the third ball removed from the urn is amber.

2.68 An urn contains 6 red balls, 7 white balls, and 8 blue balls. Three balls are drawn from the urn without replacement. What is the probability of drawing a red ball on the first draw, a white ball on the second draw, and a blue ball on the third draw?

2.69 Three cards are dealt, sequentially, from a well-shuffled deck of cards. What is the probability that the cards are, in order: a seven, a face card, and an ace?

2.70 Four cards are drawn without replacement from a well-shuffled deck. Find the probability that the fourth card drawn is a face card (jack, queen, or king), given that the first three draws are face cards.

2.71 Jim comes from a family that contains two children.

(a) What is the probability that his sibling is a girl? Answer the question in terms of conditional probability and check by enumeration.

(b) What is the probability that his younger sibling is a girl? Answer the question in terms of conditional probability and check by enumeration.

2.72 In the game "blackjack" or "21," two cards are dealt from a well-shuffled deck.

(a) Find the probability of dealing a 21 (one of the cards is an ace and the other is either a ten, jack, queen, or king) to a single player.

(b) If two players are each dealt two cards, find the probability that exactly one of the players is dealt a 21.

2.73 A fair red die and a fair green die are tossed together. Find the probability that the red die shows five spots given that the total number of spots showing on both dice is i, for $i = 6, 7, \ldots, 11$.

2.74 An urn contains n red balls and m white balls. Margot and Blake sample balls consecutively and alternatively without replacement from the urn (that is, Margot samples first, Blake samples second, Margot samples third, etc.) until a red ball is sampled. Write a computer subprogram named RedProb, which accepts positive integer arguments n and m and computes the probability that Margot samples the first red ball. Test your code on $n = 6$, $m = 7$ and on $n = 20$, $m = 30$, and report the exact results from RedProb.

2.75   Three cards are drawn from a well-shuffled deck of playing cards. Find the probability of drawing a face card, followed by an ace, followed by a numbered card.

2.76   Show that if A and B are independent events, then A' and B' are also independent events.

2.77   Consider n biased coins such that the probability that heads is flipped is i/n for coin i, for $i = 1, 2, \ldots, n$. If a coin is selected at random, flipped, and shows heads, what is the probability that the coin was coin i?

2.78   Shandelle rolls a pair of fair dice and sums the number of spots that appear on the up faces. She then flips a fair coin the number of times associated with the sum of the spots, e.g., if she rolled a 3 and a 4, then she flips the fair coin 7 times. If the coin flipping part of the random experiment yielded an equal number of heads and tails, find the probability that she rolled an 8 on the dice rolling part of the random experiment. Write a Monte Carlo simulation in R to support the analytic solution.

2.79   A bag contains two coins: one fair and one double-headed. A coin is selected at random and flipped. If the result is heads, what is the probability that the coin selected is double-headed?

2.80   Ronald has developed a 10-second test for strep throats. Wanting to capitalize on his discovery, he decides to open a drive-up McClinic, where a customer's throat is swabbed by a nurse at one window and antibiotics are dispersed at a second window if the strep test is positive. The strep test returns a positive result when a customer really does not have strep throat (a "false positive") with probability p, $0 < p < 1$. The strep test returns a negative result when a customer really has strep throat (a "false negative") with probability q, $0 < q < 1$. If the probability that a customer driving to Ronald's McClinic has strep throat is r, $0 < r < 1$, find the probability that a customer who drives away from the McClinic with antibiotics really does have strep.

2.81   Gale Sayers was one of the greatest NFL running backs of all time. He played for the Chicago Bears and was known for his ability to excel on muddy fields in the rain. The probability that the Bears win given that the field is muddy is 0.6. The probability that the Bears win given that the field is dry is 0.5. If the forecast for rain on game day has an 80% chance of rain, what is the probability that the Bears win?

2.82   An ordinary deck of playing cards has 13 ranks: 2, 3, $\ldots$, 10, jack, queen, king, and ace. Two cards are drawn in sequence from a well-shuffled deck. Find the probability that the second card drawn is of higher rank than the first card drawn if

    (a) the draws are made without replacement,

    (b) the draws are made with replacement.

2.83   Consider the following experiment involving three urns. Urn i contains 2 amber balls and i black balls, for $i = 1, 2, 3$. A ball is drawn at random from Urn 1 and transferred to Urn 2. A ball is then drawn at random from Urn 2 and transferred to Urn 3. Finally, a ball is drawn at random from Urn 3. Find the probability that the ball drawn from Urn 3 is black.

2.84   In a large state, 50% of the eligible voters are Democrats, 45% of the eligible voters are Republicans, and 5% of the eligible voters are Independents. The voter turnouts for a particular election for the three groups are 40%, 55%, and 65%, respectively.

    (a) Find the probability that an eligible voter in this state votes.

    (b) Find the probability that someone who voted in the election is a Democrat.

2.85 Some sects of Mormonism have strict dietary codes, and members abstain from alcohol, caffeine, and tobacco. Assume that historical data has shown that lung cancer is half as likely to occur in Mormons as in non-Mormons. If eight percent of the population are Mormons, what is the probability that someone who is diagnosed with lung cancer is a Mormon?

2.86 The probability that a boy is born with Down's syndrome is p, $0 < p < 1$. The probability that a girl is born with Down's syndrome is q, $0 < q < 1$. If the chances of a boy or girl being born are equal, find the probability that a baby born with Down's syndrome is a boy.

2.87 Suzette is taking an n-question multiple choice exam, where each question has m possible responses, only one of which is correct. For each question:

- with probability $1/3$ she is certain of the correct response,
- with probability $1/3$ she guesses amongst the m responses,
- with probability $1/3$ she guesses amongst 2 responses (she can eliminate all but two of the responses).

Her responses to the questions are mutually independent events.

(a) On an individual question, find the probability that she answers the question correctly.

(b) On an individual question, find the probability that she knew the answer given that she answered the question correctly.

(c) Find the probability that she gets a perfect score on the exam.

(d) Find the probability that she gets exactly x correct responses on the exam.

2.88 An urn contains n fair coins and m two-headed coins. One coin is selected from the urn at random. You see only the results of the selected coin after being flipped.

(a) If the coin is flipped once and shows heads, what is the probability that it was a two-headed coin?

(b) If the coin is flipped x times and always shows heads, what is the probability that it was a two-headed coin?

(c) If $n = m$, find the smallest number of flips, all of which show heads, required to be at least 99% certain that you have selected a two-headed coin.

2.89 A grocery store gives away stickers with the six letters V, A, L, U, E, and S, with probabilities $1/2$, $1/4$, $1/8$, $1/16$, $1/32$, and $1/32$, respectively. If Mark has just collected his nth sticker, what is the probability that it is a letter that he does not already have?

2.90 Consider 100 balls: 90 red and 10 white. A random sample of 50 of the balls are placed in an urn, while the remaining 50 balls are placed in a second urn. One ball is sampled from the first urn. It is white. This white ball is placed in the second urn. A ball is then sampled at random from the second urn. Find the probability that this second ball sampled is white.

2.91 Show that the probability of an even number of heads appearing in n tosses of a biased coin is

$$\frac{1 + (1 - 2p)^n}{2},$$

where p is the probability of tossing a head on a single toss.

2.92 The reality-based television show *Survivor* is getting too competitive. Contestants' feelings are being hurt and lawyers and animal rights activists are being drawn into the fray. The producers decide to film just three people in a remote area for several weeks, then use probability to decide a winner. Contestants 1, 2, and 3 are given coins with probabilities p_1, p_2, and p_3 of coming up heads. The contestants flip their coins simultaneously. If one contestant's outcome differs from the other two, he or she is declared the winner. If all coins come up the same, they re-flip repeatedly until they get a winner. Give probabilities of winning for the three contestants.

2.93 Consider the events A and B.

   (a) If A and B are disjoint, what is $P(A|B)$?

   (b) If A and B are independent, what is $P(A|B)$?

   (c) If A is a subset of B, what is $P(A|B)$?

   (d) If A is a subset of B, what is $P(B|A)$?

   (e) Since A and A' partition the sample space S, under what conditions is this formula valid?

$$P(A|B) = \frac{P(B|A)P(A)}{P(B|A)P(A) + P(B|A')P(A')}$$

2.94 Atlas and Bruce agree to engage in the following test of strength. They will have consecutive arm wrestling matches until one of them wins two matches in a row and is declared the winner. Atlas wins a given match with probability $3/5$. Assuming that the matches are mutually independent, express the probability that Atlas is declared the winner as an exact fraction. Also, write a Monte Carlo simulation in R to support the analytic solution.

2.95 Show that if $A_1, A_2, \ldots, A_n$ are independent events, then

$$P(A_1 \cup A_2 \cup \ldots \cup A_n) = 1 - \prod_{i=1}^{n} [1 - P(A_i)].$$

2.96 There are 20 people in a room. What is the probability that one or more of them were born on February 20? Assume that

   • all 365 days of the year are equally likely to be someone's birthday;

   • there are no twins, triplets, etc. in the class;

   • leap years don't exist.

2.97 A telephone survey is conducted on a large population to determine whether respondents are Democrats, Independents, or Republicans (these are the only three possible responses). Let p_D, p_I, and p_R denote the fraction of Democrats, Independents, and Republicans in the population, where $p_D + p_I + p_R = 1$. Find the probability exactly two Democrats and exactly three Independents are encountered before the fourth Republican is encountered.

2.98 The one and six faces are on opposite sides of a standard die. Assume that an unfair type of die is shortened or lengthened in the one-six direction so that $p = P(\text{rolling a one}) = P(\text{rolling a six})$, where $0 < p < 0.50$.

   (a) Draw pictures of the die (with the six face showing on the top of the die) corresponding to $p = 0.01$ and $p = 0.49$.

   (b) Find the probability of rolling a total of 7 when two such dice (each shortened or lengthened by the same amount) are tossed for any value of p.

2.99 The NBA season free throw record holder is Calvin Murphy of the Houston Rockets. He made 95.8% of his free throws during the 1980–81 season, hitting 206 of his 215 attempts. If his free throws during the season can be considered to be independent, what is the minimum number of free throws he would be required to take in order for you to be at least 80% certain that he will miss one or more free throws.

2.100 Amanda, Becky, and Charise toss a coin in sequence until one person "wins" by tossing the first head.

    (a) If the coin is fair, find the probability that Amanda wins.

    (b) If the coin is fair, find the probability that Becky wins.

    (c) If the coin is fair, find the probability that Charise wins.

    (d) If p is the probability that the coin comes up heads on each individual toss, plot the probability that each of the three win for $0 < p < 1$.

2.101 Joyce rolls n fair dice simultaneously and independently. What is the smallest value of n so that she can be at least 80% certain that two or more fours will appear?

2.102 An urn contains 99 red balls and one white ball. A random sample of n balls is drawn from the urn. Find the smallest value of n so that the probability that a white ball is in the sample exceeds $3/4$ if:

    (a) sampling is performed *without* replacement,

    (b) sampling is performed *with* replacement.

2.103 Deji and Iswat have three children who get married and have three children each. Find the probability that both genders are represented in each of the three sets of grandchildren. Assume that the probability of having a grandson or granddaughter equals $1/2$.

2.104 Deji and Iswat have n children who get married and have $m_1, m_2, m_3, \ldots, m_n$ children each, where $m_i > 1$, $i = 1, 2, \ldots, n$. Give an expression for the probability that both genders are represented in all sets of grandchildren. Let each grandchild's gender be independent of all others and let p be the probability that a grandchild is a girl, $0 < p < 1$.

2.105 During the "fastest finger" portion of the *Who Wants to be a Millionaire* game show, 11 contestants try to order four items chronologically. If all of the contestants guess, find the probability that all contestants get the ordering wrong.

2.106 A coin is biased so that $P(\text{tossing a head}) = p$, where $0 < p < 1$. This coin is flipped repeatedly until a head appears. Find the probability that the first head appears on an even-numbered toss.

2.107 A biased coin, with probability of tossing a head equal to p on one particular toss, is tossed $n + m$ times, where $n > m$. Find the probability of observing a string of at least n consecutive heads.

2.108 An equilateral triangle is inscribed in a circle. What is the probability that a random chord is longer than a side of the triangle? This problem has been (purposely) stated in a vague manner because the notion of a "random chord" of the circle has not been defined. Define the meaning of a random chord and solve the problem using your definition. This problem is known as Bertrand's paradox.

2.109 A coin is biased so that the probability of tossing a head is p. Find the value of p so that the probability of tossing 2 or fewer heads in 4 tosses equals $8/9$.

2.110   Nahum and Ilya have a contest where the first person to make a free throw wins. Nahum shoots first, and then they alternate shots. If Nahum is a 0.17 free throw shooter (a 17% free throw shooter), find the value of p, the probability that Ilya makes an individual free throw, so that it is equally likely that Nahum and Ilya win the contest.

2.111   Determine whether each statement below is true or false.

    (a) If events A_1, A_2, A_3 partition a sample space, then events A_1 and A_2 are independent.

    (b) If events A_1, A_2, A_3 partition a sample space, then events A_1 and A_2 are disjoint.

    (c) If A_1, A_2, A_3 are pairwise independent events, then A_1, A_2, A_3 are mutually independent events.

    (d) For any two events A and B, if $P(A) > 0$, $P(B) > 0$, and A and B are independent, then A and B cannot be disjoint.

    (e) For any two disjoint events A and B, $P(A) + P(B) = 1$.

2.112   A spinner has four equally-likely outcomes. The spinner is spun three times. Define the following three events. Let the event A_{12} be the first and second spins yielding the same outcome. Let the event A_{13} be the first and third spins yielding the same outcome. Let the event A_{23} be the second and third spins yielding the same outcome.

    (a) What is $P(A_{12})$?

    (b) Are the three events pairwise independent?

    (c) Are the three events mutually independent?

2.113   Tashika's birthday is January 17. Tashika's asks n people if they were born on January 17. What is the smallest value of n if Tashika wants to be 90% sure that someone has her birthday? Assumptions: ignore leap years, equally-likely birth dates, no twins/triplets/etc.

2.114   In the communications network depicted in Figure 2.15, a signal is sent from the source node at the far left to the sink node at the far right along the directed communication links depicted by the arrows. If the communications links are mutually independent and function with common probability p, where $0 < p < 1$, what is the probability that the signal successfully traverses the communications network? Write your final answer as a polynomial in standard form.

Figure 2.15: Communications network.

Chapter 3

Random Variables

This chapter overviews the definition of a *random variable* and its application to solving probability problems. The chapter begins by defining two fundamentally different types of random variables: discrete and continuous. The probability distribution of a discrete random variable can be characterized by its probability mass function; the probability distribution of a continuous random variable can be characterized by its probability density function. This is followed by an introduction to the cumulative distribution function, which gives the probability that a random variable assumes a value that is less than or equal to a particular value. The last two sections introduce the expectation of a random variable and some inequalities that are useful when working with random variables.

3.1   Discrete Random Variables

The problems solved in the previous chapter all seemed to have custom solutions tailored specifically for the random experiment in question. The purpose of this chapter is to define a *random variable* that will force probability problems to look more similar to one another. Random variables are, by their nature, real-valued. This means that the qualitative outcomes such as heads/tails, even/odd, red/blue, Democrat/Republican found in the sample spaces of problems in the previous chapter must be mapped into real values in order to use random variables to describe them. Although this switch requires a bit more mental effort on most problems, we will now be able to use the familiar tools of algebra and calculus to solve probability problems.

The idea behind the use of random variables is to formulate a *rule* or *function* that assigns a real number x to each element of the sample space S.

> **Definition 3.1** Given a random experiment with an associated sample space S, a *random variable* is a function X that assigns to each element $s \in S$ one and only one real number $X(s) = x$. The *support* of X is the set of real numbers $\mathcal{A} = \{x \mid x = X(s), s \in S\}$.

A random variable truly is a function. Its domain is the sample space and its range is a set of real numbers $\mathcal{A}$. A more compact and more intuitive definition for a random variable is that it is a variable whose value is subject to chance. It is common practice to drop the argument in $X(s)$ and simply write a random variable as the function name, in this case X. It is also common practice to use upper-case letters such as X, Y, or Z for random variables. If there are several of these random variables in a particular probability problem, then they are often subscripted, for example, $X_1, X_2, \ldots, X_n$.

The sequence from left to right in Figure 3.1 illustrates the concept of a random variable. The process begins with a random experiment on the left. This could be flipping a coin, rolling a die, asking someone whether they liked the service at a particular restaurant, or crashing a car into a wall. In any case, this leads to the set of all possible outcomes, the sample space S. Each element in S will

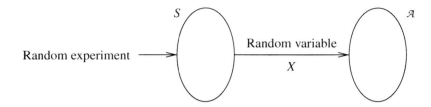

Figure 3.1: A random variable X associated with a random experiment.

be mapped to a single real value in the support set $\mathcal{A}$ by the function X, the random variable.

The same problems that were addressed in Chapter 2 can still be addressed with random variables, but the process involves an extra step. In Chapter 2, a random experiment was performed and an event of interest E was identified. After the sample space S was enumerated, the probability of the event of interest E was calculated using the various tools presented in the chapter. The result was $P(E)$, the probability of event E occurring. Using random variables, the process also begins with the same random experiment. The random variable X maps every possible outcome of that random experiment to a real number in a set $\mathcal{A}$, known as the support. Finally, some set of interest $A \subset \mathcal{A}$ is identified, and we compute $P(X \in A)$. The role of the set E, which is associated with an event, is analogous to the role of the set A, which is associated with the random variable X. This process is illustrated in the following example.

Example 3.1 A fair coin is tossed twice. Find the probability that both tosses come up heads.

The two processes are carried out in parallel below. On the left-hand side, using the Chapter 2 approach, the sample space S is given with $2^2 = 4$ elements by the multiplication rule. The event of interest E corresponding to exactly two heads is identified. Finally, because all of the elements of S are equally likely, the probability of E is calculated as $P(E) = 1/4$. On the right-hand side, the first step is to define the random variable X as the number of heads that appear in the two tosses. The support (or possible values) of X is $\mathcal{A} = \{x \mid x = 0, 1, 2\}$. The subset of $\mathcal{A}$ associated with the event of interest is identified as $A = \{x \mid x = 2\}$, and the probability of X being a member of A is computed as $P(X \in A) = 1/4$.

$$S = \{HH, HT, TH, TT\} \qquad\qquad \mathcal{A} = \{x \mid x = 0, 1, 2\}$$
$$E = \{HH\} \qquad\qquad\qquad A = \{x \mid x = 2\}$$
$$P(E) = 1/4 \qquad\qquad\qquad P(X \in A) = 1/4$$

Figure 3.2 illustrates the random experiment, the sample space S, the event E, the support of the random variable X, the support $\mathcal{A}$, and the set A for this problem.

If this process is carried out for all of the values in the support $\mathcal{A}$, then one arrives at the following three probabilities.

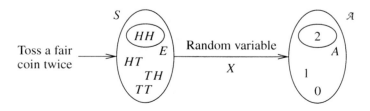

Figure 3.2: The random variable X, the number of heads in two tosses of a fair coin.

x	0	1	2
$P(X = x)$	1/4	1/2	1/4

The three probabilities listed here show how probability is "distributed" over the three values in the support of X. This is the origin of the term *probability distribution*.

The random variable X described in the previous example is known as a *discrete random variable* because of the number of elements in the support $\mathcal{A}$. In general, let X denote a random variable with one-dimensional support $\mathcal{A}$. A random variable X is discrete if the support $\mathcal{A}$ is countable, that is

- X is a discrete random variable if $\mathcal{A}$ is finite, or

- X is a discrete random variable if $\mathcal{A}$ is denumerable.

Let $f(x)$ be a function of x that satisfies

$$\sum_{\mathcal{A}} f(x) = 1 \qquad \text{and} \qquad f(x) > 0, x \in \mathcal{A}.$$

For $x \notin \mathcal{A}$, $f(x) = 0$, which will not be written explicitly in the examples that follow. For some set $A \in \mathcal{A}$, if the probability $P(A)$ can be expressed as

$$P(A) = P(X \in A) = \sum_{A} f(x),$$

then X is a discrete random variable and $f(x) = P(X = x)$ is called a *probability mass function*, often abbreviated pmf. Other common terms for $f(x)$ are the *probability density function* or just the *probability function*. For Example 3.1, the probability mass function for X, the number of heads that appears in two tosses of a fair coin, is

$$f(x) = \begin{cases} 1/4 & x = 0 \\ 1/2 & x = 1 \\ 1/4 & x = 2. \end{cases}$$

Most of the examples that follow are special cases of popular probability models that will be discussed in the next chapter.

Example 3.2 Toss a fair coin 8 times. Let the random variable X be the number of heads. Find the support of X and the probability mass function for X.

When a coin is tossed 8 times, there can be as few as 0 heads tossed to as many as 8 heads tossed. Thus the support of the random variable X is

$$\mathcal{A} = \{x \mid x = 0, 1, 2, \ldots, 8\}.$$

To determine the probability mass function, consider the $2^8 = 256$ possible equally-likely outcomes via the multiplication rule. The 2^8 is placed in the denominator of the probability mass function and the challenge is to determine how many of these outcomes correspond to getting exactly x heads for the numerator. The indices of the x positions where the heads will occur can be selected (the order is not important) in $\binom{8}{x}$ different ways, so the probability that there will be exactly x heads is given by

$$f(x) = \frac{\binom{8}{x}}{2^8} \qquad x \in \mathcal{A}$$

or

$$f(x) = \frac{8!}{256(8-x)!x!} \qquad x \in \mathcal{A}.$$

This probability mass function is plotted in Figure 3.3 by using the R statements

```
x = 0:8
f = factorial(8) / (256 * factorial(8 - x) * factorial(x))
plot(x, f, type = "h")
```

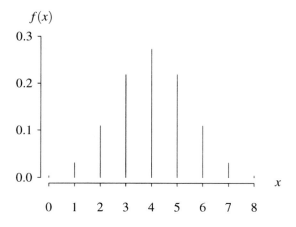

Figure 3.3: Probability mass function $f(x)$ for the number of heads.

The mass values are traditionally plotted as spikes, rather than just points, which emphasizes that the sum of the heights of the nine spikes must be 1. (The type = "h" option on the plot command tells R to plot the spikes.) The symmetry in the graph is due to the fact that the coin is fair.

Example 3.3 A 10-card hand is dealt from a well-shuffled 52-card deck. Let the random variable X be the number of jacks in the hand. Find the support of X and the probability mass function of X.

Since there are only 4 jacks in the deck, there can be as few as 0 jacks in the hand to as many as 4 jacks in the hand. Thus the support of the random variable X is

$$\mathcal{A} = \{x \mid x = 0, 1, 2, 3, 4\}.$$

Now determine the probability mass function $f(x)$. Since the sampling from the hand is performed without replacement and the order with which the cards are dealt is not relevant, there are $\binom{52}{10}$ different hands that can be dealt, which goes in the denominator of the probability mass function. The numerator must give the count of the number of hands containing exactly x jacks. There are $\binom{4}{x}$ ways to select the x jacks, which is multiplied by the $\binom{48}{10-x}$ ways to select the $10 - x$ non-jacks, yielding a probability mass function

$$f(x) = \frac{\binom{4}{x}\binom{48}{10-x}}{\binom{52}{10}} \qquad x \in \mathcal{A}.$$

As is the case with all probability mass functions, the sum of the values of $f(x)$ over $\mathcal{A}$ is 1. The R statements

```
x = 0:4
f = choose(4, x) * choose(48, 10 - x) / choose(52, 10)
plot(x, f, type = "h")
```

are used to plot the probability mass function given in Figure 3.4. Most of the mass associated with the random variable X is concentrated at $X = 0$ (no jacks in the hand) and $X = 1$ (one jack in the hand).

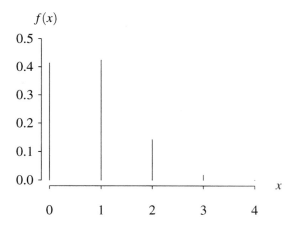

$f(x)$

Figure 3.4: Probability mass function $f(x)$ for the number of jacks.

Example 3.4 Two billiard balls are drawn with replacement from a bag containing 15 balls numbered 1, 2, ..., 15. Let X be the sum of the two numbers drawn. Find the support of X, the probability mass function of X, and the probability that X equals 17.

For the two billiard balls drawn with replacement from the bag, the sum can be as small as 2 and as large as 30. Thus, the support of the random variable X is

$$\mathcal{A} = \{x \,|\, x = 2, 3, \ldots, 30\}.$$

The sampling of the two billiard balls can be thought of conceptually as rolling a pair of fair, 15-sided dice as shown in Figure 3.5. The rows represent the number on the ball encountered on the first draw and the columns represent the number on the ball encountered on the second draw. Figure 3.5, which is analogous to Figure 2.6, contains a 15×15 grid that corresponds to the $15^2 = 225$ possible (ordered) outcomes to the experiment. The probability that the sum will be 2 is $1/225$, the probability that the sum will be 3 is $2/225$, the probability that the sum will be 4 is $3/225$, etc. Writing this probability mass function in a compact form,

$$f(x) = \frac{15 - |x - 16|}{15^2} \qquad x \in \mathcal{A}.$$

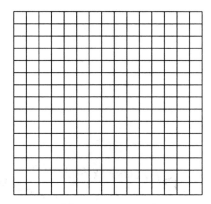

Figure 3.5: Sample space for sampling two billiard balls from a bag.

The probability mass function is plotted in Figure 3.6 using the R statements

```
x = 2:30
f = (15 - abs(x - 16)) / 225
plot(x, f, type = "h")
```

which reveals a symmetric triangular shape to the distribution.

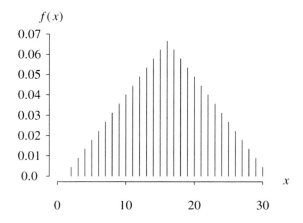

Figure 3.6: Probability mass function $f(x)$ for the sum of the numbers on the two balls.

Finally, the probability that the sum of the two numbers drawn is 17 is

$$P(X = 17) = f(17) = \frac{15 - |17 - 16|}{15^2} = \frac{14}{225} \cong 0.06222.$$

Example 3.5 Roll a fair die 60 times. Let the random variable X be the number of sixes that appear. Find the support of X, the probability mass function of X, and compute the probability that an even number of sixes are rolled.

For a die that is rolled 60 times, there can be as few as 0 sixes that appear to as many as 60 sixes that appear. Thus the support of the random variable X is

$$\mathcal{A} = \{x \mid x = 0, 1, 2, \ldots, 60\}.$$

To determine the probability mass function, the probability of rolling a six on each roll is $1/6$ and the rolls are independent. So the probability of rolling x sixes and $60 - x$ non-sixes *in a specified order* is

$$\left(\frac{1}{6}\right)^x \left(\frac{5}{6}\right)^{60-x}.$$

One example of a specified order is to roll the x sixes first and roll the subsequent $60 - x$ non-sixes last. This quantity must be multiplied by the number of orderings of the rolls with exactly x sixes. There are $\binom{60}{x}$ different ways to choose the x positions from the 60 positions where the sixes will occur. Therefore the probability mass function for the number of sixes that occur is

$$f(x) = \binom{60}{x} \left(\frac{1}{6}\right)^x \left(\frac{5}{6}\right)^{60-x} \qquad x \in \mathcal{A}.$$

This probability mass function is plotted in Figure 3.7 by using the R statements

```
x = 0:60
f = choose(60, x) * (1 / 6) ^ x * (5 / 6) ^ (60 - x)
plot(x, f, type = "h")
```

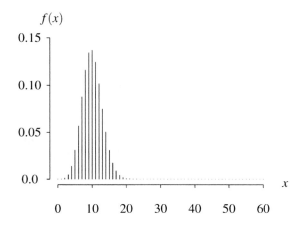

Figure 3.7: Probability mass function $f(x)$ for the number of sixes in 60 rolls.

The nonsymmetry of the probability mass function is due to the fact that the probability of rolling a six is only $1/6$, which skews the probability mass function to the right. A significant portion of the mass of the distribution is gathered around $x = 10$. This is sensible because one would expect to see 10 sixes, on average, in 60 rolls of a fair die.

The probability of getting an even number of sixes is

$$
\begin{aligned}
P(X = 0, X = 2, X = 4, \ldots, X = 60) &= f(0) + f(2) + f(4) + \cdots + f(60) \\
&= \sum_{x=0}^{30} f(2x) \\
&= \sum_{x=0}^{30} \binom{60}{2x} \left(\frac{1}{6}\right)^{2x} \left(\frac{5}{6}\right)^{60-2x}.
\end{aligned}
$$

(handwritten annotations:) odd $\quad \binom{60}{2x+1}\left(\frac{1}{6}\right)^{2x+1}\left(\frac{5}{6}\right)^{59-2x}$

The value of this sum can be computed exactly using the Maple statement

```
sum(binomial(60, 2 * x) * (1 / 6) ^ (2 * x) * (5 / 6) ^ (60 - 2 * x),
    x = 0 .. 30);
```

which, in lowest terms, yields the rather intimidating ratio

$$
\frac{423911582763691250189012801 77}{847823165504324070285888664 02}.
$$

When this fraction is evaluated as a floating point number using the Maple `evalf` function, the result is just slightly over 0.5. The result can also be calculated in R using

```
x = seq(0, 60, by = 2)
sum(choose(60, x) * (1 / 6) ^ x * (5 / 6) ^ (60 - x));
```

These statements return 0.5, which, as seen from the Maple code, is not quite the exact probability. Submitting the R `options(digits = 12)` command increases the accuracy, and yields 0.500000000014 as the probability to 12 digits. The R `seq` function creates a vector with elements 0, 2, 4, ..., 60. As will be seen in the next chapter, this probability distribution is a special case of the *binomial distribution*. Finally, a few keystrokes can be saved by typing the R statements

```
x = seq(0, 60, by = 2)
sum(dbinom(x, 60, 1 / 6))
```

There are more details concerning the dbinom function and other related functions in the next chapter.

In all of the examples presented so far, the support of the distribution of the random variable X, denoted by $\mathcal{A}$, has been listed separately from the probability mass function $f(x)$. Henceforth, the two will be listed together. The support will be listed to the right of $f(x)$. Both must be included together to have any meaning, for example $f(x) = 1/3$ is not meaningful alone; its support, say, $x = 6, 7, 8$, must be included. We make the assumption throughout the text that the probability mass function assumes a value of 0 at any x-values that are not in the support of the distribution.

Example 3.6 A spinner yields three equally-likely outcomes: 1, 2, 3. If the random variable X denotes the product of the outcomes of two spins, find the probability mass function $f(x)$, $P(X = 6)$, and $P(X \leq 6)$.

By the multiplication rule, there are $3^2 = 9$ equally-likely outcomes for the two spins. Table 3.1 shows the results of the first spin on the rows, the results of the second spin on the columns, and the products as entries in the 3×3 table.

second spin

		1	2	3
	1	1	2	3
first spin	2	2	4	6
	3	3	6	9

Table 3.1: Sample space for the product of two spins.

The probability mass function can be written as

$$
f(x) = \begin{cases}
1/9 & x = 1 \\
2/9 & x = 2 \\
2/9 & x = 3 \\
1/9 & x = 4 \\
2/9 & x = 6 \\
1/9 & x = 9.
\end{cases}
$$

This new way of writing $f(x)$ and the support $\mathcal{A}$ together conveys the same amount of information as writing them separately. The probability mass function is plotted with the R statements

```
x = c(1, 2, 3, 4, 6, 9)
f = c(1 / 9, 2 / 9, 2 / 9, 1 / 9, 2 / 9, 1 / 9)
plot(x, f, type = "h")
```

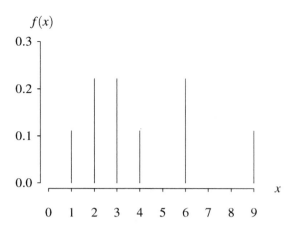

Figure 3.8: Probability mass function $f(x)$ for the product of the two spins.

and is displayed in Figure 3.8. The probability that the product of the numbers on the two spins is 6 is given by

$$P(X = 6) = f(6) = \frac{2}{9}.$$

The probability that the product of the numbers on the two spins is less than or equal to 6 is given by

$$P(X \le 6) = 1 - P(X > 6) = 1 - f(9) = 1 - \frac{1}{9} = \frac{8}{9}.$$

Example 3.7 Consider a bag of 15 billiard balls numbered $1, 2, \ldots, 15$. If 5 balls are drawn without replacement, find the probability mass function of the random variable X, the second to the largest number selected.

The order that the balls are drawn from the bag is not important, so there are $\binom{15}{5}$ different samples. The support of X ranges from 4 to 14. When $X = x$, there are $x - 1$ numbers smaller than x and $15 - x$ numbers larger than x. Thus the probability mass function of X is

$$f(x) = \frac{\binom{x-1}{3}\binom{1}{1}\binom{15-x}{1}}{\binom{15}{5}} = \frac{(x-1)(x-2)(x-3)(15-x)}{18{,}018} \qquad x = 4, 5, \ldots, 14.$$

The probability mass function is plotted with the R statements

```
x = 4:14
f = (x - 1) * (x - 2) * (x - 3) * (15 - x) / 18018
plot(x, f, type = "h")
```

The probability mass function in Figure 3.9 is skewed to the left because the random variable X is the *second highest* of the five balls sampled. If it had been the middle value (the third highest) of those sampled, it would have been a symmetric probability mass function. The R function call sum(f) can be used to assure that $\sum_{x=4}^{14} f(x) = 1$, which is indeed the case.

All of the examples of discrete random variables presented so far have had a support set $\mathcal{A}$ that is finite. We end this section with an instance of a discrete random variable that has a support set that is denumerable (that is, its elements can be placed in a one-to-one correspondence with the positive integers).

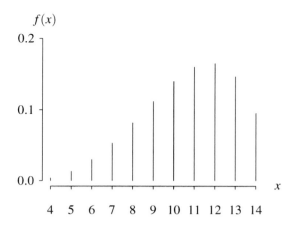

Figure 3.9: Probability mass function $f(x)$ for the second largest number sampled.

Example 3.8 Flip a fair coin repeatedly until a head appears. Let X be the number of flips required. Find the probability mass function of X (finding the support of X is implicit), and find $P(X \in A)$ where $A = \{x \mid x = 4, 5, \ldots\}$.

Since there is no upper bound on the number of flips required, the support of X is the positive integers. Using the independence of the flips, the probability mass function is

$$f(x) = \left(\frac{1}{2}\right)^x \qquad x = 1, 2, \ldots.$$

The number of elements in the support A is denumerable. The probability mass function can be plotted for the first 8 values of the support with the R statements

```
x = 1:8
f = (1 / 2) ^ x
plot(x, f, type = "h")
```

and is displayed in Figure 3.10.

To calculate the probability that it will take 4 or more flips to produce the first head:

$$
\begin{aligned}
P(X \in A) &= P(X = 4) + P(X = 5) + P(X = 6) + \cdots \\
&= f(4) + f(5) + f(6) + \cdots \\
&= \left(\frac{1}{2}\right)^4 + \left(\frac{1}{2}\right)^5 + \left(\frac{1}{2}\right)^6 + \cdots \\
&= \left(\frac{1}{2}\right)^4 \left[1 + \frac{1}{2} + \frac{1}{4} + \cdots\right] \\
&= \left(\frac{1}{2}\right)^4 \left[\frac{1}{1 - 1/2}\right] \\
&= \left(\frac{1}{2}\right)^3 \\
&= \frac{1}{8}.
\end{aligned}
$$

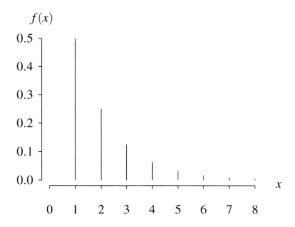

Figure 3.10: Probability mass function $f(x)$ for the number of coin flips for first head.

The series that was summed was a geometric series. As in Chapter 2, Monte Carlo simulation can be used to check this result. The R code below simulates 100,000 random experiments consisting of flipping a fair coin repeatedly until a head appears.

```
nrep  = 100000
count = 0
for (i in 1:nrep) {
  x = 1
  while (runif(1) < 0.5) x = x + 1
  if (x >= 4) count = count + 1
}
print(count / nrep)
```

Five runs of this simulation yield

| 0.12527 | 0.12505 | 0.12531 | 0.12384 | 0.12535, |

which hover around the theoretical value $1/8 = 0.125$, lending support to the correctness of the analytic solution.

The notion of a random variable has been introduced in this section, along with several illustrations of discrete random variables. The key points introduced in this section are listed below.

- A *random variable* is a rule that assigns a real number to each outcome of a random experiment.

- Upper case letters such as X, Y, and Z are traditionally used to denote random variables.

- The *support* $\mathcal{A}$ of a random variable X is the set of all possible values that the random variable can assume.

- A random variable is *discrete* if its support is a countable set (that is, a finite set or a denumerable set).

- The manner in which the probability is distributed over its support $\mathcal{A}$ for a discrete random variable X is characterized by its *probability mass function* $f(x) = P(X = x)$.

- All probability mass functions satisfy the *existence conditions*

$$\sum_{\mathcal{A}} f(x) = 1 \qquad \text{and} \qquad f(x) \geq 0 \text{ for all } x.$$

- Probabilities associated with a random variable can be found by summing appropriate members of $f(x)$.

3.2   Continuous Random Variables

All of the random variables from the examples in the previous section were discrete. We now consider another type of random variable known as a *continuous* random variable. We motivate this new type of random variable with two examples, one involving falling while ice skating on a frozen pond and another involving the limit of a discrete random variable.

Richard decides to go skating on a large frozen pond. Now let's say that the inevitable happens—he takes a fall. Richard is going to make an initial impact on that frozen pond at one particular point. Assume that the point of impact is a point in the mathematical sense—it has no area. If we assign a small probability to falling at that point, no matter how small we make the probability, there are an infinite number of points on the pond and we will exceed our limit of 1. We will have violated the third probability axiom, which states that $P(S) = 1$. The only option that we have is to define the probability of Richard falling at each point as zero, but the probability that he falls in a particular area of the frozen pond can be nonzero.

This paradox of falling at a particular point on the pond which has zero probability is much more difficult to get your head wrapped around than the notion of a discrete random variable. Suffice it to say for now that continuous random variables have a well-established theory and methodology. Even though there are some counter-intuitive aspects to continuous random variables, they are as easy to work with as discrete random variables once you have adjusted to the conventions.

Richard's point of impact on the frozen pond example can be thought of as a two-dimensional random variable. The next example considers a one-dimensional random variable.

Example 3.9 What probability distribution formalizes the notion of "equally-likely" outcomes in the unit interval $[0, 1]$?

One way of proceeding with this rather ill-defined question is to begin with a discrete random variable X with probability mass function defined on a support with $n + 1$ rational values:

$$f(x) = \frac{1}{n+1} \qquad x = \frac{0}{n}, \frac{1}{n}, \frac{2}{n}, \ldots, \frac{n-1}{n}, \frac{n}{n}.$$

The support of the random variable X ranges from 0 to 1 for all values of n. For example, when $n = 1000$,

$$f(x) = \frac{1}{1001} \qquad x = 0, \frac{1}{1000}, \frac{2}{1000}, \ldots, \frac{999}{1000}, 1.$$

There are 1001 mass values, each with equal probability. In the limit as $n \to \infty$, the support values fill in the gaps in the support and, correspondingly, the $f(x)$ values decrease. The support of the random variable X approaches

$$\mathcal{A} = \{x \mid 0 \leq x \leq 1\}$$

in the limit as $n \to \infty$. What are some reasonable properties for this limiting distribution? One is that for any interval of length, say, $1/3$, the probability of X falling in that interval should be $1/3$. So

$$P\left(0 < X < \frac{1}{3}\right), \qquad P\left(\frac{1}{9} < X < \frac{4}{9}\right), \qquad P\left(\frac{1}{3} < X < \frac{2}{3}\right), \qquad P\left(\frac{2}{3} < X < 1\right)$$

should all be exactly $1/3$. This is exactly what happens to the discrete distribution for large n. But an undesirable side effect also occurs. The probability of getting one particular value, for example

$$P\left(X = \frac{4}{9}\right)$$

is zero in the limit as $n \to \infty$. This is the trade-off that is made. Probability calculations over intervals of nonzero width on $[0, 1]$ yield nonzero probabilities; probability calculations for a particular value on $[0, 1]$ yield a probability of 0. For instance,

$$P\left(\frac{3}{5} < X < \frac{4}{5}\right) = \frac{1}{5}$$

but

$$P\left(X = \frac{2}{5}\right) = 0.$$

So a _continuous random variable_ X is simply a random variable that has a support set $\mathcal{A}$ that is uncountable. We now define a function that plays a similar role to the probability mass function for a discrete random variable. This will be referred to as a _probability density function_ that will also be denoted by $f(x)$. All probability density functions must satisfy the existence conditions:

$$\int_{\mathcal{A}} f(x)\,dx = 1 \qquad \text{and} \qquad f(x) \geq 0 \text{ for all real } x.$$

The probability density function is often abbreviated by pdf. For some set A which is a subset of $\mathcal{A}$, the probability of event A occurring is

$$P(X \in A) = \int_A f(x)\,dx.$$

In the special case of real constants $a < b$,

$$P(a < X < b) = \int_a^b f(x)\,dx.$$

The assumption that the probability of a random variable X, assuming one particular value is zero, means that the four probabilities

$$P(a < X < b) \qquad P(a \leq X < b) \qquad P(a < X \leq b) \qquad P(a \leq X \leq b)$$

are all evaluated by the same integral. In addition, for any particular real value a,

$$P(X = a) = 0.$$

Example 3.10 Define the continuous random variable X to be uniformly distributed between 0 and 1. Find the probability density function $f(x)$, the probability that X lies between $1/5$ and $1/2$, and the probability that $X = 2/3$.

There is only one probability density function $f(x)$ that: (a) is defined on the support $\mathcal{A} = \{x \mid 0 < x < 1\}$, ($b$) is constant with respect to x to satisfy the uniformity condition, (c) integrates to 1 in order to satisfy the first existence condition for probability density functions, and (d) is nonnegative over all real x to satisfy the second existence condition for probability density functions. This function is

$$f(x) = 1 \qquad 0 < x < 1.$$

The probability that X lies between $1/5$ and $1/2$ is found by integrating the probability density function between $1/5$ and $1/2$

$$P\left(\frac{1}{5} < X < \frac{1}{2}\right) = \int_{1/5}^{1/2} 1\,dx = \left[x\right]_{1/5}^{1/2} = \frac{1}{2} - \frac{1}{5} = \frac{3}{10}.$$

Since the probability that X assumes one particular value is zero,

$$P\left(X = \frac{2}{3}\right) = \int_{2/3}^{2/3} 1\,dx = 0.$$

The probability density function and the area representing $P(1/5 < X < 1/2)$ are shown in Figure 3.11. This uniform distribution is one of the most mathematically tractable distributions because of the ease of integrating $f(x)$.

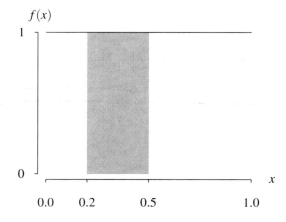

Figure 3.11: Probability density function $f(x) = 1$ on $0 < x < 1$ with $P(1/5 < X < 1/2)$ shaded.

Example 3.11 Let the continuous random variable X have probability density function

$$f(x) = \frac{x}{2} \qquad 0 < x < 2.$$

Find the probability that X is greater than 1.

The probability that X is greater than 1 can be found by integrating the probability density function from 1 to 2, yielding

$$P(X > 1) = \int_{1}^{2} \frac{x}{2}\,dx = \left[\frac{x^2}{4}\right]_{1}^{2} = 1 - \frac{1}{4} = \frac{3}{4}.$$

The R statements

```
x = c(0, 2)
f = c(0, 1)
plot(x, f, type = "l")
polygon(c(1, 1, 2, 2), c(0, 0.5, 1, 0), col = 15)
```

plot the probability density function and shade the area associated with $P(X > 1)$, which is displayed in Figure 3.12. The type = "l" parameter in the plot function connects the two points with a line. The polygon function shades the area in the polygon defined by the x and y coordinates given in the vectors that are its first two arguments. The argument col = 15 controls the shade of gray (or color) of the shaded polygon.

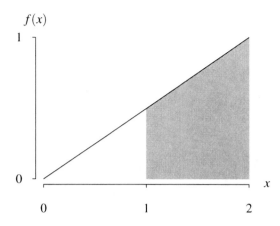

Figure 3.12: Probability density function $f(x) = x/2$ on $0 < x < 2$ with $P(X > 1)$ shaded.

Example 3.12 Let the continuous random variable X have probability density function

$$f(x) = e^{-x} \qquad x > 0.$$

Find the probability that $\lfloor X \rfloor$ is even.

The floor function truncates all of the digits to the right of the decimal point for X, so the probability that the floor of X is even is equivalent to X falling in one of the following intervals:

$$[0, 1), [2, 3), [4, 5), \ldots.$$

So integrations must be performed for an infinite number of integrals over disjoint intervals. The probability that $\lfloor X \rfloor$ is even is

$$
\begin{aligned}
P\big(\lfloor X \rfloor = 0 \text{ or } \lfloor X \rfloor = 2 \text{ or } \ldots\big) &= \int_0^1 e^{-x}\,dx + \int_2^3 e^{-x}\,dx + \int_4^5 e^{-x}\,dx + \cdots \\
&= \left[-e^{-x}\right]_0^1 + \left[-e^{-x}\right]_2^3 + \left[-e^{-x}\right]_4^5 + \cdots \\
&= \left[1 - e^{-1}\right] + \left[e^{-2} - e^{-3}\right] + \left[e^{-4} - e^{-5}\right] + \cdots \\
&= 1 - e^{-1} + e^{-2} - e^{-3} + \cdots \\
&= \frac{1}{1 + e^{-1}} \\
&\cong 0.7311.
\end{aligned}
$$

The geometric series has a common ratio $-e^{-1}$. The left-hand portion of the probability density function is shown in Figure 3.13, along with the shaded areas associated with the probability that the floor of X is even.

Discrete random variables can be distinguished from continuous random variables based on the type of support set, countable or uncountable. In a particular application, the distinction might not be so clear. If the random variable of interest involves some type of counting (for example, the number of jacks in a ten-card hand), then it is typically a discrete random variable. If the random variable of interest involves any measurement on a continuous scale, for example, volume, temperature, length, weight, brightness, or density, then it is typically a continuous random variable.

In practice, many random variables are inherently continuous, but can only be measured to finite precision. A simple example might be a person's weight. This random variable is inherently

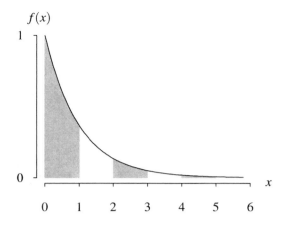

Figure 3.13: Probability density function $f(x) = e^{-x}$ on $x > 0$ with $P(\lfloor X \rfloor$ even$)$ shaded.

continuous, but a scale might only measure it to the nearest 0.1 pound. In applications like these, a continuous random variable is used for a *probability model* that approximates the weight of a person, although associated data can not be collected to infinite precision.

The vast majority of random variables are either purely discrete or purely continuous random variables. There are occasions, however, where a hybrid of these two basic types of random variables arise. These are known as *mixed discrete–continuous random variables*. Consider the following examples:

- daily rainfall, measured in inches, on the roof of the building where your probability class is being held;

- the delay time that you encounter waiting for service in a queue at an ice cream shop;

- the lifetime of a light bulb.

In all three of these cases, the random variable of interest, rainfall, waiting time, and lifetime, are all inherently continuous. However, there is a nonzero probability that it will not rain at all on one particular day, that you will not need to wait at all at the ice cream shop, and that the light bulb will burn out immediately upon being placed in the socket. In other words, if X is the random variable of interest, $P(X = 0) > 0$, so there is a discrete spike in $f(x)$ at 0.

A generic $f(x)$ for a mixed discrete–continuous distribution is shown in Figure 3.14. There is a spike of height 0.37 at $x = 0$ that corresponds to the discrete part of the distribution. In the previous three scenarios, this spike corresponds to

- no rain on 37% of the days;

- a probability of 0.37 that you will not have to wait in line for ice cream;

- a probability of 0.37 that a new bulb fails immediately (choose another brand!).

There is a continuous curve that appears to the right of the discrete spike at $x = 0$ that corresponds to the continuous part of the distribution. The area underneath this curve is 0.63. In the previous three scenarios, this continuous curve corresponds to

- the daily rainfall, in inches, for the 63% of the days when it does rain;

- the wait time, in minutes, for the 63% of the customers that need to wait for service;

- the lifetime, in years, of the 63% of the light bulbs that do not fail immediately.

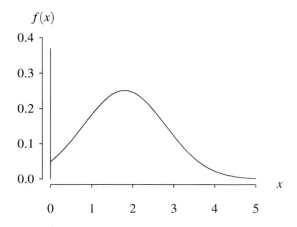

Figure 3.14: A mixed discrete–continuous distribution.

This concludes the introduction to one-dimensional continuous random variables and their distributions, which are characterized by their support and probability density function. The key points introduced in this section concerning continuous random variables are listed below.

- A *continuous* random variable has a support set $\mathcal{A}$ which is uncountable.

- The manner in which the probability is distributed over the support $\mathcal{A}$ for a continuous random variable X is characterized by its *probability density function* $f(x)$.

- All probability density functions $f(x)$ satisfy the two *existence conditions*

$$\int_{\mathcal{A}} f(x)\, dx = 1 \qquad \text{and} \qquad f(x) \geq 0 \text{ for all real } x.$$

- For some set $A \subset \mathcal{A}$, the probability that $X \in A$ can be found by integrating:

$$P(X \in A) = \int_{A} f(x)\, dx.$$

In particular,

- if a and b are real values satisfying $a < b$, then $P(a < X < b) = \int_{a}^{b} f(x)dx$,

- if a is a real value, then $P(X = a) = 0$.

continuous RV:
pdf
discrete RV
pmf

3.3 Cumulative Distribution Functions

The previous two sections have introduced discrete and continuous random variables. The distribution of a discrete random variable is characterized by its probability mass function $f(x)$ and its associated support $\mathcal{A}$; the distribution of a continuous random variable is characterized by its probability density function $f(x)$ and its associated support $\mathcal{A}$.

There is another way of characterizing the distribution of a random variable that is known as the *cumulative distribution function.* In some applications, it is of value to know whether a random variable X is less than or equal to some value x. A cumulative distribution function (also called a *distribution function* or abbreviated *cdf*) is defined the same for a discrete or continuous random variable X:

$$F(x) = P(X \leq x).$$

For a discrete random variable X, the cumulative distribution function is calculated as a sum:

$$F(x) = P(X \leq x) = \sum_{w \leq x} f(w).$$

For a continuous random variable X, the cumulative distribution function is calculated as an integral:

$$F(x) = P(X \leq x) = \int_{-\infty}^{x} f(w)\,dw.$$

The following list contains some properties of cumulative distribution functions that can be useful for solving problems.

- Since $F(x)$ is defined as a probability, $0 \leq F(x) \leq 1$.

- $F(x)$ is a nondecreasing function of x, that is, for $a < b$, $F(a) \leq F(b)$.

- $\lim_{x \to -\infty} F(x) = 0$.

- $\lim_{x \to \infty} F(x) = 1$.

- $P(a < X \leq b) = F(b) - F(a) = P(X \leq b) - P(X \leq a).$ **Important relation.**

- The random variables X and Y are identically distributed if and only if they have identical cumulative distribution functions.

- In actuarial science, biostatistics, and reliability, the related *survival function* $S(x) = P(X \geq x)$ is used more frequently than the cumulative distribution function.

- If X is a discrete random variable, $F(x)$ is a right-continuous step function. For discrete and continuous random variables, $F(x)$ is right-continuous.

- For a discrete random variable X with support $\mathcal{A} = \{x_1, x_2, \ldots, x_n\}$ where $x_1 < x_2 < \cdots < x_n$,

$$f(x_i) = F(x_i) - F(x_{i-1})$$

for $i = 2, 3, \ldots, n$, and $f(x_1) = F(x_1)$. In other words, since converting a probability mass function into a cumulative distribution function for a discrete random variable X involves summing, going in the other direction involves differencing.

- For a continuous random variable X, if $F'(x)$ exists, then

$$f(x) = F'(x).$$

In other words, since converting a probability density function into a cumulative distribution function for a continuous random variable X involves integrating, going in the other direction involves differentiating.

Example 3.13 Flip a fair coin twice. Let X be the number of heads tossed. Find $F(x)$.

Since X counts the number of heads flipped, it is a discrete random variable. Since there are $2^2 = 4$ equally-likely outcomes by the multiplication rule, the probability mass function for X is

$$f(x) = \begin{cases} 1/4 & x = 0 \\ 1/2 & x = 1 \\ 1/4 & x = 2. \end{cases}$$

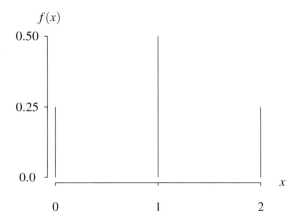

Figure 3.15: Probability mass function $f(x)$ for the number of heads in two coin tosses.

The probability mass function is plotted in Figure 3.15. The cumulative distribution function is found by summing:

$$F(x) = \begin{cases} 0 & x < 0 \\ 1/4 & 0 \le x < 1 \\ 3/4 & 1 \le x < 2 \\ 1 & x \ge 2. \end{cases}$$

Notice that the convention in this text is to assume that the probability mass function is zero anywhere that it is not defined, but the cumulative distribution function is given explicitly for all real values of x. The cumulative distribution function is right continuous because each finite interval is closed on the left and open on the right. The cumulative distribution function is plotted in Figure 3.16 using the R code shown below.

```
x = c(0, 1, 2)
fx = c(1 / 4, 1 / 2, 1 / 4)
Fx = cumsum(fx)
plot(stepfun(x, Fx), type = "l")
```

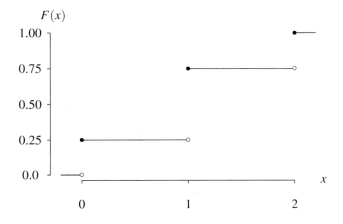

Figure 3.16: Cumulative distribution function $F(x)$ for the number of heads in two coin tosses.

The cumsum function calculates cumulative sums and the stepfun function is used to plot stepwise functions. R has the convention of including the risers when plotting a stepwise function. The plot displayed in Figure 3.16 was created by drawing the lines between the various points. The variable Fx is used rather than F because F is a reserved abbreviation for FALSE in R.

The probability mass function and the cumulative distribution function have both been given in pieces for various values of the support. This is fine for two flips of a coin, but not practical for a large number of flips. The probability mass function can be stated in a single line as

$$f(x) = \binom{2}{x} \left(\frac{1}{2}\right)^2 \qquad\qquad x = 0, 1, 2.$$

Stating the cumulative distribution function in a single line is a bit more difficult, since it assumes a constant value between the values in the support. By using the floor function in the upper limit of the summation, however, the cumulative distribution function can be written in one line as

$$F(x) = \sum_{w=0}^{\lfloor x \rfloor} \binom{2}{w} \left(\frac{1}{2}\right)^2 \qquad\qquad -\infty < x < \infty.$$

Example 3.14 Find the cumulative distribution function for X, the number of sixes that appear in 60 rolls of a fair die.

As determined in Example 3.5, the probability mass function of X is

$$f(x) = \binom{60}{x} \left(\frac{1}{6}\right)^x \left(\frac{5}{6}\right)^{60-x} \qquad\qquad x = 0, 1, 2, \ldots, 60.$$

The cumulative distribution function simply accumulates these probability mass function values as x increases:

$$F(x) = \sum_{w=0}^{\lfloor x \rfloor} \binom{60}{w} \left(\frac{1}{6}\right)^w \left(\frac{5}{6}\right)^{60-w} \qquad\qquad -\infty < x < \infty.$$

The cumulative distribution function can be plotted in R with the statements

```
x = 0:60
fx = choose(60, x) * (1 / 6) ^ x * (5 / 6) ^ (60 - x)
Fx = cumsum(fx)
plot(stepfun(x, Fx), type = "l")
```

The cumulative distribution shown in Figure 3.17 as a step function (with the risers of the steps plotted) that increases rapidly in the vicinity of $x = 10$, which is the expected number of sixes in 60 rolls of a fair die.

The previous two examples have illustrated the calculation of the cumulative distribution function $F(x)$ for discrete random variables. This function will always be a nondecreasing step function, with the steps occurring at the values in the support $\mathcal{A}$. The height of each step is given by the associated $f(x)$ value. We now illustrate the cumulative distribution function for a continuous random variable, which will be a continuous function.

Example 3.15 Find the cumulative distribution function for a continuous random variable X that is uniformly distributed between 0 and 1.

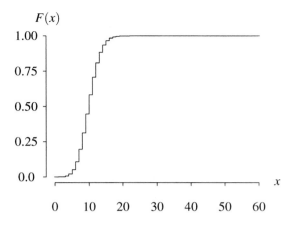

Figure 3.17: Cumulative distribution function $F(x)$ for the number of sixes in 60 rolls.

From Example 3.10, the probability density function is

$$f(x) = 1 \qquad 0 < x < 1.$$

Since X is a continuous random variable, the cumulative distribution function is found by integrating. Over the support of X, the cumulative distribution function is

$$F(x) = \int_0^x 1\, dw = x \qquad 0 < x < 1.$$

The proper way to state the cumulative distribution function is to include all real values of x, so

$$F(x) = \begin{cases} 0 & x \leq 0 \\ x & 0 < x < 1 \\ 1 & x \geq 1, \end{cases}$$

which is plotted in Figure 3.18. It is not surprising that the function is linear because the probability density function is constant.

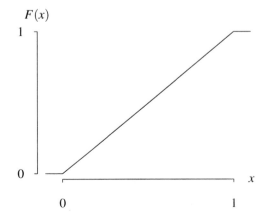

Figure 3.18: Cumulative distribution function $F(x) = x$ on $0 < x < 1$.

Example 3.16 Find the cumulative distribution function associated with

$$f(x) = \frac{x}{2} \qquad 0 < x < 2.$$

Use the cumulative distribution function to find $P(1/2 < X < 3/2)$.

The cumulative distribution function on the support is again found by integrating

$$F(x) = \int_0^x \frac{w}{2} \, dw = \frac{x^2}{4} \qquad 0 < x < 2.$$

It is not surprising that $F(x)$ is quadratic on $\mathcal{A}$ because the integral of a linear function is quadratic. Expressing the cumulative distribution function over its entire range,

$$F(x) = \begin{cases} 0 & x \le 0 \\ x^2/4 & 0 < x < 2 \\ 1 & x \ge 2. \end{cases}$$

The cumulative distribution function is plotted in Figure 3.19. To calculate $P(1/2 < X < 3/2)$ using the cumulative distribution function,

$$\begin{aligned} P(1/2 < X < 3/2) &= P(X \le 3/2) - P(X \le 1/2) \\ &= F(3/2) - F(1/2) \\ &= 9/16 - 1/16 \\ &= 1/2. \end{aligned}$$

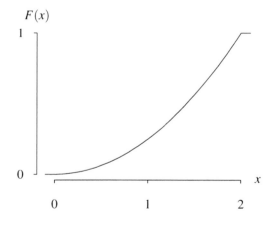

Figure 3.19: Cumulative distribution function $F(x) = x^2/4$ on $0 < x < 2$.

Example 3.17 Find the cumulative distribution function for the random variable X with probability density function

$$f(x) = \begin{cases} 1 - x & 0 < x < 1 \\ 1/4 & 5 < x < 7. \end{cases}$$

A random variable whose probability density function is defined in a piecewise fashion poses no difficulty other than a careful accounting of the probability as it is accumulated by the integral. The cumulative distribution function on $0 < x < 1$ has an integral that looks like the previous two examples:

$$F(x) = \int_0^x (1 - w)\, dw = x - x^2/2 \qquad\qquad 0 < x < 1.$$

The cumulative distribution function has accumulated an area of $F(1) = 1/2$ by $x = 1$ and the cumulative distribution function is constant at $1/2$ between $x = 1$ and $x = 5$. When computing the cumulative distribution function on $5 < x < 7$, make sure to add in the area of $1/2$ already encountered, that is

$$F(x) = \frac{1}{2} + \int_5^x \frac{1}{4}\, dw = \frac{x - 3}{4} \qquad\qquad 5 < x < 7.$$

The cumulative distribution function is plotted in Figure 3.20.

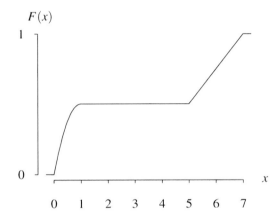

Figure 3.20: Piecewise cumulative distribution function $F(x)$ on $0 < x < 7$.

This concludes our examples of elementary cumulative distribution functions for discrete (two examples) and continuous (three examples) random variables. Mixed discrete–continuous random variables also have cumulative distribution functions. The random variable illustrated in Figure 3.14, for example, has a cumulative distribution function that takes a step upward of 0.37 at $x = 0$, then has a continuous portion to the right of 0, as illustrated in Figure 3.21.

Five topics that are somewhat related to cumulative distribution functions need to be introduced and this is a good place to introduce them. The topics are: percentiles, random variate generation, transformations of random variables, random variables in a computer algebra system, and mixtures. Each topic has a separate subheading.

Percentiles

Standardized tests for academic performance in grammar school and for college entrance often report a raw score along with a percentile. For example, a raw SAT (scholastic aptitude test) score on a college-entrance exam of 1200 might correspond to a percentile of 78. The interpretation of the percentile is that 78% of those that took the SAT exam scored 1200 or below. Likewise, a ten-year-old girl who is at the 37th percentile of her peer's heights is taller than 37% of the population of ten-year-old girls and shorter than 63% of the population of ten-year-old girls.

Percentiles are also called fractiles and quantiles in probability and statistics. Fortunately, cumulative distribution functions make it quite easy to calculate percentiles, as illustrated in the following two examples.

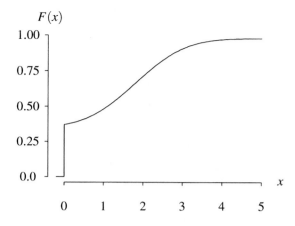

Figure 3.21: Cumulative distribution function for a mixed distribution.

Example 3.18 Flip a fair coin twice. Let X be the number of heads tossed. Find the 81st percentile of the distribution.

Since X counts the number of heads flipped, it is a discrete random variable. Since there are $2^2 = 4$ equally-likely outcomes by the multiplication rule, the probability mass function for X is

$$f(x) = \begin{cases} 1/4 & x = 0 \\ 1/2 & x = 1 \\ 1/4 & x = 2. \end{cases}$$

The cumulative distribution function is

$$F(x) = \begin{cases} 0 & x < 0 \\ 1/4 & 0 \le x < 1 \\ 3/4 & 1 \le x < 2 \\ 1 & x \ge 2, \end{cases}$$

as illustrated in Figure 3.22. In order to find the 81st percentile of this distribution

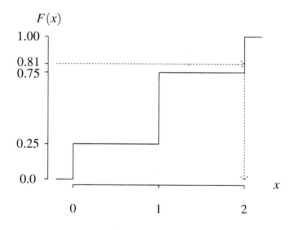

Figure 3.22: The 81st percentile of the number of heads in two coin flips.

geometrically, locate 0.81 on the vertical axis of the cumulative distribution function and draw a horizontal line at $F(x) = 0.81$. This line intersects one of the risers of the step function. The support value where this intersection occurs is the 81st percentile. In this case, the 81st percentile is

$$x_{0.81} = 2.$$

One question that might arise from the previous example concerns a horizontal line extended from the vertical axis that does not intersect a riser, such as the 75th percentile of the distribution. We follow the R convention of letting the 75th percentile of this distribution equal $x_{0.75} = 1$. In general, for a discrete distribution, the convention used here is to define the pth percentile $(0 < p < 1)$ as

$$x_p = \min \{x \,|\, F(x) \geq p\}.$$

This distinction is not necessary for continuous distributions. Let p be the fraction of the area under the probability density function to the left of the percentile of interest. We seek x_p. This quantity is found by setting $F(x_p) = p$ and solving for x_p. Assuming that the inverse of the cumulative distribution function exists, then the pth percentile is given by

$$x_p = F^{-1}(p),$$

as illustrated next.

Example 3.19 Find the 81st percentile associated with the continuous random variable with cumulative distribution function

$$F(x) = \begin{cases} 0 & x \leq 0 \\ x^2/4 & 0 < x < 2 \\ 1 & x \geq 2. \end{cases}$$

The geometry associated with the problem is shown in Figure 3.23. This problem will be solved for any general percentile, then the 81st percentile will be calculated. Equating the cumulative distribution function on its support set $0 < x < 2$ equal to p

$$x_p^2/4 = p$$

and solving for x_p yields a general expression for the pth percentile:

$$x_p = F^{-1}(p) = 2\sqrt{p}.$$

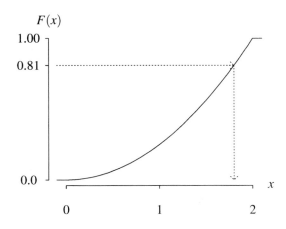

Figure 3.23: Cumulative distribution function $F(x) = x^2/4$ on $0 < x < 2$ and 81st percentile.

When $p = 0.81$, the 81st percentile is given by

$$x_{0.81} = 2\sqrt{0.81} = 1.8.$$

There are some special types of percentiles that arise so often in practice that they are given special names. The first is the *population median*, which is the 50th percentile. The population median is, in some sense, the "middle" of the distribution in that a continuous random variable is equally likely to fall above or below the population median. The *quartiles* are the 25th, 50th, and 75th percentiles of a probability distribution. (The middle of these three values is, of course, the population median.) These three values effectively partition the probability distribution into four equal portions, each associated with a probability of $1/4$. The statistical analogs of these quartiles were encountered in the discussion of box plots in Example 1.9.

Example 3.20 Find the quartiles associated with the continuous random variable with cumulative distribution function

$$F(x) = \begin{cases} 0 & x \leq 0 \\ x^2/4 & 0 < x < 2 \\ 1 & x \geq 2. \end{cases}$$

From the previous example, the pth percentile is given by

$$x_p = 2\sqrt{p}.$$

Substituting $p = 0.25, 0.50, 0.75$ yields the three quartiles:

$$x_{0.25} = 1 \qquad\qquad x_{0.50} = \sqrt{2} \cong 1.4142 \qquad\qquad x_{0.75} = \sqrt{3} \cong 1.7321.$$

The quartiles are illustrated in Figure 3.24.

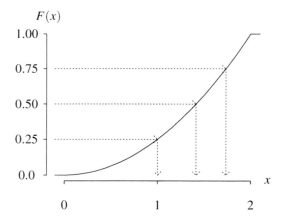

Figure 3.24: Cumulative distribution function $F(x) = x^2/4$ on $0 < x < 2$ and the quartiles.

Random variate generation

Monte Carlo simulation was introduced in Chapter 2 as a tool for confirming analytic solutions and for addressing probability problems where an analytic solution is not possible. The purpose here is to extend the method of Monte Carlo simulation to random variables.

A well-known result in probability that is known as the *probability integral transformation* states that if X is a random variable with cumulative distribution function $F(x)$, then $F(X)$ is uniformly distributed between 0 and 1. In this result, the random variable X is being transformed by its own cumulative distribution function, resulting in another random variable, which must be between 0 and 1 (because all cumulative distribution functions are between 0 and 1). The remarkable discovery associated with the probability integral transformation is that $F(X)$ is *uniformly* distributed between 0 and 1. This result allows us to generate an instance of a random variable (known as a *random variate*) by simply equating $F(x)$ to u, and solving for the random variate x, where u denotes a random variate that is uniformly distributed between 0 and 1. Symbolically, the random variate x is generated via

$$x \leftarrow F^{-1}(u)$$

when an inverse cumulative distribution function F^{-1} exists.

The previous discussion on percentiles folds nicely into random variate generation. Consider the generation of the discrete random variable X, the number of heads in two flips of a fair coin. If the value of u happens to be 0.81 (or any u value between 0.75 and 1), then $x = 2$ heads are generated, as illustrated in Figure 3.22. Likewise, consider the generation of the continuous random variable X with cumulative distribution function

$$F(x) = \begin{cases} 0 & x \leq 0 \\ x^2/4 & 0 < x < 2 \\ 1 & x \geq 2. \end{cases}$$

If the value of u happens to be 0.81, then the random variate $x = 1.8$ is generated, as illustrated in Figure 3.23.

The next example illustrates the estimation of a percentile of a continuous random variable by Monte Carlo simulation.

Example 3.21 Estimate the population median of the distribution with cumulative distribution function

$$F(x) = \begin{cases} 0 & x \leq 0 \\ x^2/4 & 0 < x < 2 \\ 1 & x \geq 2 \end{cases}$$

by Monte Carlo simulation.

The population median was calculated to be exactly

$$x_{0.5} = \sqrt{2}$$

in Example 3.20. In order to generate random variates, the cumulative distribution function must be inverted, that is

$$F(x) = \frac{x^2}{4} = u$$

must be solved for x yielding

$$F^{-1}(u) = 2\sqrt{u}$$

for $0 < u < 1$. Random variates are generated by

$$x \leftarrow 2\sqrt{u}$$

for a random variate u that is uniformly distributed between 0 and 1. The R code to estimate the population median is developed as follows. First, the `runif` function is used to generate a vector of 10,001 random numbers between 0 and 1. (An odd number of variates is used in order for there to be a true "middle" value.) These variates are transformed into a vector of random variates between 0 and 2 by plugging into the inverse cumulative distribution function. Finally, the random variates are sorted in ascending order and the middle value (the 5001st sorted value) is printed.

```
x = 2 * sqrt(runif(10001))
sort(x)[5001]
```

After a call to `set.seed(3)` to initialize the random number stream, the code is executed five times, yielding

1.4217   1.4153   1.4041   1.4248   1.4092.

These values hover around the analytic solution $x_{0.5} = \sqrt{2} \cong 1.4142$, supporting the analytic solution.

So far, Monte Carlo simulation has only been used to verify known analytic results. It is also possible to use Monte Carlo simulation on problems where the exact solution is not known.

Example 3.22 What is the 90th percentile of the distance between two points chosen at random in the interior of a unit square?

The analytic tools presently available do not allow us to solve this problem analytically, although it poses no difficulty for Monte Carlo simulation. The two points in the unit square (x_1, y_1) and (x_2, y_2) are generated using four random variates that are uniformly distributed between 0 and 1. The distance between the two points,

$$\sqrt{(x_1 - x_2)^2 + (y_1 - y_2)^2},$$

can range from 0 (when the points are identical) to $\sqrt{2}$ (when the points are at opposite corners of the square). Three realizations of random line segments in the unit square are shown in Figure 3.25.

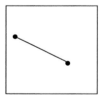 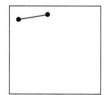 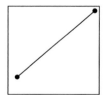

Figure 3.25: Three realizations of random line segments in a unit square.

The R code to estimate the 90th percentile of the distance between the two points is given below.

```
nrep = 1000000
x1 = runif(nrep)
x2 = runif(nrep)
y1 = runif(nrep)
y2 = runif(nrep)
d  = sqrt((x1 - x2) ^ 2 + (y1 - y2) ^ 2)
sort(d)[0.9 * nrep]
```

Of the 1,000,000 distances generated and stored in the vector d, the 900,000th ordered distance is the Monte Carlo estimate for the 90th percentile. After a call to `set.seed(3)` to initialize the random number stream, five runs of the Monte Carlo simulation code are run, yielding

0.8592064   0.8585644   0.8589713   0.8582711   0.8587005.

How many digits should be reported? Since four of the five results round to 0.859, it is reasonable to report the estimate of the 90th percentile of the distance between the points as

$$x_{0.9} \cong 0.859.$$

Transformations of random variables

Occasions arise when the distribution of a function of a random variable is of interest. If X is the radius of a solid copper pipe, for example, then the distribution of the cross-sectional area of the pipe $Y = \pi X^2$ might be of interest.

Figure 3.26 conceptualizes the progression. On the left is a random experiment. The sample space S is the set of all possible outcomes to the random experiment. The random variable X is a mapping from the sample space S to a real number in the set $\mathcal{A}$. Finally the transformation $Y = g(X)$ transforms the random variable X, with support $\mathcal{A}$, to the random variable Y, with support $\mathcal{B}$. The support of Y is the set

$$\mathcal{B} = \{y \mid y = g(x), x \in \mathcal{A}\}$$

Our goal here is to find the cumulative distribution function of Y.

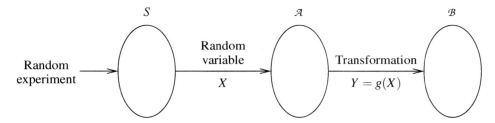

Figure 3.26: Transformation $Y = g(X)$ of a random variable X.

Now, for the first time, there are two random variables: X and Y. In order to keep track of the two random variables, yet reserve the letters f and F to maintain their current meanings, we have adopted the convention of using subscripts to distinguish these functions. For example,

- $F_X(x)$ is the cumulative distribution function for a random variable X,

- $F_Y(y)$ is the cumulative distribution function for a random variable Y,

- $f_X(x)$ is the probability mass function for a discrete random variable X,

- $f_Y(y)$ is the probability density function for a continuous random variable Y.

The purpose here is to use a method known as the *cumulative distribution function technique* to find the distribution of Y given the distribution of X. The technique always begins with the definition of the cumulative distribution function, followed by a substitution of $Y = g(X)$ as follows:

$$F_Y(y) = P(Y \le y) = P(g(X) \le y).$$

The next three examples illustrate the use of the cumulative distribution function technique to find the probability distribution of a function of a random variable. Chapter 7 is devoted entirely to the important topic of functions of random variables in much more generality.

Example 3.23 Let the random variable X be uniformly distributed between 0 and 1. Find the cumulative distribution function of $Y = g(X) = \sqrt{X}$.

As seen in Example 3.15, the cumulative distribution function of X is

$$F_X(x) = \begin{cases} 0 & x \le 0 \\ x & 0 < x < 1 \\ 1 & x \ge 1. \end{cases}$$

The transformation $Y = g(X) = \sqrt{X}$ is a 1–1 transformation from $\mathcal{A} = \{x \mid 0 < x < 1\}$ to $\mathcal{B} = \{y \mid 0 < y < 1\}$. So using the cumulative distribution function technique,

$$\begin{aligned} F_Y(y) &= P(Y \le y) \\ &= P(\sqrt{X} \le y) \\ &= P(X \le y^2) \\ &= F_X(y^2) \\ &= y^2 \qquad 0 < y < 1. \end{aligned}$$

Finally, the cumulative distribution function of Y is

$$F_Y(y) = \begin{cases} 0 & y \le 0 \\ y^2 & 0 < y < 1 \\ 1 & y \ge 1. \end{cases}$$

The geometry associated with this transformation is shown in Figure 3.27. The transformation effectively takes a random variable X that is uniformly distributed between 0 and 1 and pushes more of the probability toward 1, which is reflected in the cumulative distribution of Y given above.

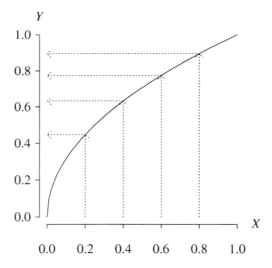

Figure 3.27: The transformation $Y = g(X) = \sqrt{X}$ on $0 < X < 1$.

Example 3.24 Let the random variable X be uniformly distributed between 0 and 1. Find the probability density function of $Y = g(X) = \arcsin X$.

The cumulative distribution function of X is

$$F_X(x) = \begin{cases} 0 & x \le 0 \\ x & 0 < x < 1 \\ 1 & x \ge 1. \end{cases}$$

The transformation $Y = g(X) = \arcsin X$ is a 1–1 transformation from $\mathcal{A} = \{x \mid 0 < x < 1\}$ to $\mathcal{B} = \{y \mid 0 < y < \pi/2\}$. So using the cumulative distribution function technique,

$$
\begin{aligned}
F_Y(y) &= P(Y \leq y) \\
&= P(\arcsin X \leq y) \\
&= P(X \leq \sin y) \\
&= F_X(\sin y) \\
&= \sin y \qquad 0 < y < \pi/2.
\end{aligned}
$$

The problem asks for the probability density function, which requires that we differentiate the cumulative distribution function:

$$
f_Y(y) = \cos y \qquad 0 < y < \pi/2.
$$

Example 3.25 Let the random variable X have the probability density function

$$
f_X(x) = e^{-x} \qquad x > 0.
$$

Find the cumulative distribution function of $Y = g(x) = X^2$.

The cumulative distribution function of X on its support is

$$
F_X(x) = \int_0^x f(w)\,dw = \int_0^x e^{-w}\,dw = 1 - e^{-x} \qquad x > 0.
$$

The transformation $Y = g(X) = X^2$ is a 1–1 transformation from $\mathcal{A} = \{x \mid x > 0\}$ to $\mathcal{B} = \{y \mid y > 0\}$. So using the cumulative distribution function technique,

$$
\begin{aligned}
F_Y(y) &= P(Y \leq y) \\
&= P(X^2 \leq y) \\
&= P(X \leq \sqrt{y}) \\
&= F_X(\sqrt{y}) \\
&= 1 - e^{-\sqrt{y}} \qquad y > 0.
\end{aligned}
$$

Random variables in a computer algebra system

The probability problems encountered thus far have been made artificially small so that they can be easily manipulated by hand. A Maple-based language known as APPL (A Probability Programming Language) has been designed to solve probability problems involving random variables. Several of the small problems encountered earlier in this chapter will now be solved using APPL.

Example 3.26 Two billiard balls are drawn with replacement from a bag containing 15 balls numbered 1, 2, ..., 15. Let X be the sum of the two numbers. Find the probability that X equals 17.

APPL has a built-in function `UniformDiscreteRV` that defines a random variable associated with an equally-likely (uniform) discrete random variable. Another APPL function `Convolution` computes the distribution of the sum of two of these random variables. Finally, `PDF` with one argument returns the probability mass function $f(x)$ of the sum of the two numbers. If the optional second argument is included, it will return the probability mass function for a specific x value. So the APPL code

```
X := Convolution(UniformDiscreteRV(1, 15), UniformDiscreteRV(1, 15));
PDF(X, 17);
```

returns

$$P(X = 17) = f(17) = \frac{14}{225}$$

as derived analytically in Example 3.4.

Example 3.27 A spinner yields three equally-likely outcomes: 1, 2, 3. If the random variable X denotes the product of the outcomes of two spins, find the probability mass function $f(x)$, $P(X = 6)$, and $P(X \leq 6)$.

APPL has a built-in function Product that is used to find the distribution of the product of the outcome of the two independent spins. The required quantities $f(x)$, $P(X = 6)$, and $P(X \leq 6)$ are found by appropriate calls to PDF, which returns the probability mass function of X, and CDF, which returns the cumulative distribution function of X. The APPL code

```
X := Product(UniformDiscreteRV(1, 3), UniformDiscreteRV(1, 3));
PDF(X);
PDF(X, 6);
CDF(X, 6);
```

returns

$$f(x) = \begin{cases} 1/9 & x = 1 \\ 2/9 & x = 2 \\ 2/9 & x = 3 \\ 1/9 & x = 4 \\ 2/9 & x = 6 \\ 1/9 & x = 9, \end{cases}$$

$$P(X = 6) = f(6) = \frac{2}{9}, \qquad \text{and} \qquad P(X \leq 6) = F(6) = \frac{8}{9},$$

consistent with Example 3.6. Although these problems were easily worked by hand, APPL makes it easy to solve a problem like this one for: (a) more spins, (b) outcomes that are not equally likely, and/or (c) more than three outcomes on the spinner.

Example 3.28 Flip a fair coin twice. Let X be the number of heads tossed. Find $F(x)$ and the 81st percentile of the distribution of X.

In this particular application, we give the data structure that APPL uses to store a discrete random variable. The number of heads that appears in the two coin tosses is stored in a *list-of-lists* format in APPL. The first list contains the probability mass function values; the second list contains the support values; the third list contains two strings that indicate that X is a discrete random variable, and that the first list contains $f(x)$ values. Finally, to solve the problem, the CDF and IDF (inverse distribution function) functions are called to find the cumulative distribution function $F(x)$ and the 81st percentile $x_{0.81}$.

```
X := [[1 / 4, 1 / 2, 1 / 4], [0, 1, 2], ["Discrete", "PDF"]];
CDF(X);
IDF(X, 0.81);
```

These statements yield

$$F(x) = \begin{cases} 0 & x < 0 \\ 1/4 & 0 \leq x < 1 \\ 3/4 & 1 \leq x < 2 \\ 1 & x \geq 2 \end{cases}$$

and
$$x_{0.81} = 2,$$

consistent with Example 3.18.

APPL has analogous functions that are designed to manipulate continuous random variables using similar data structures. The subsequent examples illustrate its use on continuous random variables encountered previously.

Example 3.29 Consider the continuous random variable X with probability density function
$$f(x) = \frac{x}{2} \qquad 0 < x < 2.$$

Find

(a) the cumulative distribution function,

(b) $P(X > 1)$,

(c) the quartiles of the distribution of X.

The APPL data structure for continuous random variables is similar to that for discrete random variables. The first list contains the probability density function; the second list contains the endpoints of the support of X; the third list contains strings indicating that X is continuous and that the first list contains a probability density function. The three questions listed above are solved using the APPL code

```
X := [[x -> x / 2], [0, 2], ["Continuous", "PDF"]];
CDF(X);
1 - CDF(X, 1);
IDF(X, 0.25);
IDF(X, 0.50);
IDF(X, 0.75);
```

which returns
$$F(x) = \begin{cases} 0 & x \le 0 \\ x^2/4 & 0 < x < 2 \\ 1 & x \ge 2, \end{cases}$$

$$P(X > 1) = \frac{3}{4},$$

and

$$x_{0.25} = 1 \qquad x_{0.50} = \sqrt{2} \cong 1.4142 \qquad x_{0.75} = \sqrt{3} \cong 1.7321$$

consistent with Example 3.20.

Example 3.30 Find the cumulative distribution function for the random variable X with probability density function
$$f(x) = \begin{cases} 1 - x & 0 < x < 1 \\ 1/4 & 5 < x < 7. \end{cases}$$

The distribution of X is keyed into APPL in the same fashion as the previous example, although there are now three distinct piecewise segments to the probability density function.

```
X := [[x -> 1 - x, x -> 0, x -> 1 / 4], [0, 1, 5, 7],
      ["Continuous", "PDF"]];
CDF(X);
```

The call to CDF correctly returns the cumulative distribution function of X as

$$F(x) = \begin{cases} 0 & x < 0 \\ x - x^2/2 & 0 \leq x < 1 \\ 1/2 & 1 \leq x < 5 \\ (x-3)/4 & 5 \leq x < 7 \\ 1 & x \geq 7, \end{cases}$$

consistent with Example 3.17.

Example 3.31 Let the random variable X be uniformly distributed between 0 and 1. Find the cumulative distribution function of $Y = g(X) = \sqrt{X}$.

To use APPL to find the distribution of the transformation of a random variable, first define the distribution of X in the usual fashion. Next, define the transformation $Y = g(X)$ as a data structure that is a list of two lists. The first list contains the transformation; the second list contains the x-values associated with the endpoints of the intervals in which the transformation is valid. Finally, a call to the APPL Transform function will calculate the distribution of Y. Since Transform typically returns the distribution of Y as a probability density function, it needs to be converted to a cumulative distribution function by the CDF function. The APPL code

```
X := [[x -> 1], [0, 1], ["Continuous", "PDF"]];
g := [[x -> sqrt(x)], [0, 1]];
Y := Transform(X, g);
CDF(Y)
```

returns

$$F_Y(y) = \begin{cases} 0 & y \leq 0 \\ y^2 & 0 < y < 1 \\ 1 & y \geq 1, \end{cases}$$

consistent with Example 3.23.

Mixtures

The last topic to be covered in this section is *mixtures*. So far, distributions of random variables have been viewed in isolation—one at a time. Probabilists and statisticians have found uses for mixing several random variables together.

Definition 3.2 Let $X_1, X_2, \ldots, X_k$ be random variables drawn from distributions with cumulative distribution functions $F_{X_1}(x), F_{X_2}(x), \ldots, F_{X_k}(x)$. Let $p_1, p_2, \ldots, p_k$ be the positive mixing probabilities, where $p_1 + p_2 + \cdots + p_k = 1$. Then X has a *finite mixture distribution* if its cumulative distribution function is

$$F_X(x) = p_1 F_{X_1}(x) + p_2 F_{X_2}(x) + \cdots + p_k F_{X_k}(x).$$

A finite mixture distribution effectively mixes k distributions according to the mixing probabilities $p_1, p_2, \ldots, p_k$ in the same fashion that a chemist mixes k solutions. The mixing probabilities correspond to the quantity of the solution mixed and the cumulative distribution functions model the different solutions being mixed.

Example 3.32 There are two factories that produce light bulbs. Factory 1 produces bulbs that have lifetimes modeled by the random variable X_1 with cumulative distribution function

$$F_{X_1}(x) = \begin{cases} 0 & x < 0 \\ 1 - e^{-x} & x \geq 0. \end{cases}$$

Factory 2 produces bulbs that have lifetimes modeled by the random variable X_2 with cumulative distribution function

$$F_{X_2}(x) = \begin{cases} 0 & x < 0 \\ 1 - e^{-2x} & x \geq 0. \end{cases}$$

Furthermore, it is known that $2/3$ of the bulbs come from Factory 1 and $1/3$ of the bulbs come from Factory 2. What is the probability distribution of the lifetime of a bulb whose factory of origin can't be identified?

Using the finite mixture notation, the mixing probabilities are

$$p_1 = \frac{2}{3} \qquad \text{and} \qquad p_2 = \frac{1}{3}.$$

Let X be the lifetime of a light bulb whose origin can't be identified. The cumulative distribution function of the mixture on the positive portion of the support is

$$F_X(x) = \frac{2}{3}\left(1 - e^{-x}\right) + \frac{1}{3}\left(1 - e^{-2x}\right) \qquad x > 0$$

or

$$F_X(x) = 1 - \frac{2}{3}e^{-x} - \frac{1}{3}e^{-2x} \qquad x > 0.$$

Finite mixtures can also be defined in terms of probability mass functions or probability density functions. Differentiating the cumulative distribution function for X in Definition 3.2 with respect to x yields

$$f_X(x) = p_1 f_{X_1}(x) + p_2 f_{X_2}(x) + \cdots + p_k f_{X_k}(x).$$

In addition to mixing product lifetimes as in the light bulb example, actuaries use mixtures to mix claim distributions. An insurer may cover $k = 3$ types of dwellings: houses, condominiums, and apartments. Each of these three types of dwellings has a distribution of the size of a claim due to damage from fire, theft, flooding, etc. The mixing probabilities p_1, p_2, and p_3 correspond to the probabilities of claims on the dwellings insured by a particular company. In this case the mixture gives the company the distribution of claim size that they can expect from the portfolio of dwellings that they are insuring.

There is a second type of mixing that is known as a *continuous mixture*. Rather than the mixing probabilities $p_1, p_2, \ldots, p_k$, you could have a continuous mixing parameter referred to generically as θ. The distribution of the random variable of interest X, which can be either a discrete or continuous random variable, given the value of the continuous mixing parameter θ is denoted by

$$f_{X \mid \Theta}(x \mid \theta).$$

In this case the unconditional distribution of X (when the value of the continuous mixing parameter θ is unknown) is

$$f_X(x) = \int_\theta f_\Theta(\theta) f_{X \mid \Theta}(x \mid \theta)\, d\theta,$$

where $f_\Theta(\theta)$ is the probability density function of the continuous mixing parameter θ. Continuous mixtures will be illustrated in subsequent chapters.

3.4   Expected Values

The mathematical expectation, or expected value, of a random variable X is what you "expect" (on the average, in the long run) for the value of X. The standard notation for the expected value of X is $\mu = E[X]$. The expected value of X is also known as the "population mean."

As a simple example, if there are 4000 sweepstakes cards returned for a single \$1000 prize, each card has an expected worth of $\frac{1}{4000}(\$1000) = \$0.25$.

Definition 3.3 Let X be a random variable defined on the support $\mathcal{A}$ with probability mass function $f(x)$ if X is discrete and probability density function $f(x)$ if X is continuous. The *expected value* of X is

$$\mu = E[X] = \begin{cases} \displaystyle\sum_{\mathcal{A}} xf(x) & X \text{ discrete} \\[2mm] \displaystyle\int_{\mathcal{A}} xf(x)\,dx & X \text{ continuous} \end{cases} \qquad \text{know!}$$

when the sum or integral exists. When the sum or integral diverges, the expected value is undefined.

The sum exists when $\sum_{\mathcal{A}} |x| f(x) < \infty$ and the integral exists when $\int_{\mathcal{A}} |x| f(x)\,dx < \infty$. The expected value of X can be thought of as a weighted average (with the $f(x)$ values as the weights) of the values in the support $\mathcal{A}$. Consider the following elementary examples.

Example 3.33 You are given the choice between

(a) a certain \$5, or

(b) \$0 or \$10 based on the results of the toss of a fair coin.

Which would you choose?

Let the random variable X denote the winnings in part (b). The probability mass function of X is

$$f(x) = \begin{cases} 1/2 & x = 0 \\ 1/2 & x = 10. \end{cases}$$

So the expected value of X is

$$\mu = E[X] = \sum_{\mathcal{A}} xf(x) = 0 \cdot \frac{1}{2} + 10 \cdot \frac{1}{2} = 5.$$

In one sense, the two options are equivalent in terms of their expected value. You are being given the choice between a certain \$5 and a gamble that has an expected value of \$5.

Example 3.34 You are given the choice between

(a) a certain \$500,000, or

(b) \$0 or \$1,000,000 based on the results of the toss of a fair coin.

Which would you choose?

Let the random variable X denote the winnings in part (b). The probability mass function of X is

$$f(x) = \begin{cases} 1/2 & x = 0 \\ 1/2 & x = 1,000,000. \end{cases}$$

So the expected value of X is

$$\mu = E[X] = \sum_{\mathcal{A}} xf(x) = 0 \cdot \frac{1}{2} + 1,000,000 \cdot \frac{1}{2} = 500,000.$$

As in the smaller scale example, the expected values of the two options are equal. In this example, however, because the stakes are much higher, a large majority of people would not consider the two options equivalent and would take the certainty of choice (a). This leads to a subject known as *utility theory* which will not be investigated here. For now, we simply conclude the two options are identical in terms of expected values.

Example 3.35 Find the expected number of spots (pips!) showing when rolling a fair die.

Let the random variable X be the number of spots showing. The probability mass function of X is

$$f(x) = \frac{1}{6} \qquad x = 1, 2, 3, 4, 5, 6.$$

The expected number of spots showing is

$$\sum_{x=1}^{n} x = \boxed{\frac{n(n+1)}{2}}$$

$$\mu = E[X] = \sum_{x=1}^{6} xf(x) = 1 \cdot \frac{1}{6} + 2 \cdot \frac{1}{6} + \cdots + 6 \cdot \frac{1}{6} = \frac{1}{6}[1 + 2 + \cdots + 6] = \frac{1}{6}\left[\frac{6 \cdot 7}{2}\right] = \frac{7}{2}.$$

This example illustrates the case where the expected value of X does not coincide with one of the mass values.

Example 3.36 The probability mass function for the random variable X is

$$f(x) = \frac{7^x e^{-7}}{x!} \qquad \text{Poisson distribution} \qquad x = 0, 1, 2, \ldots .$$

What is the expected value of X?

The expected value of X is

$$\mu = E[X] = \sum_{x=0}^{\infty} x \frac{7^x e^{-7}}{x!}$$

resulting in the non-trivial summation

$$\mu = E[X] = e^{-7}\left(\frac{1 \cdot 7^1}{1!} + \frac{2 \cdot 7^2}{2!} + \frac{3 \cdot 7^3}{3!} + \cdots\right).$$

Using Maple to evaluate the series with the statement

```
sum(x * 7 ^ x * exp(-7) / x!, x = 0 .. infinity);
```

yields

$$\mu = E[X] = 7.$$

The expected value of X is also known as the *population mean*. (The word *population* modifies the word *mean* to differentiate it from the *sample mean*.) The population mean also has a physical interpretation. When X is discrete with support values at $x_1, x_2, \ldots$, with weights $f(x_1), f(x_2), \ldots$, the population mean

$$\mu = E[X] = \sum_{A} xf(x) = x_1 f(x_1) + x_2 f(x_2) + \cdots$$

is a "weighted average" or "arithmetic mean" or "mean value" of X. A physical interpretation of the population mean is the center of gravity. If a weightless bar has weights of $f(x_1), f(x_2), \ldots$, placed at x-positions $x_1, x_2, \ldots$, then $\mu = E[X]$ is the x-position at which one can pick up the bar without it tilting to the left or the right. A similar analogy exists for continuous random variables.

Example 3.37 Find the expected winnings in a single $1 bet in a game of craps.

Recall from Example 2.40 that the probability of winning craps is $244/495 \cong 0.4929$. Define the random variable X to be the winnings. The probability mass function of X is

$$f(x) = \begin{cases} 251/495 & x = -1 \\ 244/495 & x = 1. \end{cases}$$

Thus the expected value of X is

$$\mu = E[X] = \sum_{A} x f(x) = -1 \cdot \frac{251}{495} + 1 \cdot \frac{244}{495} = -\frac{7}{495} \cong -0.01414.$$

You can expect to lose about 1.4 cents when you bet $1 on a game of craps. Using the center of gravity interpretation, if a weight of $251/495$ is placed at $x = -1$ and a weight of $244/495$ is placed at $x = 1$, then the balance point of the bar is at $\mu = -7/495 \cong -0.01414$.

Example 3.38 Jonah places a well-shuffled deck of cards face down on a table. Let the random variable X be the number of cards between the table and the heart that is furthest down in the deck.

(a) Find A, the support of X.

(b) Find $P(X = 2)$.

(c) Find $E[X]$.

Assume that Jonah is playing with a full deck and all 52! shufflings are equally likely.

(a) Since there could be a heart on the bottom of the deck ($x = 0$) or all 13 hearts could be at the top of the deck ($x = 39$), the support of X is $A = \{x \mid x = 0, 1, 2, \ldots, 39\}$.

(b) Using conditional probability,

$$P(X = 2) = \frac{39}{52} \cdot \frac{38}{51} \cdot \frac{13}{50} = \frac{19,266}{132,600} = \frac{247}{1700} \cong 0.1453.$$

(c) First calculate the probability mass function of X and then compute the population mean using

$$E[X] = \sum_{x=0}^{39} x f(x).$$

Using similar reasoning to part (b), the probability mass function of X is

$$f(x) = \frac{13 \cdot 39! \cdot (51 - x)!}{(39 - x)! \cdot 52!} \qquad x = 0, 1, \ldots, 39.$$

Due to the large factorials, Maple or APPL are the best platforms for computing $E[X]$. The APPL code that defines the probability mass function of X and invokes the Mean function

```
X := [[x -> 13 * 39! * (51 - x)! / ((39 - x)! * 52!)], [0 .. 39],
      ["Discrete", "PDF"]];
Mean(X);
```

returns $E[X] = 39/14 \cong 2.7857$. The expected number of cards between the table and the heart furthest down in the deck is approximately 2.7857 cards.

Example 3.39 An urn contains 3 red balls and 4 blue balls. Balls are drawn successively at random and without replacement from the urn. Let the random variable X be the trial number when the first red ball is drawn. Find $E[X]$.

Solution 1. Using conditional probability, the probability mass function of X is

$$
f(x) = \begin{cases}
\dfrac{3}{7} & x = 1 \\[2mm]
\dfrac{4}{7} \cdot \dfrac{3}{6} & x = 2 \\[2mm]
\dfrac{4}{7} \cdot \dfrac{3}{6} \cdot \dfrac{3}{5} & x = 3 \\[2mm]
\dfrac{4}{7} \cdot \dfrac{3}{6} \cdot \dfrac{2}{5} \cdot \dfrac{3}{4} & x = 4 \\[2mm]
\dfrac{4}{7} \cdot \dfrac{3}{6} \cdot \dfrac{2}{5} \cdot \dfrac{1}{4} \cdot \dfrac{3}{3} & x = 5
\end{cases}
$$

or

$$
f(x) = \begin{cases}
\dfrac{3}{7} & x = 1 \\[2mm]
\dfrac{2}{7} & x = 2 \\[2mm]
\dfrac{6}{35} & x = 3 \\[2mm]
\dfrac{3}{35} & x = 4 \\[2mm]
\dfrac{1}{35} & x = 5.
\end{cases}
$$

Thus, the expected value of X is

$$
E[X] = \sum_{x=1}^{5} x f(x) = 1 \cdot \frac{3}{7} + 2 \cdot \frac{2}{7} + 3 \cdot \frac{6}{35} + 4 \cdot \frac{3}{35} + 5 \cdot \frac{1}{35} = 2
$$

draws.

Solution 2. The random variable X has probability mass function

$$
f(x) = \frac{\binom{4}{x-1} \cdot \binom{3}{1}}{\binom{7}{x}} \cdot \frac{1}{x} \qquad x = 1, 2, 3, 4, 5.
$$

The first fraction of this probability mass function accounts for the probability of drawing exactly $x - 1$ blue balls and exactly 1 red ball from the urn in any order. The factor $\frac{1}{x}$ appears to account for the fact that the red draw must occur last. So the expected value of X is

$$
E[X] = \sum_{x=1}^{5} x \cdot \frac{\binom{4}{x-1} \binom{3}{1}}{\binom{7}{x}} \cdot \frac{1}{x} = 2.
$$

The following APPL code can be used to calculate $E[X]$.

```
X := [[x -> 3 * binomial(4, x - 1) / (x * binomial(7, x))],
     [1, 2, 3, 4, 5], ["Discrete", "PDF"]];
Mean(X);
```

Example 3.40 Let n be a positive integer. A cube is comprised of n^3 smaller cubes, as illustrated in Figure 3.28 for $n = 4$. If one of the n^3 smaller cubes is selected at random, give an expression for the expected number of exposed faces. (*Hint*: an interior smaller cube has no exposed faces; a corner smaller cube has three exposed faces, etc.)

Figure 3.28: A $4 \times 4 \times 4$ array of cubes.

Solution 1 (brute force). Table 3.2 classifies the number of smaller cubes of various types for $n > 1$. The sum of the elements in the second column of this table is n^3 as expected. If the random variable X models the number of exposed faces in a cube selected at random, then the probability mass function of X is

$$f(x) = \begin{cases} \dfrac{(n-2)^3}{n^3} & x = 0 \\[2mm] \dfrac{6(n-2)^2}{n^3} & x = 1 \\[2mm] \dfrac{12(n-2)}{n^3} & x = 2 \\[2mm] \dfrac{8}{n^3} & x = 3. \end{cases}$$

Thus an expression for the expected number of exposed faces is:

$$E[X] = 0 \cdot \frac{(n-2)^3}{n^3} + 1 \cdot \frac{6(n-2)^2}{n^3} + 2 \cdot \frac{12(n-2)}{n^3} + 3 \cdot \frac{8}{n^3} = \frac{6}{n}$$

for $n = 1, 2, \ldots$. As n becomes large, $\lim_{n \to \infty} E[X] = 0$ because of the overwhelming number of interior smaller cubes with no exposed faces.

Type of smaller cube	Number of smaller cubes	Number of exposed faces
Interior	$(n-2)^3$	0
Face	$6(n-2)^2$	1
Edge	$12(n-2)$	2
Corner	8	3

Table 3.2: Classifying the cubes.

Solution 2 (finesse). The total number of *faces* on the smaller cubes is $6n^3$. The total number of *exposed faces* on the smaller cubes is $6n^2$. If a *face* of a smaller cube is selected at random, the probability that the face is exposed is

$$\frac{6n^2}{6n^3} = \frac{1}{n}$$

which is also the expected number of exposed faces. Since there are six faces on a smaller cube selected at random, the expected number of exposed faces is

$$\frac{6}{n}.$$

Example 3.41 A 10-card hand is dealt from a well-shuffled 52-card deck. What is the expected number of jacks in the hand? What would be a "fair" amount to pay to play a game that paid you $25 for each jack in a 10-card hand?

Let the random variable X be the number of jacks in the hand. From Example 3.3, the probability mass function of X is

$$f(x) = \frac{\binom{4}{x}\binom{48}{10-x}}{\binom{52}{10}} \qquad x = 0, 1, 2, 3, 4.$$

The expected number of jacks in a hand is

$$E[X] = \sum_{x=0}^{4} x \cdot \frac{\binom{4}{x}\binom{48}{10-x}}{\binom{52}{10}} = \frac{10}{13} \cong 0.7692.$$

Thus it would be fair to pay

$$25 \cdot \frac{10}{13} = \frac{250}{13} \cong \$19.23$$

to play a game that pays $25 for each jack that appears in 10 cards dealt from a well-shuffled deck.

We now shift gears and calculate expected values for continuous random variables, where integrals replace summations in the calculations.

Example 3.42 Find the population mean of the random variable X with probability density function

$$f(x) = x/2 \qquad 0 < x < 2.$$

The expected value of X is

$$\mu = E[X] = \int_0^2 x \cdot \frac{x}{2} \, dx = \frac{4}{3}.$$

The center of gravity interpretation is as follows. If a barbell of length 2 (the width of the support) is made of continuously varying material such that its weight at position x is given by $x/2$, then the balance point for such a bar is $\mu = 4/3$.

Example 3.43 For the random variable X with probability density function

$$f(x) = \begin{cases} 1 - x & 0 < x < 1 \\ 1/4 & 5 < x < 7, \end{cases}$$

calculate $\mu = E[X]$.

The population mean of X is

$$\mu = E[X] = \int_0^1 x(1-x)\,dx + \int_5^7 \frac{x}{4}\,dx = \frac{1}{6} + 3 = \frac{19}{6}.$$

Example 3.44 Find $E[X]$ for the random variable X with probability density function

$$f(x) = \frac{1}{x^2} \qquad x > 1.$$

This distribution is a special case of the *Pareto* distribution. The population mean is

$$E[X] = \int_1^\infty x \cdot \frac{1}{x^2}\,dx = \int_1^\infty \frac{1}{x}\,dx,$$

which does not exist. This is the first example we have encountered of a distribution with a population mean that does not exist.

Example 3.45 Find $E[X]$ for the random variable X with probability density function

$$f(x) = \frac{1}{\pi(1+x^2)} \qquad -\infty < x < \infty.$$

This distribution is known as the *Cauchy* distribution, and it is an oft-cited example of a distribution with an undefined population mean. The population mean is

$$E[X] = \int_{-\infty}^{\infty} \frac{x}{\pi(1+x^2)} \, dx = \frac{1}{2\pi} \left[\ln(1+x^2) \right]_{-\infty}^{\infty},$$

which does not exist. Hence $E[X]$ does not exist for the Cauchy distribution. Both the Pareto and Cauchy distributions do, however, have medians, as do all continuous distributions.

So far, we have encountered two so-called "measures of central tendency," namely the population median $x_{0.5}$ and the population mean μ. There is a third measure that can be considered a measure of central tendency for some unimodal distributions, which is known as the *population mode*. We take a brief diversion from expected values to discuss and illustrate the calculation of the population mode.

The population mode of a discrete probability distribution is the value in the support $\mathcal{A}$ where the probability mass function $f(x)$ assumes its maximum value. The population mode is the value that is the most frequently occurring value for a discrete distribution. The population mode of a continuous probability distribution is the value in the support $\mathcal{A}$ where the probability density function $f(x)$ achieves its maximum value.

Example 3.46 Find the population mode, mean, and median of the continuous random variable X with probability density function

$$f(x) = xe^{-x} \qquad x > 0.$$

The population mode can be found by differentiating the probability density function with respect to x. Using the product rule

$$f'(x) = -xe^{-x} + e^{-x} = e^{-x}(1-x).$$

Setting the derivative to zero and solving for x yields $x = 1$. Since $f(0) = 0$ and $\lim_{x \to \infty} f(x) = 0$, there is only a single value x which maximizes $f(x)$, the mode occurs at $x = 1$. The population mean and median can be calculated by computing the integral

$$\mu = \int_0^{\infty} x^2 e^{-x} \, dx$$

and solving the equation

$$\int_0^{x_{0.5}} xe^{-x} \, dx = \frac{1}{2}$$

for $x_{0.5}$. They can also be calculated with the APPL commands

```
X := [[x -> x * exp(-x)], [0, infinity], ["Continuous", "PDF"]];
Mean(X);
IDF(X, 0.5);
```

which yield $\mu = 2$ and $x_{0.5} \cong 1.6783$. Figure 3.29 shows that the skewed nature of the probability density function results in three different values for these measures of central tendency. The population mode is the value in the support where $f(x)$ achieves

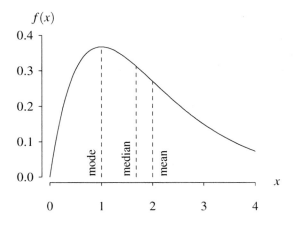

Figure 3.29: Population mode, median and mean for $f(x) = xe^{-x}$ on $x > 0$.

its global maximum. The population median is the "middle" value of the probability distribution with area $1/2$ under $f(x)$ to the left and right of $x_{0.5}$. The population mean, or expected value of X, is the average value obtained for an X value sampled from the probability distribution.

A more careful definition of the mode is motivated by the discrete random variable X with probability mass function

$$f(x) = \frac{1}{2} \qquad x = 0, 1$$

and the continuous random variable X with probability density function

$$f(x) = 1 \qquad 0 < x < 1.$$

Both of these distributions maximize $f(x)$ at multiple values of x, so it is best to define a mode as a unique value as follows.

Definition 3.4 If there is only one value x such that $f(x)$ is maximized at x, it is the *population mode* of the distribution.

In an awkward use of terminology, probabilists and statisticians use the term "bimodal" for a random variable with $f(x)$ having two local maximums. Likewise the term "multimodal" describes the distribution of a random variable with $f(x)$ having more than two local maximums. It remains the case that the mode, however, is at the highest of the peaks.

This concludes the introduction of the expected value of a univariate discrete or continuous random variable X. The population mean $\mu = E[X]$ is a measure of central tendency that gives a "best guess" at the average value that X will attain. The population mean, however,

(a) gives no indication of the dispersion or variance of the probability distribution,

(b) gives no indication of the symmetry of the probability distribution,

(c) is influenced by heavy-tailed distributions.

Measures of dispersion and symmetry will be addressed later in this section.

There are three remaining miscellaneous topics that are related to expected values: properties of expected values, moments, and moment generating functions. Each topic has a separate subheading.

Properties of expected values

There are several mathematical properties that the expected value operator satisfies that are stated and proven here. The proofs are given for either continuous random variables or discrete random variables only. The analogous proofs for the other type of random variable typically involves swapping integrals and summations.

Theorem 3.1 Given a random variable X and a real constant c,

$$E[c] = c.$$

Proof (continuous random variable X) Since all probability density functions integrate to one, the expected value of c is

$$E[c] = \int_{-\infty}^{\infty} c f(x)\, dx = c \int_{-\infty}^{\infty} f(x)\, dx = c \cdot 1 = c. \qquad \square$$

Not surprisingly, the expected value of a non-random quantity c is simply the value c. The next result highlights the fact that the expected value operator $E[\cdot]$ is a linear operator.

Theorem 3.2 Given a random variable X and a real constant c,

$$E[cX] = cE[X].$$

Proof (continuous random variable X) The expected value of cX is

$$E[cX] = \int_{-\infty}^{\infty} cx f(x)\, dx = c \int_{-\infty}^{\infty} x f(x)\, dx = cE[X]. \qquad \square$$

This result implies that constants found inside the expected value operator can simply be pulled outside of the operator. We now consider an important result concerning the expected value of a function of a random variable. We first motivate the result with an example.

Example 3.47 Let the continuous random variable X be uniformly distributed between 0 and 1 with probability density function

$$f_X(x) = 1 \qquad 0 < x < 1.$$

Find $E\left[\sqrt{X}\right]$.

There are two different ways of thinking about the meaning of $E\left[\sqrt{X}\right]$ that are presented here.

Approach 1. Think of generating a large number of X values uniformly distributed between 0 and 1 using Monte Carlo simulation, then taking the square root of each value (the transformation), then averaging the results (the expected value operator). Five runs of the R code

```
nrep = 100000
x = runif(nrep)
mean(sqrt(x))
```

yield

| 0.6674065 | 0.6649523 | 0.6657345 | 0.6662218 | 0.6663353. |

The results seem to hover around $2/3$. This will be checked using the analytic approach next.

Approach 2. In Example 3.23, the distribution of $Y = g(X) = \sqrt{X}$ was determined using the cumulative distribution function technique, where X is uniformly distributed between 0 and 1. The probability density function of Y, found by differentiating the cumulative distribution function with respect to y, is

$$f_Y(y) = 2y \qquad 0 < y < 1.$$

Since $Y = \sqrt{X}$, it seems reasonable to proceed to find $E[Y] = E\left[\sqrt{X}\right]$ using the definition of the expected value as

$$E[Y] = \int_0^1 y \cdot 2y \, dy = \frac{2}{3},$$

which is consistent with the Monte Carlo simulation results from the first approach.

The next result involving the expected value of a function of a random variable allows us to calculate the expected value without first finding the distribution of $Y = g(X)$.

Theorem 3.3 Let X be a random variable defined on the support $\mathcal{A}$ with probability mass function $f(x)$ if X is discrete and probability density function $f(x)$ if X is continuous. The *expected value* of $g(X)$ is

$$E[g(X)] = \begin{cases} \displaystyle\sum_{\mathcal{A}} g(x)f(x) & X \text{ discrete} \\ \displaystyle\int_{\mathcal{A}} g(x)f(x) \, dx & X \text{ continuous} \end{cases}$$

when the sum or integral exists. When the sum or integral diverges, the expected value (also known as the "mathematical expectation") is undefined.

Proof (discrete random variable X) Let $Y = g(X)$ be a 1–1 transformation from $\mathcal{A}$ to $\mathcal{B}$. The proof when $Y = g(X)$ is not 1–1 is similar but involves more tedious bookkeeping. The random variable Y has probability mass function $f_Y(y)$. The random variable X has probability mass function $f_X(x)$. By definition, the expected value of $Y = g(X)$ is

$$\begin{aligned} E[Y] &= \sum_{\mathcal{B}} y f_Y(y) \\ &= \sum_{\mathcal{B}} y P(Y = y) \\ &= \sum_{\mathcal{B}} y P(g(X) = y) \\ &= \sum_{\mathcal{A}} g(x) P(X = x) \\ &= \sum_{\mathcal{A}} g(x) f_X(x). \end{aligned}$$

This proves the theorem in the case of a discrete random variable X because $E[Y] = E[g(X)]$. $\qquad\square$

Example 3.48 Let the continuous random variable X be uniformly distributed between 0 and 1 with probability density function

$$f_X(x) = 1 \qquad 0 < x < 1.$$

Find $E\left[\sqrt{X}\right]$.

This is the same problem posed in the previous example, but this time it will be worked using Theorem 3.3. The expected value of the square root of X is

$$E\left[\sqrt{X}\right] = \int_0^1 \sqrt{x} \cdot 1 \, dx = \left[\frac{2x^{3/2}}{3}\right]_0^1 = \frac{2}{3},$$

which is consistent with the other two approaches to solving the problem.

Example 3.49 Divide a line segment of unit length randomly into two parts. Find the expected value of the quotient of the lengths of the two segments.

The first task is to determine the meaning of randomly dividing a line segment. It is not unreasonable to assume that the break point is uniformly distributed along the length of the line segment. Let X be the break point with probability density function

$$f(x) = 1 \qquad 0 < x < 1.$$

Each segment has expected length of $1/2$ because

$$E[X] = \frac{1}{2}$$

by symmetry, and

$$E[1 - X] = 1 - E[X] = 1 - \frac{1}{2} = \frac{1}{2}.$$

Using Theorem 3.3, the expected value of the ratio of the two lengths is

$$E\left[\frac{X}{1 - X}\right] = \int_0^1 \frac{x}{1 - x} \cdot 1 \, dx = \left[-x - \ln(x - 1)\right]_0^1,$$

which does not exist. The distribution of the random variable $X/(1 - X)$ has a support set which is the positive real numbers and a heavy enough right-hand tail so that the population mean of this random variable is not defined, as was the case of the Pareto and Cauchy distributions. In spite of the fact that the expected value is undefined, Monte Carlo simulation can still be used to estimate the expected value. After a call to set.seed(1) to initialize the random number stream, five runs of the following R statements to estimate the expected value of the quotient

```
x = runif(500000)
mean(x / (1 - x))
```

result in

13.05924	16.71324	19.40304	15.93129	43.56639.

Other random number streams yield similar results. In spite of the large number of replications, the averages of the 500,000 replications of the experiment are not converging. This is consistent with the undefined expected value. The fifth run, for example, produced a sample mean that was more than double the others. It must have been the case that during this run, one (or more) large random variate X was generated from the right-hand tail that resulted in a huge value for $X/(1 - X)$.

Theorem 3.3 can be applied to finding the expected value of arbitrarily complex functions of a random variable X, as seen in the next example.

Example 3.50 During the early part of the 20th century, newsboys sold newspapers on U.S. city streets. Each newsboy would get one chance to purchase newspapers at the beginning of the day to sell to his customers. At the end of the day, the unsold newspapers must be discarded. If he purchases too few newspapers, then potential sales are lost, representing an opportunity cost. If he purchases too many newspapers, then some of the newspapers must be discarded. If the daily demand for a newspaper is a discrete random variable X with known probability mass function $f(x)$ that is based on previous experience, how many newspapers should the newsboy purchase to maximize his expected profit?

This well-known "newsboy problem" is also known as the "Christmas tree problem" in that it involves purchasing a particular quantity of a product when there is only one opportunity to do so. There are many products (for example, magazines, flowers) that fall into this category. Sometimes there is a small recovery value for unsold items; sometimes there is a disposal cost for unsold items. In this model we assume that there is no recovery value and no disposal cost—the newsboy discards the unsold newspapers.

The following notation will be used to solve the newsboy problem.

- Let p be the newsboy's *profit* on each newspaper sold.

- Let l be the newsboy's *loss* on each newspaper unsold (that is, the price he pays to purchase the newspaper).

- Let q be the *quantity of newspapers purchased* by the newsboy at the beginning of the day.

- Let the discrete random variable X, with nonnegative support, probability mass function $f(x)$, and cumulative distribution function $F(x)$, be the daily *demand* for newspapers sold by this newsboy.

- Let the random variable P_q be the newsboy's profit when the newsboy stocks q newspapers on one particular day.

The goal here is to find the optimal value of q in order to maximize the newsboy's expected profit. The random variable P_q can be written in terms of the random variable X as

$$P_q = \min\{pX - l(q - X), pq\}.$$

When q newspapers are purchased, the profit can't exceed pq, the second term in the minimum expression. This accounts for the excess demand case. On the other hand, when the demand X is less than the quantity ordered q, the profit is the sum of the profit from the X newspapers sold, pX, minus the loss from the $q - X$ newspapers not sold $l(q - X)$.

The expected profit when the newsboy purchases q newspapers is

$$
\begin{aligned}
E[P_q] &= E\left[\min\{pX - l(q - X), pq\}\right] \\
&= \sum_{x=0}^{q} (px - l(q - x))f(x) + \sum_{x=q+1}^{\infty} pqf(x) \\
&= (p + l)\sum_{x=0}^{q} xf(x) - lq\sum_{x=0}^{q} f(x) + pq\left[1 - \sum_{x=0}^{q} f(x)\right] \\
&= pq + (p + l)\sum_{x=0}^{q} xf(x) - (p + l)q\sum_{x=0}^{q} f(x) \\
&= pq + (p + l)\sum_{x=0}^{q} (x - q)f(x).
\end{aligned}
$$

We take a brief break in the general derivation to look at a specific case. Using early 20th century numbers, the newsboy purchases newspapers from the publisher for 5 cents each and sells them for 9 cents each. Using the notation in the mathematical model, the newsboy makes a profit of $p = 4$ cents for each newspaper sold and incurs a loss of $l = 5$ cents for every newspaper that he can't sell. Assume that customer demand is equally likely between 1 and 20, that is

$$f(x) = \frac{1}{20} \qquad x = 1, 2, \ldots, 20.$$

The expected profit $E[P_q]$ can now be plotted for various values of q, shown in Figure 3.30. This graph is created with the R statements

```
p = 4
l = 5
x = 1:20
f = 1 / 20
profit = rep(0, 20)
for (q in 1:20) {
   profit[q] = p * q + (p + l) * sum(x[1:q] - q) * f
}
plot(x, profit, type = "h")
```

When the (conservative) newsboy purchases just $q = 1$ newspaper, he is certain of a sale, so his expected profit is $E[P_1] = 4$ cents. When the newsboy purchases $q = 2$ newspapers, he receives a profit of 8 cents with probability $19/20$ and a loss of 1 cent with probability $1/20$, so his expected profit is

$$E[P_2] = 8 \cdot \frac{19}{20} - 1 \cdot \frac{1}{20} = 7.55$$

cents. At the other extreme, the (reckless) newsboy purchases $q = 20$ newspapers, which results in a high probability of discarding many unsold newspapers, and an expected profit of $E[P_{20}] = -5.5$ cents. The optimal value of q is to purchase $q^* = 9$ newspapers for an expected profit of $E[P_9] = 19.8$ cents.

Returning to the general derivation, we desire to maximize $E[P_q]$, but the discrete nature of q makes this more difficult than if q were continuous. One way to proceed is to

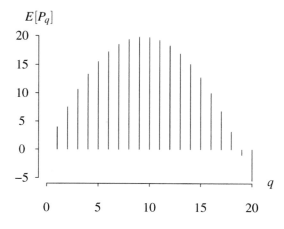

Figure 3.30: Expected profit $E[P_q]$ as a function of quantity ordered q.

increase q sequentially by one until the difference between the expected profit when $q + 1$ newspapers are purchased and q newspapers are purchased becomes negative. The expected profit when $q + 1$ newspapers are purchased is

$$E[P_{q+1}] = p(q+1) + (p+l) \sum_{x=0}^{q+1} (x - q - 1) f(x)$$

$$= p(q+1) + (p+l) \sum_{x=0}^{q} (x - q - 1) f(x).$$

So the difference in the expected profit when stocking $q + 1$ and q items is

$$E[P_{q+1}] - E[P_q] = p(q+1) + (p+l) \sum_{x=0}^{q} (x - q - 1) f(x) - pq - (p+l) \sum_{x=0}^{q} (x - q) f(x)$$

$$= p - (p+l) \sum_{x=0}^{q} f(x)$$

$$= p - (p+l) F(q).$$

The purchase quantity q should be increased as long as the difference $p - (p+l)F(q)$ remains positive. In other words

$$p - (p+l) F(q) > 0$$

or, equivalently, purchasing $q + 1$ newspapers is preferred to purchasing q newspapers as long as

$$F(q) < \frac{p}{p+l}.$$

In our numerical example with $p = 4$ and $l = 5$, the newsboy prefers stocking $q + 1$ newspapers to q newspapers as long as

$$F(q) < \frac{4}{9}.$$

The newsboy problem in this particular instance can be solved using the APPL code

```
p := 4;
l := 5;
X := UniformDiscreteRV(1, 20);
IDF(X, p / (p + 1));
```

which yields $q^* = 9$.

We end the discussion on properties of expectation with two results concerning the expected value of a function of a random variable.

Theorem 3.4 Given a random variable X, a real constant c, and a function $g(X)$,

$$E[cg(X)] = cE[g(X)]$$

when the expected values exist.

Proof (continuous random variable X) The expected value of $cg(X)$ is

$$E[cg(X)] = \int_{-\infty}^{\infty} cg(x) f(x) \, dx = c \int_{-\infty}^{\infty} g(x) f(x) \, dx = cE[g(X)]$$

when the expected values exist. $\qquad \square$

As expected, constants within the expected value operator can be pulled outside of the operator. The next result concerns the expected value of the sum of the expected value of two functions, which generalizes to the sum of more than two functions.

Theorem 3.5 Given a random variable X and functions $g_1(X)$ and $g_2(X)$,

$$E[g_1(X) + g_2(X)] = E[g_1(X)] + E[g_2(X)]$$

when the expected values exist.

Proof (continuous random variable X) The expected value of $g_1(X) + g_2(X)$ is

$$
\begin{aligned}
E[g_1(X) + g_2(X)] &= \int_{-\infty}^{\infty} (g_1(x) + g_2(x)) f(x)\, dx \\
&= \int_{-\infty}^{\infty} g_1(x) f(x)\, dx + \int_{-\infty}^{\infty} g_2(x) f(x)\, dx \\
&= E[g_1(X)] + E[g_2(X)]
\end{aligned}
$$

when the expected values exist.   □

Moments

The mean, median, and mode are popular measures of central tendency. There are other expected values that can be useful for characterizing certain aspects of a probability distribution. We generically refer to these expected values as *moments*.

The first such expected value considered is the *population variance*. We motivate the use of the population variance with the following example.

Example 3.51 Suppose we decide to measure the distance between a random variable X and constant a with $(X - a)^2$. (The squaring gives more weight to X values far away from a and makes the side of a that X falls on irrelevant.) Find the value of a that minimizes $E\left[(X - a)^2\right]$.

This is a standard minimization problem from calculus. The solution below is for a continuous random variable X. When X is discrete, integrals are replaced by summations. By Theorem 3.3, the expected value of $(X - a)^2$ is

$$E\left[(X - a)^2\right] = \int_{-\infty}^{\infty} (x - a)^2 f(x)\, dx.$$

Differentiating with respect to a and equating to zero yields

$$-\int_{-\infty}^{\infty} 2(x - a) f(x)\, dx = 0$$

or

$$\int_{-\infty}^{\infty} x f(x)\, dx - \int_{-\infty}^{\infty} a f(x)\, dx = 0$$

or

$$a = \mu.$$

The expected value is minimized at $a = \mu$.

From Example 3.51, the expected value at the minimum, namely

$$E\left[(X - \mu)^2\right],$$

can be interpreted as a measure of dispersion, or spread, of the distribution of a random variable X. It measures the expected squared distance between a random variable and its population mean. This expected value is known as the population variance of the random variable X, as defined below.

Definition 3.5 For a random variable X with mean μ, the *population variance* of X is

$$\sigma^2 = V[X] = E\left[(X-\mu)^2\right]$$

when the expected value exists.

The following facts can be useful when working with the population variance.

- The units on the population variance are the square of the units of the random variable X.

- The positive square root of the population variance is the *population standard deviation* σ. One reason for the popularity of σ is that its units are the same units as the random variable X.

- If the discrete random variable X has a support $\mathcal{A}$ that contains only one x-value, then $\sigma^2 = 0$. This distribution is often known as a *degenerate* distribution. ⟶ only 1 *x-value*

- Some authors use Var$[X]$ for the population variance.

- The population variance is not the only measure of dispersion for a random variable X. Two other, less popular, measures of dispersion are

  - the population range $R = \sup\mathcal{A} - \inf\mathcal{A}$,
  - the population mean absolute deviation $E[|X - \mu|]$.

The next two results will be used throughout the text when working with population variance. The definition of the population variance given in Definition 3.5 is known as a *defining formula*. The next result gives a *shortcut* or *computational* formula that is often easier to calculate for a given random variable X.

Theorem 3.6 For the random variable X with population mean μ and population variance σ^2,

$$V[X] = E\left[X^2\right] - \mu^2,$$

which is known as the *shortcut formula* for computing the population variance.

Proof For a discrete or continuous random variable X,

$$
\begin{aligned}
E\left[(X-\mu)^2\right] &= E\left[X^2 - 2\mu X + \mu^2\right] \\
&= E\left[X^2\right] - 2\mu E[X] + \mu^2 \\
&= E\left[X^2\right] - \mu^2,
\end{aligned}
$$

where all expectations exist because the population mean and variance exist. □

Example 3.52 Using the defining and computational formulas, find the population variance of the number of spots showing when rolling a fair die.

Let the random variable X be the number of spots showing. The probability mass function of X is

$$f(x) = \frac{1}{6} \qquad x = 1, 2, 3, 4, 5, 6.$$

The expected number of spots showing is

$$\mu = E[X] = \sum_{x=1}^{6} x f(x) = 1\cdot\frac{1}{6} + 2\cdot\frac{1}{6} + \cdots + 6\cdot\frac{1}{6} = \frac{1}{6}[1+2+\cdots+6] = \frac{1}{6}\left[\frac{6\cdot 7}{2}\right] = \frac{7}{2}.$$

Using the defining formula, the population variance of the number of spots showing is

$$
\begin{aligned}
\sigma^2 &= E\left[(X-\mu)^2\right] \\
&= \sum_{x=1}^{6}\left(x-\frac{7}{2}\right)^2 f(x) \\
&= \left(1-\frac{7}{2}\right)^2\frac{1}{6}+\left(2-\frac{7}{2}\right)^2\frac{1}{6}+\cdots+\left(6-\frac{7}{2}\right)^2\frac{1}{6} \\
&= \frac{25}{4}\cdot\frac{1}{6}+\frac{9}{4}\cdot\frac{1}{6}+\cdots+\frac{25}{4}\cdot\frac{1}{6} \\
&= \frac{35}{12}.
\end{aligned}
$$

Using the shortcut formula, the population variance of the number of spots showing is

$$
\begin{aligned}
\sigma^2 &= E\left[X^2\right]-\mu^2 \\
&= \sum_{x=1}^{6} x^2 f(x)-\left(\frac{7}{2}\right)^2 \\
&= 1^2\cdot\frac{1}{6}+2^2\cdot\frac{1}{6}+\cdots+6^2\cdot\frac{1}{6}-\frac{49}{4} \\
&= \frac{1}{6}\left(1^2+2^2+\cdots+6^2\right)-\frac{49}{4} \\
&= \frac{1}{6}\left(\frac{6\cdot7\cdot13}{6}\right)-\frac{49}{4} \\
&= \frac{91}{6}-\frac{49}{4} \\
&= \frac{35}{12}
\end{aligned}
$$

using the formula for the sum of squares for the first six integers. The population mean and variance of X can also be calculated with the APPL statements

```
X := UniformDiscreteRV(1, 6);
Mean(X);
Variance(X);
```

which also yield

$$\mu=\frac{7}{2} \qquad \text{and} \qquad \sigma^2=\frac{35}{12}.$$

Example 3.53 Calculate the population variance of the random variable X with probability density function

$$f(x)=\frac{x}{2} \qquad 0<x<2$$

using the defining formula and the computational formula.

The expected value of X is

$$\mu=E[X]=\int_0^2 x\cdot\frac{x}{2}\,dx=\frac{4}{3}.$$

Using the defining formula, the population variance is

$$\begin{aligned}
\sigma^2 &= E\left[(X-\mu)^2\right] \\
&= \int_0^2 \left(x - \frac{4}{3}\right)^2 \frac{x}{2}\, dx \\
&= \frac{2}{9}.
\end{aligned}$$

Using the shortcut formula, the population variance is

$$\begin{aligned}
\sigma^2 &= E\left[X^2\right] - \mu^2 \\
&= \int_0^2 x^2 \frac{x}{2}\, dx - \left(\frac{4}{3}\right)^2 \\
&= \frac{2}{9}.
\end{aligned}$$

The population mean and variance of X can also be calculated with the APPL statements

```
X := [[x -> x / 2], [0, 2], ["Continuous", "PDF"]];
Mean(X);
Variance(X);
```

which also yield

$$\mu = \frac{4}{3} \qquad \text{and} \qquad \sigma^2 = \frac{2}{9}.$$

Theorem 3.7 For the random variable X with population mean μ and population variance σ^2,

$$V[aX+b] = a^2 V[X]$$

for real constants a and b.

Proof For a discrete or continuous random variable X, using the definition of the population variance,

$$\begin{aligned}
V[aX+b] &= E\left[((aX+b) - E[aX+b])^2\right] \\
&= E\left[(aX - aE[X])^2\right] \\
&= E\left[a^2(X-\mu)^2\right] \\
&= a^2 E\left[(X-\mu)^2\right] \\
&= a^2 V[X],
\end{aligned}$$

where all expectations exist because the population mean and variance exist. $\qquad\square$

The intuition associated with Theorem 3.7 is that shifting the distribution of a random variable by a constant b does not influence the variance. Two corollaries that result from Theorem 3.7 can be found by setting $b = 0$ and $a = 0$:

$$V[aX] = a^2 V[X] \qquad \text{and} \qquad V[b] = 0.$$

The equation on the left indicates that a constant inside of the variance operator can be extracted as the square of that constant. In this sense, the population variance is a quadratic operator. Taking the square roots of both sides of this equation, the standard deviation of aX must include the absolute value of a, that is $\sigma_{aX} = |a|\sigma_X$. The equation on the right, $V[b] = 0$, indicates that the population variance of a constant is 0. This is reasonable because a constant is the equivalent of a degenerate probability distribution, which also has population variance 0.

This ends the discussion of the population variance σ^2 and population standard deviation σ which are the primary measures of the dispersion or spread of the distribution of a random variable. There are higher order moments that are also of interest from time to time. We begin with general definitions of population moments about the origin and moments about the mean.

Definition 3.6 For a random variable X and positive integer r, the rth population moment of X about the origin is

$$E[X^r]$$

when the expectation exists.

We have already encountered the population mean $E[X]$ with $r = 1$, and $E[X^2]$ with $r = 2$ in the shortcut formula for the population variance.

Definition 3.7 For a random variable X and positive integer r, the rth population moment of X about the population mean is

$$E[(X - \mu)^r]$$

when the expectation exists.

The rth population moment of X about the population mean is sometimes called the rth central population moment of X. The population variance $E[(X - \mu)^2]$, for example, is the 2nd population moment about the mean. To define higher order population moments, probabilists often find that it is easier to work with a *standardized random variable*.

Definition 3.8 For a random variable X with population mean μ and positive population standard deviation σ, the random variable

$$\frac{X - \mu}{\sigma}$$

is called a *standardized random variable*, which has mean 0 and standard deviation 1 when the expectations exist.

Some authors prefer to call these random variables *standardized, centralized random variables* because subtracting the mean centralizes the random variable (so its population mean is 0) and dividing by σ standardizes the random variable (so its population standard deviation is 1).

Example 3.54 Show that the standardized random variable has population mean 0 and population standard deviation 1 when the expectations exist.

The population mean of the standardized random variable is

$$E\left[\frac{X - \mu}{\sigma}\right] = \frac{1}{\sigma}E[X - \mu] = \frac{1}{\sigma}[E[X] - E[\mu]] = \frac{1}{\sigma} \cdot (\mu - \mu) = 0.$$

The population variance of the standardized random variable is

$$V\left[\frac{X - \mu}{\sigma}\right] = \frac{1}{\sigma^2}V[X - \mu] = \frac{1}{\sigma^2}V[X] = \frac{\sigma^2}{\sigma^2} = 1.$$

The population mean μ is the first moment about the origin. It is the primary measure of the central tendency of the probability distribution. The population variance σ^2 is the second moment about the mean. It is the primary measure of the dispersion of the probability distribution. We now introduce a new moment which is the population *skewness*, which is the expected value of the cube of the standardized random variable. .

> **Definition 3.9** For a random variable X with population mean μ and positive population variance σ^2, the *population skewness* is
>
> $$E\left[\left(\frac{X-\mu}{\sigma}\right)^3\right]$$
>
> when the expectation exists.

The population skewness is a measure of the symmetry of a probability distribution. If $f(x)$ is symmetric, the skewness is 0; if $f(x)$ is skewed to the right, the skewness is positive; if $f(x)$ is skewed to the left, the skewness is negative.

Although it is pushing the limits of interpretation, the fourth standardized moment also gets its own name: the population *kurtosis*. It is a measure of the heaviness of the tails of a probability density function as well as measuring the peakedness of a probability density function. Thankfully, there is little discussion of population moments higher than order 4.

> **Definition 3.10** For a random variable X with population mean μ and positive population variance σ^2, the *population kurtosis* is
>
> $$E\left[\left(\frac{X-\mu}{\sigma}\right)^4\right]$$
>
> when the expectation exists.

Example 3.55 Find the population skewness and kurtosis of a continuous random variable X that is uniformly distributed between 0 and 1.

By using the symmetry of $f(x)$ and the center-of-gravity interpretation of μ, or by using calculus, the population mean is

$$\mu = \int_0^1 x \cdot 1 \, dx = \frac{1}{2}.$$

Using the defining formula, the population variance is

$$\sigma^2 = \int_0^1 \left(x - \frac{1}{2}\right)^2 \cdot 1 \, dx = \frac{1}{12}.$$

The population skewness is

$$E\left[\left(\frac{X-\mu}{\sigma}\right)^3\right] = 12^{3/2} \int_0^1 \left(x - \frac{1}{2}\right)^3 \cdot 1 \, dx = 0.$$

The population kurtosis is

$$E\left[\left(\frac{X-\mu}{\sigma}\right)^4\right] = 12^2 \int_0^1 \left(x - \frac{1}{2}\right)^4 \cdot 1 \, dx = \frac{9}{5}.$$

The first four population moments can also be calculated using the following APPL statements.

```
X := [[x -> 1], [0, 1], ["Continuous", "PDF"]];
Mean(X);
Variance(X);
Skewness(X);
Kurtosis(X);
```

We end the discussion of population moments with a random variable with a nonsymmetric probability density function, which typically results in a nonzero skewness.

Example 3.56 Find the population skewness and kurtosis of a continuous random variable X with probability density function

$$f(x) = \frac{x}{2} \qquad 0 < x < 2.$$

The population mean is

$$\mu = \int_0^2 x \cdot \frac{x}{2} \, dx = \frac{4}{3}.$$

Using the defining formula, the population variance is

$$\sigma^2 = \int_0^2 \left(x - \frac{4}{3} \right)^2 \cdot \frac{x}{2} \, dx = \frac{2}{9}.$$

The population skewness is

$$E\left[\left(\frac{X - \mu}{\sigma} \right)^3 \right] = \left(\frac{9}{2} \right)^{3/2} \int_0^2 \left(x - \frac{4}{3} \right)^3 \cdot \frac{x}{2} \, dx = -\frac{2\sqrt{2}}{5}.$$

The negative skewness is an indication that the probability density function of X is "skewed to the left" because of the long left-hand tail of the ramp-shaped probability density function. The population kurtosis is

$$E\left[\left(\frac{X - \mu}{\sigma} \right)^4 \right] = \left(\frac{9}{2} \right)^2 \int_0^2 \left(x - \frac{4}{3} \right)^4 \cdot \frac{x}{2} \, dx = \frac{12}{5}.$$

The first four population moments can also be calculated using the following APPL statements.

```
X := [[x -> x / 2], [0, 2], ["Continuous", "PDF"]];
Mean(X);
Variance(X);
Skewness(X);
Kurtosis(X);
```

The $f(x)$ and $F(x)$ functions completely define the distribution of a discrete or continuous random variable X. The population moments, however, do not completely specify the distribution of a random variable X. The next subsection defines the moment generating function which effectively holds all of the population moments for a random variable X in one function. The moment generating function also completely defines the distribution of a random variable X.

Moment generating functions

This subsection explores some elementary properties of the moment generating function $M(t)$. The moment generating function is a special expected value, indexed by t, that has several nice properties that can be exploited in solving probability problems. As with $f(x)$ and $F(x)$, we attach the random variable as a subscript to the moment generating function, that is $M_X(t)$, whenever there might be any confusion about which random variable is under consideration.

Definition 3.11 Let X be a random variable. The *moment generating function* of X is

$$M(t) = E\left[e^{tX} \right]$$

provided that the expected value exists on the interval $-h < t < h$ for some positive real number h.

So if this expected value exists on a neighborhood about the origin, the random variable X has a moment generating function. Moment generating functions have three applications. First, by virtue of their name, they are good at generating moments. We will focus our attention entirely upon this application here. Second, in Chapter 7, they will also be used to find the distribution of sums of independent random variables. Finally, in Chapter 8, they will also be used to find the limiting distribution of a random variable.

Example 3.57 Find the moment generating function for a continuous random variable X that is uniformly distributed between 0 and 1.

The probability density function of X is

$$f(x) = 1 \qquad 0 < x < 1.$$

The moment generating function of X is the expected value of e^{tX}

$$
\begin{aligned}
M(t) &= E\left[e^{tX}\right] \\
&= \int_0^1 e^{tx} \cdot 1 \, dx \\
&= \left[\frac{1}{t} e^{tx}\right]_0^1 \\
&= \frac{1}{t}\left(e^t - 1\right).
\end{aligned}
$$

This moment generating function does exist in a neighborhood about 0. Use l'Hôpital's rule to find the moment generating function at 0, which is

$$\lim_{t \to 0} M(t) = \lim_{t \to 0} \frac{e^t - 1}{t} = \lim_{t \to 0} \frac{e^t}{1} = 1.$$

So $M(t)$ is a continuous function that exists on $-\infty < t < \infty$.

This paragraph outlines several properties of moment generating functions. First, not all distributions have moment generating functions. The Cauchy distribution, which was introduced in Example 3.45, does not have a moment generating function because the expected value does not exist. Second, the moment generating function at $t = 0$ must be 1, that is $M(0) = 1$. This is because $M(0) = E\left[e^0\right] = E[1] = 1$. Third, the moment generating function is unique and completely defines the distribution of X when it exists. Because of this property, if two random variables have the same moment generating function, they necessarily have the same distribution. Fourth, the existence of a moment generating function implies the existence of derivatives of all orders of the moment generating function.

Example 3.58 Consider the random variable X with moment generating function

$$M(t) = 0.7e^t + 0.2e^{2t} + 0.1e^{3t} \qquad -\infty < t < \infty.$$

Is X discrete or continuous? What is the probability mass function or probability density function of X?

Since the moment generating function is written as a sum, there is a suspicion that the random variable X is discrete because the general formula for the moment generating function of a discrete random variable X is

$$M(t) = E\left[e^{tX}\right] = \sum_{\mathcal{A}} e^{tx} f(x)$$

on some neighborhood about 0. It is clear from this formula that the support of X is $A = \{x \mid x = 1, 2, 3\}$ and the probability mass function is

$$f(x) = \begin{cases} 0.7 & x = 1 \\ 0.2 & x = 2 \\ 0.1 & x = 3. \end{cases}$$

The following theorem provides the basis for using moment generating functions to calculate population moments of a distribution about the origin. The process is straightforward: to calculate the rth population moment about the origin, take r derivatives of the moment generating function with respect to t, then plug $t = 0$ into the resulting function.

Theorem 3.8 If X has moment generating function $M(t)$ then for some positive integer r

$$E[X^r] = M^{(r)}(0) = \frac{d^r}{dt^r} M(t) \Big|_{t=0}.$$

Proof (continuous random variable X for $r = 1$, other cases analogous) The derivative of the moment generating function with respect to t is

$$\begin{aligned} \frac{d}{dt} M(t) &= \frac{d}{dt} \int_{-\infty}^{\infty} e^{tx} f(x) \, dx \\ &= \int_{-\infty}^{\infty} \left(\frac{d}{dt} e^{tx} \right) f(x) \, dx \\ &= \int_{-\infty}^{\infty} \left(x e^{tx} \right) f(x) \, dx \\ &= E[X e^{tX}]. \end{aligned}$$

Using $t = 0$ as an argument in this derivative yields

$$\frac{d}{dt} M(t) \Big|_{t=0} = E[X e^{tX}]_{t=0} = E[X]. \qquad \square$$

Theorem 3.8 can also be proved for any positive integer r by using series expansions. The alternative proof is presented next.

Proof Expand e^{tX} in a series about 0:

$$e^{tX} = 1 + tX + \frac{(tX)^2}{2!} + \frac{(tX)^3}{3!} + \cdots.$$

Taking the expected value of both sides of this equation yields

$$M(t) = 1 + tE[X] + \frac{t^2}{2!} E[X^2] + \cdots$$

when the expected values exist. Expanding $M(t)$ about zero,

$$M(t) = M(0) + M'(0)t + M''(0) \frac{t^2}{2!} + \cdots.$$

Equating like terms in the two previous expressions provides the desired result:

$$E[X^r] = \frac{d^r}{dt^r} M(t) \Big|_{t=0}. \qquad \square$$

Example 3.59 Use the moment generating function to find $E[X]$, $E[X^2]$, and $E[X^3]$ for the continuous random variable X with probability density function

$$f(x) = e^{-x} \qquad x > 0.$$

The moment generating function of X is

$$
\begin{aligned}
M(t) &= E\left[e^{tX}\right] \\
&= \int_0^\infty e^{tx} e^{-x} dx \\
&= \int_0^\infty e^{(t-1)x} dx \\
&= \frac{1}{t-1} \left[e^{(t-1)x} \right]_0^\infty \\
&= \frac{1}{1-t} \qquad t - 1 < 0.
\end{aligned}
$$

The integral in the expression only converges when $t - 1 < 0$, or equivalently, when $t < 1$. Using Definition 3.11, this means that a constant h must be selected so that the moment generating function exists in an interval about 0. Choosing any h between 0 and 1 accomplishes this goal. Therefore, the moment generating function exists. Taking the first three derivatives of the moment generating function:

$$M'(t) = (1-t)^{-2} \qquad M''(t) = 2(1-t)^{-3} \qquad M'''(t) = 6(1-t)^{-4}$$

for $t < 1$. Finally, using $t = 0$ as an argument in these three expressions yields

$$E[X] = 1 \qquad E[X^2] = 2 \qquad E[X^3] = 6.$$

Working this problem using moment generating functions is much easier than calculating the expected values using the definition of an expected value. For example,

$$E[X^3] = \int_0^\infty x^3 e^{-x} dx$$

requires repeated integration by parts which is tedious and prone to error.

Example 3.60 Use the moment generating function to find $E[X]$, $E[X^2]$, and $E[X^3]$ for the discrete random variable X with probability mass function

$$
f(x) = \begin{cases}
0.7 & x = 1 \\
0.2 & x = 2 \\
0.1 & x = 3.
\end{cases}
$$

The moment generating function is

$$M(t) = \sum_{x=1}^3 e^{tx} f(x) = 0.7e^t + 0.2e^{2t} + 0.1e^{3t}$$

for $-\infty < t < \infty$. Taking the first three derivatives of the moment generating function:

$$M'(t) = 0.7e^t + 0.4e^{2t} + 0.3e^{3t} \qquad M''(t) = 0.7e^t + 0.8e^{2t} + 0.9e^{3t}$$

$$M'''(t) = 0.7e^t + 1.6e^{2t} + 2.7e^{3t}$$

for $-\infty < t < \infty$. Using $t = 0$ as an argument in these three expressions yields

$$E[X] = 1.4 \qquad E[X^2] = 2.4 \qquad E[X^3] = 5.$$

APPL can be used to check the calculation of the moment generating function and the three expected values using the statements

```
X := [[0.7, 0.2, 0.1], [1, 2, 3], ["Discrete", "PDF"]];
MGF(X);
subs(t = 0, diff(MGF(X), t));
subs(t = 0, diff(MGF(X), [t, t]));
subs(t = 0, diff(MGF(X), [t, t, t]));
```

These are the same values obtained using the definition of expected value with the APPL statements

```
X := [[0.7, 0.2, 0.1], [1, 2, 3], ["Discrete", "PDF"]];
ExpectedValue(X, x -> x);
ExpectedValue(X, x -> x ^ 2);
ExpectedValue(X, x -> x ^ 3);
```

Many distributions (for example, the Cauchy and Pareto distributions) don't have moment generating functions. The *characteristic function*, defined as

$$\phi(t) = E\left[e^{itX}\right],$$

where $i = \sqrt{-1}$, exists for all distributions. As with the moment generating function, the characteristic function completely defines a distribution. The two functions are related via $\phi(t) = M(it)$. The characteristic function has applications in spectral analysis, where random signals are broken into trigonometric functions. Characteristic functions can also be used to generate moments via

$$E\left[X^r\right] = \frac{\phi^{(r)}(0)}{i^r} = \frac{1}{i^r}\left[\frac{d^r}{dt^r}\phi(t)\right]_{t=0},$$

where r is a positive integer.

3.5   Inequalities

Sometimes it is useful to have bounds on probabilities when only partial information (for example one or two moments) is known about a distribution. We present two inequalities, attributed to Markov and Chebyshev, in this section.

Theorem 3.9 (Markov's inequality) Let X be a random variable with nonnegative support (that is, $P(X \geq 0) = 1$) for which $E[X]$ exists. Then for any positive constant a

$$P(X \geq a) \leq \frac{E[X]}{a}.$$

Proof (continuous random variable X) Due to the assumption of nonnegative support, the integrals below evaluate to nonnegative values. Thus the expected value of X can be written as

$$
\begin{aligned}
E[X] &= \int_0^\infty xf(x)\,dx \\
&= \int_0^a xf(x)\,dx + \int_a^\infty xf(x)\,dx \\
&\geq \int_a^\infty xf(x)\,dx \\
&\geq \int_a^\infty af(x)\,dx \\
&= a\int_a^\infty f(x)\,dx \\
&= aP(X \geq a),
\end{aligned}
$$

which, after rearranging terms, proves Markov's inequality. □

This inequality states that the probability of getting a value of X that is much larger than $E[X]$ will be small, regardless of the distribution being considered. Markov's inequality gives an upper bound for the probability of X exceeding a.

Example 3.61 Roll a fair die 60 times. Let the random variable X be the number of sixes that appear. Use Markov's inequality to find an upper bound on the probability of rolling 30 or more sixes.

From Example 3.5, the probability mass function of X is

$$f(x) = \binom{60}{x} \left(\frac{1}{6}\right)^x \left(\frac{5}{6}\right)^{60-x} \qquad x = 0, 1, 2, \ldots, 60.$$

The expected value of X is

$$E[X] = \sum_{x=0}^{60} x \binom{60}{x} \left(\frac{1}{6}\right)^x \left(\frac{5}{6}\right)^{60-x} = 10.$$

So Markov's inequality gives the upper bound of $P(X \geq 30)$ as

$$P(X \geq 30) \leq \frac{E[X]}{30}$$

or

$$P(X \geq 30) \leq \frac{1}{3}.$$

As is the case with most inequalities in probability theory, since very little information about the distribution of X is required (only its expected value), very little information is delivered. The exact value of $P(X \geq 30)$ is, in fact, *much* less than $1/3$, as illustrated in Figure 3.7.

Before stating Chebyshev's inequality, we prove a generalization of Chebyshev's inequality. Compare the statement of this result to Markov's inequality.

Theorem 3.10 Let $g(X)$ be a nonnegative function of the random variable X. If $E[g(X)]$ exists, then, for every positive, real constant a,

$$P(g(X) \geq a) \leq \frac{E[g(X)]}{a}.$$

Proof (continuous random variable X) Let $A = \{x \mid g(x) \geq a\}$ and $f(x)$ be the probability density function of X. The set A is illustrated in Figure 3.31.

$$E[g(X)] = \int_{-\infty}^{\infty} g(x)f(x)\,dx = \int_A g(x)f(x)dx + \int_{A'} g(x)f(x)dx$$

Since $g(x) \geq 0$ (by assumption) and $f(x) \geq 0$ for all x

$$E[g(X)] \geq \int_A g(x)f(x)\,dx$$

Since $x \in A$ implies that $g(x) \geq a$, $E[g(X)] \geq \int_A af(x)dx$. Also, since $\int_A f(x)\,dx = P(X \in A) = P(g(X) \geq a)$,

$$E[u(X)] \geq aP(u(X) \geq a).$$

Dividing by a yields the result in the theorem. □

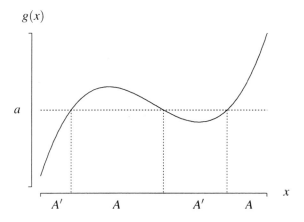

Figure 3.31: Definition of the set A.

Markov's inequality bounds probabilities when only the first moment, $\mu = E[X]$, is known. Chebyshev's inequality, which is presented next, bounds probabilities when the first two moments, $\mu = E[X]$ and $\sigma^2 = V[X]$, are known.

Theorem 3.11 (Chebyshev's inequality) Let the random variable X have a finite population mean μ and a finite population variance σ^2. For every $k > 0$

$$P(|X - \mu| \geq k\sigma) \leq \frac{1}{k^2}.$$

Proof In Theorem 3.10, let $g(X) = (X - \mu)^2$ and $a = k^2\sigma^2$. Then

$$P\left((X - \mu)^2 \geq k^2\sigma^2\right) \leq \frac{E\left[(X - \mu)^2\right]}{k^2\sigma^2}$$

or

$$P\left(|X - \mu| \geq k\sigma\right) \leq \frac{1}{k^2}. \qquad \square$$

Chebyshev's inequality, which is often called *Chebyshev's theorem*, can be stated in several equivalent ways.

(a) Switching the sense of the inequality in the probability statement yields

$$P\left(|X - \mu| < k\sigma\right) \geq 1 - \frac{1}{k^2}.$$

(b) Dropping the absolute values results in a slightly less compact formulation:

$$P(\mu - k\sigma < X < \mu + k\sigma) \geq 1 - \frac{1}{k^2}.$$

(c) In words, the previous statement of Chebyshev's inequality states that the probability that a random variable X assumes a value within k "σ-units" of its population mean is at least $1 - \frac{1}{k^2}$.

(d) Another statement lets $\varepsilon = k\sigma$, which leads to:

$$P\left(|X - \mu| < \varepsilon\right) > 1 - \frac{\sigma^2}{\varepsilon^2}.$$

Chebyshev's inequality is valuable when only μ and σ^2 are known about a distribution. It is of use only for $k > 1$. The next example illustrates the notion of σ-units and Chebyshev's inequality.

Example 3.62 Let X be the number of screws delivered to a box by an automatic filling device. Assume $\mu = 1000$ and $\sigma^2 = 25$. There are problems associated with having too many (giving away free product) or too few (potential irritated customers) screws in a box.

(a) How many σ-units to the right of μ is 1009?

(b) What X value is 2.6 σ-units to the left of μ?

(c) Use Chebyshev's inequality to find a bound on $P[994 < X < 1006]$.

(d) For what value s can the manufacturer be certain that

$$P(1000 - s < X < 1000 + s) \geq 0.80?$$

The formula for converting a specific value of the random variable X to σ-units is

$$k = \frac{X - \mu}{\sigma}$$

and the formula for converting a σ-unit to a specific value of the random variable X is

$$X = \mu + k\sigma,$$

where k is the number of σ-units.

(a) Using the formula above, $X = 1009$ is

$$k = \frac{1009 - 1000}{5} = \frac{9}{5} = 1.8$$

σ-units to the right of μ.

(b) Using the formula above, $k = -2.6$ σ-units corresponds to the X value

$$X = 1000 - (2.6)(5) = 1000 - 13 = 987.$$

(c) Since this interval is symmetric about $\mu = 1000$, the endpoints are

$$k = \frac{1006 - 1000}{5} = \frac{6}{5} = 1.2$$

σ-units from μ, and Chebyshev's inequality implies that

$$P\big(1000 - (1.2)(5) < X < 1000 + (1.2)(5)\big) \geq 1 - \frac{1}{(1.2)^2}$$

or

$$P(994 < X < 1006) \geq 0.3056.$$

(d) To find the value s can the manufacturer be certain that:

$$P(1000 - s < X < 1000 + s) \geq 0.80$$

requires solving

$$1 - \frac{1}{k^2} = 0.80$$

for k yielding $k = \sqrt{5}$. So Chebyshev's inequality indicates that $k\sigma = \sqrt{5} \cdot 5 \cong 11.18$ or

$$P(988.82 < X < 1011.18) \geq 0.8.$$

Chebyshev's inequality gives only a *lower bound* on the probability of interest. The bound is typically not very tight (close to the true probability), as illustrated in the next example.

Example 3.63 Let the continuous random variable X be uniformly distributed between 0 and 1. Find a bound on and the exact value of $P\left(\frac{1}{4} < X < \frac{3}{4}\right)$.

As seen in Example 3.55, the population mean and variance of a random variable X that is uniformly distributed between 0 and 1 are

$$\mu = \frac{1}{2} \qquad \text{and} \qquad \sigma^2 = \frac{1}{12}.$$

The standard deviation is $\sigma = \frac{1}{2\sqrt{3}}$. The interval of interest is symmetric about μ, so Chebyshev's inequality can be applied with

$$k = \frac{X - \mu}{\sigma} = \frac{\frac{3}{4} - \frac{1}{2}}{\frac{1}{2\sqrt{3}}} = \frac{\sqrt{3}}{2}$$

yielding

$$P\left(\frac{1}{2} - \left(\frac{\sqrt{3}}{2}\right)\left(\frac{1}{2\sqrt{3}}\right) < X < \frac{1}{2} + \left(\frac{\sqrt{3}}{2}\right)\left(\frac{1}{2\sqrt{3}}\right)\right) \geq 1 - \frac{1}{\left(\frac{\sqrt{3}}{2}\right)^2}$$

or

$$P\left(\frac{1}{4} < X < \frac{3}{4}\right) \geq -\frac{1}{3}.$$

This interval gives us absolutely no information at all, because we know that probabilities can't be negative. On the other hand, the true probability can be calculated exactly using the probability density function $f(x) = 1$ for $0 < x < 1$ which yields

$$P\left(\frac{1}{4} < X < \frac{3}{4}\right) = \int_{1/4}^{3/4} 1 \, dx = \left[x\right]_{1/4}^{3/4} = \frac{3}{4} - \frac{1}{4} = \frac{1}{2}.$$

So far, Chebyshev's inequality has not produced spectacularly helpful bounds. There are occasions, however, when all of the stars align properly, and the bound can be quite helpful. Here is one such example.

Example 3.64 Find a bound for $P(-10 < X < 10)$ for the discrete random variable X with probability mass function

$$f(x) = \begin{cases} 1/8 & x = -10 \\ 3/4 & x = 0 \\ 1/8 & x = 10. \end{cases}$$

For this distribution, $\mu = 0$ and $\sigma = 5$. Chebyshev's inequality with $k = \frac{10-0}{5} = 2$ is

$$P\big(0 - (2)(5) < X < 0 + (2)(5)\big) \geq 1 - \frac{1}{2^2}$$

or

$$P(-10 < X < 10) \geq \frac{3}{4}$$

and the exact probability is $\frac{3}{4}$. In this case, Chebyshev's inequality is a tight bound in the sense that it gives us substantial information about the probability based on just μ and σ.

3.6 Exercises

3.1 An urn contains r red balls and w white balls, where r and w are positive integers and $r \geq 3$. Balls are drawn successively and without replacement. Let the random variable X be the draw number when the third red ball is drawn. Find the probability mass function of X.

3.2 Atlas and Bruce agree to engage in the following test of strength. They will have consecutive arm wrestling matches until one of them wins two matches in a row and is declared the winner. Atlas wins a given match with probability $3/5$. Assuming that the matches are independent, give the probability mass function of the number of matches required to declare a winner.

3.3 Five fair dice are tossed simultaneously. Find the probability that the total number of spots showing is less than or equal to 11.

3.4 A fair die is rolled n times. Let X_i denote the number of spots that are on the up face on roll i, for $i = 1, 2, \ldots, n$. Find the probability mass function of $Y = \max\{X_1, X_2, \ldots, X_n\}$.

3.5 The Maple function `bubblesort` is listed below.

```
bubblesort := proc(a :: list)
local n, i, j, tmp, T:
n := nops(a):
T := a:
for i from 1 to (n - 1) do
  for j from 1 to (n - i) do
    if (T[j + 1] < T[j]) then
      tmp := T[j]:
      T[j] := T[j+1]:
      T[j+1] := tmp:
    fi:
  od:
od:
T;
end proc:
```

A *swap* occurs when the inside `if` statement is executed. If a list containing a random permutation of three distinct integers is passed to `bubblesort`, give the probability mass function of the number of swaps required by the algorithm to sort the random permutation.

3.6 A fair green die and a fair red die are tossed together. Let X denote the number of spots showing on the green die and Y denote the number of spots showing on the red die. Find the probability mass function for:

    (a) the sum of the spots showing on the two dice, $X + Y$,

    (b) the number of spots showing on the green die minus the number of spots showing on the red die, $X - Y$,

    (c) the difference between the spots showing on the two dice, $|X - Y|$,

    (d) the maximum number of spots showing on a single die, $\max\{X, Y\}$,

    (e) the minimum number of spots showing on a single die, $\min\{X, Y\}$.

3.7 A family of six consisting of a mother, father, and four children line up (shoulder-to-shoulder) for a family portrait in a random order. Find the probability mass function of the random variable X, the number of children located between the father and mother in the photograph.

3.8   A bag contains 20 balls numbered 10, 11, ..., 29. A single ball is drawn at random from the bag. Let the random variable X be the sum of the digits on the ball. Find the probability mass function of X.

3.9   Let the random variable X have probability density function

$$f(x) = 2x \qquad 0 < x < 1.$$

Find $P\left(X^2 < \frac{1}{4}\right)$.

3.10   A continuous random variable X has a single parameter a. The probability density function of X is

$$f(x) = c\left(1 - x^2\right) \qquad -a < x < a.$$

for some real constant $c > 0$.

(a) What are the allowable values of the parameter a?

(b) What is the value of the constant c?

3.11   Consider the random variable X with cumulative distribution function

$$F(x) = \begin{cases} 0 & x \le 0 \\ 1 - e^{-(\lambda x)^\kappa} & x > 0, \end{cases}$$

where λ is a positive scale parameter and κ is a positive shape parameter. Assuming that $\lambda = 0.001$ and $\kappa = 2$, find $P(X > 80 \,|\, X > 50)$.

3.12   A graph of the piecewise-linear probability density function of X is shown below. Draw a graph of the cumulative distribution function.

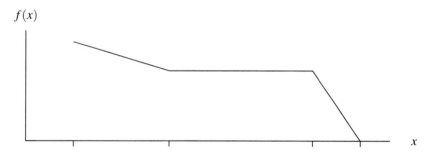

Figure 3.32: A piecewise-linear probability density function.

3.13   If the random variable X has cumulative distribution function $F_X(x)$, what is the cumulative distribution function of the random variable $Y = a + bX$, where a and b are real constants?

3.14   Let X be a continuous random variable with 25th, 50th, and 75th percentiles of 2, 7, and 17, respectively. Give an expression for the 75th percentile of $Y = g(X)$, where $g(\cdot)$ is a continuous, monotone decreasing function.

3.15   Find the cumulative distribution function for the random variable X with probability mass function

$$f(x) = x/10 \qquad x = 3, 7.$$

3.16   Find the cumulative distribution function for the random variable X with probability density function

$$f(x) = 2x \qquad 0 < x < 1.$$

3.17 Find the 64th percentile, $x_{0.64}$, of the distribution of the random variable X with probability mass function

$$f(x) = x/10 \qquad x = 3, 7.$$

3.18 Find the 64th percentile, $x_{0.64}$, of the distribution of the random variable X with probability density function

$$f(x) = 2x \qquad 0 < x < 1.$$

3.19 Find the cumulative distribution function associated with the probability density function

$$f(x) = \begin{cases} x & 0 < x < 1 \\ 2 - x & 1 \leq x < 2. \end{cases}$$

3.20 (Probability integral transformation). Let the continuous random variable X have cumulative distribution function $F_X(x)$. Show that the random variable $Y = F_X(X)$ is a continuous random variable that is uniformly distributed between 0 and 1.

3.21 Let X be a continuous random variable with probability density function

$$f(x) = \frac{1}{3} \qquad -1 < x < 2.$$

Find the probability density function of $Y = X^2$.

3.22 If the income in the United States (normalized so that it is expressed as an hourly wage) is modeled by a random variable X with cumulative distribution function

$$F(x) = \begin{cases} 0 & x \leq 7.25 \\ 1 - (7.25/x)^{5/4} & x > 7.25, \end{cases}$$

find the 95th percentile of the hourly wage.

3.23 Let the random variable X have hazard function

$$h(x) = \begin{cases} \lambda & 0 < x < 1 \\ \lambda x & x \geq 1 \end{cases}$$

for $\lambda > 0$, where $h(x) = f(x)/(1 - F(x))$. What is $F(x)$?

3.24 An n-sided fair die is rolled m times. Find the probability mass function for the maximum number of spots showing on the m rolls.

3.25 A random variable X has cumulative distribution function

$$F(x) = \begin{cases} 0 & x \leq 0 \\ x^\theta & 0 < x < 1 \\ 1 & x \geq 1, \end{cases}$$

where θ is a positive parameter. What is the population median of this distribution?

3.26 Seven cards are dealt from a well-shuffled deck. Let the random variable X denote the number of jacks in the hand.

(a) Find $f(-1)$.

(b) Find $F(-1)$.

(c) Find $F(5)$.

(d) Find $F(1)$.

3.27 Does $F(x)$ correspond to a legitimate cumulative distribution function for each of the following five functions?

(a) $F(x) = \begin{cases} 0 & x \leq 0 \\ x^3/8 & 0 < x < 2 \\ 1 & x \geq 2 \end{cases}$

(b) $F(x) = \begin{cases} 0 & x \leq 0 \\ x^2 + 2x & x > 0 \end{cases}$

(c) $F(x) = \begin{cases} 0 & x \leq 0 \\ e^{-1/x} & x > 0 \end{cases}$

(d) $F(x) = \begin{cases} 0 & x \leq -1 \\ x^2 - 1 & -1 < x < \sqrt{2} \\ 1 & x \geq \sqrt{2} \end{cases}$

(e) $F(x) = \frac{e^x}{1+e^x}$ $-\infty < x < \infty$

3.28 Find $E[X]$ for the discrete random variable X with probability mass function

$$f(x) = \begin{cases} 1/3 & x = 0 \\ 2/3 & x = 3. \end{cases}$$

3.29 Find $E[X(3-X)]$ for the discrete random variable X with probability mass function

$$f(x) = \begin{cases} 1/3 & x = 0 \\ 2/3 & x = 3. \end{cases}$$

3.30 Find $E[X^\pi]$ for the continuous random variable X with probability density function

$$f(x) = \frac{1}{4} \qquad 2 < x < 6.$$

3.31 A fair die has 1 and 6 on opposite sides, 2 and 5 on opposite sides, and 3 and 4 on opposite sides. If this die is tossed, find the expected *product* of the number of spots on the up and down sides.

3.32 Let the continuous random variable X have probability density function

$$f(x) = cx \qquad 0 < x < 4,$$

where c is a positive real constant.

(a) Find the value of c so that $f(x)$ is a legitimate probability density function.
(b) Find $P(X > 1)$.
(c) Find $E[X]$.

3.33 Consider the random variable X with cumulative distribution function

$$F(x) = \begin{cases} 0 & x \leq 0 \\ x^2/36 & 0 < x < 6 \\ 1 & x \geq 6. \end{cases}$$

What is the expected value of X?

3.34 Three points are chosen at random on the circumference of a circle of radius 1. Use Monte Carlo simulation to estimate the expected area of the triangle created by the three points. Report your estimate to three-digit accuracy. *Hint:* Heron's formula can be used to calculate the area of a triangle from the lengths of its three sides.

3.35 The random variable X has moment generating function

$$M(t) = 0.2e^{4t} + 0.7e^{7t} + 0.1e^{9t} \qquad -\infty < t < \infty.$$

Find $P(X = 7)$.

3.36 Write APPL code to find the population kurtosis of the random variable X with probability density function

$$f(x) = c \cdot \sin x \qquad 0 < x < \pi,$$

where c is a real constant that is chosen to satisfy the existence conditions for a probability density function. Key these APPL statements into Maple and write the resulting population kurtosis as an exact value and the floating point value to four digits. Write a Monte Carlo simulation in R to support the analytic solution. *Hint:* Use `Pi` rather than `pi` for π in your APPL code.

3.37 The probability of winning a game of craps is $244/495$. If Vladi places $1 bets on two consecutive games of craps, find his expected winnings. (Note: if he wins a game, his winnings are $1; if he loses a game, his winnings are $-$1.)

3.38 Let X be the number of *black jacks* (that is, jack of ♣ or jack of ♠) in a five-card poker hand dealt from a well-shuffled deck. Find $E[X]$.

3.39 Let the distribution of the random variable X be described by the probability density function

$$f(x) = \frac{1}{5} \qquad 0 < x < 5.$$

Find the population mean and variance of

$$3\lceil 2X \rceil + 4.$$

3.40 Let X be a continuous random variable. Find the population mean and variance of $Y = \lceil X \rceil - \lfloor X \rfloor$.

3.41 For the probability mass function defined by

$$f(x) = \begin{cases} 4/5 & x = 0 \\ 1/5 & x = 2, \end{cases}$$

what is the coefficient of variation σ/μ?

3.42 Consider a discrete random variable X with cumulative distribution function $F(x)$ defined on the positive integers. If $F(17) < 1/2$ and $F(18) > 1/2$, what is the population median of the distribution?

3.43 Write the probability density function of a continuous random variable X that has population median 0 and $f_X(0) = 0$.

3.44 Buses leave a particular bus stop at 0 and 20 minutes after the hour around the clock. Find the expected wait time of a patron who arrives to the bus stop at a random point in time.

3.45 Two points are chosen at random on the circumference of a circle of radius 1.

(a) Find the 50th percentile (that is, the population median) of the length of the chord connecting the two points (this can be done by inspection—no math needed).

(b) Write a Monte Carlo simulation using 10,001 random pairs of points that supports your solution to part (a).

3.46   Find the population mean of the random variable X with cumulative distribution function

$$F(x) = 1 - e^{1-e^{\lambda x}} \qquad x > 0,$$

where λ is a positive scale parameter.

3.47   Find the population mean of a *Benford* random variable X with probability mass function

$$f(x) = \frac{\ln(1 + 1/x)}{\ln(10)} \qquad x = 1, 2, \ldots, 9.$$

3.48   Consider the three fair (balanced) dice A, B, and C with the following numbers on the six faces:

- Die A: 3, 3, 5, 5, 7, 7
- Die B: 2, 2, 4, 4, 9, 9
- Die C: 1, 1, 6, 6, 8, 8.

Find the population mean number that will appear in a single roll of each one of the dice individually. In a head-to-head roll between all pair of dice (for example, A vs. B, B vs. C, and C vs. A) where the higher number wins, find the probabilities that die A beats die B, die B beats die C, and die C beats die A.

3.49   Consider the following alternative definition for the population median. The *population median* of a univariate distribution is a value x such that $P(X < x) \le \frac{1}{2}$ and $P(X \le x) \ge \frac{1}{2}$. If there is only one such x, then it is called the population median of the distribution. Consider the random variable X with probability mass function

$$f(x) = \binom{3}{x} p^x (1-p)^{3-x} \qquad x = 0, 1, 2, 3$$

for $0 < p < 1$. Find the population median of the distribution as a function of p.

3.50   Let the random variable X have cumulative distribution function

$$F(x) = \begin{cases} 0 & x < 2 \\ 1 - e^{-\lambda x} & x \ge 2, \end{cases}$$

for a positive real constant λ. Find $\mu = E[X]$.

3.51   A bag contains five balls numbered 1, 2, 3, 4, and 5. A sample of three balls is drawn without replacement. Let the random variable X denote the largest number drawn minus the smallest number drawn. What is the expected value of X?

3.52   Let X be a continuous random variable with probability density function $f(x)$ defined on $\mathcal{A} = \{x \mid -\pi/2 < x < \pi/2\}$. Give an expression for $V[\sin X]$.

3.53   An urn contains 30 red balls and 40 blue balls. Balls are drawn successively at random from the urn. Let the random variable X be the trial number when the *third* red ball is drawn. Find $E[X]$ when

(a) sampling is performed with replacement,

(b) sampling is performed without replacement.

3.54   Let X be a random variable with finite population mean μ and finite population variance σ^2. What is the population mean and standard deviation of $Y = a + bX$, where a and $b \ne 0$ are real constants?

3.55   The expression $E[|X - m|]$, where m is the population median, is often used to measure variability or dispersion in a probability distribution.

(a) Find $E[|X - m|]$ for a population with probability density function

$$f(x) = \frac{1}{b - a} \qquad a < x < b,$$

for $a < b$.

(b) Find $E[|X - m|]$ for a population with probability mass function

$$f(x) = \begin{cases} p^2 & x = 0 \\ 2p(1 - p) & x = 1 \\ (1 - p)^2 & x = 2, \end{cases}$$

for $0 < p < 1$.

3.56   Bridget is a bridge player. She has devised a point system for scoring the hand she has been dealt. She assigns four points for every ace in her hand, three points for every king in her hand, two points for every queen in her hand, and one point for every jack in her hand. If she is dealt a 13-card hand (without replacement) from a well-shuffled deck, find the expected number of points in her hand.

3.57   The game of *Battleship*™consists of a 10 by 10 matrix of positions which are initially populated by five ships: a "Carrier" (5 positions), a "Battleship" (4 positions), a "Cruiser" (3 positions), a "Submarine" (3 positions), and a "Destroyer" (2 positions), arranged in nonoverlapping horizontal or vertical positions on the matrix. Sequential guesses by a player who does not know the locations of the ships are made with only the information "hit" or "miss" conveyed. The game ends when all five of the ships are sunk (that is, 17 "hits"). Find the expected number of guesses necessary to sink all of the ships when random guesses are made among the positions in the matrix which have not been previously guessed.

3.58   The random variable X has moment generating function

$$M(t) = 0.2 + 0.5e^{3t} + 0.3e^{8t}$$

for $-\infty < t < \infty$. Give the probability mass function of X.

3.59   Find the population mean of the distribution defined by the probability density function

$$f(x) = \theta x^{\theta - 1} \qquad 0 < x < 1,$$

where θ is a positive parameter.

3.60   Consider the random variable X with probability density function

$$f(x) = \theta x^{\theta - 1} \qquad 0 < x < 1,$$

where θ is a positive parameter. For what value of θ is the population variance of X maximized?

3.61   A bag contains 100 bills, 80 of which are authentic bills and 20 of which are counterfeit bills. Bills are drawn sequentially and without replacement from the bag. Let X be the number of counterfeit bills drawn prior to the *third* authentic bill being drawn.

(a) Find the probability mass function of X.

(b) Write and execute a computer program that calculates the population mean of X.

3.62   Tanujit spies a square sheet of paper that measures 1 foot by 1 foot. He cuts four identical squares out of each corner with a random side length X (measured in feet), where X is a continuous random variable with probability density function

$$f(x) = 8x \qquad 0 < x < \frac{1}{2}.$$

Find the expected volume of the box, illustrated in Figure 3.33, that Tanujit has created when he folds up the sides.

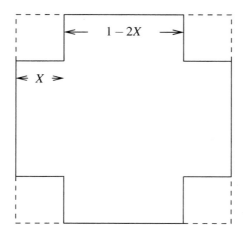

Figure 3.33: Tanujit's creation.

3.63   Two cards are dealt from a well-shuffled, 52-card deck. Let X be the sum of the ranks of the two cards dealt, where aces count as one and face cards count as ten.

   (a) What is $E[X]$?
   (b) What is the population median and 85th percentile of the distribution of X?

3.64   An urn contains n red balls and m white balls. Balls are sampled randomly until the rth white ball is encountered. Give an expression for the expected number of balls sampled when

   (a) sampling is performed with replacement,
   (b) sampling is performed without replacement.

3.65   A balanced die is tossed. Let the random variable X denote the number of spots showing, and x a particular realization of this random variable. After the die has been tossed, x bills are selected at random and without replacement from a bag containing 6 five dollar bills and 4 twenty dollar bills. Find the expected monetary value withdrawn from the bag.

3.66   A kindergarten class consists of 12 boys and 4 girls. The children are arranged from tallest to shortest. Assume that all 16! rankings are equally likely, and no two children are exactly the same height. Let the random variable X be the rank of the *second* tallest boy. Assume that the tallest person in the class is rank 1.

   (a) Find $f(x)$.
   (b) Calculate $E[X]$ and $V[X]$.

3.67   A jigsaw puzzle is a rectangular array of $n \times m$ pieces. There are three disjoint categories of pieces: edge pieces, corner pieces and interior pieces. If two jigsaw puzzle pieces are sampled without replacement from the pieces, find the expected number of edge pieces sampled.

3.68 Consider a three-point discrete distribution

$$f_X(x) = \begin{cases} p_1 & x = x_1 \\ p_2 & x = x_2 \\ p_3 & x = x_3, \end{cases}$$

where $x_1 < x_2 < x_3$ and $p_1 + p_2 + p_3 = 1$. Find equations governing the choices of x_1, x_2, x_3, $p_1, p_2,$ and p_3 such that

$$\mu = E[X] = 0, \quad \sigma^2 = V[X] = 1, \quad E\left[\left(\frac{X-\mu}{\sigma}\right)^3\right] = 0.$$

3.69 Consider the random variable X with probability density function

$$f(x) = mx + b \qquad 0 < x < 1,$$

where m and b are real-valued constants.

(a) What are the constraints on m and b so that $f(x)$ is a legitimate probability density function?

(b) What is $\mu = E[X]$? What values can μ assume?

3.70 If $E[X] = 2$, $E[X^2] = 5$, $E[X^3] = 0$, and $E[X^4] = 30$, find

(a) $E[(X - \pi)^3]$,

(b) $V[17 - 4X]$,

(c) $V[X^2]$.

3.71 Consider the continuous random variable X with probability density function

$$f(x) = \frac{3}{2}x^2 \qquad -1 < x < 1.$$

(a) Find the population median of this distribution.

(b) Find the interquartile range of this distribution.

(c) Find $P(X > 1/3)$.

3.72 Consider the continuous random variable X with probability density function

$$f(x) = cx^2 \qquad -1 < x < 1,$$

where c is a real-valued constant.

(a) Find c.

(b) Find $F(x)$. Plot $f(x)$ and $F(x)$.

(c) Find $E[X]$ and $V[X]$.

(d) Find $E\left[\left(\frac{X-\mu}{\sigma}\right)^3\right]$.

(e) Find $P\left(X > \frac{1}{2} \mid X > \frac{1}{4}\right)$.

3.73 Consider the random variable X with probability density function

$$f(x) = \frac{2}{9}x \qquad 0 < x < 3.$$

Find $E[X]$.

3.74   A series of baseball games is played between teams from the American League and the National League. Each game is independent of the others and p is the probability that the American League wins a single game. The series ends when one of the teams wins n games. Let X be the number of games played in the series.

   (a) Find $f(x)$.

   (b) Give expressions for $E[X]$ and $V[X]$.

   (c) Plot $E[X]$ on $0 < p < 1$ for $n = 1, 2, 3$ and 4.

3.75   An urn contains three red balls and two white balls. Balls are removed sequentially without replacement until a white ball is withdrawn. Let X be the number of balls withdrawn.

   (a) Find the probability mass function of the number of balls withdrawn, $f(x)$.

   (b) Find the expected number of balls withdrawn, $E[X]$.

   (c) Find the population variance of the number of balls withdrawn, $V[X]$.

3.76   A coupon collector obtains one of n distinct coupons in succession. Coupon type i is obtained with probability p_i, $i = 1, 2, \ldots, n$. Assume that the sequence of coupon types are independent trials with n possible outcomes. Let X be the number of coupons required to obtain at least one coupon of each type.

   (a) Find an expression for $f(x)$.

   (b) Compute $E[X]$ when $n = 6$ and $p_1 = p_2 = \cdots = p_6 = 1/6$. Write a sentence interpreting $E[X]$ in terms of repeated rolls of a fair die.

3.77   The survivor function of a random variable X is typically defined as

$$S(x) = P(X \geq x).$$

If the discrete random variable X with survivor function $S(x)$ has support on the positive integers, show that

$$E[X] = \sum_{x=1}^{\infty} S(x).$$

3.78   A bag contains balls numbered $1, 2, \ldots, 365$. A ball is drawn at random from the bag. Let the random variable X be the number on that ball.

   (a) Find $E[X]$.

   (b) Find $E[1/X]$ to five significant digits.

3.79   The probability density function for a Weibull random variable is given by

$$f(x) = \kappa \lambda^{\kappa} x^{\kappa-1} e^{-(\lambda x)^{\kappa}} \qquad\qquad x > 0,$$

where λ and κ are positive parameters.

   (a) Find expressions for the population mean, median, and mode. (*Hint:* they might not all be closed-form.)

   (b) Find parameter values associated with each of the following three cases: the population median and mode of the distribution are equal; the population mean and median of the distribution are equal; the population mean and mode of the distribution are equal.

3.80   Let the random variable A be the number of heads that appear in two tosses of a fair coin. Find the expected (Euclidean) distance between the x- and y-intercepts of the equation

$$x + y = A.$$

3.81 An urn contains n balls. A single ball is drawn at random with replacement repeatedly until a ball previously sampled is withdrawn. Let X be the number of samples necessary.

(a) Find the probability mass function $f(x)$.

(b) Give an expression for the expected number of draws $E[X]$.

(c) Calculate the expected number of draws $E[X]$ when $n = 10$.

3.82 A piecewise-linear cumulative distribution function associated with the data values 1, 2, and 9 can be found by connecting the points $(1, 0)$, $(2, 1/2)$, and $(9, 1)$ with lines. Find

(a) the population mean,

(b) the population variance, and

(c) the moment generating function

of the random variable associated with this piecewise-linear cumulative distribution.

3.83 The waiting time in a queue is a mixed discrete–continuous random variable X with cumulative distribution function

$$F(x) = \begin{cases} 0 & x < 0 \\ 1 - 0.7e^{-x} & x \geq 0. \end{cases}$$

Find $E[X]$ and $V[X]$.

3.84 Find the population mean of the random variable X with probability density function

$$f(x) = \begin{cases} \dfrac{2(x-a)}{(b-a)(m-a)} & a < x < m \\[2mm] \dfrac{2(b-x)}{(b-a)(b-m)} & m \leq x < b, \end{cases}$$

where $a < m < b$ are real numbers.

3.85 For a random variable X with cumulative distribution function $F(x)$, identify each of the following as a *constant*, *random variable*, *event*, *undefined*, or *none of the above*.

(a) $\cos X < 0.3$

(b) $E[\cos X < 0.3]$

(c) $V\left[\sqrt{X}\right]$

(d) X^2

(e) $F(3)$

3.86 A bag contains three balls numbered -1, 0, and 2. Two balls are sampled at random. Find the expected value of the product of the numbers on the two balls that are sampled when

(a) sampling is performed without replacement,

(b) sampling is performed with replacement.

3.87 A continuous random variable X has probability density function

$$f(x) = c\left(x^2 + 2x\right) \qquad 0 < x < 1.$$

(a) Find the constant c so that $f(x)$ is a legitimate probability density function.

(b) Find $E[X]$.

3.88 Consider a random variable X with moment generating function

$$M(t) = e^{3t} \qquad -\infty < t < \infty.$$

  (a) Find $E[X]$.

  (b) Find $V[X]$.

  (c) Find $f(x)$.

3.89 Consider an arbitrary rectangle. A second, smaller rectangle, is created by taking the first rectangle and scaling both sides by a single realization of a random variable X with probability density function

$$f(x) = 2x \qquad 0 < x < 1.$$

  (a) Find the expected ratio of the area of the smaller rectangle to the larger rectangle.

  (b) Support your result from part (a) via Monte Carlo simulation for an arbitrary rectangle of your choosing.

3.90 Consider the following game. A die is rolled repeatedly. You receive a dollar for each spot showing on the up face of the last roll. If a one is rolled you must stop rolling the die. You may stop the rolling earlier, however.

  (a) If the die is fair, what is the optimal strategy to maximize your expected winnings?

  (b) If the die is biased so that the number of spots X appearing on a single roll has probability mass function

$$f(x) = x/21 \qquad x = 1, 2, 3, 4, 5, 6,$$

  what is the optimal strategy to maximize your expected winnings?

3.91 Three married couples attend a dance. The women are paired with the men at random for one particular song. Let the random variable X be the number of couples with the spouses dancing together for the song.

  (a) Find the probability mass function of X.

  (b) Find $E[X]$.

  (c) Find $V[X]$.

3.92 A fair die is rolled. Let the random variable X denote the number of spots that appear on the up face. Find $E[1/X]$.

3.93 Consider a random variable X with moment generating function

$$M_X(t) = \frac{1}{2} e^{-ct} + \frac{1}{2} e^{ct}$$

for some positive constant c. Find the value of c so that $V[X] = 1$.

3.94 Suppose that the only information known about the random variable X is that $E[X] = 60$ and $V[X] = 100$. What probability statement can be made about $P(40 < X < 80)$?

3.95 Let the discrete random variable X be the number of heads that appear in 16 tosses of a fair coin. Find the exact value of $P(4 < X < 12)$ and the Chebyshev lower bound on this probability.

Chapter 4

Common Discrete Distributions

A number of discrete random variables have probability distributions that occur so frequently in practice that they have been named and given *parameters* to enhance their flexibility in solving probability problems. This chapter outlines common discrete distributions; the next chapter outlines common continuous distributions.

Many of the probability distributions described here arise naturally or in applications. Just as we study parabolas because of trajectories and reflectors, or the hyperbolic cosine because of hanging wires and catenary arches, we study distributions of discrete random variables because they arise in natural or man-made systems. For example,

- the Bernoulli distribution arises in clinical trials, athletics, polling, genomics, etc.,

- the Poisson distribution arises in radioactive decay, arrival processes, quality control, etc.,

- the normal distribution arises in agricultural yields, IQ scores, heights of children, etc.

For the distributions presented in this chapter, we will consider the support $\mathcal{A}$, the distribution's parameters, the probability mass function $f(x) = P(X = x)$, the cumulative distribution function $F(x) = P(X \leq x)$, the moment generating function $M(t) = E\left[e^{tX}\right]$, the population mean μ, and the population variance σ^2. Applications of each of the distributions will also be presented.

4.1   Bernoulli Distribution

The Bernoulli distribution is the simplest of the probability distributions introduced in this chapter. Its support consists of just two members: $\mathcal{A} = \{0, 1\}$.

Definition 4.1 A discrete random variable X with probability mass function

$$f(x) = \begin{cases} 1 - p & x = 0 \\ p & x = 1 \end{cases}$$

for $0 < p < 1$ is a Bernoulli(p) random variable.

The Bernoulli distribution has a single parameter p that allows the distribution to be applied to a variety of situations that might arise. The probability mass function can also be written in a single line as

$$f(x) = p^x (1 - p)^{1-x} \qquad x = 0, 1.$$

A single observation of a Bernoulli random variable is often known as a "Bernoulli trial." Tradition dictates that the two possible outcomes of a Bernoulli trial are generically referred to as "success" $(X = 1)$ and "failure" $(X = 0)$. Any random experiment that has two outcomes can be modeled as a Bernoulli trial. Examples include

- whether the registered voter supported a particular candidate,

- whether the coin came up heads when flipped,

- whether the child caught the flu last winter,

- whether the guest found the service at the hotel satisfactory,

- whether the high jumper cleared the bar on her last jump,

- whether the cancer patient survived five years after diagnosis.

In all cases, there is a single trial and two outcomes. Choosing which outcome constitutes a success and which outcome constitutes a failure is a decision made by the modeler.

The moment generating function for the Bernoulli distribution is

$$M(t) = E\left[e^{tX}\right] = \sum_{\mathcal{A}} e^{tx} f(x) = \sum_{x=0}^{1} e^{tx} f(x) = 1 - p + pe^{t} \qquad -\infty < t < \infty.$$

All derivatives of $M(t)$ with respect to t are identical, that is, for any positive integer r

$$M^{(r)}(t) = pe^{t} \qquad -\infty < t < \infty,$$

so the expected value of X^r is

$$E[X^r] = M^{(r)}(0) = p.$$

Letting $r = 1$, the population mean is $\mu = E[X] = p$, which can be checked by using the definition of the expected value:

$$\mu = E[X] = \sum_{\mathcal{A}} x f(x) = \sum_{x=0}^{1} x f(x) = (0)(1 - p) + (1)(p) = p.$$

The population variance is given by

$$\sigma^2 = V[X] = E\left[(X - \mu)^2\right] = \sum_{\mathcal{A}} (x - \mu)^2 f(x) = (0 - p)^2(1 - p) + (1 - p)^2 p = p(1 - p).$$

Finally, the population skewness and kurtosis, after simplification, are

$$E\left[\left(\frac{X - \mu}{\sigma}\right)^3\right] = \frac{1 - 2p}{\sqrt{p(1 - p)}} \qquad \text{and} \qquad E\left[\left(\frac{X - \mu}{\sigma}\right)^4\right] = \frac{3p^2 - 3p + 1}{p(1 - p)}.$$

These quantities can be calculated with the following APPL code.

```
X := BernoulliRV(p);
MGF(X);
Mean(X);
Variance(X);
Skewness(X);
Kurtosis(X);
```

All of the distributions introduced in this chapter are built into the APPL language. The function name is usually the name of the distribution appended with the letters RV, for random variable, to avoid any conflict with existing built-in Maple functions. The call to the function `BernoulliRV` simply creates the list of three lists

```
[[1 - p, p], [0, 1], ["Discrete", "PDF"]]
```

with the added assumption that p must lie between 0 and 1.

An interesting special case of the Bernoulli distribution occurs when $p = 1/2$. This special case arises in modeling one flip of a fair coin. The two outcomes of the random experiment, heads and tails, are mapped to $x = 0$ and $x = 1$ by the random variable X. (The choice between the two possible mappings is arbitrary.) Since there is equal mass at $x = 0$ and $x = 1$, the population mean is $\mu = 1/2$ using the center-of-gravity interpretation of the population mean. The population variance is $\sigma^2 = 1/4$, which is the largest population variance possible for a Bernoulli random variable. Since the probability mass function is symmetric when $p = 1/2$, the population skewness is 0. Finally, the population kurtosis is 1 when $p = 1/2$, which is the smallest population kurtosis possible for *any* random variable. The corresponding probability mass function has the widest peak about the population mean of any random variable.

Before leaving the Bernoulli distribution behind, a small piece of new notation is introduced. When a random variable has one of these common distributions, probabilists like to use the shorthand

$$X \sim \text{Bernoulli}(p)$$

to indicate that X has the Bernoulli distribution with parameter p. The $\sim$ should be read as "is distributed as." This notation applies to all of the distributions introduced in this chapter and the next chapter.

Bernoulli trials, with their binary outcomes, form the basis for the distributions described in the next three sections: binomial, geometric, and negative binomial.

4.2   Binomial Distribution

Most random experiments involving Bernoulli trials include more than just a single trial. The *binomial distribution* models the number of successes in n independent Bernoulli trials, each with probability of success p, where n is a positive integer. When n Bernoulli trials are conducted, each with an identical probability of success p, the entire experiment is known as a *binomial random experiment*, as defined below.

Definition 4.2 A *binomial random experiment* satisfies the following criteria.

(a) The random experiment consists of n identical Bernoulli trials.

(b) There are two possible outcomes for each Bernoulli trial.

(c) The Bernoulli trials are mutually independent.

(d) The probability of success on each Bernoulli trial is identical.

The two outcomes of each Bernoulli trial are again generically referred to here as "success" (denoted by S and corresponding to an outcome of $x = 1$ for the Bernoulli trial) and "failure" (denoted by F and corresponding to an outcome of $x = 0$ for the Bernoulli trial). The probability of success on each trial is p, where $0 < p < 1$. The discrete random variable of interest in a binomial random experiment is X, the count of the number of successes in the n Bernoulli trials. We now derive the support of X and the functional form of the probability mass function $f(x) = P(X = x)$.

The support of X is straightforward. Since there are n trials, there can be as few as 0 successes and as many as n successes, so the support is $\mathcal{A} = \{0, 1, 2, \ldots, n\}$. What is the probability that the first x Bernoulli trials are successes and the last $n - x$ Bernoulli trials are failures? Because the Bernoulli trials are independent, the sequence

$$\underbrace{S\,S \ldots S}_{x}\underbrace{F\,F \ldots F}_{n-x}$$

occurs with probability $p^x(1-p)^{n-x}$. Similarly, what is the probability that the first $n-x$ Bernoulli trials are failures and the last x Bernoulli trials are successes? Because the trials are independent, the sequence

$$\underbrace{F\,F\ldots F}_{n-x}\underbrace{S\,S\ldots S}_{x}$$

also occurs with probability $p^x(1-p)^{n-x}$. More generally, how many of the 2^n possible sequences of successes and failures result in exactly x successes? This number will be multiplied by $p^x(1-p)^{n-x}$ to arrive at the probability mass function $f(x) = P(X = x)$. Consider the indices where the x successes occur. There are

$$\binom{n}{x}$$

different ways of selecting x positions for the successes. Combinations are used because a binomial random variable is a count of the number of successes and therefore the order in which the indices are selected is not important. This observation leads to the probability mass function defined next.

Definition 4.3 A discrete random variable X with probability mass function

$$f(x) = \binom{n}{x} p^x(1-p)^{n-x} \qquad\qquad x = 0, 1, 2, \ldots, n$$

for some positive integer n and $0 < p < 1$ is a binomial(n, p) random variable.

Order of indices not important. [handwritten annotation]

Here are some important tidbits about binomial random variables. First, a binomial(n, p) random variable is the number of successes in a binomial random experiment. Second, we use the same shorthand, $X \sim$ binomial(n, p), to indicate that a random variable has the binomial distribution with parameters n and p. Third, the origins of the binomial distribution are successive terms in the expansion of $\big((1-p)+p\big)^n$ using the binomial theorem. These terms sum to $1^n = 1$, which shows that the binomial distribution satisfies the existence conditions for a probability mass function. Fourth, the moment generating function, population mean, and population variance are most easily derived by treating a binomial random variable X as the sum of mutually independent Bernoulli random variables $X_1, X_2, \ldots, X_n$:

$$X = X_1 + X_2 + \cdots + X_n.$$

Mutually independent random variables will be defined in Chapter 6. Fifth, the binomial(n, p) distribution reduces to the Bernoulli(p) distribution when $n = 1$.

The population mean of a binomial(n, p) random variable can be calculated as

$$
\begin{aligned}
\mu &= \sum_{x=0}^{n} x \binom{n}{x} p^x(1-p)^{n-x} \\[4pt]
&= \sum_{x=1}^{n} x \binom{n}{x} p^x(1-p)^{n-x} \\[4pt]
&= \sum_{x=1}^{n} \frac{n!}{(x-1)!(n-x)!} p^x(1-p)^{n-x} \\[4pt]
&= np \sum_{x=1}^{n} \frac{(n-1)!}{(x-1)!(n-x)!} p^{x-1}(1-p)^{n-x} \\[4pt]
&= np \sum_{y=0}^{n-1} \frac{(n-1)!}{y!(n-y-1)!} p^{y}(1-p)^{n-y-1} \\[4pt]
&= np.
\end{aligned}
$$

The last summation corresponds to summing the probability mass function of a binomial($n-1$, p) random variable, which must be 1 by the existence conditions. The substitution $y = x-1$ was used to place the summation in this particular form.

Using similar methodology, the population variance of a binomial(n, p) random variable is

$$\sigma^2 = np(1-p)$$

and the population skewness and kurtosis are

$$E\left[\left(\frac{X-\mu}{\sigma}\right)^3\right] = \frac{1-2p}{\sqrt{np(1-p)}} \qquad \text{and} \qquad E\left[\left(\frac{X-\mu}{\sigma}\right)^4\right] = 3 + \frac{1-6p(1-p)}{np(1-p)}.$$

The population skewness and kurtosis converge to 0 and 3, respectively, in the limit as $n \to \infty$. Finally, the moment generating function for a binomial(n, p) random variable is

$$M(t) = \left(1 - p + pe^t\right)^n \qquad -\infty < t < \infty.$$

The shape of the probability mass function for a binomial(n, p) random variable typically follows a bell shape. Consider the following three binomial random variables.

- The number of fours in 60 rolls of a fair die: $X \sim \text{binomial}(60, 1/6)$.

- The number of even numbers in 60 rolls of a fair die: $X \sim \text{binomial}(60, 1/2)$.

- The number of non-fours in 60 rolls of a fair die: $X \sim \text{binomial}(60, 5/6)$.

Plots of the bell-shaped probability mass functions are shown in Figure 4.1, with identical vertical scales on the three probability mass functions. The left-hand probability mass function is centered around $\mu = 60 \cdot \frac{1}{6} = 10$ and is skewed to the right; the middle probability mass function is centered around $\mu = 60 \cdot \frac{1}{2} = 30$ and is symmetric; the right-hand probability mass function is centered around $\mu = 60 \cdot \frac{5}{6} = 50$ and is skewed to the left. The R commands that create these plots are given below.

```
par(mfrow = c(1, 3))
x = 0:60
plot(x, dbinom(x, 60, 1 / 6), type = "h")
plot(x, dbinom(x, 60, 1 / 2), type = "h")
plot(x, dbinom(x, 60, 5 / 6), type = "h")
```

The `mfrow` argument in `par` indicates that a 1×3 array of plots is to be displayed. The `dbinom` function returns the probability mass function for the binomial distribution.

The binomial distribution is one of the pillars in applied probability because it arises so often in applications. Applications of the distribution are now considered in the following sequence of examples.

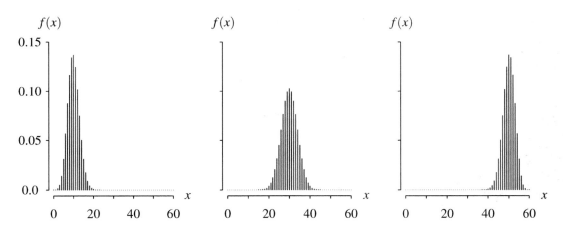

Figure 4.1: Three binomial probability mass functions.

Example 4.1 Emma is a 70% free throw shooter. If she takes 3 shots and X is the number that she makes, find $f(x)$, μ, σ^2, and $P(X = 2)$. *[handwritten: $\mu = np$ $\sigma^2 = pn(1-p)$.]*

Assuming that each of the shots constitutes an independent Bernoulli trial, the probability mass function for $X \sim$ binomial$(3, 0.7)$ is

$$f(x) = \binom{3}{x}(0.7)^x(0.3)^{3-x} \qquad x = 0, 1, 2, 3.$$

[handwritten: $f(x) = \binom{3}{x}(.7)^x(.3)^{3-x}$]

The population mean number of shots that she makes is

$$\mu = np = (3)(0.7) = 2.1.$$

The population variance of the number of shots that she makes is

$$\sigma^2 = np(1 - p) = (3)(0.7)(0.3) = 0.63.$$

Finally, the probability that Emma makes exactly two shots of the three shots is

$$P(X = 2) = f(2) = \binom{3}{2}(0.7)^2(0.3) = 0.441.$$

These quantities can be calculated with the following APPL code.

```
X := BinomialRV(3, 0.7);
PDF(X);
Mean(X);
Variance(X);
PDF(X, 2);
```

[handwritten: # of trials]

Example 4.2 A dozen eggs contain 3 defectives. If a sample of 5 is taken *with replacement*, find the probability that exactly 2 of the eggs sampled are defective. Also, find the probability that 2 or fewer are defective. *[handwritten: success = defective.]*

Since eggs are sampled with replacement, each of the 5 eggs taken from the dozen is a Bernoulli trial. Define "success" on a trial to be drawing a defective egg. So the parameters for the binomial distribution are $n = 5$ and $p = 3/12 = 1/4$. The probability mass function for $X \sim$ binomial$(5, 1/4)$ is

[handwritten margin: -- if it was w/out repl. then the probability for each egg would be different which violates the definition of the bernoulli and thus binomial distribution. No longer ber noulli trials.]

$$f(x) = P(X = x) = \binom{5}{x}\left(\frac{1}{4}\right)^x\left(\frac{3}{4}\right)^{5-x} \qquad x = 0, 1, 2, 3, 4, 5.$$

The probability that exactly 2 of the eggs in the sample of 5 are defective is

$$P(X = 2) = f(2) = \binom{5}{2}\left(\frac{1}{4}\right)^2\left(\frac{3}{4}\right)^3 = \frac{270}{1024} \cong 0.2637.$$

The probability of sampling 2 or fewer defective eggs in the sample of 5 is

$$
\begin{aligned}
F(2) &= P(X \leq 2) \\
&= P(X = 0) + P(X = 1) + P(X = 2) \\
&= \binom{5}{0}\left(\frac{1}{4}\right)^0\left(\frac{3}{4}\right)^5 + \binom{5}{1}\left(\frac{1}{4}\right)^1\left(\frac{3}{4}\right)^4 + \binom{5}{2}\left(\frac{1}{4}\right)^2\left(\frac{3}{4}\right)^3 \\
&= \frac{243}{1024} + \frac{405}{1024} + \frac{270}{1024} \\
&= \frac{918}{1024} \\
&\cong 0.8965.
\end{aligned}
$$

R has four functions associated with the binomial distribution that are useful in performing the calculations listed in Table 4.1. In the previous example, the probability of exactly 2 defective eggs is calculated with

```
dbinom(2, 5, 1 / 4)
```
pmf / pdf.

which returns $P(X = 2) = f(2) \cong 0.2637$. Likewise, the probability of getting 2 or fewer defective eggs in the sample of 5 eggs is calculated with

```
pbinom(2, 5, 1 / 4)
```
cdf

which returns $P(X \leq 2) = F(2) \cong 0.8965$. The 95th percentile of the distribution of X is calculated with

```
qbinom(0.95, 5, 1 / 4)
```
quartile

which returns $x_{0.95} = 3$. A vector of 30 binomial(5, 1/4) random variates is generated with

```
rbinom(30, 5, 1 / 4)
```
random variate.

which yields

> 1 1 1 1 4 1 2 2 2 1 2 3 1 1 3 1 1 2 3 3 1 1 0 1 1 1 0 2 0 1.

Each one of these random binomial variates represents the count of defective eggs in a random sample of 5 eggs drawn with replacement from a dozen eggs containing 3 defective eggs.

In order to use Monte Carlo simulation to check the probability of getting exactly 2 defective eggs in the sample, conduct one million replications of the random experiment with the following R statement

```
sum(rbinom(1000000, 5, 1 / 4) == 2) / 1000000
```

which, after a call to `set.seed(14)`, yields

> 0.263616   0.263377   0.263614   0.263177   0.263431.

All of these values are less than the true value $270/1024 \cong 0.2637$. Is there a problem with the analytic solution? In this case, there is no need to panic. Just as flipping a coin and getting heads five times in a row does not imply that the coin is biased, we can't conclude that the analytic solution is wrong from these five values. Running the statement five more times with another random number seed results in output that falls on both sides of the analytic value.

R follows this same pattern for other distributions—a good example of logical program design. For any distribution, the first letter of the function name determines the action: d for computing $f(x)$; p for computing $F(x)$; q for computing a percentile (quantile); r for generating random variates. The letters that follow the first letter give the name of the distribution. The other discrete distributions covered in this chapter are the geometric (`geom`), negative binomial (`nbinom`), Poisson (`pois`), and hypergeometric (`hyper`). The continuous distributions covered in the next chapter are the uniform (`unif`), exponential (`exp`), gamma (`gamma`), normal (`norm`), and chi-square (`chisq`).

R code cheat sheet.

Function name and parameters	Returned value for $X \sim \text{binomial}(n, p)$
`dbinom(x, n, p)`	calculates the probability mass function $f(x) = P(X = x)$
`pbinom(x, n, p)`	calculates the cumulative distribution function $F(x) = P(X \leq x)$
`qbinom(u, n, p)`	calculates the percentile (quantile) $F^{-1}(u)$
`rbinom(m, n, p)`	generates m random variates

Table 4.1: R functions for the binomial distribution.

Example 4.3 Give an expression for the probability of breaking even or coming out ahead in bets on *red* in roulette (which wins with probability $18/38 = 9/19$ per spin) if

[handwritten in margin: Equals Winning.]

(a) one bet of $1000 is made on red, *[handwritten: $P(win) = 9/19$.]*

(b) one thousand repeated bets on red of $1 are made.

Each spin of the roulette wheel constitutes an independent Bernoulli trial with probability of success $p = 9/19$.

(a) The probability of winning with one bet on red is just $9/19 \cong 0.4737$.

(b) Let the random variable X be the number of wins in the 1000 independent Bernoulli trials. Since the 1000 spins constitute a binomial random experiment, $X \sim \text{binomial}(1000, 9/19)$. Breaking even or coming out ahead in 1000 bets is equivalent to the event $X \geq 500$. So the probability of breaking even or coming out ahead in 1000 repeated bets on red is

$$P(X \geq 500) = \sum_{x=500}^{1000} \binom{1000}{x} \left(\frac{9}{19}\right)^x \left(\frac{10}{19}\right)^{1000-x} \cong 0.05110,$$

which is calculated with the R statement 1 - pbinom(499, 1000, 9 / 19).

Betting $1000 in these two different fashions highlights the fact that the optimal strategy for the bettor is to place one large bet rather than several smaller bets. The casino, on the other hand, hopes that many small bets, rather than a few large bets are made at their casino. The *expected* winnings associated with the strategies in (a) and (b) are equal, but the two probabilities that the gambler comes out ahead differ greatly. The reader is encouraged to sketch the probability mass functions of the winnings for parts (a) and (b).

Example 4.4 Emma is shooting free throws again. If she is a 70% shooter, how many shots should she take to be 99% certain that she will make at least 2?

The goal of this problem is to find the smallest value of n such that $P(X \geq 2) > 0.99$, where $X \sim \text{binomial}(n, 0.7)$. So

$$
\begin{aligned}
P(X \geq 2) &= 1 - P(X \leq 1) \\
&= 1 - [P(X = 0) + P(X = 1)] \\
&= 1 - \binom{n}{0}(0.7)^0(0.3)^n - \binom{n}{1}(0.7)^1(0.3)^{n-1} \\
&= \underbrace{1 - (0.3)^n - n(0.7)(0.3)^{n-1}}_{r} > 0.99
\end{aligned}
$$

This inequality has no analytic solution, so it must be solved by trial and error. The R statement 1 - pbinom(1, n, 0.7) for several values of n gives the following values for r.

n	2	3	4	5	6	7
r	0.4900000	0.7840000	0.9163000	0.9692200	0.9890650	0.9962092

Emma should take $n = 7$ shots so that the probability of making two or more shots exceeds 0.99. Figure 4.2 contains a graph of the probability mass function of the binomial distribution with $n = 7$ and $p = 0.7$. The probability mass function with $n = 7$ shots is the smallest value of n such that the sum of the mass values between 2 and n exceeds 0.99.

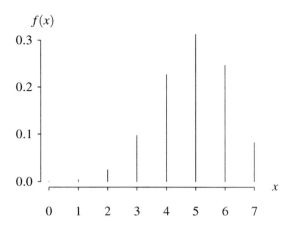

Figure 4.2: The probability mass function for $X \sim \text{binomial}(7, 0.7)$.

Example 4.5 An airplane has 100 seats. The airline "overbooks" a flight (that is, sells more tickets than available seats) in order to maximize their profit. Assume that each ticket holder's decision to show up for a flight is an independent Bernoulli trial with a probability of showing up for the flight of 0.92. (Corporate passengers make the independent trials assumption more plausible; families make the independent trials assumption less plausible.) If the airline profit is \$10 for each seat sold and the airline loses \$40 (and considerable customer goodwill) for each "bumped" passenger, what is the expected profit if 103 seats are sold?

Let the random variable X be the number of passengers that show up for a flight. Since each passenger's decision to show up for the flight (defined here as "success") is an independent Bernoulli trial,

$$X \sim \text{binomial}(103, 0.92),$$

which has probability mass function.

$$f(x) = \binom{103}{x} 0.92^x 0.08^{103-x} \qquad x = 0, 1, 2, \ldots, 103.$$

The profit generated from selling 103 tickets for the flight is

$$(103)(\$10) = \$1030.$$

The probabilities of 101, 102, and 103 customers arriving for the flight are

$$P(X = 101) = f(101) = \binom{103}{101} 0.92^{101} 0.08^2 \cong 0.007399,$$

$$P(X = 102) = f(102) = \binom{103}{102} 0.92^{102} 0.08^1 \cong 0.001668,$$

and

$$P(X = 103) = f(103) = \binom{103}{103} 0.92^{103} 0.08^0 \cong 0.0001863.$$

When these probabilities are multiplied by the associated losses due to the bumped customers of \$40, \$80, and \$120, and subtracted from the profit from sales, the overall expected profit is $1029.55. The expected profit for overbooking is greater than the expected profit for not overbooking. The fact that only $0.45 was lost from the bumped customers is due to the fact that the associated probabilities are so small.

The R code shown below calculates the expected profit when 103 seats are sold.

```
103 * 10 - 40  * dbinom(101, 103, 0.92) -
            80  * dbinom(102, 103, 0.92) -
           120 * dbinom(103, 103, 0.92)
```

Ignoring customer ill will, the airline should do some overbooking because the profit is $1000 on a full flight without overbooking.

This is an example of an application from operations research known as *yield management*, where prices and overbooking are manipulated based on demand in order to maximize profit. Yield management is commonly used in the airline and hotel industries. The optimal number of customers to overbook can be found by following the above procedure for several different levels of overbooking. The R code shown below calculates the expected profit for several levels of overbooking with n tickets sold.

```
profit = rep(0, 110)
for (n in 100:110) {
  sales = 10 * n
  loss  = 0
  for (j in 100:n) loss = loss + 40 * (j - 100) * dbinom(j, n, 0.92)
  profit[n] = sales - loss
}
x = 100:110
plot(x, profit[x], type = "h")
```

The plot that is generated from this code is shown in Figure 4.3. The expected profit on the flight is $1000 when 100 tickets are sold. The expected profit increases until 107 tickets are sold, where it peaks at $1051.20. After this point, the effect of the losses associated with bumped customers overwhelms the additional profit from ticket sales, and the expected profit drops.

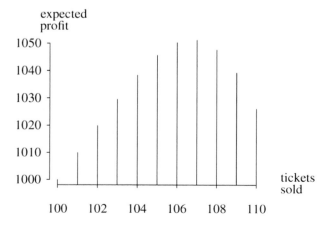

Figure 4.3: The expected profit as a function of number of tickets sold.

Example 4.6 Let $X \sim$ binomial(n, p). Define the random variable X/n as the *relative frequency of success*. Find the population mean of X/n and the population variance of X/n. Also, can any statements be made concerning X/n and p in the limit as $n \to \infty$?

The expected value of X/n is

$$E\left[\frac{X}{n}\right] = \frac{1}{n}E[X] = p.$$

The expected value of the relative frequency of success is equal to the population probability of success. This can be interpreted as the fraction X/n is a good estimator for p because it is aimed directly at the target, on average. The population variance of X/n is

$$V\left[\frac{X}{n}\right] = \frac{1}{n^2}V[X] = \frac{1}{n^2}np(1-p) = \frac{p(1-p)}{n}.$$

The n in the denominator of this formula indicates that

$$\lim_{n \to \infty} V\left[\frac{X}{n}\right] = 0,$$

which is to say as the number of Bernoulli trials increases, the variability of the relative frequency goes to zero. These two properties, being aimed at the right target and having a variability that goes to zero, are good properties for X/n, which is used to estimate p. We use this particular form of Chebyshev's inequality:

$$P(|X - \mu| < k\sigma) \geq 1 - \frac{1}{k^2},$$

to make statements about the relationship between X/n and p. Replacing X with X/n and μ and σ with the appropriate population mean and variance from the formulas above,

$$P\left(\left|\frac{X}{n} - p\right| < \underbrace{k\sqrt{\frac{p(1-p)}{n}}}_{\varepsilon}\right) \geq 1 - \frac{1}{k^2} \quad \Longrightarrow \quad P\left(\left|\frac{X}{n} - p\right| < \varepsilon\right) \geq 1 - \frac{p(1-p)}{n\varepsilon^2}.$$

The quantity

$$\left|\frac{X}{n} - p\right|$$

is the difference between the estimator X/n and the true probability of success p. In this sense it can be interpreted as the error in estimation. So as the number of Bernoulli trials increases,

$$\lim_{n \to \infty} P\left(\left|\frac{X}{n} - p\right| < \varepsilon\right) = 1.$$

If some small positive constant ε is chosen, this result says that we can be certain (with probability 1) that the error between the estimate X/n and the true probability of success p is less than ε in the limit as $n \to \infty$. This result is one form of the "weak law of large numbers." It was first published in 1713 by Swiss mathematician Jakob Bernoulli's *Ars Conjectandi* (The Art of Conjecturing). One important application area for this is in Monte Carlo simulation. Almost all of the Monte Carlo simulations performed so far in the text have involved repeated Bernoulli trials in an effort to estimate a probability. The result indicates that as the number of replications, usually denoted by the R variable `nrep`, goes to infinity, the estimate converges to the true probability of interest.

Example 4.7 Grant makes 4 out of 5 free throw shots. Find the value of p that maximizes $f(4)$, where $X \sim$ binomial$(5, p)$.

The probability mass function of a random variable X that has a binomial$(5, p)$ distribution is

$$f(x) = \binom{5}{x} p^x (1-p)^{5-x} \qquad x = 0, 1, 2, 3, 4, 5.$$

Our goal is to maximize $f(4)$ with respect to p. The probability mass function at $x = 4$ is

$$f(4) = \binom{5}{4} p^4 (1-p) = 5p^4 - 5p^5.$$

In order to maximize this function with respect to p, differentiate with respect to p

$$\frac{d}{dp} f(4) = 20p^3 - 25p^4$$

and equate the derivative to 0 yielding

$$20p^3 - 25p^4 = 5p^3(4 - 5p) = 0.$$

Solving this equation for p yields $p = 0$ and $p = 4/5$. The more interesting root of the equation, $p = 4/5$, happens to be the relative frequency from the previous example. If Grant would like to maximize $f(4)$, the probability of obtaining exactly four out of five free throws, he should use the relative frequency $p = 4/5$. This notion of maximizing the probability mass function (or probability density function if X is continuous) is known as *maximum likelihood estimation* in statistics.

Example 4.8 A company produces biased coins that come up heads when flipped with probability 0.7. You are not sure whether you have one of these biased coins or whether you have a fair coin, so you devise the following experiment: (1) flip the coin 100 times, (2) if there are 62 or more heads conclude that the coin is biased, otherwise, conclude that the coin is fair. The choice of 62 as a cutoff is arbitrary. Place four probabilities in the table below giving the probability of the four outcomes. Note that the rows concern the conclusion of the experiment and the columns concern the true state of the coin.

		True coin status	
		Coin fair	Coin biased
Experiment	Coin fair		
conclusion	Coin biased		

Since each coin flip is a Bernoulli trial,

$$X \sim \text{binomial}(100, 0.5)$$

when the coin is fair, and

$$X \sim \text{binomial}(100, 0.7)$$

when the coin is biased. This means that the four probabilities of interest can all be expressed as summations of the probability mass functions of the binomial distribution as follows.

True coin status

		Coin fair	Coin biased
Experiment	Coin fair	$\displaystyle\sum_{x=0}^{61}\binom{100}{x}(0.5)^x(0.5)^{100-x}$	$\displaystyle\sum_{x=0}^{61}\binom{100}{x}(0.7)^x(0.3)^{100-x}$
conclusion	Coin biased	$\displaystyle\sum_{x=62}^{100}\binom{100}{x}(0.5)^x(0.5)^{100-x}$	$\displaystyle\sum_{x=62}^{100}\binom{100}{x}(0.7)^x(0.3)^{100-x}$

The column sums are one. The values in the 2×2 matrix can be calculated with the following R commands.

True coin status

		Coin fair	Coin biased
Experiment	Coin fair	pbinom(61, 100, 0.5)	pbinom(61, 100, 0.7)
conclusion	Coin biased	1 - pbinom(61, 100, 0.5)	1 - pbinom(61, 100, 0.7)

Likewise, after the APPL statements

```
X := BinomialRV(100, 1 / 2);
Y := BinomialRV(100, 7 / 10);
```

the four probabilities can be calculated with the following APPL commands.

True coin status

		Coin fair	Coin biased
Experiment	Coin fair	CDF(X, 61)	CDF(Y, 61)
conclusion	Coin biased	1 - CDF(X, 61)	1 - CDF(Y, 61)

Executing these statements in either language yields the following four probabilities.

True coin status

		Coin fair	Coin biased
Experiment	Coin fair	0.9895	0.0340
conclusion	Coin biased	0.0105	0.9660

There is no perfect test to determine if the coin is fair or biased. There is always the possibility of an error. The random experiment conducted here is known in statistics as a "hypothesis test." The two smaller probabilities are associated with what is called a "Type I" and "Type II" error. In these two cases, the wrong conclusion would be drawn about the coin from the flipping experiment.

The Bernoulli distribution is a fundamental discrete distribution that arises frequently in probability problems. Independent Bernoulli trials lead to the binomial distribution, but they also lead to the geometric distribution, which is described next.

4.3   Geometric Distribution

The geometric distribution is another probability distribution that is based on Bernoulli trials. A geometric random variable X is the number of failures before the first success in repeated independent and identically distributed Bernoulli trials. Since there can be as few as zero failures before the first success and no upper bound on the number of trials before the first success, the support of a geometric random variable is the denumerable set $\mathcal{A} = \{0, 1, 2, \ldots\}$. Table 4.2 lists the possible sequences of Bernoulli trials, associated values of x, and the associated probabilities.

Bernoulli trial outcomes	x value	$P(X = x)$
S	0	p
F S	1	$p(1-p)$
F F S	2	$p(1-p)^2$
F F F S	3	$p(1-p)^3$
$\vdots$	$\vdots$	$\vdots$

Table 4.2: Repeated Bernoulli trials ending with a success.

Generalizing from this table, the probability that the geometric random variable X assumes the value x is $P(X = x) = p(1-p)^x$ for $x \in \mathcal{A}$, which leads to the following definition.

Definition 4.4 A discrete random variable X with probability mass function

$$f(x) = p(1-p)^x \qquad x = 0, 1, 2, \ldots$$

for $0 < p < 1$ is a geometric(p) random variable.

We again use the shorthand $X \sim \text{geometric}(p)$ to indicate that the random variable X has the geometric distribution with parameter p. The geometric distribution occurs in any application where the random variable of interest is the number of failures before the first success in independent Bernoulli(p) trials. Applications include gambling, polling, hunting, cold-call sales, etc.

It is unusually easy to show that $f(x)$ is a legitimate probability mass function. The first existence condition is that $f(x)$ must sum to one over $\mathcal{A}$:

$$\sum_{\mathcal{A}} f(x) = \sum_{x=0}^{\infty} p(1-p)^x = p\left[1 + (1-p) + (1-p)^2 + \cdots\right] = \frac{p}{1-(1-p)} = 1. \quad \checkmark$$

The second existence condition, that $f(x) \geq 0$, is clear by inspection.

The origin of the geometric distribution is that values of the probability mass function $f(x)$ are successive terms in the expansion of $p(1 - (1-p))^{-1}$. More specifically, using the summation formula for a geometric series

$$p(1 - (1-p))^{-1} = p\left[1 + (1-p) + (1-p)^2 + \cdots\right] = p + p(1-p) + p(1-p)^2 + \cdots$$

which is the reverse of proving the first existence condition of the probability mass function for the geometric distribution.

The previous two paragraphs have shown that the right-hand tail of the geometric distribution's probability mass function is easy to sum because it is a geometric series. Therefore, unlike most of the discrete distributions introduced in this chapter, the geometric distribution has a closed-form cumulative distribution function:

$$
\begin{aligned}
F(x) &= P(X \leq x) \\
&= 1 - P(X > x) \\
&= 1 - \sum_{w=x+1}^{\infty} f(w) \\
&= 1 - \sum_{w=x+1}^{\infty} p(1-p)^{w} \\
&= 1 - p(1-p)^{x+1} \left[1 + (1-p) + (1-p)^2 + \cdots \right] \\
&= 1 - p(1-p)^{x+1} \left[\frac{1}{1-(1-p)} \right] \\
&= 1 - (1-p)^{x+1}
\end{aligned}
$$

for any x in $\mathcal{A}$.

   The geometric distribution is unique in the sense that it is the only discrete distribution that has the *memoryless property*. The only continuous distribution that has the memoryless property is the exponential distribution, which will be introduced in the next chapter.

Theorem 4.1 (memoryless property) For $X \sim$ geometric(p) and any two nonnegative integers x and y,
$$
P(X \geq x+y \,|\, X \geq x) = P(X \geq y).
$$

Proof The conditional probability is

$$
\begin{aligned}
P(X \geq x+y \,|\, X \geq x) &= \frac{P(X \geq x+y, \, X \geq x)}{P(X \geq x)} \\
&= \frac{P(X \geq x+y)}{P(X \geq x)} \\
&= \frac{(1-p)^{x+y}}{(1-p)^{x}} \\
&= (1-p)^{y} \\
&= P(X \geq y),
\end{aligned}
$$

which proves the memoryless property. $\qquad\qquad\qquad\qquad\qquad\square$

   The memoryless property can be interpreted as follows. Consider repeated, independent, identical Bernoulli trials and a random variable X that is the number of failures before the first success. If you know that X is greater than or equal to x, then the distribution of the *remaining* number of Bernoulli trials before the first success has the same distribution as if the original x trials had never occurred. This interpretation is consistent with intuition. The previous history of the sequence of Bernoulli trials has no effect on the outcomes of future Bernoulli trials.

   The memoryless property also has a geometric interpretation. Consider the probability mass function for a geometric(p) random variable from some value x to infinity. The sum of the mass values from x to infinity does not equal 1; rather it equals $(1-p)^{x}$. If each of the mass values from x to infinity is divided by $(1-p)^{x}$, then the resulting conditional probability mass function looks identical to the original geometric(p) probability mass function; it is just shifted to the right. No other discrete distribution has this property. For all other discrete probability distributions, the original (unconditional) probability mass function and the conditional probability mass function have different shapes.

The moment generating function for a geometric(p) random variable X is

$$
\begin{aligned}
M(t) &= E\left[e^{tX}\right] \\
&= \sum_{x=0}^{\infty} e^{tx} p(1-p)^x \\
&= p \sum_{x=0}^{\infty} \left(e^t(1-p)\right)^x \\
&= \frac{p}{1-(1-p)e^t}
\end{aligned}
$$

for $(1-p)e^t < 1$ or $t < -\ln(1-p)$, which is required for the geometric series to converge. The moment generating function exists in a neighborhood about $t = 0$.

The population mean of a geometric(p) random variable can be found in three different ways. First, one can use the definition of the expected value

$$
\begin{aligned}
E[X] &= \sum_{\mathcal{A}} x f(x) \\
&= \sum_{x=0}^{\infty} x p(1-p)^x \\
&= 1 \cdot p(1-p) + 2 \cdot p(1-p)^2 + 3 \cdot p(1-p)^3 + 4 \cdot p(1-p)^4 + \cdots.
\end{aligned}
$$

This non-trivial summation can be simplified by writing its terms in the following fashion

$$
\begin{aligned}
E[X] = \ p(1-p) \ &+ \ p(1-p)^2 \ + \ p(1-p)^3 \ + \ p(1-p)^4 \ + \ \cdots \\
&+ \ p(1-p)^2 \ + \ p(1-p)^3 \ + \ p(1-p)^4 \ + \ \cdots \\
& + \ p(1-p)^3 \ + \ p(1-p)^4 \ + \ \cdots \\
& + \ p(1-p)^4 \ + \ \cdots \\
& + \ \ddots
\end{aligned}
$$

Each row is a geometric series with common multiplier $(1-p)$, so taking the row sums yields

$$
E[X] = (1-p) + (1-p)^2 + (1-p)^3 + (1-p)^4 + \cdots
$$

which is itself a geometric series with common multiplier $(1-p)$ that sums to

$$
E[X] = \frac{1-p}{p}.
$$

A second way to find the population mean of $X \sim$ geometric(p) is to use the moment generating function. The first derivative of the moment generating function with respect to t is

$$
M'(t) = \frac{p(1-p)e^t}{\left(1-(1-p)e^t\right)^2}
$$

for $|(1-p)e^t| < 1$. Using $t = 0$ as an argument yields the population mean

$$
E[X] = M'(0) = \frac{1-p}{p}.
$$

The third way to find the population mean of a geometric random variable is to use conditional expectation. This important topic and the derivation of the population mean of a geometric random variable will be presented in Chapter 6.

Using the moment generating function or conditioning, the population variance of a geometric(p) random variable X is

$$
V[X] = E\left[(X-\mu)^2\right] = \frac{1-p}{p^2}.
$$

The population skewness and kurtosis of a geometric(p) random variable X are

$$E\left[\left(\frac{X-\mu}{\sigma}\right)^3\right] = \frac{2-p}{\sqrt{1-p}} \qquad \text{and} \qquad E\left[\left(\frac{X-\mu}{\sigma}\right)^4\right] = \frac{p^2-9p+9}{1-p}.$$

As $p \to 0$, the population skewness and kurtosis approach 2 and 9, respectively.

Example 4.9 Roll a pair of fair dice repeatedly until a "double six" appears. Let X be the number of rolls *prior* to the appearance of the first double six. Find $f(x)$, $E[X]$, $V[X]$, and $P(X < 24)$.

Letting "success" be rolling a double six so that $p = 1/36$, the random experiment is a sequence of repeated independent, identical Bernoulli trials. Therefore the number of rolls prior to the first double six is geometric($1/36$). The probability mass function of X is

$$f(x) = \frac{1}{36}\left(\frac{35}{36}\right)^x \qquad x = 0, 1, 2, \ldots$$

and is graphed in Figure 4.4.

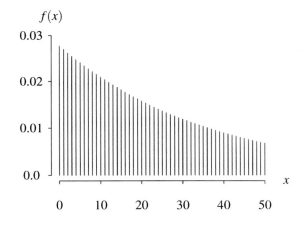

Figure 4.4: The probability mass function for $X \sim$ geometric($1/36$).

The population mean and variance of X are

$$E[X] = 35 \qquad \text{and} \qquad V[X] = 36 \cdot 35 = 1260.$$

The units on the population mean and variance are rolls and squared rolls. The probability that X is less than 24 is

$$
\begin{aligned}
P(X < 24) &= 1 - P(X \geq 24) \\
&= 1 - \sum_{x=24}^{\infty}\left(\frac{1}{36}\right)\left(\frac{35}{36}\right)^x \\
&= 1 - \left(\frac{1}{36}\right)\left(\frac{35}{36}\right)^{24}\left[1 + \left(\frac{35}{36}\right) + \left(\frac{35}{36}\right)^2 + \cdots\right] \\
&= 1 - \left(\frac{1}{36}\right)\left(\frac{35}{36}\right)^{24}\left(\frac{1}{1 - 35/36}\right) \\
&= 1 - \left(\frac{35}{36}\right)^{24} \\
&\cong 0.4914.
\end{aligned}
$$

This probability matches the solution that relied on independence from Example 2.14. Since $P(X < 24) = P(X \leq 23) = F(23)$, this probability can be calculated in R with the statement

```
pgeom(23, 1 / 36)
```

This particular probability question is of historical significance because it was posed in a seventeenth century written correspondence between Blaise Pascal and The Chevalier de Méré which formed the genesis of modern probability theory. There was a gambling game which was won when the gambler rolled a double six in 24 rolls of a pair of fair dice. Gamblers had noticed that they won slightly less than half of the time, and Pascal was able to work through the mathematics to confirm that this was indeed the case. The fact that the probability of winning the game is so close to $1/2$ highlights the importance that the dice are fair.

To review, a geometric(p) random variable X models the *number of failures before the first success* in repeated independent, identical Bernoulli trials. The probability mass function for X is

$$f(x) = p(1 - p)^x \qquad\qquad x = 0, 1, 2, \ldots$$

where $0 < p < 1$. The population mean and variance are

$$\mu = \frac{1 - p}{p} \qquad\text{and}\qquad \sigma^2 = \frac{1 - p}{p^2}.$$

Unfortunately, not everyone parameterizes the geometric distribution in this fashion. A geometric random variable X can also model the *trial number of the first success* in repeated independent, identical Bernoulli trials. This effectively shifts the distribution one x-value to the right, resulting in the following definition.

Definition 4.5 A discrete random variable X with probability mass function

$$f(x) = p(1 - p)^{x-1} \qquad\qquad x = 1, 2, \ldots$$

for $0 < p < 1$ is a Geometric(p) random variable.

The upper-case G in Geometric(p) is used in this text (this notation is not standard) to distinguish between the two versions of the geometric distribution. The cumulative distribution function for a Geometric(p) random variable parameterized from 1 on its support is shifted to the right by one unit:

$$F(x) = 1 - (1 - p)^x \qquad\qquad x = 1, 2, \ldots.$$

The population mean of the Geometric(p) random variable is also shifted up by one, but the population variance is unaffected by the shift:

$$\mu = \frac{1}{p} \qquad\text{and}\qquad \sigma^2 = \frac{1 - p}{p^2}.$$

The population skewness and kurtosis remain the same because they are the expected values of powers of the standardized geometric random variable.

The two definitions of a geometric random variable are annoying but unavoidable. The R language, for example, parameterizes the geometric distribution with a support that begins at 0; the APPL language parameterizes the geometric distribution with a support that begins at 1. All future references to the geometric(p) distribution will specify whether the support begins at 0 or 1. Some probability questions are more naturally modeled in terms of one parameterization or the other.

Example 4.10 How many tosses of a pair of fair dice are necessary to be 99% certain that a double six will appear?

The random experiment again consists of repeated, independent, identical Bernoulli trials, with an interest in the toss number of the first double six; therefore, using a Geometric(1/36) distribution is an appropriate probability model. Let X be the trial number of the first appearance of a double six, which has probability mass function

$$f(x) = \left(\frac{1}{36}\right)\left(\frac{35}{36}\right)^{x-1} \qquad x = 1, 2, \ldots .$$

The goal here is to find the number of tosses c that satisfies

$$P(X \le c) \ge 0.99,$$

which can be written as

$$\sum_{x=1}^{c} \left(\frac{1}{36}\right)\left(\frac{35}{36}\right)^{x-1} \ge 0.99$$

or

$$1 - \sum_{x=c+1}^{\infty} \left(\frac{1}{36}\right)\left(\frac{35}{36}\right)^{x-1} \ge 0.99.$$

Summing the geometric series in the usual fashion results in

$$\left(\frac{35}{36}\right)^{c} < 0.01.$$

The left-hand side of this inequality is the probability of seeing no double sixes in the first c tosses. Taking the natural logarithm of both sides of this equation yields

$$c \ln\left(\frac{35}{36}\right) < \ln 0.01$$

or

$$c > \frac{\ln 0.01}{\ln(35/36)} \cong 163.4727.$$

The smallest value of c satisfying this inequality is $c = 164$. It requires 164 tosses to be 99% certain of seeing a double six.

This problem can be restated as finding the 99th percentile of a Geometric(1/36) random variable. This can be found with the APPL statements

```
X := GeometricRV(1 / 36);
IDF(X, 0.99);
```

because APPL parametrizes its geometric distribution with a support beginning at 1. It can also be found with the R statement

```
qgeom(0.99, 1 / 36)
```

as long as you remember to increase the returned value by one to account for the fact that R parametrizes the geometric distribution with a support beginning at 0.

4.4   Negative Binomial Distribution

The geometric distribution models the number of failures before the first success in repeated, independent Bernoulli trials, each with probability of success p. The negative binomial distribution is a generalization of the geometric distribution. The negative binomial distribution models the number of failures before the rth success in repeated, independent Bernoulli trials, each with probability of success p. Since there can be as few a zero failures before the rth success and no upper bound on the number of trials before the rth success, the support of a negative binomial random variable is the denumerable set $A = \{0, 1, \ldots\}$.

We now turn to the more difficult question of determining the probability mass function for a negative binomial random variable X. The probability of r successes and x failures *in a specified order,* for example

$$FFSFFFSS$$

associated with $r = 3$ and $x = 5$, is

$$p^r(1-p)^x.$$

The rth success occurs at position $x + r$ in the sequence. This means that there are $x + r - 1$ positions prior to the rth success, where the prior $r - 1$ successes must be distributed. Since the order is not important as to which positions are selected for the $r - 1$ initial successes, there are

$$\binom{x+r-1}{r-1}$$

different sequences of failures and successes associated with x failures prior to the rth success. This leads to the definition of a negative binomial random variable.

Definition 4.6 A discrete random variable X with probability mass function   *r successes*

$$f(x) = \binom{x+r-1}{r-1} p^r(1-p)^x \qquad x = 0, 1, 2, \ldots \quad \textit{x failures.}$$

for some positive integer r and $0 < p < 1$ is a negative binomial(r, p) random variable.

The negative binomial is also known as the Pascal distribution. We denote a negative binomial distribution with parameters r and p by $X \sim$ negative binomial(r, p). The origins of this distribution are that values of $f(x)$ are successive terms in the expansion of $p^r(1-(1-p))^{-r}$. The geometric distribution is a special case of the negative binomial distribution when $r = 1$.

A negative binomial random variable can be thought of as the concatenation of r random experiments associated with the geometric distribution in the following fashion. A geometric random variable involves repeated Bernoulli(p) trials until the first success. A negative binomial random variable is r of these random experiments placed back-to-back. Stated another way, if $X_1, X_2, \ldots, X_r$ are independent geometric(p) random variables then $X_1 + X_2 + \cdots + X_r$ is a negative binomial(r, p) random variable. The notion of independence for random variables will be defined in Chapter 6.

The moment generating function for a negative binomial(r, p) random variable is

$$M(t) = \left[\frac{p}{1-(1-p)e^t}\right]^r$$

for $(1-p)e^t < 1$ or $t < -\ln(1-p)$. Using $t = 0$ as an argument in $M'(t)$ yields the population mean

$$\mu = E[X] = \frac{r(1-p)}{p}.$$

Similarly, the population variance is

$$\sigma^2 = V[X] = \frac{r(1-p)}{p^2}$$

and the population skewness and kurtosis are

$$E\left[\left(\frac{X-\mu}{\sigma}\right)^3\right] = \frac{2-p}{\sqrt{r(1-p)}} \qquad \text{and} \qquad E\left[\left(\frac{X-\mu}{\sigma}\right)^4\right] = \frac{p^2 - 6p - 3pr + 3r + 6}{r(1-p)}.$$

The population skewness and kurtosis approach 0 and 3 in the limit as $r \to \infty$. These two values, along with the central limit theorem proven in Chapter 8, indicate that the probability mass function for a negative binomial distribution approaches a symmetric bell shape as $r \to \infty$.

Example 4.11 Raghu is making cold sales calls. The probability of a sale on each call is 0.4. The calls may be considered independent Bernoulli trials.

(a) What is the probability that he has exactly five failed calls before his second successful sales call?

(b) What is the probability that he has fewer than five failed calls before his second successful sales call?

Let the random variable X be the number of failures before Raghu's second success. So $X \sim$ negative binomial$(2, 0.4)$. The probability mass function for X is

$$f(x) = \binom{x+2-1}{2-1}(0.4)^2(0.6)^x \qquad x = 0, 1, 2, \ldots$$

or

$$f(x) = (x+1)(0.16)(0.6)^x \qquad x = 0, 1, 2, \ldots .$$

The probability mass function is plotted in Figure 4.5 using the R statements

```
x = 0:9
f = dnbinom(x, 2, 0.4)
plot(x, f, type = "h")
```

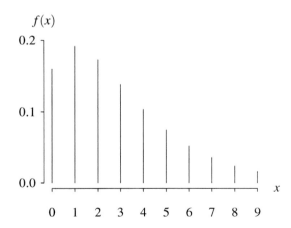

Figure 4.5: The probability mass function for $X \sim$ negative binomial$(2, 0.4)$.

(a) The probability that Raghu has exactly five failures before his second successful sales call is $P(X = 5)$, that is

$$P(X = 5) = f(5) = 6(0.16)(0.6)^5 \cong 0.07465,$$

which can be calculated in R with the statement `dnbinom(5, 2, 0.4)`. This probability is the height of $f(5)$ in Figure 4.5.

[handwritten margin notes: "b/c discrete, we must pay attention to detail. < 5 = ≤ 4 definition ≤ of cdf is ≤ not ≤ for discrete."]

(b) The probability that Raghu has fewer than five failures before his second successful sales call is $P(X < 5) = P(X \leq 4)$, that is

$$
\begin{aligned}
P(X < 5) &= f(0) + f(1) + \cdots + f(4) \\
&= (0.16)(0.6)^0 + 2(0.16)(0.6)^1 + \cdots + 5(0.16)(0.6)^4 \\
&= 0.16 + 0.192 + \cdots + 0.10368 \\
&= 0.76672,
\end{aligned}
$$

which can be calculated in R with the statement pnbinom(4, 2, 0.4). This probability is the sum of the heights of $f(0)$ through $f(4)$ in Figure 4.5.

Example 4.12 Ravi has r coins. Scott has s coins. They agree to perform a Bernoulli trial with probability of success p to determine whether Scott or Ravi will place the next coin in the jukebox. "Success" corresponds to Ravi placing the next coin in the jukebox. Find the probability that Ravi runs out of coins before Scott.

Let the random variable X be the number of failures before the rth success. The rth success corresponds to Ravi running out of coins. The probability mass function of X is

$$
f(x) = \binom{x + r - 1}{r - 1} p^r (1 - p)^x \qquad x = 0, 1, 2, \ldots.
$$

The question is asking for the probability that X is less than s, which is

$$
P(X < s) = \sum_{x=0}^{s-1} \binom{x + r - 1}{r - 1} p^r (1 - p)^x.
$$

When Ravi and Scott begin with equal fortunes, that is $r = s$, this probability can be plotted for various values of p using the R code

```
p  = seq(0, 1, by = 0.001)
r  = 1
y1 = pnbinom(r - 1, r, p)
r  = 3
y2 = pnbinom(r - 1, r, p)
r  = 10
y3 = pnbinom(r - 1, r, p)
matplot(p, cbind(y1, y2, y3), type = "l")
```

The plot appears in Figure 4.6 for initial fortunes of $r = s = 1, 3, 10$ coins. The curves become steeper for larger initial fortunes. When Ravi and Scott have $r = s = 1$ coin each, the probability of Ravi running out first corresponds to a single flip, resulting in the line connecting $(0, 0)$ and $(1, 1)$. When each begin with an initial fortune of $r = s = 10$ coins, the competition is essentially a "best out of 19" series, resulting in the steeper probability curve. It is worth imagining what this curve would do in the limit as the size of the equal fortunes goes to infinity.

To review, a negative binomial random variable X models the *number of failures* before the rth success in repeated independent, identical Bernoulli trials. The probability mass function for X is

$$
\boxed{f(x) = \binom{x + r - 1}{r - 1} p^r (1 - p)^x} \qquad x = 0, 1, 2, \ldots
$$

where $0 < p < 1$. The population mean and variance are

$$
\mu = \frac{r(1 - p)}{p} \qquad \text{and} \qquad \sigma^2 = \frac{r(1 - p)}{p^2}.
$$

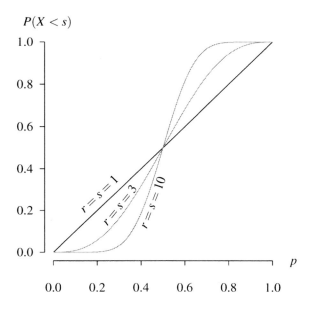

$P(X < s)$

Figure 4.6: The probability that Ravi runs out of coins first.

As was the case with the geometric distribution, not everyone parameterizes the negative binomial distribution in this fashion. A negative binomial random variable X can also model the *trial number* of the rth success in repeated independent, identical Bernoulli trials. This new parameterization means that we must define another negative binomial distribution whose support begins at r. As with the geometric distribution, we capitalize the first letter to distinguish this particular parameterization.

Definition 4.7 A discrete random variable X with probability mass function

$$f(x) = \binom{x-1}{r-1} p^r (1-p)^{x-r} \qquad x = r, r+1, \ldots$$

for some positive integer r and $0 < p < 1$ is a Negative binomial(r, p) random variable.

The binomial coefficient $\binom{x-1}{r-1}$ corresponds to the number of ways to distribute the $r - 1$ initial successes in the available $x - 1$ positions in the sequence of Bernoulli trials. The rth success occurs on the xth trial. The population mean of a Negative binomial(r, p) random variable is

$$\mu = E[X] = \frac{r}{p}.$$

The population variance, skewness, and kurtosis remain the same as in the original parameterization.

The fact that there are multiple manners in which the negative binomial(r, p) distribution can be defined means that we must be careful to indicate which parameterization is being used and we must check any computer software to see which parameterization the software developers used. The R language parameterizes the negative binomial distribution from 0 in its functions

```
dnbinom        pnbinom        qnbinom        rnbinom
```

which calculate the probability mass function, calculate the cumulative distribution function, calculate a percentile (quantile), and generate random variates. The APPL function `NegativeBinomialRV`, on the other hand, parameterizes the negative binomial distribution from r. APPL uses the associated functions `PDF`, `CDF`, `IDF`, `Mean`, `Variance`, etc. to calculate quantities of interest.

We end this section with an example that is parameterized using the negative binomial distribution with a support beginning at r.

Example 4.13 Consider two random experiments:

- Let X be the roll number of the third double six in repeated rolls of a pair of fair dice.
- Let Y be the flip number of the 54th head in repeated flips of a fair coin.

Plot the probability mass functions of these two random variables.

These two random variables are both negative binomial:

$$X \sim \text{Negative binomial}(3, 1/36) \qquad \text{and} \qquad Y \sim \text{Negative binomial}(54, 1/2),$$

where the negative binomial distribution is parameterized beginning at r. In addition, each has a mean population $\mu = r/p$ equal to $\mu = 108$. The probability mass function for X is plotted in Figure 4.7 for $x = 3, 4, \ldots, 200$ with the R statements shown below.

```
r = 3
p = 1 / 36
x = r:200
f = choose(x - 1, r - 1) * p ^ r * (1 - p) ^ (x - r)
plot(x, f, type = "h")
```

The probability mass function for Y is plotted in Figure 4.8 using similar code. Although the means of the two distributions are the same, the shapes of the two probability mass functions differ significantly. The probability mass function for X is skewed to the right, whereas the probability mass function for Y is nearly symmetric about its population mean. This is consistent with the observation that the population skewness of the negative binomial distribution approaches 0 as $r \to \infty$. Also, the distribution of Y is clustered more tightly about its population mean, consistent with the population variances:

$$V[X] = \frac{3(1 - 1/36)}{(1/36)^2} = 3 \cdot 35 \cdot 36 = 3780$$

and

$$V[Y] = \frac{54(1 - 1/2)}{(1/2)^2} = 54 \cdot 2 = 108.$$

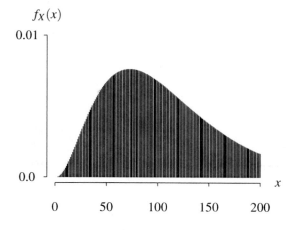

Figure 4.7: The probability mass function for $X \sim \text{Negative binomial}(3, 1/36)$.

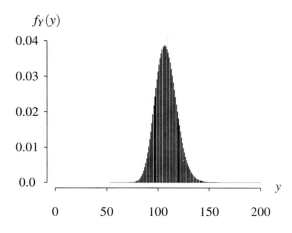

Figure 4.8: The probability mass function for $Y \sim$ Negative binomial$(54, 1/2)$.

4.5 Poisson Distribution

The Poisson distribution was introduced by French mathematician Simeon Poisson (1781–1840) in a book that he published in 1837. There are two ways to develop the Poisson distribution. First, the Poisson distribution can be used as an approximation to the binomial distribution, which is effective for large n and small p. Second, the Poisson distribution can be used to model the number of events that occur over space or time in what is known as a *Poisson process*. We begin with the Poisson approximation to the binomial distribution.

Approximation to the binomial distribution

The binomial distribution, introduced in Section 4.2, is one of the key discrete distributions in probability. In many applications, n can be quite large, and calculating the factorials in the binomial coefficient that appear in the probability mass function can be difficult. When n is large and p is small, it is possible to arrive at a distribution that approximates the binomial distribution. Recall that the population mean of a binomial(n, p) random variable X is $\mu = np$. Replacing p with μ/n and doing some strategic rearrangement of terms yields the following probability mass function:

$$
\begin{aligned}
f(x) &= \binom{n}{x} p^x (1-p)^{n-x} \\
&= \frac{n(n-1)(n-2)\ldots(n-x+1)}{x!} \left(\frac{\mu}{n}\right)^x \left(1-\frac{\mu}{n}\right)^{n-x} \\
&= \frac{n}{n} \cdot \frac{n-1}{n} \cdot \ldots \cdot \frac{n-x+1}{n} \cdot \frac{\mu^x}{x!} \cdot \left(1-\frac{\mu}{n}\right)^n \left(1-\frac{\mu}{n}\right)^{-x} \qquad x = 0, 1, 2, \ldots, n.
\end{aligned}
$$

Taking the limit as $n \to \infty$, the $\mu^x/x!$ term can be factored out of the limit because it does not involve n. Furthermore, all of the other factors go to 1 in the limit as $n \to \infty$ except $\left(1-\frac{\mu}{n}\right)^n$, which approaches $e^{-\mu}$ using a result from calculus. Therefore

$$
\lim_{n \to \infty} f(x) = \lim_{n \to \infty} \binom{n}{x} p^x (1-p)^{n-x} = \frac{\mu^x e^{-\mu}}{x!} \qquad x = 0, 1, 2, \ldots.
$$

This limiting distribution in n is known as the *Poisson distribution*. Tradition dictates that the parameter λ is used for the population mean μ, resulting in the following definition.

Definition 4.8 A discrete random variable X with probability mass function

$$f(x) = \frac{\lambda^x e^{-\lambda}}{x!} \qquad x = 0, 1, 2, \ldots$$

for $\lambda > 0$ is a Poisson(λ) random variable.

In order to check to see that the probability mass function satisfies the two existence conditions, first see that $f(x)$ sums to one over its support. Using the exponential series expansion,

$$\sum_{x=0}^{\infty} \frac{\lambda^x e^{-\lambda}}{x!} = e^{-\lambda} \sum_{x=0}^{\infty} \frac{\lambda^x}{x!} = e^{-\lambda} e^{\lambda} = 1.$$

The probability mass function is positive for all values in its support, so it satisfies both existence conditions.

The moment generating function for $X \sim$ Poisson(λ) is

$$\begin{aligned}
M(t) &= E\left[e^{tX}\right] \\
&= \sum_{x=0}^{\infty} e^{tx} \frac{\lambda^x e^{-\lambda}}{x!} \\
&= e^{-\lambda} \sum_{x=0}^{\infty} \frac{(\lambda e^t)^x}{x!} \\
&= e^{-\lambda} e^{\lambda e^t} \\
&= e^{\lambda(e^t - 1)}
\end{aligned}$$

moment generating function

for $-\infty < t < \infty$. The moments can be calculated by taking derivatives of the moment generating function. The first two derivatives of $M(t)$ with respect to t are

$$M'(t) = e^{\lambda(e^t - 1)} \lambda e^t \qquad \text{and} \qquad M''(t) = e^{\lambda(e^t - 1)} \lambda e^t + (\lambda e^t)^2 e^{\lambda(e^t - 1)}$$

for $-\infty < t < \infty$. When these two derivatives are evaluated at $t = 0$, the first two moments about the origin are

$$E[X] = M'(0) = \lambda \qquad \text{and} \qquad E\left[X^2\right] = M''(0) = \lambda + \lambda^2.$$

So using the shortcut formula for the population variance $V[X] = E\left[X^2\right] - E[X]^2$, the population mean and variance of the Poisson distribution are equal:

$$\mu = E[X] = \lambda \qquad \text{and} \qquad \sigma^2 = V[X] = \lambda.$$

Again using the moment generating function, the population skewness and kurtosis can be calculated, yielding

$$E\left[\left(\frac{X - \mu}{\sigma}\right)^3\right] = \frac{1}{\sqrt{\lambda}} \qquad \text{and} \qquad E\left[\left(\frac{X - \mu}{\sigma}\right)^4\right] = \frac{1 + 3\lambda}{\lambda}.$$

As was the case with the negative binomial distribution, the skewness and kurtosis approach 0 and 3 in the limit as $\lambda \to \infty$. The probability mass function for the Poisson distribution approaches a symmetric bell shape as $\lambda \to \infty$.

In terms of computations involving the Poisson distribution, there is, thankfully, only the single parameterization given in Definition 4.8. This means that the APPL code

```
X := PoissonRV(lambda);
Mean(X);
Variance(X);
Skewness(X);
Kurtosis(X);
MGF(X);
```

for example, can be used to compute the moments discussed so far. Likewise, the R functions

$$\text{dpois} \qquad \text{ppois} \qquad \text{qpois} \qquad \text{rpois}$$

can be used to calculate the probability mass function, calculate the cumulative distribution function, calculate a percentile (quantile), and generate random variates from the Poisson distribution. The Poisson approximation to the binomial distribution works best for large n and small p, as illustrated in the following examples.

Example 4.14 A rare disease affects 0.2% of the population. Find the probability that a city of 500,000 people has 1040 or fewer people with the disease.

Solution 1 (exact solution using the binomial distribution). Each person in the city is a Bernoulli trial with $p = 0.002$ as the probability of "success," which in this case is being infected with the disease. Let the random variable X be the number in the city that are infected with the disease, which is a binomial random variable with $n = 500000$ and $p = 0.2$. The exact probability is

$$P(X \leq 1040) = \sum_{x=0}^{1040} \binom{500000}{x} 0.002^x 0.998^{500000-x}.$$

Calculating this probability can be difficult due to the large factorials. Nevertheless, the R language has numerical procedures to overcome the overflow potential, and the probability can be computed with

```
pbinom(1040, 500000, 0.002)
```

which yields $P(X \leq 1040) \cong 0.8995$.

Solution 2 (approximate solution using the Poisson approximation to the binomial distribution). This problem certainly supports the criteria of a large n and small p, so the Poisson approximation to the binomial distribution with $\lambda = (500000)(0.002) = 1000$ can also be used to solve the problem. The approximate probability is

$$P(X \leq 1040) = \sum_{x=0}^{1040} \frac{1000^x e^{-1000}}{x!}.$$

The factorials can again be an issue, but R calculates this probability with the statement

```
ppois(1040, 500000 * 0.002)
```

which yields $P(X \leq 1040) \cong 0.8993$. The difference between the exact and approximate methods occurs out in the fourth digit, so the two solutions are essentially identical.

Example 4.15 As a continuation of Example 4.5, an airplane that holds 100 passengers is overbooked in order to increase the revenue generated by a flight. The probability an individual ticketed passenger shows up for a flight is 0.92, and all passenger decisions to show up for a flight constitute independent Bernoulli trials. If the airline books 107 seats for a particular flight (the optimum overbooking level from Example 4.5), what is the probability that one or more ticketed passenger will be bumped from the flight?

This problem will be solved exactly using the binomial distribution, then approximated using the Poisson distribution. The approximation works well when n is large and p is small, so we define "success" on each Bernoulli trial to be *not* showing up for a flight so that there are $n = 107$ Bernoulli trials, each with probability of success $p = 0.08$. Let the discrete random variable X be the number of successes (no-shows) in the $n = 107$

Bernoulli trials. We want to calculate the probability that one or more ticketed passenger will be bumped from the flight, which is $P(X < 7)$.

Solution 1 (exact solution using the binomial distribution). The distribution of X is binomial(107, 0.08), with probability mass function

$$f(x) = \binom{107}{x} (0.08)^x (0.92)^{107-x} \qquad x = 0, 1, 2, \ldots, 107.$$

The probability mass function is plotted in Figure 4.9. The probability of one or more ticketed passenger being bumped from the flight is

$$P(X < 7) = \sum_{x=0}^{6} \binom{107}{x} (0.08)^x (0.92)^{107-x} \cong 0.2387.$$

Approximately 24% of the sold-out flights will require ticketed passengers to be bumped to a later flight. This may prove acceptable if the airline can produce incentives for the bumped passengers to voluntarily move to a later flight. This probability can be calculated using the R statement

```
pbinom(6, 107, 0.08)
```

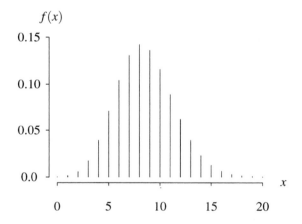

Figure 4.9: The probability mass function for $X \sim$ binomial(107, 0.08).

Solution 2 (approximate solution using the Poisson approximation to the binomial distribution). First calculate the population mean of the binomial(107, 0.08) distribution, which is

$$\mu = np = (107)(0.08) = 8.56.$$

We expect, on average, 8.56 ticketed passengers to not show up for the flight. The probability mass function of the associated Poisson(8.56) distribution is

$$f(x) = \frac{8.56^x e^{-8.56}}{x!} \qquad x = 0, 1, 2, \ldots .$$

The probability mass function is plotted in Figure 4.10. It is nearly identical to the corresponding binomial probability mass function. The probability of one or more ticketed passenger being bumped from the flight is

$$P(X < 7) = \sum_{x=0}^{6} \frac{8.56^x e^{-8.56}}{x!} \cong 0.2498.$$

This probability can be calculated using the R statement

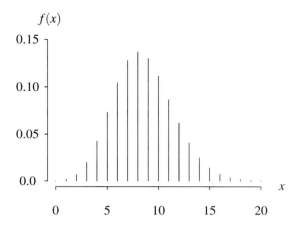

Figure 4.10: The probability mass function for $X \sim \text{Poisson}(8.56)$.

```
ppois(6, 8.56)
```

Comparison. There is a 5% difference between the exact and approximate solutions for this particular problem. There are two points of view on this difference. The first point of view is the garbage-in, garbage-out perspective. Since p is approximate and the model concerning independent decisions by the flyers isn't perfect either, an error of 5% is not a significant problem. The second point of view questions whether it is appropriate to heap an additional 5% error on top of the error already present by some dubious assumptions.

Poisson processes

A second way to introduce the Poisson distribution is through the development of a *Poisson process*, where events occur over time or space in a random fashion. We begin the introduction to Poisson processes with an example.

Sven & Barry's Very Berry Yogurt Factory is doing a booming business. During their peak arrival time, they average 30 customer arrivals per hour. They hire you as a consultant to create a discrete-event simulation model of their Very Berry Yogurt Factory during the peak time. (A discrete-event simulation model differs from a Monte Carlo simulation model in that the passage of time plays a significant role.) Your first decision is how to model customer arrival times to the Yogurt Factory.

The easiest way to model the arrival of 30 customers in an hour is to have them arrive every 2 minutes. Generating arrivals at times $2, 4, \ldots, 60$ is shown in the time axis in Figure 4.11. Instead of looking random, these arrival times look very deterministic. You immediately abandon this approach. The customers do not arrive on a conveyor belt.

0   60

Figure 4.11: Deterministic arrivals.

As a second attempt at generating more realistic random arrival times during one of their rush hours, each one-minute interval along the time axis can be considered to be an equally-likely opportunity for a customer arrival. A fair coin is flipped, and if it comes up heads, an arrival is generated in the middle of that particular time interval. This will give an *average* of 30 arrivals during an hour, which more accurately reflects the true arrival pattern. The distribution of the number of arrivals during the hour is binomial(60, 0.5). Five realizations of this process are shown in Figure 4.12. This manner of generating arrival times captures the randomness of the arrivals much more accurately than the deterministic arrivals every 2 minutes. The five realizations contain 26, 36, 29, 32, and 26 arrivals, capturing the natural random sampling variability that would occur in practice. So you take the diagram of the realizations to Sven and Barry. Their only complaint about the simulated arrival times is that the arrivals are all separated by an integer number of minutes, which is not the case in practice. You then realize that your choice of using time intervals with a width of one minute was arbitrary. That feature can be effectively eliminated by changing the interval width from minutes to seconds.

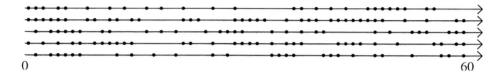

Figure 4.12: Random arrivals during each minute.

In your final arrival time model, you divide the hour into $60 \cdot 60 = 3600$ time slices, each one second wide. During each second, you flip a biased coin with probability of success $p = 1/120$ (where success corresponds to an arrival), which assures that an average of 30 arrivals will occur during the hour. The distribution of the number of arrivals is binomial(3600, 1/120). Five realizations of this process are plotted in Figure 4.13.

Figure 4.13: Random arrivals during each second.

Sven and Barry are now happy with your arrival process. You have captured the clustering of arrivals successfully and the number of arrivals varies from one realization to the next, but averages to 30 customers per hour. The five realizations contain a variable number of arrivals (27, 23, 31, 37, and 17 for the five realizations) clustered around 30, which effectively captures the random sampling variability that would occur in practice. Looking back at the binomial random variables for the one minute and one second intervals, however, you realize that n, the number of intervals, is increasing, and p, the probability of an arrival during an interval, is decreasing in a fashion such that $\mu = np$ is being held constant at 30. This means that in the limit as $n \to \infty$, the number of arrivals during the hour will have a Poisson distribution with $\lambda = 30$ because of the earlier derivation involving the Poisson approximation to the binomial distribution. In fact, the number of arrivals in *any* time interval will be Poisson, with a parameter that depends on the width of the interval.

The overall arrival rate for your process is $\lambda = 30$ customers per hour. Using hours as the time scale and interval widths of one second, each interval of time has width $w = 1/3600$ hour, or one second. Here are three key properties of this arrival process model:

1. The probability of an arrival during an interval is $\lambda w = (30)(1/3600) = 1/120$.

2. The probability of more than one arrival during a time interval is close to 0.

3. The probability of an arrival during a time interval is independent of the previous history of arrivals.

We now begin to abstract notions from the specific case of the Yogurt Factory to the more general case. A *random process* evolves over time and has some sort of chance mechanism controlling the occurrence of events. Random processes are used to control "events," which could be customer arrivals, machine breakdowns, or earthquakes, that occur randomly over time or space. The next two examples illustrate what is meant by the occurrence of events over *time* and *space*.

Example 4.16 The following examples describe events occurring over *time*.

particles	*arriving to*	a Geiger counter
customers	*arriving to*	a bank
earthquakes	*arriving to*	California
web hits	*arriving to*	a website
phone calls	*arriving to*	a call center
insurance claims	*arriving to*	an insurance company

Example 4.17 The following examples describe events occurring over *space*.

potholes	*arriving to*	a highway
defects	*arriving to*	a magnetic tape
typografical errors	*arriving to*	a page
painting defects	*arriving to*	a car door

The painting defects arriving to a car door can be viewed as a two-dimensional random process, whereas the positions of the defects on a magnetic tape are measured in one dimension.

The detailed derivations of the properties of a Poisson process are best left for a first class in *stochastic processes*, which is highly recommended. The approach taken here is to simply state the assumptions, notation, and results associated with Poisson processes with minimal mathematical rigor. Time, rather than space, is used when describing the assumptions, notation, and results.

A *Poisson process* models the occurrence of events that occur at random over time with rate λ and is governed by the following three assumptions:

1. The probability that an event occurs in a narrow time interval of width w is λw.

2. The probability that two events will occur simultaneously is 0.

3. The number of events in nonoverlapping time intervals are independent.

In order to state the significant results associated with a Poisson process, some additional notation is required. First, let

$$T_0 = 0$$

be the time origin for the Poisson process. Second, let the ordered continuous random variables

$$T_1, T_2, \ldots$$

denote the event times for the Poisson process. Third let

$$X_1, X_2, \ldots$$

denote the times between the event times in the Poisson process. The nth event time can be written as

$$T_n = X_1 + X_2 + \cdots + X_n.$$

This relationship between the event times and the time between events is illustrated in Figure 4.14 for the first three events in a Poisson process. Fourth, define the *counting function* $N(t)$ as the number of events that occur in the interval $(0, t]$. In other words,

$$N(t) = \max\{k \mid T_k \le t\}$$

for $t > 0$. When an event occurs, $N(t)$ increases by 1. The nondecreasing, integer-valued process described by $\{N(t), t > 0\}$ is often called a *counting process* and satisfies these two properties:

1. If $t_1 < t_2$, then $N(t_1) \le N(t_2)$.

2. If $t_1 < t_2$, then $N(t_2) - N(t_1)$ is the number of events in the interval $(t_1, t_2]$.

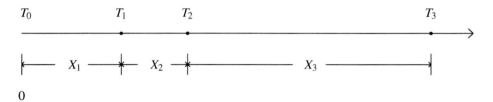

Figure 4.14: Poisson process notation.

Theorem 4.2 (Poisson number of events in any time interval) For a Poisson process with rate λ, the number of events in the interval $(a, b]$ is Poisson with probability mass function

$$f(x) = \frac{(\lambda(b-a))^x e^{-\lambda(b-a)}}{x!} \qquad x = 0, 1, 2, \dots.$$

Theorem 4.2 replaces the rate per unit time λ by $\lambda(b-a)$, which is the rate over the interval of length $b - a$. This result allows us to calculate probabilities associated with the number of arrivals in a time interval, as illustrated in the example below.

Example 4.18 For a Poisson process with arrival rate $\lambda = 30$ customers per hour, find

(a) the expected number of arrivals in the first 10 minutes of an hour, $30\left(\frac{1}{6}\right) = \dfrac{5^x e^{-5}}{x!}$

(b) the probability of exactly 4 arrivals in the first 10 minutes of an hour,

(c) the probability of 4 or fewer arrivals in the first 10 minutes of an hour,

(d) the probability of 35 or more arrivals in an hour given that there were 8 arrivals in $30\left(\frac{5}{6}\right)$
the first 10 minutes of that hour. $P(x > 27) \approx$ $f(x) = \dfrac{25^x e^{-25}}{x!}$ $= 25$

These questions can be addressed using Theorem 4.2 and the third Poisson assumption regarding independence.

(a) Let the random variable X denote the number of arrivals in the first 10 minutes (which is $1/6$ of an hour). Theorem 4.2 indicates that X has the Poisson distribution with rate $30(1/6 - 0) = 5$ customer arrivals over the 10-minute period. The population mean of this distribution is 5 customers.

(b) The probability mass function of X, the number of customer arrivals in the first 10 minutes of the hour, is

$$f(x) = \frac{5^x e^{-5}}{x!} \qquad x = 0, 1, 2, \dots,$$

so the probability of exactly 4 arrivals in the first 10 minutes is

$$P(X = 4) = f(4) = \frac{5^4 e^{-5}}{4!} = \frac{625 e^{-5}}{24} \cong 0.1755,$$

which is calculated in R with dpois(4, 5).

(c) The probability of 4 or fewer arrivals in the first 10 minutes of an hour is

$$P(X \leq 4) = \sum_{x=0}^{4} \frac{5^x e^{-5}}{x!} \cong 0.4405,$$

which is calculated in R with ppois(4, 5).

(d) Using the third assumption of a Poisson process, the number of arrivals in the first 10 minutes of an hour is independent of the number of arrivals in the nonoverlapping last 50 minutes of an hour, so the question can be restated as finding the probability of $35 - 8 = 27$ or more arrivals in the last 50 minutes of an hour. Let X be the number of arrivals in the last 50 minutes of the hour. By Theorem 4.2, X is Poisson with rate $30(1 - 1/6) = 25$ arrivals in 50 minutes. The probability mass function of X is

$$f(x) = \frac{25^x e^{-25}}{x!} \qquad x = 0, 1, 2, \ldots$$

so the probability of 27 or more arrivals during the last 50 minutes is

$$P(X \geq 27) = \sum_{x=27}^{\infty} \frac{25^x e^{-25}}{x!} \cong 0.3706,$$

which is calculated in R with 1 - ppois(26, 25).

Theorem 4.3 (exponential times between events) For a Poisson process with rate λ, the times between events are independent and identically distributed with probability density function

$$f(x) = \lambda e^{-\lambda x} \qquad x > 0.$$

As will be seen in the next chapter, this distribution is known as the *exponential* distribution, and it plays a central role in probability theory, particularly in stochastic processes, queueing theory, reliability, and survival analysis. The mode of the exponential distribution is at 0, which explains the clustering that occurs in the events of a Poisson process. Although the rate of the process is constant, events tend to clump together because of the likelihood of very short times between events. An advantage to having exponential times between arrivals in a Poisson process is that they are easy to simulate, as illustrated in the following example.

Example 4.19 Devise an algorithm for generating the event times in a Poisson process with rate λ.

Using Theorem 4.3, a clock should be initialized to zero, then incremented by an exponential random variable with parameter λ to move to the next event time. So it is necessary to generate random variates from the exponential distribution. First, find the cumulative distribution function for the exponential distribution.

$$F(x) = \int_0^x \lambda e^{-\lambda w} \, dw = 1 - e^{-\lambda x} \qquad x > 0.$$

Next, equate the cumulative distribution function to u and solve for x:

$$u = 1 - e^{-\lambda x}$$

or
$$x = -\frac{1}{\lambda}\ln(1-u).$$

Combining this into an algorithm for generating the event times of a Poisson process on $(0, s]$, yields the pseudocode given below. Indentation denotes nesting.

$T_0 \leftarrow 0$
$i \leftarrow 0$
while $T_i \leq s$
    $i \leftarrow i + 1$
    generate $U_i \sim U(0,1)$
    $T_i \leftarrow T_{i-1} - \ln(1-U_i)/\lambda$
return $T_1, T_2, \ldots, T_{i-1}$

The U_i values are uniformly distributed between 0 and 1 and are provided by a random number generator. Using Sven & Barry's arrival rate of $\lambda = 30$ customers per hour, The R code for generating arrival times during an hour is given below.

```
lambda = 30
s      = 1
event  = rexp(1, lambda)
time   = event
while (event < s) {
  event = event + rexp(1, lambda)
  if (event < s) time = c(time, event)
}
print(time)
```

After a call to set.seed(9), the following 22 event times (read row wise and measured in hours) are returned.

0.0057	0.0361	0.0512	0.0638	0.1434	0.1561	0.1881
0.2063	0.2876	0.3573	0.4026	0.4722	0.5167	0.5354
0.5625	0.5934	0.6409	0.6840	0.6964	0.7253	0.9476
0.9570.						

These arrival times are plotted in Figure 4.15 using a scale measured in minutes. Three observations can be made concerning this particular realization of the Poisson process. First, this is a particularly light arrival hour. The expected number of arrivals is 30, but this hour only generated 22 arrivals. Second, the clustering due to the exponential time between arrivals is present. Third, there is the occasional long gap, such as that between customers 20 and 21 in Figure 4.15.

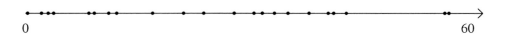

0   60

Figure 4.15: Poisson process realization.

Theorem 4.4 (Erlang time to the nth event) For a Poisson process with rate λ, the time of the nth event has probability density function

$$f(x) = \frac{\lambda(\lambda x)^{n-1}e^{-\lambda x}}{(n-1)!} \qquad x > 0.$$

As will be seen in the next chapter, this distribution is known as the *Erlang* distribution, which is the probability distribution of the sum of n independent exponential(λ) random variables.

Theorem 4.5 (superpositioning) Two independent Poisson processes with rates λ_1 and λ_2 whose event times are combined together (superpositioned) form a Poisson process with rate $\lambda_1 + \lambda_2$.

Theorem 4.5 generalizes to more than just two Poisson processes. The superposition of several Poisson processes is also a Poisson process whose rate is the sum of the constituent rates. Figure 4.16 shows the events from three Poisson processes being combined into a single Poisson process on the axis at the bottom of the figure. Calls to a switchboard (911 emergency service calls, for example), are in fact the superposition of the calls from each telephone in the service area.

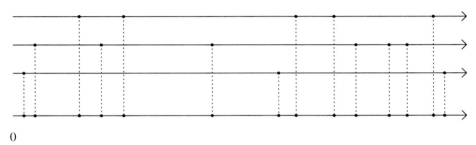

0

Figure 4.16: The superposition of the realizations of three Poisson processes.

Theorem 4.6 (decomposition) A Poisson process with rate λ whose events are subdivided into two processes by the results of independent Bernoulli(p) trials yields two Poisson processes with rates $(1 - p)\lambda$ and $p\lambda$.

Theorem 4.6 allows a modeler to break a Poisson process into sub-processes by using the flip of a biased coin to classify to which process to send an event. Sven and Barry, for example, could subdivide their customers by gender (male or female), age (child or adult), or product type (cup or cone). A decomposition of a Poisson process can also be generalized to more than just the outcomes success and failure. A biased n-sided die can be cast which subdivides each event into one of n Poisson processes with rates that sum to λ. Just as superpositioning is used combine Poisson processes together to form another Poisson process, decomposition is used to break a Poisson process into multiple Poisson processes with lower arrival rates.

Theorem 4.7 (equivalence to uniform order statistics) For a Poisson process with rate λ, the distribution of the event times on the time interval $(0, s]$ is identical to a Poisson(λs) number of event times uniformly distributed between 0 and s.

Theorem 4.7 allows a realization of a Poisson process to be generated in the following fashion: (a) generate a Poisson(λs) random variate N, the random number of arrivals on the interval $(0, s]$, (b) generate N continuous random variables that are uniformly distributed between 0 and s, then (c) sort the continuous random variables in ascending order. The sorted values are the event times in the Poisson process.

Example 4.20 Use Theorem 4.7 to generate the customer arrival times in a Poisson process over $s = 1$ hour at $\lambda = 30$ arrivals per hour.

The arrival times can be generated with just two R statements. The first statement generates the number of arrivals during the hour; the second statement generates the arrival times.

```
n = rpois(1, 30)
time = sort(runif(n))
```

The code is executed after a call to set.seed(3). The 33 arrivals during the hour are plotted in Figure 4.17.

0   60

Figure 4.17: A realization of a Poisson process.

To review the results associated with a Poisson process, there are three ways to view a Poisson process with rate λ.

(a) The number of events in the interval $(a, b]$ is Poisson with parameter $\lambda(b - a)$.

(b) The time between events is exponential(λ).

(c) The event times have the same distribution as a Poisson number of random variables that are uniformly distributed over the interval.

In addition, Poisson processes can be combined (superpositioning) to form a Poisson process or divided (decomposition) to form several Poisson processes.

We end this subsection on Poisson processes with a well-known data set involving deaths of Prussian soldiers by horse kicks.

Example 4.21 A late nineteenth century data set gathered by Prussian officials on hazards that horses posed to cavalry soldiers has been used to illustrate the fit of a Poisson process to a data set. Officials monitored 10 cavalry corps over 20 years, creating the equivalent of 200 corps-years worth of data. The number of annual deaths due to horse kicks x and the associated number of observations are given in Table 4.3. Does the Poisson distribution provide an adequate description for this data set?

x	observed
0	109
1	65
2	22
3	3
4	1
	200

Table 4.3: Observed frequency of deaths from horse kicks.

The first step is to assess the plausibility of the Poisson assumptions. First, do horse-kick deaths occur at a rate that is constant over time? Since the horses do not conspire against the soldiers in any manner, this is a reasonable assumption assuming that the number of horses and soldiers in a corps remains stable. Second, is the process "memoryless" in the sense that the number of deaths in one time interval is independent of the number of deaths in another nonoverlapping time interval. Again, this assumption seems consistent with a Poisson process. In order to see if the Poisson distribution fits the observed data well, the Poisson distribution needs to be fitted to the data set. Although it is a topic from statistics that is beyond the scope of this text, it would not be unreasonable to use the sample mean number of annual fatalities

$$\hat{\lambda} = \frac{(109)(0) + (65)(1) + (22)(2) + (3)(3) + (1)(4)}{200} = \frac{122}{200} = 0.61$$

fatalities per corps-year to estimate λ, the underlying population mean number of annual horse-kick fatalities. The R statements

```
x = 0:4
200 * dpois(x, 0.61)
```

can be used to add a third column to the table containing the data. Table 4.4 includes the fitted Poisson probabilities that are predicted by the Poisson model. The predicted values are rounded to the nearest integer. This is a spectacular agreement between the prediction from the Poisson probability model and the observed data. We can indeed conclude that the Poisson distribution is an appropriate model—horse kick deaths are likely to be a Poisson process over time.

x	observed	predicted
0	109	109
1	65	66
2	22	20
3	3	4
4	1	1
	200	200

Table 4.4: Observed vs. predicted frequency of deaths from horse kicks.

4.6 Hypergeometric Distribution

The binomial distribution can be applied to probability problems involving sampling n items with replacement from an urn containing two different types of items. What happens if the sampling is performed without replacement?

Example 4.2 concerned a sample of five taken from a dozen eggs. The binomial distribution was the appropriate model because the sample of five eggs was taken with replacement. But this is a strange way to sample the eggs. The hypergeometric distribution concerns sampling without replacement, which is more natural in this context. If the dozen eggs contain three bad eggs, the probability of obtaining exactly x defective eggs in a sample of size 5 is

$$f(x) = P(X = x) = \frac{\binom{3}{x}\binom{9}{5-x}}{\binom{12}{5}} \qquad x = 0, 1, 2, 3.$$

We now extend this example to the more general case. Using Bernoulli trial terminology, an urn contains N balls, m of which are "successes" and $N - m$ of which are "failures." A random sample of n of these balls is drawn without replacement from the urn. The meaning of a *random sample* in the case of sampling without replacement is that any of the balls that have not been sampled previously has an equal chance of being sampled. The probability of exactly x successes in the sample of size n is given by the associated value of the probability mass function.

Definition 4.9 A discrete random variable X with probability mass function *random of sample of n. N-m failures m successes N total.*

$$f(x) = \frac{\binom{m}{x}\binom{N-m}{n-x}}{\binom{N}{n}} \qquad x = 0, 1, 2, \ldots, n$$

for some nonnegative integer parameter N, $n = 0, 1, 2, \ldots, N$, and $m = 0, 1, 2, \ldots, N$, is a hypergeometric(m, N, n) random variable.

Some authors prefer to write the support of a hypergeometric random variable as follows:

$$f(x) = P(X = x) = \frac{\binom{m}{x}\binom{N-m}{n-x}}{\binom{N}{n}} \qquad x = \max\{0, m+n-N\}, \ldots, \min\{n, m\}.$$

The thinking associated with the upper bound of this definition of the support is that you can't get more successes than n in the sample of size n, but you also can't get more successes than the number of successes m in the population. Therefore the upper bound of the support is $\min\{n, m\}$. The thinking associated with the lower bound of this definition of the support is that you can't get fewer successes than 0 in the sample, but you are guaranteed to get at least $m + n - N$ successes if the population size is small enough. Therefore the lower bound of the support is $\max\{0, m+n-N\}$. The statement of the support in its much simpler form in Definition 4.9 is still correct, however, because we have assumed that $\binom{n}{x}$ is 0 whenever $x > n$ or $x < 0$ when the binomial coefficient was defined. The support given in Definition 4.9 is preferred for its simplicity and the fact that it is identical to the support for the binomial distribution.

The population mean for a hypergeometric random variable is

$$\mu = E[X] = \sum_{x=0}^{n} x \frac{\binom{m}{x}\binom{N-m}{n-x}}{\binom{N}{n}} = \frac{nm}{N}.$$

The summation can be calculated by using some combinatorial identities. Similarly, the population variance for a hypergeometric random variable is

$$\sigma^2 = V[X] = \frac{nm(N-n)(N-m)}{N^2(N-1)}.$$

The population skewness and kurtosis can also be calculated, but result in closed-form, rather messy expressions.

A thought experiment that proves to be helpful with the hypergeometric distribution is to consider a *very large* number of balls in the urn N. In this case, whether or not sampling is performed with or without replacement is largely irrelevant. So the formulas developed so far for the hypergeometric distribution should reduce to the corresponding formulas for the binomial distribution. Write the formula for the population mean as

$$\mu = n\frac{m}{N}.$$

This corresponds to the population mean of a binomial random variable $\mu = np$ because the ratio m/N is the probability of drawing a ball from the urn associated with success. Likewise, write the formula for the population variance as

$$\sigma^2 = n\left(\frac{m}{N}\right)\left(1 - \frac{m}{N}\right)\left(\frac{N-n}{N-1}\right).$$

The last term in this product goes to 1 in the limit as $n \to \infty$, and the remaining terms approach the associated population variance of the binomial random variable $\sigma^2 = np(1-p)$.

One of the primary application areas of the hypergeometric distribution is in an area within quality control known as *acceptance sampling*. The setting for acceptance sampling involves two entities: a producer and a consumer. The consumer purchases a lot consisting of N identical products from the producer. Before paying for the products, the consumer would like to test to see if the items are functioning properly. If it is too expensive to test all of the items, the consumer randomly selects n of the items to test prior to paying for the products. The hypergeometric distribution is appropriate in this setting because the random sampling of the products is performed without replacement. The parameter N is known as the *lot size* and the parameter n is known as the *sample size* in this setting. If the lot contains m defective products, then the hypergeometric random variable X models the probability of obtaining exactly x defective items in a sample of size n taken without replacement from a lot of N products.

We end this section on the hypergeometric distribution with two applications.

Example 4.22 A five-card hand is dealt from a well-shuffled 52-card deck. Let X be the number of diamonds in the hand. Find $P(X \leq 3)$.

The dealing of the five cards corresponds to sampling without replacement, so the use of the hypergeometric distribution is appropriate. The parameters in the hypergeometric distribution are $m = 13$, $N = 52$, and $n = 5$. So the probability mass function for the number of diamonds in the hand is

$$f(x) = \frac{\binom{13}{x}\binom{39}{5-x}}{\binom{52}{5}} \qquad x = 0, 1, 2, 3, 4, 5.$$

This probability mass function can be plotted with the R statements

```
x = 0:5
f = dhyper(x, 13, 39, 5)
plot(x, f, type = "h")
```

which is shown in Figure 4.18. The parameters in the R function dhyper are m, $N - m$, and n. The probability of dealing three or fewer diamonds is

$$P(X \leq 3) = \sum_{x=0}^{3} f(x) = \sum_{x=0}^{3} \frac{\binom{13}{x}\binom{39}{5-x}}{\binom{52}{5}} = \frac{969}{980} \cong 0.9888,$$

which is calculated as a floating point with the R statement phyper(3, 13, 39, 5) or as an exact fraction with the APPL statements

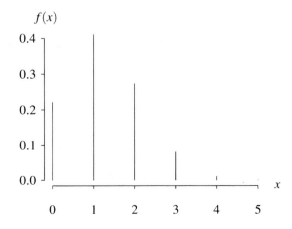

Figure 4.18: The probability mass function for $X \sim$ hypergeometric(13, 52, 5).

```
X := HypergeometricRV(52, 13, 5);
1 - (PDF(X, 4) + PDF(X, 5));
```

The meaning and order of the parameters in these two languages is not consistent, so it is worthwhile checking for consistency between the two.

Example 4.23 A biologist uses a "catch and release" program to estimate the population size of a particular animal in a region. During the catch phase, 20 animals are tagged. Several months later, 30 animals are captured, and 7 of them have tags. What is the most likely population size?

In order to use a hypergeometric model for this problem, several assumptions need to be in place.

- Enough time passes between the two catches so that the animals disperse throughout the region.
- No tags fall off between the two catches.
- No new animals enter the population by birth between the two catches.
- No animals exit the population by death between the two catches.
- No animals migrate into or out of the region.
- Both catches involve sampling without replacement. This is easier during the first catch because of the tags. During the second catch, numbered tags prevent resampling tagged animals, and untagged animals must be identified so as to prevent being captured more than once.
- The capturing of the animals is done in a fashion so that each uncaptured animal is equally likely to be captured.

If these assumptions are satisfied reasonably well, we can proceed to solve the problem using the hypergeometric distribution as a probability model. This particular setting is unique in that the population size N is the unknown. This problem is another example of the principle of "maximum likelihood," where we find the population size N that maximizes the probability of the result that we observed.

The second catch of $n = 30$ animals selected at random from a population of unknown size N results in X tagged animals taken from the population of $m = 20$ tagged animals (modeled as "successes" from the generic model) and $N - m = N - 20$ untagged animals. Using the hypergeometric model, the probability of capturing x tagged animals is

$$f(x) = P(X = x) = \frac{\binom{20}{x}\binom{N-20}{30-x}}{\binom{N}{30}} \qquad x = 0, 1, 2, \ldots, 20.$$

In the recapture experiment, the biologist captured $x = 7$ tagged animals, which occurs with probability

$$f(7) = P(X = 7) = \frac{\binom{20}{7}\binom{N-20}{23}}{\binom{N}{30}}.$$

We desire to find the value of N that makes this probability as large as possible. Intuitively, this means find the population size that makes our catching 7 tagged animals the most likely. The R code below plots the probability of capturing 7 animals for values of the population size N between 80 and 100. The graph is shown in Figure 4.19, which indicates that the most likely population size based on the results of this particular catch and release program is $N = 86$.

```
N = 80:100
f = dhyper(7, 20, N - 20, 30)
plot(N, f, type = "h")
```

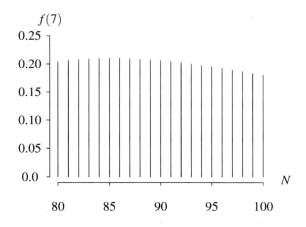

Figure 4.19: The probability mass function at $x = 7$ for $N = 80, 81, \ldots, 100$.

4.7 Other Distributions

There are several other discrete distributions that occasionally arise in probability problems, although they are not as widespread to warrant a separate section in this book. We outline four such distributions in this section: the discrete uniform distribution, the Benford distribution, the Zipf distribution, and mixture distributions. Choosing these four distributions is arbitrary. There are plenty of other good choices.

Discrete uniform distribution

Occasions arise when it is useful to have a discrete distribution that is equally likely over a sequence of consecutive integers. This distribution is known as the *discrete uniform* distribution with integer parameters a and b that denote the lower and upper limits of the support. The probability mass function of a random variable X with the discrete uniform distribution is

$$f(x) = \frac{1}{b-a+1} \qquad x = a, a+1, \ldots, b$$

for integers $a \le b$. When $a = 1$ and $b = 6$, this corresponds to the outcome of rolling a fair die. When $a = 1$ and $b = 15$, this corresponds to the number on a single draw from a bag containing 15 billiard balls numbered 1, 2, ..., 15. The only distribution from the previous section that coincides with the discrete uniform distribution is the Bernoulli distribution when $a = 0$, $b = 1$, and $p = 1/2$, which corresponds to the toss of a fair coin. The moment generating function of the discrete uniform distribution is

$$M(t) = E\left[e^{tX}\right] = \sum_{x=a}^{b} e^{tx} \frac{1}{b-a+1}$$

for $-\infty < t < \infty$. The rth derivative of the moment generating function with respect to t is

$$M^{(r)}(t) = \frac{1}{b-a+1} \sum_{x=a}^{b} x^r e^{tx}$$

for $-\infty < t < \infty$. Evaluating the rth derivative of the moment generating function at $t = 0$ yields the rth moment about the origin

$$E\left[X^r\right] = M^{(r)}(0) = \frac{1}{b-a+1} \sum_{x=a}^{b} x^r.$$

After some algebra, this leads to the population mean and variance

$$\mu = \frac{a+b}{2} \qquad \text{and} \qquad \sigma^2 = \frac{(b-a+1)^2 - 1}{12}.$$

The population skewness of the discrete uniform distribution is 0 because the probability mass function is symmetric. Finally, after some algebra, the population kurtosis is

$$E\left[\left(\frac{X-\mu}{\sigma}\right)^4\right] = 3 - \frac{6}{5}\left(\frac{(b-a+1)^2 + 1}{(b-a+1)^2 - 1}\right).$$

The kurtosis approaches 1.8 in the limit as $a \to -\infty$ and $b \to \infty$.

Example 4.24 A fair die is rolled three times. Find the probability that the sum of the spots on the up faces that appear is 5 or less.

By the multiplication rule, there are $6^3 = 216$ equally-likely outcomes, where the order of the outcomes is considered to be significant. Of these equally-likely outcomes, these 10 correspond to a total of 5 or fewer spots showing in the three rolls:

$$(1,1,1), (1,1,2), (1,2,1), (2,1,1), (1,1,3), (1,3,1), (3,1,1), (2,2,1), (2,1,2), (1,2,2).$$

So the probability that the total number of spots that appear in the three rolls is 5 or less is

$$P(\text{rolling 5 or less}) = \frac{10}{216} = \frac{5}{108}.$$

The discrete uniform distribution is preloaded into APPL. The previous probability can be calculated with a call to the Convolution function, which calculates the distribution of the sum of the spots showing on the three tosses.

```
X := UniformDiscreteRV(1, 6);
Y := Convolution(X, X, X);
CDF(Y, 5);
```

Although this problem was easily worked by hand, what if the fair die was rolled 15 times and the interest was in the probability of a total of 49 or fewer spots appearing on all of the rolls? There must be an better way than counting which of the $6^{15} = 470184984576$ possible outcomes correspond to 49 or fewer spots showing. In this case, a call to ConvolutionIID, which internally calls Convolution in a loop, can be used to solve the problem.

```
X := UniformDiscreteRV(1, 6);
Y := ConvolutionIID(X, 15);
CDF(Y, 49);
```

which yields

$$\frac{17059583999}{52242776064} \cong 0.3265.$$

This problem is much more difficult to solve by hand. The analytic solution is supported by the R Monte Carlo simulation code given below.

```
nrep  = 100000
count = 0
for (i in 1:nrep) {
  total = 0
```

```
      for (j in 1:15) total = total + ceiling(runif(1, 0, 6))
      if (total <= 49) count = count + 1
   }
count / nrep
```

Five runs of this code after a call to `set.seed(3)` yields

0.3253	0.3269	0.3260	0.3262	0.3262.

These values hover around the analytic solution of 0.3265, supporting the analytic solution via APPL.

Benford distribution

There has been increasing attention focused on the distribution of the leading digit of the values in a data set. The leading digit is the first nonzero digit, which falls between 1 and 9. For example, 365 has leading digit 3 and 0.0098 has leading digit 9. General Electric physicist Frank Benford published an article in 1938 where he analyzed the leading digits of $n = 20,229$ data values that he had gathered from a divergent set of sources (for example, populations of counties, American League baseball statistics, numbers appearing in *Reader's Digest*, areas of rivers, death rates, atomic weights). The proportions associated with each of the leading digits are given in Table 4.5.

1	2	3	4	5	6	7	8	9
0.306	0.185	0.124	0.094	0.080	0.064	0.051	0.049	0.047

Table 4.5: Benford's leading digit frequencies.

It may come as a surprise to you that the leading digits are so strongly amassed toward the lower digits. None of the discrete distributions introduced in this chapter adequately model this particular data set. First of all, except for the discrete uniform distribution (which would *not* model this data well), none of them have a support that ranges from 1 to 9. Second, this is a rather unique distribution of probability on this support and is not well modeled by a general-purpose distribution.

Benford conjectured that the distribution of the leading digit X has a distribution with probability mass function

$$f(x) = \log_{10}\left(1 + \frac{1}{x}\right) \qquad x = 1, 2, \ldots, 9$$

which we call the "Benford distribution." Any data set that follows this particular pattern for its leading digits satisfies what has become known as "Benford's law." This distribution has over 30% of its mass at $x = 1$ and less than 5% of its mass at $x = 9$. The probability mass function is monotonically decreasing from $x = 1$ to $x = 9$, and the values of $f(1), f(2), \ldots, f(9)$ are shown in Table 4.6. The values of the probability mass function associated with X show astonishing agreement with the huge data set that Benford collected.

Benford apparently independently arrived at the same conclusion in 1938 as Simon Newcomb did in 1881. Newcomb was an astronomer and mathematician who noticed "how much faster the first pages (of tables of logarithms) wear out than the last ones" leading to the counter-intuitive conclusion that the first significant digit in the values in a logarithm table is not uniformly distributed

1	2	3	4	5	6	7	8	9
0.301	0.176	0.125	0.097	0.079	0.067	0.058	0.051	0.046

Table 4.6: Probability mass values for the Benford distribution.

between 1 and 9. Using a heuristic argument, he developed the "logarithm law," which was exactly Benford's probability mass function, which he published in the *American Journal of Mathematics*.

Example 4.25 Develop an efficient random variate generation algorithm for the Benford distribution.

Using the fact that the logarithm of a quotient can be expressed as a difference of logarithms, the probability mass function can be written as a difference

$$f(x) = \log_{10}\left(1 + \frac{1}{x}\right) = \log_{10}\left(\frac{x+1}{x}\right) = \log_{10}(x+1) - \log_{10}(x)$$

for $x = 1, 2, \ldots, 9$. This means that the cumulative distribution function of X is

$$F(x) = P(X \le x) = \log_{10}(x+1)$$

for $x = 1, 2, \ldots, 9$. So a Benford random variate X can be generated by

$$X \leftarrow \lfloor 10^U \rfloor$$

where U is a random number that is uniformly distributed between 0 and 1. The R code below generates 500,000 Benford random variates and tabulates the fraction associated with the digits $x = 1, 2, \ldots, 9$.

```
nrep = 500000
f = rep(0, 9)
x = floor(10 ^ runif(nrep))
for (i in 1:nrep) f[x[i]] = f[x[i]] + 1
print(f / nrep)
```

One run of this simulation results in the estimated values for the probability mass function in Table 4.7. These values appear to be converging to the associated values for the population Benford probability mass function, so the random variate generation algorithm is considered to be verified.

1	2	3	4	5	6	7	8	9
0.302	0.176	0.124	0.097	0.078	0.067	0.058	0.051	0.046

Table 4.7: Simulated Benford random variate frequencies.

The Benford distribution is the first discrete distribution that we have encountered in this chapter that has no parameters. It is used solely for describing the distribution of the leading digit, which it has adequately modeled for a wide variety of data sets.

Benford's law has found applications in detecting accounting fraud and election fraud. If numbers are fudged because of human intervention, a lack of fit to Benford's law might become apparent. This lack of fit can be detected by subjecting the leading digits to a statistical goodness-of-fit test.

Zipf distribution

The Zipf distribution was created to account for the relative popularity of a few members of a population and the relative obscurity of other members of a population. Examples include the following.

- There are a few websites that get lots of hits, a greater number of websites that get a moderate number of hits, and a vast number of websites that hardly get any hits at all.

- A library has a few books that everyone wants to borrow (best sellers), a greater number of books that get borrowed occasionally (classics), and a vast number of books that hardly ever get borrowed.

- A natural language has a few words that are used with high frequency ("the" and "of" rank first and second in English), a greater number of words that get used with lower frequency (like "butter" and "joke"), and a vast number of words that hardly ever get used at all (like "defenestrate" which means to throw out of a window, "lucubration" which means a study or composition lasting late into the night, or "mascaron" which means a grotesque face on a door-knocker).

- The world population lives in several large cities, a greater number of medium-sized cities, and a vast number of small towns.

The probability mass function of the Zipf distribution is

$$f(x) = \frac{c}{x^s} \qquad x = 1, 2, \ldots, n$$

for the parameters $s > 0$ and a positive integer n. A normalizing constant c is chosen so that the $f(x)$ values sum to 1 over the support, that is

$$c = \left[\sum_{i=1}^{n} \left(\frac{1}{i} \right)^s \right]^{-1}.$$

Example 4.26 Assume that the English language has $n = 1,000,000$ words. Using the classic form of the Zipf distribution with $s = 1$ and assuming that the Zipf distribution adequately describes word frequency, find

(a) the frequency of occurrence of the top eight words in the language,

(b) the number of words necessary to capture half of the language.

The probability mass function of the random variable X, which represents the rank of words in terms of frequency, is

$$f(x) = \frac{c}{x} \qquad x = 1, 2, \ldots, 1000000.$$

The value of the constant c can be determined with the Maple statement

```
c := evalf(1 / sum(1 / i, i = 1 .. 1000000));
```

which yields $c \cong 0.06947953779$.

(a) The frequency of occurrence of the top eight words in the English language are given by $f(x)$ for $x = 1, 2, \ldots, 8$, and are listed in Table 4.8. The Zipf distribution predicts that the most frequent word, "the," appears about 7% of the time. The second most frequent word, "of," appears half as often as "the." The third most frequent word, "and," appears one third as often as "the."

1	2	3	4	5	6	7	8
the	of	and	a	to	in	is	you
0.0695	0.0347	0.0232	0.0174	0.0139	0.0116	0.0099	0.0087

Table 4.8: Word frequencies predicted by the Zipf distribution.

(b) The R code shown below computes the number of words necessary to capture half of the language. The variable cc is used for the constant c because c is reserved for the collect function, which is used to define vectors.

```
cc  = 0.06947953779
sum = 0
x   = 1
while(sum < 0.5) {
   sum = sum + cc / x
   x = x + 1
}
print(x - 1)
```

This code returns 749. It takes only 749 of the one million words in the English language to cover 50% of the words that appear in text if Zipf's law holds with the parameters given. This is an example of the significant few and the insignificant many when it comes to word frequency.

Mixture distribution

The last of the four distributions covered in this section is a specific example of a mixture distribution. There are dozens of possible mixtures that could be considered, but we illustrate just one here by example. As mentioned in Chapter 3, mixture distributions are an important part of statistical theory known as Bayesian statistics.

Example 4.27 Consider a binomial random variable X associated with a fixed number of Bernoulli trials n. You are so unsure of the value of p that you can only conclude that it is uniformly distributed between 0 and 1. What is the probability mass function of the number of successes when p is itself a random variable?

In this case, the distribution of X given a particular value of p is binomial with probability mass function

$$f_{X\mid P}(x\mid p) = \binom{n}{x} p^x (1-p)^{n-x} \qquad x = 0, 1, 2, \ldots, n.$$

The probability density function of the continuous random variable P, which is the probability of success on each Bernoulli trial, is

$$f_P(p) = 1 \qquad 0 < p < 1.$$

Using the continuous mixtures formula from Section 3.3, the probability mass function of X is

$$\begin{aligned}
f_X(x) &= \int_0^1 \binom{n}{x} p^x (1-p)^{n-x} \cdot 1 \, dp \\
&= \binom{n}{x} \int_0^1 p^x (1-p)^{n-x} \, dp \\
&= \frac{n!}{x!(n-x)!} \cdot \frac{x!(n-x)!}{(n+1)!} \\
&= \frac{1}{n+1} \qquad x = 0, 1, 2, \ldots, n.
\end{aligned}$$

This indicates that if the probability of success on each Bernoulli trial is a random variable P that is uniformly distributed between 0 and 1, then the unconditional distribution of X is a discrete uniform distribution between 0 and n.

To check the analytic result by Monte Carlo simulation for a fixed n value, say $n = 5$, use the following R code.

```
nrep = 1000000
f = rep(0, 6)
for (i in 1:nrep) {
  p = runif(1)
  x = rbinom(1, 5, p)
  f[x + 1] = f[x + 1] + 1
}
print(f / nrep)
```

Running just one iteration of this simulation results in the following estimate, with mass values rounded to three digits, for the probability mass function of X:

$$\hat{f}_X(x) = \begin{cases} 0.166 & x = 0 \\ 0.167 & x = 1 \\ 0.167 & x = 2 \\ 0.168 & x = 3 \\ 0.166 & x = 4 \\ 0.167 & x = 5. \end{cases}$$

The hat on f is used to show that this is an *estimate* for the true probability mass function. These simulation results lend plausibility to the analytic result that the six outcomes are from a discrete uniform distribution

$$f_X(x) = \frac{1}{6} \qquad x = 0, 1, 2, 3, 4, 5.$$

A sampling of discrete distributions that commonly arise in probability applications has been introduced in this chapter. The key discrete distributions presented in this chapter are summarized in Table 4.9, including both parameterizations of the geometric and negative binomial distributions.

Distribution	$f(x)$	Support	Mean	Variance
Bernoulli(p)	$p^x(1-p)^{1-x}$	$x = 0, 1$	p	$p(1-p)$
binomial(n, p)	$\binom{n}{x} p^x(1-p)^{n-x}$	$x = 0, 1, \ldots, n$	np	$np(1-p)$
geometric(p)	$p(1-p)^x$	$x = 0, 1, 2, \ldots$	$\dfrac{1-p}{p}$	$\dfrac{1-p}{p^2}$
Geometric(p)	$p(1-p)^{x-1}$	$x = 1, 2, \ldots$	$\dfrac{1}{p}$	$\dfrac{1-p}{p^2}$
negative binomial(r, p)	$\binom{x+r-1}{r-1} p^r(1-p)^x$	$x = 0, 1, 2, \ldots$	$\dfrac{r(1-p)}{p}$	$\dfrac{r(1-p)}{p^2}$
Negative binomial(r, p)	$\binom{x-1}{r-1} p^r(1-p)^{x-r}$	$x = r, r+1, \ldots$	$\dfrac{r}{p}$	$\dfrac{r(1-p)}{p^2}$
Poisson(λ)	$\dfrac{\lambda^x e^{-\lambda}}{x!}$	$x = 0, 1, 2, \ldots$	λ	λ
hypergeometric(m, N, n)	$\dfrac{\binom{m}{x}\binom{N-m}{n-x}}{\binom{N}{n}}$	$x = 0, 1, \ldots, n$	$\dfrac{nm}{N}$	$\dfrac{nm(N-n)(N-m)}{N^2(N-1)}$

Table 4.9: Common discrete distributions.

4.8   Exercises

4.1   Let X be a Bernoulli(p) random variable. Find $E[\cos(\pi X)]$.

4.2   There are 31 days in October. In a room of 35 people, what is the probability mass function of X, the number of October birthdays? Ignore leap years. Assume that there are no twins, triplets, etc. in the room and that each birthday is equally likely.

4.3   Aaron is giving a cake to anyone who can roll k or more sixes when rolling 32 fair dice. Find the smallest k so that the probability of winning a cake is less than 0.01.

4.4   David rolls 8 fair dice. Find the probability that he rolls 7 or more of the same number.

4.5   There are 35 unrelated people in a room. Assuming that all 365 birthdays are equally likely and ignoring leap years, what is the probability that today is the birthday for one or more of the people in the room?

4.6   The probability of winning at the game craps is $244/495$. Find the probability of coming out ahead when you make five individual \$1 bets on five games of craps.

4.7   A TV tube has a random lifetime X with probability density function

$$f(x) = \lambda e^{-\lambda x} \qquad x > 0,$$

where λ is a positive parameter. If n TV tubes are sampled at random, find the probability that two or more tubes survive to time c, where $c > 0$.

4.8   The phrase "may the best team win" is often used prior to a professional sporting event. But oftentimes the best team loses. Particularly in a single-game, such as the Super Bowl, the underdog, or weaker team, may prevail. The national basketball association and major league baseball have partially alleviated this problem by playing a "best of seven" game series, where the first team to win four games wins the series. Consider an n-game series, where n is an odd, positive integer. Let p be the probability that Team I wins a single game, $0 < p < 1$. If the games are independent, give an expression for and draw a graph of the probability that the best team wins (on the vertical axis) versus p (on the horizontal axis) for $n = 1, 3, 5, 7$.

4.9   If the moment generating function for a random variable X is

$$M_X(t) = \left(\frac{1}{10} + \frac{9}{10} e^t \right)^n$$

for $-\infty < t < \infty$ and for some positive integer n, find $E[X]$ and $V[X]$.

4.10   Plot $E\left[(X - c)^2 \right]$ and $E\left[|X - c| \right]$ for $X \sim$ binomial$(3, 1/2)$ for $-1 < c < 4$.

4.11   Let X be the number of fives that appear in 100 rolls of a fair die.

(a) Write a mathematical expression for $P(10 \leq X \leq 20)$.

(b) Write an R statement to calculate $P(10 \leq X \leq 20)$.

4.12   Chet rolls a single die 600 times and gets only 81 ones. Based on these results, you suspect that his die is biased (as opposed to fair), yielding ones with some probability that is less than $1/6$. One way to test your hypothesis is to calculate the probability of getting 81 or fewer ones in 600 rolls assuming that the die is fair. If this probability is small, then there is statistically significant evidence that the die is biased. Calculate the probability of getting 81 or fewer ones in 600 rolls of a fair die and write a sentence or two interpreting the result.

4.13 Ian rides across the country on his bicycle in 60 days. Every day on his trip, he purchases two bottles of "Purple Passion" grape soda. If the soda company claims that the probability that a consumer will win a prize on one of their bottle cap contests is $1/12$, find the probability that Ian will win one or more prizes on his trip.

4.14 The World Series is being played by the Cubs and Sox. Let p be the probability that the Cubs win on a particular game, $0 < p < 1$. Assume that the outcome of each game is independent. If instead of the current "play until you win 4 games," the Commissioner is considering a marathon series consisting of "play until you win 20 games." Plot the probability that the Cubs win the series for both scenarios on $0 < p < 1$ on a single set of axes using R.

4.15 The 2004 United States Presidential election was predicted to be quite close. Assume that the incumbent, President George W. Bush, has 260 electoral votes secure. Assume that the challenger, Senator John Kerry, also has 260 electoral votes secure. The only "swing" states where the outcome of the election hangs in the balance are: Colorado (9 electoral votes), New Mexico (5 electoral votes), and New Hampshire (4 electoral votes). Assume that the outcome of each of these three swing states is a Bernoulli trial with parameter p, where p is the probability that President Bush wins the state. Use R to plot:

(a) the probability of a win for President Bush, a win for Senator Kerry, and a tie,

(b) the expected number of electoral votes for President Bush and Senator Kerry,

for $0 < p < 1$.

4.16 Consider an m-member jury that requires n or more votes to convict a defendant. Let p be the probability that a juror votes a guilty person innocent and let q be the probability that a juror votes an innocent person guilty, $0 < p < 1, 0 < q < 1$. Assuming that r is the fraction of guilty defendants and that jurors vote independently, what is

(a) the probability a defendant is convicted?

(b) the probability a defendant is convicted when $n = 9$, $m = 12$, $p = 1/4$, $q = 1/5$, and $r = 5/6$? Use R to calculate the result.

4.17 Let X be a binomial random variable with population mean $\mu = 3$ and population variance $\sigma^2 = 2$. Find $P(X \geq 2)$.

4.18 Ten fair coins are tossed. Each coin showing tails is tossed a second time. Finally, each coin still showing tails is tossed a third time. Let the random variable X be the number of heads showing after the third set of tosses.

(a) Find $f(x)$.

(b) Find $E[X]$.

(c) Write and execute a Monte Carlo simulation to give numerical support to your solution to part (b).

4.19 Some consider the digits after the decimal in $\pi = 3.14159265\ldots$ to be mutually independent random digits. If this conjecture is assumed to be true, write an R statement to calculate the probability that there will be 8 or fewer fours in the first 100 digits of after the decimal point in π.

4.20 A coin has a probability p of turning up heads when tossed. The coin is tossed repeatedly until a head appears. Let X denote the number of tails that occur before the first head is tossed. Find the probability mass function of Y, the remainder when X is divided by 4.

4.21 Let X denote the number of tails prior to the first occurrence of a head in repeated tosses of a fair coin. Find $P(X \bmod 5 = 2)$.

4.22   There is a bear in the woods. Some think the bear is dangerous, others do not. A hunter can hit the bear on each shot with $p = 0.4$. Each shot is independent (slow bear). What is the minimum number of bullets that the hunter should carry to be at least 75% certain that he can hit the bear?

4.23   Katie has decided to sell her car, so she places an ad in the local newspaper. She will sell the car upon the first offer that exceeds some constant c. If the offer amounts are independent, each with cumulative distribution function $F(x)$, what is the expected number of offers that it takes for Katie to sell her car?

4.24   Let the random variable X denote the number of independent rolls of a fair die required to obtain the *second* occurrence of a "five." Find $P(X \geq 4)$.

4.25   The probability of winning a game of craps is $244/495$. If Barbara plays craps repeatedly, find the probability that her third win will occur on the fifth game she plays.

4.26   Mr. Legonaton bets repeatedly on double-zero in roulette until he wins three times. If double-zero occurs with probability $1/38$, find the probability mass function, expected value, and population standard deviation of the number of bets that Mr. Legonaton will make.

4.27   Consider repeated independent Bernoulli trials with probability of success p. What is the probability mass function of X, the *number of trials before* the rth success, where r is a positive integer?

4.28   An automobile manufacturer is implementing a "zero-defects" drive in order to improve the quality of their automobiles. They have found that the number of manufacturing defects on each car follows a Poisson distribution. If their goal is to have at least 98 percent of their cars defect-free, find the largest population mean μ that achieves their goal.

4.29   Perform the following calculations using a calculator and by using the appropriate R function for the distributions listed below.

     (a) Binomial:
$$\sum_{x=0}^{3} \binom{8}{x} \left(\frac{3}{10}\right)^x \left(\frac{7}{10}\right)^{8-x}.$$

     (b) Geometric:
$$\sum_{x=2}^{4} \left(\frac{1}{3}\right) \left(\frac{2}{3}\right)^x.$$

     (c) Poisson:
$$\frac{9^3 e^{-9}}{3!}.$$

4.30   The number of parking tickets issued daily is a Poisson random variable with population mean 15. Write the mathematical expression and use R to find the probability that

     (a) exactly 17 tickets will be issued in one day,

     (b) 17 or more tickets will be issued in one day,

     (c) exactly 17 tickets will be issued in one day given that the population mean number of tickets was exceeded on that particular day,

     (d) 100 or fewer tickets will be issued in any five-day period.

4.31   Power failures in the state of Virginia during the month of March are well modeled by a Poisson process with a rate of $\lambda = 7$ failures per month. Given that there have been a total of 4 failures during the first 10 days of March, what is the probability that Virginia will have more than 12 power outages during the entire month of March?

4.32 The probability that the four aces will be the top four cards in a well-shuffled deck is

$$\frac{\binom{4}{4}\binom{48}{0}}{\binom{52}{4}} = \frac{4 \cdot 3 \cdot 2 \cdot 1}{52 \cdot 51 \cdot 50 \cdot 49} = \frac{24}{6,497,400} = \frac{1}{270,725}.$$

If one million well-shuffled decks are examined, find the probability that 3 or fewer of the decks will have the four aces as the top four cards in R using:

(a) the binomial distribution,

(b) the Poisson approximation to the binomial distribution.

Report your results to enough accuracy to show the number of digits of accuracy attained by the Poisson approximation.

4.33 Let $X \sim \text{Poisson}(\mu)$, where $f(0) = f(1)$. Find $P\left((X-3)^2 = 1\right)$.

4.34 Let X be a Poisson random variable with a population mean λ. Find the value of λ that satisfies $P(X = 0 \,|\, X \leq 2) = 1/8$.

4.35 Daily *demand* for Purple Passion grape soda is a Poisson random variable with population mean 18. If a vending machine can hold just 24 cans of the popular soda, find the expected daily *sales*. Assume that the machine is restocked daily at noon. Use R to perform the calculations.

4.36 Let X be a Poisson random variable with population mean λ. Find the probability that X is even. *Hint*: consider the expansion of $e^{-\lambda} + e^{\lambda}$.

4.37 Let X be a Poisson(λ) random variable. Find the value of λ that maximizes $P(X = x)$.

4.38 Show that a Poisson (λ) random variable X has a mode at $x = \lfloor \lambda \rfloor$ for noninteger λ and no mode when λ is an integer.

4.39 The number of customers arriving daily to TVs-R-Us is a Poisson(λ) random variable. If each customer's decision to buy a TV is an independent Bernoulli trial, where p is the probability of making a purchase, give an expression for the probability mass function of the number of TVs purchased daily.

4.40 The *probability generating function* of a discrete random variable X having support on the nonnegative integers is

$$G(t) = E\left[t^X\right].$$

Find the probability generating function of $X \sim \text{Poisson}(\lambda)$. *Hint:* The probability mass function for the Poisson distribution is

$$f(x) = \frac{\lambda^x e^{-\lambda}}{x!} \qquad x = 0, 1, 2, \dots.$$

4.41 Mrs. Sindy gives a spelling test to confirm that her students can spell basic words. Her test consists of n words, and a student passes if he or she makes 0, 1, or 2 errors.

(a) Let p be the true proportion of words that a student can spell correctly. Give a general expression for $P(\text{passing})$ for $0 < p < 1$ and $n = 2, 3, \dots$ assuming that the population of basic words is infinite. Make a plot of p on the horizontal axis versus $P(\text{passing})$ on the vertical axis for $n = 5$, 10, and 20.

(b) Give a general expression for $P(\text{passing})$ for $n = 2, 3, \dots$ assuming that there are 100 basic words and the students can spell r of these words correctly. Mrs. Sindy chooses, of course, to sample her words without replacement. Make a plot of the true proportion of words a student can spell correctly versus $P(\text{passing})$ for $n = 5$, 10, and 20.

4.42   Consider a "powerball" lottery, where five balls are drawn at random and without replacement from an urn containing 25 balls numbered $1, 2, \ldots, 25$. (Aside: many state lotteries use 85 rather than 25 balls.) A player pays \$1 to play the game, which is to guess the five numbers drawn without replacement from the urn. (Assume that the player is clever enough to choose five *different* numbers and that they are all in the range $1, 2, \ldots, 25$.)

(a) Calculate the probability that a single player will match exactly x of the numbers, for $x = 0, 1, 2, 3, 4, 5$.

(b) If the lottery pays out only for matching all five of the guesses, what is the appropriate prize value to make the lottery "fair."

(c) If the lottery pays out only for matching four or five of the guesses, and the payout for matching exactly four of the numbers is $1/3$ of the payout for matching exactly five of the numbers, what are the appropriate prize values to make the lottery "fair."

(d) If one individual purchases ten tickets and makes ten independent permutations of five guesses to play the game described in part (c), what is the 1st, 50th (median), and 99th percentiles of the player's winnings (negative values for losses) rounded to the nearest penny.

4.43   Consider an infinite sequence of random variables $X_1, X_2, \ldots$, each drawn from a common probability density function $f_X(x)$.

(a) Find the probability that the leading digit is 8 (for example, 0.8629 or 0.008435) of any element of the sequence if $f(x) = 1$ for $0 < x < 1$.

(b) Find the probability that the leading digit is 8 (for example, 0.8629 or 0.008435) of any element of the sequence if $f(x) = 2x$ for $0 < x < 1$.

(c) Let k be a positive integer. Let the random variable Y be the sequence number of the kth random variable that has leading digit 8. Find $E[Y]$ when $f_X(x) = 1$ for $0 < x < 1$.

4.44   Let X have the Benford distribution with probability mass function

$$f(x) = \log_{10}\left(1 + \frac{1}{x}\right) \qquad x = 1, 2, \ldots, 9.$$

(a) What value is returned by the APPL statement

```
IDF(BenfordRV(), 0.95);
```

(b) What value is returned by the APPL statement

```
HF(BenfordRV(), 8);
```

where the hazard function $h(x)$ is defined by $h(x) = P(X = x)/P(X \geq x) = f(x)/S(x)$, and $S(x)$ is the survivor function.

Chapter 5

Common Continuous Distributions

A number of continuous random variables have probability distributions that occur so frequently in practice that they have been named and given *parameters* to enhance their flexibility in solving probability problems. This chapter outlines common continuous distributions. For the distributions presented in this chapter, we will consider the support $\mathcal{A}$, the distribution's parameters, the probability density function $f(x)$, the cumulative distribution function $F(x) = P(X \leq x)$, the moment generating function $M(t) = E\left[e^{tX}\right]$, the population mean μ, and the population variance σ^2. Applications of each of the distributions will also be presented.

The distribution parameters are constants that determine the specific functional form of $f(x)$ or $F(x)$. Parameters in a continuous distribution allow modeling of such diverse applications as a light bulb failure time, a patient's post-surgery survival time, and a union strike duration by a single distribution. Parameters can be classified into three types for continuous random variables. *Location* (or *shift*) parameters are used to shift $f(x)$ or $F(x)$ to the left or right along the x-axis. If c_1 and c_2 are two values of a location parameter with cumulative distribution function $F(x; c)$, then there exists a real constant a such that $F(x; c_1) = F(a + x; c_2)$. *Scale* parameters are used to expand or contract the x-axis by a factor a. If c_1 and c_2 are two values for a scale parameter for a lifetime distribution with cumulative distribution function $F(x; c)$, then there exists a real constant a such that $F(ax; c_1) = F(x; c_2)$. *Shape* parameters are appropriately named because they affect the shape of the probability density function. In summary, location parameters *translate* distributions along the x-axis; scale parameters *expand* or *contract* the scale for distributions; shape parameters alter the shape of $f(x)$.

5.1 Uniform Distribution

A continuous random variable X is "uniformly distributed" between a and b if X is equally likely to assume any value on the interval (a, b). Consider, for example, a bicycle tire of radius r. If a puncture is equally likely to occur anywhere around the circumference of the tire, then the probability density function of the puncture position X is given by

$$f(x) = \frac{1}{2\pi r} \qquad 0 < x < 2\pi r.$$

This probability model requires that the circumference of the tire be flattened into one dimension and that the points along the circumference be measured from a fixed point in one particular direction (say from the stem measured clockwise). The uniform distribution is generalized in the following definition to model a random variable assuming a value with uniform likelihood between a minimum value a and a maximum value b.

Definition 5.1 A continuous random variable X with probability density function

$$f(x) = \frac{1}{b-a} \qquad a < x < b$$ *don't forget the support.*

for real constants a and b satisfying $a < b$ is a $U(a, b)$ random variable.

As in Chapter 4, we again use the $\sim$ as shorthand in $X \sim U(a, b)$ to indicate that the random variable X is a continuous random variable that is uniformly distributed between a and b. The constants a and b are the parameters in the uniform distribution.

The cumulative distribution function for $X \sim U(a, b)$ is mathematically tractable. On $a < x < b$, the cumulative distribution function is

$$F(x) = \int_a^x \frac{1}{b-a}\,dw = \left[\frac{w}{b-a}\right]_a^x = \frac{x-a}{b-a}.$$ *cdf.*

So the cumulative distribution function of X over the entire range of x-values is

$$F(x) = \begin{cases} 0 & x \le a \\ \dfrac{x-a}{b-a} & a < x < b \\ 1 & x \ge b. \end{cases}$$

The moment generating function for $X \sim U(a, b)$ is

$$
\begin{aligned}
M(t) &= E\left[e^{tX}\right] \\
&= \int_a^b e^{tx}\frac{1}{b-a}\,dx \\
&= \frac{1}{b-a}\left[\frac{1}{t}e^{tx}\right]_a^b \\
&= \frac{e^{bt} - e^{at}}{t(b-a)}.
\end{aligned}
$$

This moment generating function gives the indeterminate form $0/0$ for $t = 0$, but using L'Hôpital's rule results in $M(0) = 1$ (to assure continuity of $M(t)$ at $t = 0$). So the moment generating function of X is

$$M(t) = \begin{cases} 1 & t = 0 \\ \dfrac{e^{bt} - e^{at}}{t(b-a)} & t \ne 0, \end{cases}$$

which exists in a neighborhood about $t = 0$.

The population mean and variance can be computed by taking derivatives of the moment generating function or by using the definition of the expected value. Using the latter approach, the population mean is

$$
\begin{aligned}
\mu &= E[X] \\
&= \int_a^b x f(x)\,dx \\
&= \int_a^b \frac{x}{b-a}\,dx \\
&= \left[\frac{x^2}{2(b-a)}\right]_a^b \\
&= \frac{b^2 - a^2}{2(b-a)} \\
&= \frac{a+b}{2}
\end{aligned}
$$

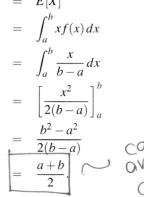 *can see this as an analogy of the center of mass.*

$a \xmapsto{\hspace{2cm}} b$

$\dfrac{a+b}{2}$

$\uparrow$

C. m

This result is consistent with the center-of-gravity interpretation of the population mean; the balance point is the average of the minimum value a and the maximum value b. Similarly, but omitting the integration steps, the population variance is

$$\sigma^2 = V[X] = E\left[(X - \mu)^2\right] = \int_a^b \left(x - \frac{a+b}{2}\right)^2 \frac{1}{b-a} dx = \frac{(b-a)^2}{12}.$$

Finally, the population skewness and kurtosis are

$$E\left[\left(\frac{X - \mu}{\sigma}\right)^3\right] = 0 \qquad \text{and} \qquad E\left[\left(\frac{X - \mu}{\sigma}\right)^4\right] = \frac{9}{5}.$$

The population skewness of zero is consistent with the fact that the uniform distribution has a symmetric probability density function. The population skewness and kurtosis do not involve the parameters a and b because X is standardized. The moments can also be calculated via the APPL code shown below.

```
X := UniformRV(a, b);
Mean(X);
Variance(X);
Skewness(X);
Kurtosis(X);
```

Example 5.1 A shuttle train at a busy airport completes a circuit between two terminals every five minutes. What is the probability that a passenger will wait more than three minutes for a shuttle train?

It is reasonable to assume that a passenger selected at random has an arrival time to the shuttle train that is uniformly distributed between 0 and 5 minutes measured in time elapsed since the last shuttle train departure. So the probability density function of a passenger waiting time X is

$$f(x) = \frac{1}{5} \qquad 0 < x < 5$$

and the probability of a wait time that is longer than three minutes, which is shaded in Figure 5.1, is

$$P(X > 3) = \int_3^5 \frac{1}{5} dx = \left[\frac{x}{5}\right]_3^5 = 1 - \frac{3}{5} = \frac{2}{5}.$$

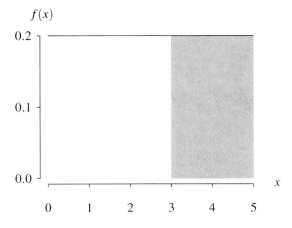

Figure 5.1: Probability density function for $X \sim U(0, 5)$ with $P(3 < X < 5)$ shaded.

Example 5.2 What is the probability that the quadratic equation $x^2 + Bx + 1 = 0$ has two real roots, where $B \sim U(0, 3)$?

The quadratic equation $ax^2 + bx + c = 0$ has two real roots if the discriminant $b^2 - 4ac$ is positive. Thus the quadratic equation $x^2 + Bx + 1 = 0$ has two real roots if $B^2 - 4 > 0$. Since the probability density function of B is

$$f(b) = \frac{1}{3} \qquad\qquad 0 < b < 3,$$

the probability that $x^2 + Bx + 1 = 0$ has two real roots is

$$
\begin{aligned}
P\left(B^2 - 4 > 0\right) &= P\left(B^2 > 4\right) \\
&= P(B > 2) \\
&= \int_2^3 \frac{1}{3}\, db \\
&= \frac{1}{3}.
\end{aligned}
$$

Example 5.3 Let $X \sim U(0, 1)$. Find $V\left[3\lfloor 2X\rfloor + 4\right]$.

The expression inside of the variance operator is complicated, so we work from the inside out. Common sense indicates that if $X \sim U(0, 1)$, then $2X \sim U(0, 2)$. But since common sense can sometimes be wrong, we use the cumulative distribution function technique to verify the hunch. Let $Y = 2X$. The cumulative distribution function of Y is

$$
\begin{aligned}
F_Y(y) &= P(Y \le y) \\
&= P(2X \le y) \qquad\qquad \text{\emph{Good Example}} \\
&= P(X \le y/2) \\
&= F_X(y/2) \\
&= y/2 \qquad\qquad 0 < y < 2.
\end{aligned}
$$

This cumulative distribution function can be recognized as that of a $U(0, 2)$ random variable, so our common sense is, in this case, confirmed. Next, what is the distribution of the floor of a $U(0, 2)$ random variable? If Y falls between 0 and 1, the floor is 0; if Y falls between 1 and 2, the floor is 1. So $\lfloor 2X\rfloor$ assumes the value 0 with probability $1/2$ and assumes the value 1 with probability $1/2$. Thus $\lfloor 2X\rfloor$ is a Bernoulli$(1/2)$ random variable. Using the properties of the variance operator and the fact that the variance of a Bernoulli(p) random variable is $p(1 - p)$, the desired population variance is

$$V\left[3\lfloor 2X\rfloor + 4\right] = V\left[3\lfloor 2X\rfloor\right] = 9V\left[\lfloor 2X\rfloor\right] = 9 \cdot \frac{1}{2} \cdot \frac{1}{2} = \frac{9}{4}.$$

To verify that we have correctly analyzed this problem, we check the solution by Monte Carlo simulation. The R code to check the solution is given below.

```
nrep = 100000
var(3 * floor(2 * runif(nrep)) + 4)
```

After a call to `set.seed(3)` to initialize the random number generator stream, five runs of this code yields the following estimates of the population variance:

   2.25002   2.24993   2.25002   2.25001   2.24999.

These values support the analytic result $V\left[3\lfloor 2X\rfloor + 4\right] = 9/4 = 2.25$.

Example 5.4 Divide a line segment of unit length randomly into two parts. Find the expected value of the product of the lengths of the two segments.

This problem has a physical interpretation. Select a point at a random position X on a fence of unit length. The random area of a rectangular pig pen that can be made by using the two pieces of fencing (one of length X and the other of length $1 - X$) is $X(1 - X)$ when placed against two walls that meet at a right angle, as shown in Figure 5.4. We want the expected value of this random area. When X is close to 0 or 1, the pigs would need to be rather skinny. The random area for the pig pen ranges from 0 to $1/4$.

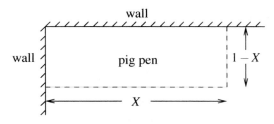

Figure 5.2: Pig pen geometry.

The probability density function of the breakpoint X is

$$f(x) = 1 \qquad 0 < x < 1.$$

Note that $E[X] = E[1 - X] = 1/2$. The expected value of $X(1 - X)$ is

$$
\begin{aligned}
E\big[X(1 - X)\big] &= \int_0^1 x(1 - x)\,dx \\
&= \left[\frac{x^2}{2} - \frac{x^3}{3} \right]_0^1 \\
&= \frac{1}{6},
\end{aligned}
$$

which is *not* equal to $E[X]E[1 - X] = (\frac{1}{2})^2 = \frac{1}{4}$, the largest possible pig pen area. This result can be confirmed with the APPL code:

```
X := UniformRV(0, 1);
g := x -> x * (1 - x);
ExpectedValue(X, g);
```

which also yields $E[X(1 - X)] = 1/6$. The Monte Carlo simulation code in R given below can also provide an estimate of the expected value.

```
nrep = 100000
x = runif(nrep)
mean(x * (1 - x))
```

Five replications of this simulation yield the following estimates for the population mean area of the pig pen:

0.16671	0.16630	0.16649	0.16657	0.16666.

These results are consistent with the analytic result $E[X(1 - X)] = 1/6$. The R statement `mean(runif(nrep) * (1 - runif(nrep)))` would *not* be the correct way to estimate $E[X(1 - X)]$ because it ignores the dependence between X and $1 - X$.

Example 5.5 Barney retires as deputy and opens First National Bank of Mayberry. On all of his savings accounts that provide monthly interest, Barney rounds the interest amount to the nearest penny. When the interest amount is rounded up to the nearest penny, the interest amount is transferred to the depositor's account. When the interest amount is rounded down to the nearest penny, the interest amount is transferred to the depositor's account and Barney "skims" (this is illegal) the fraction of a penny truncated into his own personal account. What is the distribution of the amount of money that Barney makes per month from this scheme for the four depositors (Bea, Opie, Andy, and Floyd)?

Let's begin with an example. If Opie has \$9.87 in an interest-bearing account paying interest at 5.25% annually (compounded and paid monthly), how much interest should he earn this month? Since the monthly interest rate is $0.0525/12$, Opie should be paid

$$9.87 \cdot \frac{0.0525}{12} = 0.04318125$$

or approximately 4.318 cents at the end of the month. Barney transfers 4 cents to Opie's account, then transfers 0.318 cents to his own account. This is an insignificant amount with just four accounts, but would be significant for a large bank.

What is the distribution of the fraction of a penny that is transferred to Barney's account from Opie? Since neither the deposit amounts nor the interest rates are specified in the question, it is reasonable to assume that the amount transferred to Barney's account is close to uniformly distributed between 0 and $1/2$ cent. For half of the months, on average, Opie's account generates nothing for Barney (when the interest amount is rounded up to the nearest penny), and half of the months generate a $U(0, 1/2)$ fraction of a penny for Barney (when the interest amount is rounded down to the nearest penny).

Now to determine how much revenue is generated by the four accounts. The number of accounts that generate revenue for Barney is binomial$(4, 1/2)$ because whether each account generates revenue for Barney is an independent Bernoulli trial. Each account that generates revenue yields Barney a fraction of a penny that is uniformly distributed between 0 and $1/2$. Leaving the details to APPL, the amount of revenue generated to Barney from the four accounts Y is a mixture of a degenerate distribution at 0 (when all four accounts are rounded up) and various sums of $U(0, 1/2)$ random variables.

```
X  := UniformRV(0, 1 / 2);
X0 := [[x -> 1], [0, 0], ["Continuous", "PDF"]];
X1 := X;
X2 := ConvolutionIID(X, 2);
X3 := ConvolutionIID(X, 3);
X4 := ConvolutionIID(X, 4);
p  := [1 / 16, 4 / 16, 6 / 16, 4 / 16, 1 / 16];
Y  := Mixture(p, [X0, X1, X2, X3, X4]);
```

The distribution of the mixed discrete–continuous random variable Y is described by

$$f_Y(y) = \begin{cases} \dfrac{1}{16} & y = 0 \\[2mm] \dfrac{1}{6}y^3 + y^2 + \dfrac{3}{2}y + \dfrac{1}{2} & 0 < y \le 1/2 \\[2mm] -\dfrac{1}{2}y^3 - y^2 + y + \dfrac{5}{6} & 1/2 < y \le 1 \\[2mm] \dfrac{1}{2}y^3 - y^2 - \dfrac{1}{2}y + \dfrac{4}{3} & 1 < y \le 3/2 \\[2mm] -\dfrac{1}{6}y^3 + y^2 - 2y + \dfrac{4}{3} & 3/2 < y \le 2. \end{cases}$$

The additional APPL statements `Mean(Y)` and `Variance(Y)` calculate the population mean and variance:

$$\mu = \frac{1}{2} \qquad \text{and} \qquad \sigma^2 = \frac{5}{48}.$$

The population mean is consistent with the fact that each account generates an expected $1/8$ cent revenue (the population mean of the $U(0, 1/2)$ distribution is $1/4$ which is generated by the flip of a fair coin).

The following R code provides a Monte Carlo simulation check of the population mean and variance of the analytic model. The code simulates $100,000$ months of operation of the bank with four accounts.

```
nrep = 100000
n = 4
p = 1 / 2
barney = rep(0, nrep)
for (i in 1:nrep) {
  naccts = rbinom(1, n, p)
  barney[i] = sum(runif(naccts, 0, 1 / 2))
}
mean(barney)
var(barney)
```

Five runs of this simulation yield the following estimates for the population mean

0.4982168	0.5021850	0.5005846	0.4987352	0.5017606

and the population variance

0.1037619	0.1047048	0.1041821	0.1042304	0.1048081.

These values support the analytic population mean $\mu = 1/2$ and population variance $\sigma^2 = 5/48 \cong 0.1042$.

A special case of the uniform distribution, namely the $U(0, 1)$ distribution, plays a central role in probability theory. As introduced in Chapter 3, the probability integral transformation can be used to generate a random variate via

$$x \leftarrow F^{-1}(u)$$

where u is a realization of a $U(0, 1)$ random variable, commonly known as a *random number*.

Example 5.6 Find a random variate generation algorithm for generating $X \sim U(a, b)$.

As indicated earlier, the cumulative distribution function of a $U(a, b)$ random variable is

$$F(x) = \frac{x - a}{b - a}$$

for $a < x < b$. So the inverse cumulative distribution function is

$$F^{-1}(u) = a + (b - a)u$$

for $0 < u < 1$. So a one-line variate generation algorithm is

$$x \leftarrow a + (b - a)u$$

where u is a random number. This is consistent with intuition. Multiplying a $U(0, 1)$ by $(b - a)$ results in a $U(0, b - a)$ random variate. This is a stretching step if $b - a > 1$ or a shrinking step if $b - a < 1$. Adding a to this random variate results in the desired $U(a, b)$ random variate.

5.2   Exponential Distribution

A continuous random variable with positive support $A = \{x \mid x > 0\}$ is useful in a variety of applications. Examples include

- patient survival time after the diagnosis of a particular cancer,
- the lifetime of a light bulb,
- the sojourn time (waiting time plus service time) for a customer purchasing a ticket at a box office,
- the time between births at a hospital,
- the number of gallons purchased at a gas pump,
- the time to construct an office building.

The common element associated with these random variables is that they can only assume positive values.

One fundamental distribution with positive support is the exponential distribution, which is the subject of this section. The gamma distribution also has positive support and is considered in the next section. The exponential distribution has a single scale parameter λ, as defined below.

Definition 5.2 A continuous random variable X with probability density function

$$f(x) = \lambda e^{-\lambda x} \qquad x > 0$$

for some real constant $\lambda > 0$ is an exponential(λ) random variable.

In many applications, λ is referred to as a "rate," for example the arrival rate or the service rate in queueing, the death rate in actuarial science, or the failure rate in reliability. The exponential distribution plays a pivotal role in modeling random processes that evolve over time that are known as "stochastic processes."

The exponential distribution enjoys a particularly tractable cumulative distribution function:

$$
\begin{aligned}
F(x) &= P(X \leq x) \\
&= \int_0^x f(w)\,dw \\
&= \int_0^x \lambda e^{-\lambda w}\,dw \\
&= \left[-e^{-\lambda w} \right]_0^x \\
&= 1 - e^{-\lambda x}
\end{aligned}
$$

for $x > 0$. Thus for all values of x, the cumulative distribution function is

$$
F(x) = \begin{cases} 0 & x \leq 0 \\ 1 - e^{-\lambda x} & x > 0. \end{cases}
$$

The geometric distribution, which was introduced in Section 4.3, is the only discrete distribution to possess the *memoryless property*. The only continuous distribution to possess this property is the exponential distribution.

Theorem 5.1 (memoryless property) For $X \sim$ exponential(λ) and any two positive real numbers x and y,

$$P(X \geq x + y \mid X \geq x) = P(X \geq y).$$

Proof The conditional probability is

$$P(X \geq x+y \,|\, X \geq x) = \frac{P(X \geq x+y, X \geq x)}{P(X \geq x)}$$

$$= \frac{P(X \geq x+y)}{P(X \geq x)}$$

$$= \frac{e^{-\lambda(x+y)}}{e^{-\lambda x}}$$

$$= e^{-\lambda y}$$

$$= P(X \geq y),$$

which proves the memoryless property. □

The memoryless property also has a geometric interpretation. Consider the probability density function of an exponential(λ) random variable from some value x to infinity. The area under the probability density function from x to infinity does not equal 1; rather it equals $e^{-\lambda x}$. The conditional probability density function given that the random variable X is greater than or equal to x is found by rescaling:

$$f_{X \,|\, X \geq x}(w) = \frac{\lambda e^{-\lambda w}}{e^{-\lambda x}} = \lambda e^{-\lambda(w-x)} \qquad\qquad w > x.$$

This conditional distribution, if shifted x units to the left, is identical to the original exponential(λ) distribution. The geometry associated with the memoryless property is shown in Figure 5.3. The left-hand curve is an exponential(λ) probability density function; the right-hand curve is the conditional probability density function of an exponential(λ) random variable that is greater than x.

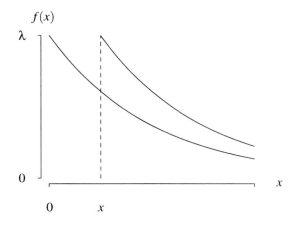

Figure 5.3: Memoryless property illustration for the exponential distribution.

The memoryless property is very strong. Consider whether the exponential distribution should be used to model the lifetime of a candle with an expected burning time of 5 hours. If several candles are sampled, you might envision a bell-shaped histogram for candle lifetimes, centered around 5 hours. The exponential lifetime model is certainly not appropriate in this case because a candle that has burned for 4 hours does not have the same lifetime distribution as a brand new candle. This same argument can be used to reason that the exponential lifetime model should not be applied to mechanical components that undergo wear (for example, bearings) or fatigue (structural supports) or electrical components that contain an element that burns away (filaments) or degrade with time (batteries). An electrical component for which the exponential lifetime assumption may be justified is a fuse. A fuse is designed to fail when there is a power surge that causes the fuse to burn out. Assuming that the fuse does not undergo any weakening or degradation over time and that power surges that cause failure occur with equal likelihood over time (that is, power surges constitute a Poisson process), the exponential lifetime assumption for fuse lifetime is appropriate, and a used fuse that has not failed is as good as a new one.

The exponential distribution should be applied judiciously because the memoryless property restricts its applicability. It is usually misapplied for the sake of simplicity because the statistical techniques for the exponential distribution are particularly tractable, or because small sample sizes do not support more than a one-parameter distribution.

The moment generating function for $X \sim$ exponential(λ) is

$$
\begin{aligned}
M(t) &= E\left[e^{tX}\right] \\
&= \int_0^\infty e^{tx} \lambda e^{-\lambda x} \, dx \\
&= \lambda \int_0^\infty e^{(t-\lambda)x} \, dx \\
&= \boxed{\frac{\lambda}{\lambda - t}}
\end{aligned}
$$

when $t - \lambda < 0$, or equivalently, when $t < \lambda$. Thus, the moment generating function exists for t in a neighborhood of 0 (just use the interval $-\lambda/2 < t < \lambda/2$ about $t = 0$ to satisfy the definition of the moment generating function).

The moments associated with $X \sim$ exponential(λ) can be calculated in four different ways. The first way is to use the definition of the expected value. For example, the expected value of X is

$$
\mu = E[X] = \int_0^\infty x \lambda e^{-\lambda x} \, dx = \cdots = \frac{1}{\lambda},
$$

which is calculated by using integration by parts. The second way is to differentiate the moment generating function. For example, the first derivative of the moment generating function is

$$
M'(t) = \frac{\lambda}{(\lambda - t)^2}
$$

and the expected value of X is

$$
E[X] = M'(0) = \frac{1}{\lambda}.
$$

The third way to calculate the moments of the exponential distribution is via APPL. The statements

```
X := ExponentialRV(lambda);
Mean(X);
Variance(X);
Skewness(X);
Kurtosis(X);
```

calculate the population mean and variance

$$
E[X] = \frac{1}{\lambda} \qquad \text{and} \qquad V[X] = \frac{1}{\lambda^2}
$$

and the population skewness and kurtosis

$$
E\left[\left(\frac{X-\mu}{\sigma}\right)^3\right] = 2 \qquad \text{and} \qquad E\left[\left(\frac{X-\mu}{\sigma}\right)^4\right] = 9,
$$

which are free of the parameter λ. The fourth way to calculate the moments is to prove the following general result for the sth moment of an exponential random variable.

Theorem 5.2 If $X \sim$ exponential(λ), then

$$E[X^s] = \frac{\Gamma(s+1)}{\lambda^s} \qquad s > -1,$$

where $\Gamma(\kappa) = \int_0^\infty x^{\kappa-1} e^{-x} dx$.

Proof The expected value of X^s is

$$
\begin{aligned}
E[X^s] &= \int_0^\infty x^s \lambda e^{-\lambda x} dx \\
&= \lambda^{-s} \int_0^\infty t^s e^{-t} dt \\
&= \lambda^{-s} \Gamma(s+1) \qquad s > -1
\end{aligned}
$$

by using the substitution $t = \lambda x$, which proves the result. $\qquad\square$

The *gamma function*

$$\Gamma(\kappa) = \int_0^\infty x^{\kappa-1} e^{-x} dx \qquad \kappa > 0$$

appears frequently in probability. When κ is an integer, $\Gamma(\kappa) = (\kappa - 1)!$, so the factorials are effectively smoothed by the gamma function, as illustrated in Figure 5.4. The gamma function is minimized at $\kappa \cong 1.4616$. Two other useful results concerning the gamma function include

- $\Gamma(\kappa + 1) = \kappa \Gamma(\kappa)$ for $\kappa > 0$,

- $\Gamma(1/2) = \sqrt{\pi}$.

The gamma function is calculated in R with the gamma function and in Maple with the GAMMA function. Returning to the moments of the exponential distribution, when s is a nonnegative integer, this expression reduces to $E[X^s] = s!/\lambda^s$. By setting $s = 1, 2, 3$, and 4, the population mean, variance, skewness, and kurtosis are again obtained as

$$E[X] = \frac{1}{\lambda} \qquad V[X] = \frac{1}{\lambda^2} \qquad E\left[\left(\frac{X-\mu}{\sigma}\right)^3\right] = 2 \qquad E\left[\left(\frac{X-\mu}{\sigma}\right)^4\right] = 9.$$

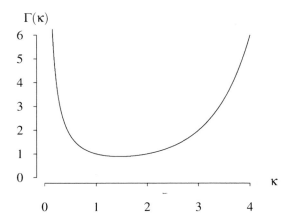

Figure 5.4: The gamma function.

Example 5.7 Arrival times to a busy yogurt shop during their peak arrival time are well modeled by a Poisson process with a rate of 30 customers per hour.

(a) Find the median time between arrivals.

(b) Find the probability that the time between the arrivals of two customers is between 2 and 4 minutes.

Since the arrival rate is measured in customers per hour and the question is posed in minutes, the first step in solving this problem is to choose the time units. Choosing minutes for the time units, the arrival rate is $\lambda = 1/2$ customer per minute. Using Theorem 4.3, the time between arrivals X is exponential($1/2$) with probability density function

$$f(x) = (1/2)e^{-x/2} \qquad x > 0$$

and cumulative distribution function

$$F(x) = \begin{cases} 0 & x \le 0 \\ 1 - e^{-x/2} & x > 0. \end{cases}$$

(a) The median time between arrivals m can be found by solving

$$F(m) = 1/2,$$

which is

$$1 - e^{-m/2} = 1/2$$

for m. Solving this equation yields the median, in minutes, as

$$m = -2\ln(1/2) \cong 1.3863.$$

(b) The probability that the time between the arrivals is between 2 and 4 minutes is the area under the probability density function between $x = 2$ and $x = 4$ (as illustrated in Figure 5.5):

$$\begin{aligned} P(2 < X < 4) &= \int_2^4 (1/2)e^{-x/2}\,dx \\ &= \left[-e^{-x/2}\right]_2^4 \\ &= e^{-1} - e^{-2} \\ &\cong 0.2325. \end{aligned}$$

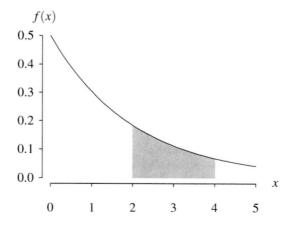

Figure 5.5: Probability density function for $X \sim$ exponential($1/2$) with $P(2 < X < 4)$ shaded.

Example 5.8 Devise a random variate generation algorithm for the exponential distribution.

The cumulative distribution function for an exponential random variable X is

$$F(x) = \begin{cases} 0 & x \leq 0 \\ 1 - e^{-\lambda x} & x > 0. \end{cases}$$

This means that the inverse cumulative distribution function is

$$F^{-1}(u) = -\frac{1}{\lambda} \ln(1 - u)$$

for $0 < u < 1$. Since the inverse cumulative distribution function can be expressed in closed form, the one-line random variate generation algorithm is

$$x \leftarrow -\frac{1}{\lambda} \ln(1 - u),$$

where u is a random number. The geometry associated with this variate generation algorithm is illustrated for $\lambda = 3$, $u = 0.9279$, and $x = 0.8766$ in Figure 5.6. Implementing this variate generation algorithm in R with $\lambda = 3$, the statements

```
nrep = 10000
lambda = 3
x = -log(1 - runif(nrep)) / lambda
mean(x)
var(x)
```

generate 10,000 exponential random variates that are placed in the vector x. The sample mean and variance are calculated using the last two statements, and the values printed hover around the analytic values $E[X] = 1/3$ and $V[X] = 1/9$ as expected. There is also a built-in random variate generator in R named rexp. The R statements

```
nrep = 10000
lambda = 3
x = rexp(nrep, lambda)
mean(x)
var(x)
```

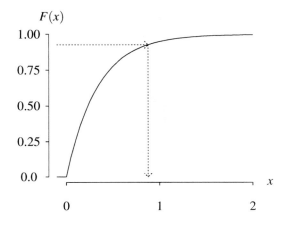

Figure 5.6: Exponential random variate generation algorithm illustration.

also generate 10,000 exponential random variates and compute their sample mean and variance. The two code segments given above do not generate identical results after a call to set.seed. The reason for the discrepancy is that the R developers have sped up their algorithm by generating their random variates via

$$x \leftarrow -\frac{1}{\lambda} \ln(u).$$

The slight difference created by replacing $1 - u$ by u [which is legitimate because $U \sim U(0, 1) \iff 1 - U \sim U(0, 1)$] keeps the algorithm probabilistically correct, but saves CPU cycles because a subtraction is saved for each variate generated.

Example 5.9 Let $X \sim$ exponential(λ). What is the distribution of $Y = \lfloor X \rfloor$?

The floor function applied to X means that the random variable Y effectively ignores all digits to the right of the decimal point, so Y is a discrete random variable with support $\mathcal{A} = \{y \mid y = 0, 1, 2, \ldots\}$. Calculate the probabilities associated with the initial values of Y:

$$P(Y = 0) = P(0 \le X < 1) = \int_0^1 \lambda e^{-\lambda x} dx = \left[-e^{-\lambda x} \right]_0^1 = 1 - e^{-\lambda},$$

$$P(Y = 1) = P(1 \le X < 2) = \int_1^2 \lambda e^{-\lambda x} dx = \left[-e^{-\lambda x} \right]_1^2 = e^{-\lambda} - e^{-2\lambda},$$

$$P(Y = 2) = P(2 \le X < 3) = \int_2^3 \lambda e^{-\lambda x} dx = \left[-e^{-\lambda x} \right]_2^3 = e^{-2\lambda} - e^{-3\lambda}.$$

Generalizing this pattern,

$$P(Y = y) = P(y \le X < y + 1) = \int_y^{y+1} \lambda e^{-\lambda x} dx = \left[-e^{-\lambda x} \right]_y^{y+1} = e^{-y\lambda} - e^{-(y+1)\lambda}$$

for $y = 0, 1, 2, \ldots$. This means that the probability mass function of Y is

$$f_Y(y) = e^{-y\lambda} - e^{-(y+1)\lambda} \qquad y = 0, 1, 2, \ldots$$

or

$$f_Y(y) = e^{-y\lambda} \left(1 - e^{-\lambda} \right) \qquad y = 0, 1, 2, \ldots.$$

This is a distribution we have already encountered: $Y \sim$ geometric $\left(1 - e^{-\lambda} \right)$. This provides a second link between the exponential and geometric distributions. Not only are they the only distributions with the memoryless property, but the floor of an exponential random variable is a geometric random variable.

Example 5.10 A truck requires a particular nonrepairable electrical component that has an exponential lifetime with a positive failure rate λ failures per hour. A site supports a large fleet of n trucks, each operating 24 hours a day, 7 days a week. A parts manager can make an order for spare parts once a week. Assuming that the lead time is 0 (that is, immediate delivery), what is the minimum inventory level of spare parts that should be maintained at the beginning of each week to ensure that there is a probability of at least 0.9999 that a truck is not down for lack of this particular nonrepairable electrical component?

This is another inventory problem, similar in nature to the newsboy problem encountered in Example 3.50. Instead of ordering newspapers, the parts manager is ordering spare parts. It is reasonable to assume that the failure of any one part during the week is an independent Bernoulli trial. Using a result that will be proved in the next chapter, the minimum of n independent exponential(λ) random variables is exponential($n\lambda$). So the time to the first failure during the week is exponential($n\lambda$). Since each operating part is as good as new at the time of the first failure by the memoryless property, the time to each subsequent failure is also exponential($n\lambda$). This means that the number of failures during the week constitutes a Poisson process with rate $n\lambda$. Over a one-week period, the number of failures has the Poisson distribution with rate $24 \cdot 7 \cdot n\lambda = 168n\lambda$. The parts manager should place an order to put the inventory level at the 99.99th percentile of this Poisson distribution.

We apply this part ordering algorithm for $n = 20$ trucks and $\lambda = 0.001$ failure per hour. The number of failures per week has a Poisson distribution with population mean

$$24 \cdot 7 \cdot 20 \cdot 0.001 = 3.36.$$

To find the 99.99th percentile of the Poisson distribution with a population mean 3.36, use the R statement qpois(0.9999, 3.36) which yields 12 parts. The parts manager should order so as to begin each week with 12 spare parts in inventory. The following code implements the inventory scheme of always having 12 spare parts on hand at the beginning of each of 100,000 weeks with a Poisson(3.36) number of failures per week. The R code

```
set.seed(3)
failures = rpois(100000, 3.36)
sum(failures <= 12) / 100000
```

returns 0.99993, which (barely) exceeds 0.9999, as desired.

To summarize, the exponential distribution is the fundamental continuous probability distribution with positive support. It has a single positive parameter λ, is the only continuous distribution with the memoryless property, and has probability density function

$$f(x) = \lambda e^{-\lambda x} \qquad x > 0.$$

The population mean and variance are

$$\mu = \frac{1}{\lambda} \qquad \text{and} \qquad \sigma^2 = \frac{1}{\lambda^2}.$$

Not everyone parameterizes the exponential distribution in this fashion. There are certain applications where it is advantageous to parameterize the exponential distribution by its population mean $\theta = 1/\lambda$. In this case, the probability density function is

$$f(x) = \frac{1}{\theta} e^{-x/\theta} \qquad x > 0$$

and the population mean and variance are

$$\mu = \theta \qquad \text{and} \qquad \sigma^2 = \theta^2.$$

The memoryless property of the exponential distribution limits its applicability. There are several two-parameter distributions that include the exponential distribution as a special case that are more flexible. We consider one such distribution in the next section.

5.3   Gamma Distribution

Adding a shape parameter to the exponential distribution increases its flexibility. We derive the gamma distribution by using the gamma function, which was defined in the last section as

$$\Gamma(\kappa) = \int_0^\infty x^{\kappa-1} e^{-x} dx \qquad \kappa > 0.$$

Making the substitution $y = x/\lambda$, where λ is a positive real constant, the gamma function becomes

$$\Gamma(\kappa) = \int_0^\infty (\lambda y)^{\kappa-1} e^{-\lambda y} \lambda \, dy \qquad \kappa > 0.$$

Dividing both sides of this equation by $\Gamma(\kappa)$ yields

$$1 = \int_0^\infty \frac{\lambda^\kappa y^{\kappa-1} e^{-\lambda y}}{\Gamma(\kappa)} \, dy.$$

Since the integrand is positive and integrates to one, it is a legitimate probability density function. The positive parameter λ is a scale parameter; the positive parameter κ is a shape parameter. This probability density function corresponds to the gamma distribution as defined below.

Definition 5.3 A continuous random variable X with probability density function

$$f(x) = \frac{\lambda^\kappa x^{\kappa-1} e^{-\lambda x}}{\Gamma(\kappa)} \qquad x > 0$$

for some real constants $\lambda > 0$ and $\kappa > 0$ is a gamma(λ, κ) random variable.

The cumulative distribution function for $X \sim$ gamma(λ, κ) on its support is

$$
\begin{aligned}
F(x) &= \int_0^x \frac{\lambda^\kappa w^{\kappa-1} e^{-\lambda w}}{\Gamma(\kappa)} \, dw \\
&= \frac{\lambda^\kappa}{\Gamma(\kappa)} \int_0^x w^{\kappa-1} e^{-\lambda w} \, dw \\
&= \frac{\lambda^\kappa}{\Gamma(\kappa)} \int_0^{\lambda x} \left(\frac{y}{\lambda}\right)^{\kappa-1} e^{-y} \frac{1}{\lambda} \, dy \\
&= \frac{1}{\Gamma(\kappa)} \int_0^{\lambda x} y^{\kappa-1} e^{-y} \, dy
\end{aligned}
$$

for $x > 0$. This is the first continuous distribution that we have encountered that does not have a closed-form cumulative distribution function. The integrand is identical to the integrand in the definition of the gamma function, but the upper limit of the integral is not ∞. For this reason, the cumulative distribution function of the gamma distribution is proportional to what is known as the *incomplete gamma function*. The cumulative distribution function of a gamma random variable is coded into R in the function pgamma(x, kappa, lambda). The probability that $P(X < 2)$, for example, when $X \sim$ gamma$(3, 4)$ is found with the statement pgamma(2, 4, 3). (R swaps the order of the parameters from our convention.)

The moment generating function for $X \sim \text{gamma}(\lambda, \kappa)$ is

$$
\begin{aligned}
M(t) &= E\left[e^{tX}\right] \\
&= \int_0^\infty e^{tx} \frac{\lambda^\kappa x^{\kappa-1} e^{-\lambda x}}{\Gamma(\kappa)}\, dx \\
&= \frac{\lambda^\kappa}{\Gamma(\kappa)} \int_0^\infty x^{\kappa-1} e^{-(\lambda-t)x}\, dx \\
&= \frac{\lambda^\kappa}{\Gamma(\kappa)} \int_0^\infty \left(\frac{y}{\lambda-t}\right)^{\kappa-1} e^{-y} \frac{1}{\lambda-t}\, dy \\
&= \frac{\lambda^\kappa}{(\lambda-t)^\kappa \Gamma(\kappa)} \int_0^\infty y^{\kappa-1} e^{-y}\, dy \\
&= \left(\frac{\lambda}{\lambda-t}\right)^\kappa \qquad t < \lambda.
\end{aligned}
$$

nice closed form.

The fact that the moment generating function is mathematically tractable means that there will be mathematically tractable moments. The APPL statements

```
X := GammaRV(lambda, kappa);
MGF(X);
Mean(X);
Variance(X);
Skewness(X);
Kurtosis(X);
```

yield the following values for the population mean, variance, skewness, and kurtosis

$$
E[X] = \frac{\kappa}{\lambda} \qquad V[X] = \frac{\kappa}{\lambda^2} \qquad E\left[\left(\frac{X-\mu}{\sigma}\right)^3\right] = \frac{2}{\sqrt{\kappa}} \qquad E\left[\left(\frac{X-\mu}{\sigma}\right)^4\right] = \frac{3(\kappa+2)}{\kappa}.
$$

The limiting values of the population skewness and kurtosis as $\kappa \to \infty$ are 0 and 3.

In addition to its flexibility for modeling continuous distributions with positive support, the gamma distribution is important for its special cases. The three special cases outlined in the paragraphs below are the exponential, Erlang, and chi-square distributions.

The gamma distribution collapses to the *exponential distribution* when $\kappa = 1$. In this case, the probability density function is

$$
f(x) = \lambda e^{-\lambda x} \qquad x > 0
$$

and the population moments are

$$
E[X] = \frac{1}{\lambda} \qquad V[X] = \frac{1}{\lambda^2} \qquad E\left[\left(\frac{X-\mu}{\sigma}\right)^3\right] = 2 \qquad E\left[\left(\frac{X-\mu}{\sigma}\right)^4\right] = 9.
$$

The R functions dexp, pexp, qexp, and rexp are used to calculate $f(x)$, $F(x)$, percentiles, and random variates. The exponential distribution is the only continuous distribution with the memoryless property and the minimum of independent exponential random variables is also exponential. Additional details associated with the exponential distribution were given in the previous section.

The gamma distribution collapses to the *Erlang distribution* when κ is a positive integer, say k. The probability density function of $X \sim \text{Erlang}(\lambda, k)$ random variable is

$$
f(x) = \frac{\lambda^k x^{k-1} e^{-\lambda x}}{(k-1)!} \qquad x > 0.
$$

The population moments are

$$E[X] = \frac{k}{\lambda} \qquad V[X] = \frac{k}{\lambda^2} \qquad E\left[\left(\frac{X-\mu}{\sigma}\right)^3\right] = \frac{2}{\sqrt{k}} \qquad E\left[\left(\frac{X-\mu}{\sigma}\right)^4\right] = \frac{3(k+2)}{k}.$$

The cumulative distribution function of an Erlang(λ, k) random variable can be expressed as a summation

$$F(x) = \begin{cases} 0 & x \le 0 \\ 1 - \sum_{y=0}^{k-1} \dfrac{(\lambda x)^y e^{-\lambda x}}{y!} & x > 0. \end{cases}$$

Finally, the sum of mutually independent exponential random variables is Erlang. More specifically, if $X_1, X_2, \ldots, X_k$ are mutually independent and identically distributed exponential(λ) random variables, then $X_1 + X_2 + \cdots + X_k \sim$ Erlang (λ, k). More details associated with this result will be given in a subsequent chapter. Because of the previous property, the time of the kth event for a Poisson process with a rate λ is Erlang (λ, k).

The gamma distribution collapses to the *chi-square distribution* when the shape parameter κ assumes the value $k/2$, where k is a positive integer parameter known as the *degrees of freedom*, and $\lambda = 1/2$. The resulting probability density function of a chi-square random variable X is

$$f(x) = \frac{x^{k/2-1} e^{-x/2}}{\Gamma(k/2) 2^{k/2}} \qquad x > 0.$$

The shorthand for a chi-square random variable X with k degrees of freedom is $X \sim \chi^2(k)$. The moments are

$$E[X] = k \qquad V[X] = 2k \qquad E\left[\left(\frac{X-\mu}{\sigma}\right)^3\right] = \frac{2\sqrt{2}}{\sqrt{k}} \qquad E\left[\left(\frac{X-\mu}{\sigma}\right)^4\right] = \frac{3(k+4)}{k}.$$

The R functions `dchisq`, `pchisq`, `qchisq`, and `rchisq` are used to calculate $f(x)$, $F(x)$, percentiles, and random variates.

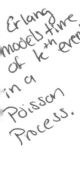

Example 5.11 Arrival times to a busy yogurt shop during the peak arrival time are well modeled by a Poisson process with a rate of 30 customers per hour. Find the probability that 8 or more customers arrive in a 10-minute interval.

Solution 1. Using minutes as the time scale, the arrival rate is $\lambda = 1/2$ customer per minute and the arrival time of the 8th customer is $X \sim$ Erlang($1/2, 8$). The probability density function of the 8th customer's arrival time is shown in Figure 5.7. The probability that 8 or more customers arrive in a 10-minute interval is equivalent to the probability that the 8th customer arrives before time 10:

$$P(X \le 10) = F(10) = 1 - \sum_{y=0}^{k-1} \frac{(\lambda x)^y e^{-\lambda x}}{y!} = 1 - \sum_{y=0}^{7} \frac{5^y e^{-5}}{y!} \cong 0.1334.$$

The area associated with the above probability is shaded in Figure 5.7. This probability can be calculated with the R statement

```
pgamma(10, 8, 1 / 2)
```

which returns

$$P(X \le 10) \cong 0.1334.$$

Solution 2. For a Poisson process, the number of events in any time interval has the Poisson distribution. In this case, the number of customer arrivals (the events) in the

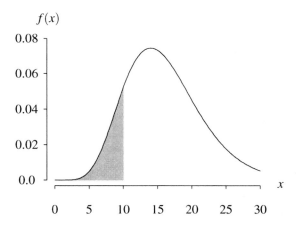

Figure 5.7: Erlang$(1/2, 8)$ probability density function with $P(X < 10)$ shaded.

10-minute interval has a Poisson distribution with $\lambda = (30)(10/60) = 5$. Let Y denote the number of customer arrivals in a 10-minute interval. The probability mass function of Y is

$$f(y) = \frac{5^y e^{-5}}{y!} \qquad y = 0, 1, 2, \ldots$$

and the probability of interest is

$$P(Y \geq 8) = \sum_{y=8}^{\infty} \frac{5^y e^{-5}}{y!}.$$

This probability is calculated with the R statement

```
1 - ppois(7, 5)
```

which returns

$$P(Y \geq 8) \cong 0.1334$$

which matches the previous solution.

To summarize, the gamma distribution is a generalization of the exponential distribution with positive support. It has a positive scale parameter λ, a positive shape parameter κ, and a probability density function

$$f(x) = \frac{\lambda^\kappa x^{\kappa-1} e^{-\lambda x}}{\Gamma(\kappa)} \qquad x > 0.$$

The population mean and variance are

$$\mu = \frac{\kappa}{\lambda} \qquad \text{and} \qquad \sigma^2 = \frac{\kappa}{\lambda^2}.$$

There are three special cases of the gamma distribution that often arise in applications: (a) the exponential distribution when $\kappa = 1$, (b) the Erlang distribution when κ is a positive integer k, and (c) the chi-square distribution when $\lambda = 1/2$ and $\kappa = k/2$, where k is the degrees of freedom.

5.4   Normal Distribution

Many random variables that arise in practice have bell-shaped distributions. Examples include

- cholesterol levels of 50-year-old men,

- heights of adult women,

- newborn baby weights,

- crop yields,

- ball bearing diameters.

None of the distributions encountered so far have a symmetric bell-shaped probability density function (although the gamma distribution approaches a bell-shaped distribution as $\kappa \to \infty$). We begin our search for such a distribution by considering the *potential* probability density function

$$f^*(x) = e^{-x^2/2} \qquad -\infty < x < \infty.$$

The $*$ superscript on f indicates that we are not sure whether this is a legitimate probability density function. This function is (a) symmetric about $x = 0$, (b) positive, and (c) bell-shaped. The only question that remains is whether it integrates to 1. Define the area under $f^*(x)$ as

$$a = \int_{-\infty}^{\infty} e^{-x^2/2} dx.$$

This integral can't be evaluated analytically. Surprisingly, however, its square can be evaluated analytically. Using polar coordinates with $x = r\cos\theta$ and $y = r\sin\theta$,

$$
\begin{aligned}
a^2 &= \left(\int_{-\infty}^{\infty} e^{-x^2/2} dx \right) \left(\int_{-\infty}^{\infty} e^{-y^2/2} dy \right) \\
&= \int_{-\infty}^{\infty} \int_{-\infty}^{\infty} e^{-(x^2+y^2)/2} dx dy \\
&= \int_{0}^{2\pi} \int_{0}^{\infty} e^{-r^2/2} r dr d\theta \\
&= \int_{0}^{2\pi} \left[-e^{-r^2/2} \right]_{0}^{\infty} d\theta \\
&= \int_{0}^{2\pi} d\theta \\
&= 2\pi.
\end{aligned}
$$

This means that $a = \sqrt{2\pi}$. So we have a legitimate probability density function by simply dividing $f^*(x)$ by $\sqrt{2\pi}$ so that it integrates to one:

$$f(x) = \frac{1}{\sqrt{2\pi}} e^{-\frac{1}{2}x^2} \qquad -\infty < x < \infty.$$

We now have a symmetric, bell-shaped probability density function $f(x)$. Its only remaining drawback is that it is centered about 0 and has no parameters to enhance its flexibility. To enhance this distribution's generality, let $y = (x - \mu)/\sigma$ in the integral

$$\int_{-\infty}^{\infty} \frac{1}{\sqrt{2\pi}} e^{-\frac{1}{2}y^2} dy = 1$$

yielding

$$\int_{-\infty}^{\infty} \frac{1}{\sqrt{2\pi}\sigma} e^{-\frac{1}{2}\left(\frac{x-\mu}{\sigma}\right)^2} dx = 1.$$

The integrand is the probability density function of the *normal* distribution defined below.

Definition 5.4 A continuous random variable X with probability density function

$$f(x) = \frac{1}{\sqrt{2\pi}\sigma} e^{-\frac{1}{2}\left(\frac{x-\mu}{\sigma}\right)^2} \qquad -\infty < x < \infty$$

for some real constants μ and $\sigma > 0$ is a $N\left(\mu, \sigma^2\right)$ random variable.

The normal distribution is the most important distribution because (a) many random variables measuring natural and man-made phenomena tend to come from bell-shaped probability density functions and (b) of the central limit theorem, which will be proved in Chapter 8. The normal distribution was introduced by Abraham DeMoivre in the 18th century, and applied by Carl Friedrich Gauss in the 19th century, which is why it is sometimes called the *Gaussian distribution*. Statisticians called data sets that had bell-shaped histograms "normally" distributed, and the term stuck.

Using μ and σ for the parameters is no coincidence—these are the population mean and population standard deviation of the distribution; μ is a location parameter and σ is a scale parameter. There are a number of properties and applications associated with the normal distribution which follow.

- The applications listed at the beginning of this section (cholesterol levels, heights, weights, crop yields, diameters) are all best modeled by a distribution with positive support, yet the normal distribution is defined on $(-\infty, \infty)$. Therefore the normal distribution must be considered an approximate model. As the well-known statistician George Box stated: "All models are wrong; some models are useful." If a bell-shaped distribution is called for in a case of data that is inherently positive, the normal distribution should be used when the area under $f(x)$ to the left of 0 is negligible.

- The normal distribution has the most applications in a field generically referred to as "classical statistics," where data values are often assumed to arise from a bell-shaped population.

- One of the most productive uses of applied statistics is a field known as "quality control." Control charts were invented by Walter Shewhart in 1923, then exported to Japan by W. Edwards Deming and Joseph Juran. Most control charts are based on the normal distribution.

- The nearly-universal shorthand for X having a normal distribution with population mean μ and population variance σ^2 is $X \sim N(\mu, \sigma^2)$.

- Since the probability density function is symmetric, the population mean, median, and mode of the normal distribution all equal μ.

- The probability density function has inflection points at $\mu \pm \sigma$.

- A special case of the normal distribution is the *standard normal distribution* which occurs when $\mu = 0$ and $\sigma = 1$. Many authors use Z, rather than X, for a standard normal random variable.

- The cumulative distribution function of $X \sim N(\mu, \sigma^2)$,

$$F(x) = \int_{-\infty}^{x} \frac{1}{\sqrt{2\pi}\sigma} e^{-\frac{1}{2}\left(\frac{w-\mu}{\sigma}\right)^2} dw \qquad -\infty < x < \infty,$$

  can't be evaluated in closed form. To calculate $F(x)$ for a specific value of x, we must resort to numerical methods, tables, and functions. Some authors use $\Phi(x)$, rather than $F(x)$, for the standard normal cumulative distribution function (we will not adopt this notation).

- The normal distribution follows the 68—95—99.7 rule. The probabilities of falling within 1, 2, and 3 population standard deviations of the population mean are approximately 0.68, 0.95, and 0.997, respectively. This is illustrated for a standard normal random variable in Figure 5.8.

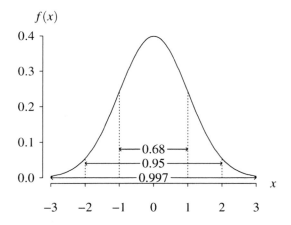

Figure 5.8: Probabilities associated with the standard normal distribution.

- The normal distribution can be derived as the limit of the binomial distribution as $n \to \infty$.

- The limiting distribution of a gamma(λ, κ) distribution is normally distributed as $\kappa \to \infty$.

- The R functions dnorm, pnorm, qnorm, and rnorm are used to calculate $f(x)$, $F(x)$, percentiles, and random variates. The third argument in these functions is the population standard deviation σ rather than the population variance σ^2.

- The APPL statements

```
X := NormalRV(mu, sigma);
Mean(X);
Variance(X);
Skewness(X);
Kurtosis(X);
```

calculate the first four population moments as

$$E[X] = \mu \qquad V[X] = \sigma^2 \qquad E\left[\left(\frac{X-\mu}{\sigma}\right)^3\right] = 0 \qquad E\left[\left(\frac{X-\mu}{\sigma}\right)^4\right] = 3.$$

- The Maple erf function can be used to calculate the cumulative distribution function:

$$F(x) = \frac{1}{2} + \frac{1}{2}\text{erf}\left(\frac{\sqrt{2}}{2}\left(\frac{x-\mu}{\sigma}\right)\right).$$

The moment generating function for $X \sim N(\mu, \sigma^2)$ is

$$
\begin{aligned}
M(t) &= E\left[e^{tX}\right] \\
&= \int_{-\infty}^{\infty} e^{tx} \frac{1}{\sqrt{2\pi}\sigma} e^{-\frac{1}{2}\left(\frac{x-\mu}{\sigma}\right)^2} dx \\
&= \int_{-\infty}^{\infty} \frac{1}{\sqrt{2\pi}\sigma} e^{-\frac{1}{2\sigma^2}\left(-2tx\sigma^2 + (x-\mu)^2\right)} dx \\
&= e^{\mu t + \frac{1}{2}t^2\sigma^2} \left[\int_{-\infty}^{\infty} \frac{1}{\sqrt{2\pi}\sigma} e^{-\frac{1}{2}\left(\frac{x-(\mu+t\sigma^2)}{\sigma}\right)^2} dx\right] \\
&= e^{\mu t + \frac{1}{2}\sigma^2 t^2} \qquad -\infty < t < \infty.
\end{aligned}
$$

The transition from the third to the fourth line in this derivation involves completing the square of the expression

$$
\begin{aligned}
-2tx\sigma^2 + (x-\mu)^2 &= x^2 - 2\mu x + \mu^2 - 2tx\sigma^2 \\
&= x^2 - 2\left(\mu + t\sigma^2\right)x + \mu^2 \\
&= x^2 - 2\left(\mu + t\sigma^2\right)x + \left(\mu + t\sigma^2\right)^2 - \left(\mu + t\sigma^2\right)^2 + \mu^2 \\
&= \left[x - \left(\mu + t\sigma^2\right)\right]^2 - 2\mu t\sigma^2 - t^2\sigma^4.
\end{aligned}
$$

The population mean of X, for example, is found by taking the derivative of $M(t)$:

$$
M'(t) = \left(e^{\mu t + \frac{1}{2}\sigma^2 t^2}\right)(\mu + t\sigma^2)
$$

for $-\infty < t < \infty$. Using $t = 0$ as an argument results in

$$
E[X] = M'(0) = \mu.
$$

Taking a second derivative of $M(t)$ yields

$$
M''(t) = \left(e^{\mu t + \frac{1}{2}\sigma^2 t^2}\right)(\sigma^2) + \left(\mu + t\sigma^2\right)^2\left(e^{\mu t + \frac{1}{2}\sigma^2 t^2}\right)
$$

for $-\infty < t < \infty$. Using $t = 0$ as an argument results in

$$
E\left[X^2\right] = M''(0) = \sigma^2 + \mu^2.
$$

Using the shortcut formula for the population variance,

$$
V[X] = E\left[X^2\right] - E[X]^2 = \sigma^2.
$$

We now prove three elementary results concerning the normal distribution. The first states that a linear function of a normal random variable is normally distributed.

Theorem 5.3 If $X \sim N(\mu, \sigma^2)$ then $Y = aX + b \sim N(a\mu + b, a^2\sigma^2)$.

Proof The proof of this result uses moment generating functions. The moment generating function of X is

$$
M_X(t) = E\left[e^{tX}\right] = e^{\mu t + \frac{1}{2}\sigma^2 t^2} \qquad -\infty < t < \infty.
$$

The moment generating function of Y is

$$
\begin{aligned}
M_Y(t) &= E\left[e^{tY}\right] \\
&= E\left[e^{t(aX+b)}\right] \\
&= E\left[e^{atX+bt}\right] \\
&= E\left[e^{bt}e^{atX}\right] \\
&= e^{bt}E\left[e^{atX}\right] \\
&= e^{bt}e^{a\mu t + \frac{1}{2}\sigma^2(at)^2} \\
&= e^{(a\mu+b)t + \frac{1}{2}(a\sigma)^2 t^2} \qquad -\infty < t < \infty.
\end{aligned}
$$

A careful inspection of this moment generating function shows that this is the moment generating function of a $N(a\mu + b, a^2\sigma^2)$ random variable. Since moment generating functions completely define a distribution, the result is proved. $\square$

Theorem 5.4 If $X \sim N(\mu, \sigma^2)$ then $Z = \dfrac{X - \mu}{\sigma} \sim N(0, 1)$.

Proof One way to prove this result is to simply use $a = 1/\sigma$ and $b = -\mu/\sigma$ in Theorem 5.3. A second way to prove the result is to use the cumulative distribution function technique:

$$
\begin{aligned}
F_Z(z) &= P(Z \leq z) \\
&= P\left(\frac{X - \mu}{\sigma} \leq z\right) \\
&= P(X \leq \mu + \sigma z) \\
&= \int_{-\infty}^{\mu + \sigma z} \frac{1}{\sqrt{2\pi}\sigma} e^{-\frac{1}{2}\left(\frac{x-\mu}{\sigma}\right)^2} dx \\
&= \int_{-\infty}^{z} \frac{1}{\sqrt{2\pi}} e^{-\frac{1}{2}y^2} dy.
\end{aligned}
$$

Since the integrand is the probability density function of a standard normal random variable $Z \sim N(0, 1)$, the result is proved. □

Theorem 5.5 If $X \sim N(\mu, \sigma^2)$ then $Y = \left(\dfrac{X - \mu}{\sigma}\right)^2 \sim \chi^2(1)$.

Proof From Theorem 5.4, $Z = \frac{X - \mu}{\sigma} \sim N(0, 1)$. Exploiting the symmetry of the probability density function of the standard normal distribution, the cumulative distribution function of Y is

$$
\begin{aligned}
F_Y(y) &= P(Y \leq y) \\
&= (Z^2 \leq y) \\
&= P(-\sqrt{y} < Z < \sqrt{y}) \\
&= 2\int_0^{\sqrt{y}} \frac{1}{\sqrt{2\pi}} e^{-z^2/2} dz \\
&= \int_0^{y} \frac{1}{\sqrt{2\pi}\sqrt{w}} e^{-w/2} dw \qquad y > 0.
\end{aligned}
$$

Differentiating with respect to y yields the probability density function of Y

$$
\begin{aligned}
f_Y(y) &= \frac{1}{\sqrt{\pi}\sqrt{2}\sqrt{y}} e^{-y/2} \\
&= \frac{1}{\Gamma(1/2)2^{1/2}} y^{1/2 - 1} e^{-y/2} \qquad y > 0,
\end{aligned}
$$

which is recognized as the probability density function of a $\chi^2(1)$ random variable. □

Example 5.12 IQ scores are normally distributed with population mean 100 and population standard deviation 15.

(a) Find the probability that an IQ score is less than 130.

(b) Find the probability that an IQ score falls between 80 and 110.

(c) Find the 99th percentile of the distribution of IQ scores.

We assume that IQ scores are reported on a continuous scale (which is certainly not the case in practice but a reasonable approximation) by letting $X \sim N(100, 15^2)$.

(a) The historical approach to solving this problem is to first standardize the random variable X by subtracting its population mean and dividing by its population standard deviation in order to convert it to a standardized random variable Z. By Theorem 5.4, this random variable has a standard normal distribution. The probabilities associated with a standard normal random variable are found in a normal table because the cumulative distribution function is not closed form. Normal tables are not included in this text because of the availability of calculators and software to determine the required probabilities. For this problem

$$P(X < 130) = P\left(\frac{X - 100}{15} < \frac{130 - 100}{15}\right)$$
$$= P(Z < 2)$$
$$\cong 0.9772.$$

$z \quad \mu = 0$
$\sigma^2 = 1$

The standardizing is done internally in R. The statement

```
pnorm(130, 100, 15)
```

calculates the probability directly. Likewise, the APPL statements

```
X := NormalRV(100, 15);
CDF(X, 130);
```

return the same probability after a call to `evalf` to evaluate the integral as a floating point value.

(b) The probability that an IQ score falls between 80 and 110 is shaded in Figure 5.9 and can be calculated with the R statement

```
pnorm(110, 100, 15) - pnorm(80, 100, 15)
```

which returns

$$P(80 < X < 110) \cong 0.6563.$$

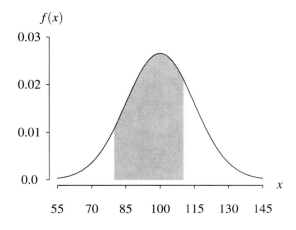

Figure 5.9: Probability density function for $X \sim N\left(100, 15^2\right)$ with $P(80 < X < 110)$ shaded.

(c) The 99th percentile of X is found with the R statement

```
qnorm(0.99, 100, 15)
```

which returns
$$x_{0.99} \cong 134.8952.$$

Example 5.13 If $X \sim N(6, 1)$, find $P(|X - 5| < 2)$.

Using algebra to simplify the absolute value, the desired probability is

$$
\begin{aligned}
P(|X - 5| < 2) &= P(-2 < X - 5 < 2) \\
&= P(3 < X < 7) \\
&= P\left(\frac{3-6}{1} < \frac{X-6}{1} < \frac{7-6}{1}\right) \\
&= P(-3 < Z < 1) \\
&= P(Z < 1) - P(Z < -3) \\
&\cong 0.8413 - 0.001350 \\
&\cong 0.8400.
\end{aligned}
$$

This probability can be calculated, without standardizing X, with the R statement

```
pnorm(7, 6, 1) - pnorm(3, 6, 1)
```

Example 5.14 If $X \sim N(\mu, \sigma^2)$, find $E\big[|X - \mu|\big]$.

Exploiting the symmetry of the probability density function of the normal distribution about μ, the expected value can be calculated as

$$
\begin{aligned}
E\big[|X - \mu|\big] &= \int_{-\infty}^{\infty} |x - \mu| \frac{1}{\sqrt{2\pi}\sigma} e^{-\frac{1}{2}\left(\frac{x-\mu}{\sigma}\right)^2} dx \\
&= 2\int_{\mu}^{\infty} \frac{x-\mu}{\sqrt{2\pi}\sigma} e^{-\frac{(x-\mu)^2}{2\sigma^2}} dx \\
&= \frac{2\sigma}{\sqrt{2\pi}} \int_{\mu}^{\infty} \frac{x-\mu}{\sigma^2} e^{-\frac{(x-\mu)^2}{2\sigma^2}} dx \\
&= \sqrt{\frac{2}{\pi}}\sigma \left[-e^{-\frac{(x-\mu)^2}{2\sigma^2}}\right]_{\mu}^{\infty} \\
&= \sqrt{\frac{2}{\pi}}\sigma.
\end{aligned}
$$

This derivation can be confirmed in APPL with the statements

```
X := NormalRV(mu, sigma);
ExpectedValue(X, x -> abs(x - mu));
```

For one set of parameters, say $\mu = 17$ and $\sigma^2 = 36$, this result can be supported via Monte Carlo simulation with the R statement

```
mean(abs(rnorm(1000000, 17, 6) - 17))
```

Five runs of this simulation yield

4.78294	4.79239	4.78721	4.78996	4.78318,

which hover around the analytic value $E\big[|X - \mu|\big] = \sqrt{\dfrac{2}{\pi}} \cong 4.7873$.

Example 5.15 Find the value of σ (to four-digit accuracy) so that a $N(0, \sigma^2)$ random variable has an interquartile range of 10.

Let $X \sim N(0, \sigma^2)$. The 25th percentile $x_{0.25}$ and the 75th percentile $x_{0.75}$ satisfy

$$P(x_{0.25} < X < x_{0.75}) = \frac{1}{2}.$$

The problem requires the value of σ such that $x_{0.75} - x_{0.25} = 10$. One way to solve this problem is to use trial and error in R with the statement

```
qnorm(0.75, 0, sigma) - qnorm(0.25, 0, sigma)
```

until the difference between the percentiles equals 10. A more direct solution is to recognize that because the distribution of X is centered about 0, the 75th percentile must equal $x_{0.75} = 5$. So use the APPL statements

```
X := NormalRV(0, sigma);
solve(CDF(X, 5) = 3 / 4, sigma);
```

which return $\sigma \cong 7.4130$.

This solution can be checked using Monte Carlo simulation. R has a function named `IQR` which calculates the sample interquartile range associated with a vector of data values passed to the function as an argument. The R statements

```
nrep = 10000000
sigma = 7.43011090
IQR(rnorm(nrep, 0, sigma))
```

return

9.99566	10.00533	10.00268	10.00442	9.99966.

These values are hovering around the population interquartile range (which is 10), so the analytic solution is supported.

5.5 Other Distributions

There are dozens of other continuous probability distributions that would be worthwhile introducing in this section. Only a sampling of these distributions are introduced in this section due to space limitations: the beta distribution, the triangular distribution, and the Weibull distribution. Two other important distributions that arise in applied statistics, the t and F distributions, will be introduced in a subsequent chapter.

Beta distribution

The $U(0, 1)$ distribution is useful because its support is $\mathcal{A} = \{x \mid 0 < x < 1\}$. The beta distribution is a generalization of the $U(0, 1)$ distribution that has two shape parameters, α and β, that provide a variety of shapes for the probability density function. This flexible distribution defined on $(0, 1)$ is useful for modeling batting averages of baseball players, availability of aircraft, interest rates, probabilities, percentages, etc.

A continuous random variable X with probability density function

$$f(x) = \frac{\Gamma(\alpha + \beta)}{\Gamma(\alpha)\Gamma(\beta)} x^{\alpha-1}(1-x)^{\beta-1} \qquad 0 < x < 1$$

for some real positive constants α and β is a beta(α, β) random variable. The beta distribution includes the $U(0, 1)$ distribution as a special case when $\alpha = \beta = 1$. As is the case with all continuous random variables, the probability density function of X integrates to one, that is

$$\int_0^1 \frac{\Gamma(\alpha+\beta)}{\Gamma(\alpha)\Gamma(\beta)} x^{\alpha-1} (1-x)^{\beta-1} \, dx = 1$$

or

$$\int_0^1 x^{\alpha-1} (1-x)^{\beta-1} \, dx = \frac{\Gamma(\alpha)\Gamma(\beta)}{\Gamma(\alpha+\beta)}.$$

The right-hand side of this equation defines what is known as the *beta function*

$$B(\alpha, \beta) = \frac{\Gamma(\alpha)\Gamma(\beta)}{\Gamma(\alpha+\beta)}.$$

The cumulative distribution function of a beta random variable X is

$$F(x) = \int_0^x \frac{\Gamma(\alpha+\beta)}{\Gamma(\alpha)\Gamma(\beta)} w^{\alpha-1} (1-w)^{\beta-1} \, dw$$

for $0 < x < 1$. Ignoring the constant involving the gamma functions that can be pulled outside of the integral, this is known as the *incomplete beta function*. Like the gamma and normal distributions, there is no closed form solution for $F(x)$. Numerical methods must be relied on for calculating probabilities. The R function pbeta calculates values of the cumulative distribution function numerically. In the special case when both α and β are positive integers, the integrand is a polynomial and easy to integrate. Furthermore, when both parameters are positive integers, the cumulative distribution function can be expressed as

$$F(x) = \int_0^x \frac{\Gamma(\alpha+\beta)}{\Gamma(\alpha)\Gamma(\beta)} w^{\alpha-1} (1-w)^{\beta-1} \, dw = \sum_{y=\alpha}^n \binom{n}{y} x^y (1-x)^{n-y}$$

for $0 < x < 1$, where $n = \alpha + \beta - 1$. This result is proved using integration by parts. The right-hand side of this expression is a summation of probabilities associated with the binomial distribution, where x is playing the role of p. So when $\alpha = 2$ and $\beta = 3$, for example, the cumulative distribution function evaluated at $x = 4/5$ can be found via the R statement

```
pbeta(4 / 5, 2, 3)
```
$$pbeta(x, \alpha, \beta).$$

or, using the right-hand side of the previous equation

```
1 - pbinom(1, 4, 4 / 5)
```

both of which yield 0.9728.

Using the definition of the expected value, the population mean of $X \sim$ beta(α, β) is

$$\begin{aligned}
E[X] &= \int_0^1 x \frac{\Gamma(\alpha+\beta)}{\Gamma(\alpha)\Gamma(\beta)} x^{\alpha-1} (1-x)^{\beta-1} \, dx \\
&= \frac{\Gamma(\alpha+\beta)}{\Gamma(\alpha)\Gamma(\beta)} \int_0^1 x^{\alpha} (1-x)^{\beta-1} \, dx \\
&= \frac{\Gamma(\alpha+\beta)}{\Gamma(\alpha)\Gamma(\beta)} \int_0^1 x^{\alpha+1-1} (1-x)^{\beta-1} \, dx \\
&= \frac{\Gamma(\alpha+\beta)}{\Gamma(\alpha)\Gamma(\beta)} \cdot \frac{\Gamma(\alpha+1)\Gamma(\beta)}{\Gamma(\alpha+\beta+1)} \\
&= \frac{\alpha\Gamma(\alpha)\Gamma(\alpha+\beta)}{(\alpha+\beta)\Gamma(\alpha+\beta)\Gamma(\alpha)} \\
&= \frac{\alpha}{\alpha+\beta}.
\end{aligned}$$

The population moments of the beta distribution can also be calculated with the APPL statements

```
X := BetaRV(alpha, beta);
Mean(X);
Variance(X);
Skewness(X);
Kurtosis(X);
```

which yields population mean and variance

$$\mu = E[X] = \frac{\alpha}{\alpha+\beta} \qquad\qquad \sigma^2 = V[X] = \frac{\alpha\beta}{(\alpha+\beta)^2(\alpha+\beta+1)}$$

and population skewness and kurtosis

$$E\left[\left(\frac{X-\mu}{\sigma}\right)^3\right] = \frac{2(\beta-\alpha)\sqrt{\alpha+\beta+1}}{(\alpha+\beta+2)\sqrt{\alpha\beta}}$$

$$E\left[\left(\frac{X-\mu}{\sigma}\right)^4\right] = \frac{3(\alpha+\beta+1)(2\alpha^2+2\beta^2-2\alpha\beta+\alpha^2\beta+\alpha\beta^2)}{\alpha\beta(\alpha+\beta+2)(\alpha+\beta+3)}.$$

Example 5.16 Batting 0.400 for a major league baseball season is a rare and extraordinary feat. A 0.400 season was seen just 13 times from 1900–1941, a feat accomplished by 8 players. No one has had a 0.400 season since. One theory for the demise of the 0.400 hitter is that while the batting average for all players is fairly constant season-to-season, the variability has been decreasing over time, that is, batters are becoming more consistent. Rule changes, such as the size of the strike zone, also play a role. The beta distribution can be used to model batting averages because all batting averages must fall between 0 and 1. Find the parameters of the beta distribution so that the population mean batting average is 0.260 and the population standard deviation of the batting averages is 0.04. With this particular choice of α and β, what is the probability a batting average exceeds 0.400?

The parameters α and β must satisfy

$$\mu = \frac{\alpha}{\alpha+\beta} = 0.26 \qquad\qquad \sigma = \sqrt{\frac{\alpha\beta}{(\alpha+\beta)^2(\alpha+\beta+1)}} = 0.04.$$

This 2×2 set of simultaneous equations can be solved for α and β in APPL with the statements

```
X := BetaRV(alpha, beta);
solve({Mean(X) = 26 / 100, sqrt(Variance(X)) = 4 / 100}, {alpha, beta});
```

Leaving the population mean and standard deviation values as exact fractions forces Maple to solve for α and β as exact fractions. These values yield

$$\alpha = \frac{6201}{200} = 31.005 \qquad\text{and}\qquad \beta = \frac{17649}{200} = 88.245.$$

This probability density function of this beta random variable is shown in Figure 5.10. If this particular probability model was an adequate description of seasonal major league baseball batting averages, then a histogram of the batting averages of all of the players in a season would be a close approximation to this probability density function. The probability that a batting average exceeds 0.400 can be calculated with the R command

```
1 - pbeta(0.400, 31.005, 88.245)
```

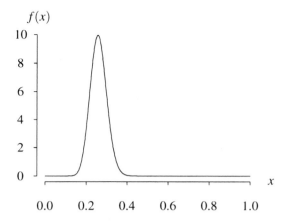

Figure 5.10: Probability density function for $X \sim \text{beta}(31.005, 88.245)$.

which returns $P(X > 0.400) \cong 0.0006077$.

The beta distribution is a very flexible two-parameter distribution. It includes the hump shape from the previous example, but also a "U" shape or "J" shape. The variety of potential probability density functions are difficult to capture in a single plot. Figure 5.11 contains a 3×3 matrix of probability density functions, where

- the rows correspond to $\alpha = 1/2$, $\alpha = 1$, and $\alpha = 2$,

- the columns correspond to $\beta = 1/2$, $\beta = 1$, and $\beta = 2$.

Due to space constraints the axis labels are suppressed. The horizontal axis on each of the 9 plots is x and the vertical axis is $f(x)$. The mirror image about the line $x = 1/2$ for non-diagonal elements on the opposite sides of the 3×3 matrix illustrates the symmetric role of the α and β parameters. These plots highlight the variety of potential shapes for the probability density function.

We end our discussion of the beta distribution with an example illustrating a distribution that has a parameter that is itself a random variable.

Example 5.17 The binomial distribution was presented in the last chapter with parameters n, a fixed positive integer, and a fixed probability $0 < p < 1$. But what if the probability p is unknown? More specifically, if $X \sim \text{binomial}(n, p)$ and p is a random variable which has the beta distribution, that is, $P \sim \text{beta}(\alpha, \beta)$, what is the probability mass function of X?

This is a continuous mixture situation as described in Section 3.3. The unconditional probability mass function of X is

$$
\begin{aligned}
f_X(x) &= \int_0^1 f_P(p) f_{X \mid p}(x \mid p) \, dp \\
&= \int_0^1 \left[\frac{\Gamma(\alpha+\beta)}{\Gamma(\alpha)\Gamma(\beta)} \, p^{\alpha-1}(1-p)^{\beta-1} \right] \left[\binom{n}{x} p^x (1-p)^{n-x} \right] dp \\
&= \binom{n}{x} \frac{\Gamma(\alpha+\beta)}{\Gamma(\alpha)\Gamma(\beta)} \int_0^1 p^{\alpha+x-1}(1-p)^{n-x+\beta-1} \, dp \\
&= \binom{n}{x} \frac{\Gamma(\alpha+\beta)}{\Gamma(\alpha)\Gamma(\beta)} \cdot \frac{\Gamma(\alpha+x)\Gamma(\beta+n-x)}{\Gamma(\alpha+\beta+n)} \\
&= \frac{\Gamma(n+1)\Gamma(\alpha+\beta)\Gamma(\alpha+x)\Gamma(\beta+n-x)}{\Gamma(x+1)\Gamma(n-x+1)\Gamma(\alpha)\Gamma(\beta)\Gamma(\alpha+\beta+n)} \qquad x = 0, 1, 2, \ldots.
\end{aligned}
$$

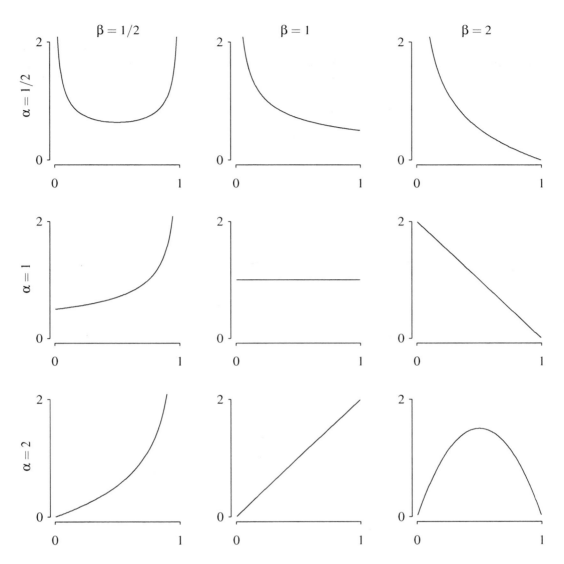

Figure 5.11: Probability density functions for various beta distributions.

This probability mass function corresponds to what is known as the *beta–binomial distribution*. This solution can be verified in APPL with the statements

```
Y := BetaRV(alpha, beta);
X := BinomialRV(n, p);
int(Y[1][1](p) * X[1][1](x), p = 0 .. 1);
```

which confirm the analytic derivation above.

Triangular distribution

Situations arise when there is no data available on a random quantity of interest. The triangular distribution has three parameters: the minimum a, the mode b and the maximum c. I encountered an application of the triangular distribution when working at a NASA research center in 1985. I was constructing a discrete-event simulation model of the proposed space station which required the time to assemble a particular truss structure in space. No data or estimates existed on this assembly time, so I asked one of the more experienced engineers the following three questions: (*a*) If everything went right, how quickly could the truss structure get assembled? (*b*) What is the most likely amount of time to assemble the truss structure? (*c*) If everything went wrong, how long would it take to assemble the truss structure? From the engineer's responses to these three questions, I was able to construct a crude approximate probability model using the triangular distribution for the truss structure assembly time.

A continuous random variable X with probability density function

$$f(x) = \begin{cases} \dfrac{2(x-a)}{(b-a)(c-a)} & a < x \leq b \\[2ex] \dfrac{2(c-x)}{(c-a)(c-b)} & b < x < c, \end{cases}$$

for some real constants satisfying $a < c$ and $a \leq b \leq c$ is a triangular(a, b, c) random variable. The parameter a is the *minimum*; the parameter b is the *mode*; the parameter c is the *maximum*. A generic triangular(a, b, c) probability density function is shown in Figure 5.12.

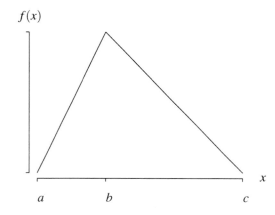

Figure 5.12: Probability density function for $X \sim$ triangular(a, b, c).

The cumulative distribution function of $X \sim$ triangular(a, b, c) on $a < x \leq b$ is

$$F(x) = \int_a^x \frac{2(t-a)}{(b-a)(c-a)} \, dt = \frac{(x-a)^2}{(b-a)(c-a)}.$$

The cumulative distribution function on $b < x < c$ is

$$F(x) = \frac{b-a}{c-a} + \int_b^x \frac{2(c-t)}{(c-a)(c-b)} \, dt = \frac{-x^2 + 2cx - ac + ab - bc}{(c-a)(c-b)} = 1 - \frac{(c-x)^2}{(c-a)(c-b)}.$$

Thus, the cumulative distribution function is the piecewise quadratic function

$$F(x) = \begin{cases} 0 & x \leq a \\[1ex] \dfrac{(x-a)^2}{(b-a)(c-a)} & a < x \leq b \\[2ex] 1 - \dfrac{(c-x)^2}{(c-a)(c-b)} & b < x < c \\[2ex] 1 & x \geq c. \end{cases}$$

The integrations can be avoided by computing the cumulative distribution function with the APPL statement CDF(TriangularRV(a, b, c)).

The population mean of $X \sim \text{triangular}(a, b, c)$ is

$$
\begin{aligned}
E[X] &= \int_{-\infty}^{\infty} x f(x) \, dx \\
&= \int_{a}^{b} x \frac{2(x-a)}{(b-a)(c-a)} \, dx + \int_{b}^{c} x \frac{2(c-x)}{(c-a)(c-b)} \, dx \\
&= \frac{a+b+c}{3}.
\end{aligned}
$$

Similarly, the population variance is

$$
V[X] = \frac{a^2 + b^2 + c^2 - ab - ac - bc}{18}.
$$

Finally, the population skewness and kurtosis are

$$
E\left[\left(\frac{X-\mu}{\sigma}\right)^3\right] = \frac{\sqrt{2}(a-2b+c)(2a-b-c)(a-b-2c)}{5(a^2+b^2+c^2-ab-ac-bc)^{3/2}}
$$

and

$$
E\left[\left(\frac{X-\mu}{\sigma}\right)^4\right] = \frac{12}{5}.
$$

These quantities can be calculated with the APPL statements

```
X := TriangularRV(a, b, c);
Mean(X);
Variance(X);
Skewness(X);
Kurtosis(X);
```

Example 5.18 Let $X \sim \text{triangular}(0, 2, 4)$.

(a) Find $E\left[|X - 3|\right]$.

(b) If X is the lifetime of a personal computer in years, find the expected remaining lifetime of a personal computer that is three years old.

(c) Check the results of parts (a) and (b) using APPL.

The probability density function of X is

$$
f_X(x) = \begin{cases} x/4 & 0 < x \leq 2 \\ (4-x)/4 & 2 < x < 4. \end{cases}
$$

(a) The expected value of $|X - 3|$ is

$$
\begin{aligned}
E\left[|X - 3|\right] &= \int_{-\infty}^{\infty} |x - 3| f_X(x) \, dx \\
&= \int_{-\infty}^{3} (3 - x) f_X(x) \, dx + \int_{3}^{\infty} (x - 3) f_X(x) \, dx \\
&= \int_{0}^{2} (3 - x) \cdot \frac{x}{4} \, dx + \int_{2}^{3} (3 - x) \cdot \frac{4 - x}{4} \, dx + \int_{3}^{4} (x - 3) \cdot \frac{4 - x}{4} \, dx \\
&= \frac{5}{6} + \frac{5}{24} + \frac{1}{24} \\
&= \frac{13}{12}.
\end{aligned}
$$

(b) The probability density function of the failure time W of the personal computer that is still operating after three years is

$$f_W(w) = \frac{f_X(w)}{\int_3^4 f_X(x)\,dx} = \frac{(4-w)/4}{1/8} = 8 - 2w \qquad 3 < w < 4.$$

So the expected time of failure is

$$E[W] = \int_3^4 w \cdot (8-2w)\,dw = \left[4w^2 - \frac{2}{3}w^3\right]_3^4 = \frac{10}{3}.$$

So the expected *remaining* lifetime is $1/3$ of a year.

(c) The APPL code to solve these problems is:

```
X := TriangularRV(0, 2, 4);
ExpectedValue(X, x -> abs(x - 3));
W := Truncate(X, 3, 4);
Mean(W) - 3;
```

Weibull distribution

The gamma distribution was introduced in Section 5.3 as a generalization of the exponential distribution. A second generalization of the exponential distribution, which is quite popular in modeling the lifetimes of products in reliability analysis and survival times in biostatistics, is the *Weibull distribution*.

As mentioned earlier, the exponential distribution is limited in applicability because of the memoryless property, so probability distributions that generalize the exponential are useful. A continuous random variable X with probability density function

$$f(x) = \kappa\lambda^\kappa x^{\kappa-1} e^{-(\lambda x)^\kappa} \qquad x > 0$$

for real positive constants λ and κ is a Weibull(λ, κ) random variable. Like the gamma distribution, λ and κ are the scale and shape parameters of the distribution, respectively. The exponential distribution is a special case of the Weibull distribution when $\kappa = 1$. Another special case occurs when $\kappa = 2$, commonly known as the *Rayleigh distribution*. When $3 < \kappa < 4$, the probability density function resembles that of a normal probability density function, and the mode and median of the distribution are equal when $\kappa \cong 3.26$. Graphs of the probability density functions of several Weibull distributions for $\lambda = 1$ and several values of κ are given in Figure 5.13.

The cumulative distribution function of the Weibull distribution has a closed-form expression. For $x > 0$,

$$F(x) = \int_0^x \kappa\lambda^\kappa w^{\kappa-1} e^{-(\lambda w)^\kappa}\,dw = \left[e^{-(\lambda w)^\kappa}\right]_0^x = 1 - e^{-(\lambda x)^\kappa}.$$

So for all values of x, the cumulative distribution function is

$$F(x) = \begin{cases} 0 & x \le 0 \\ 1 - e^{-(\lambda x)^\kappa} & x > 0. \end{cases}$$

Population moments for the Weibull distribution are not as mathematically tractable as those for the exponential distribution. Using the expression

$$E[X^s] = \frac{s}{\kappa\lambda^s}\Gamma\left(\frac{s}{\kappa}\right),$$

for $s = 1, 2, \ldots$, which is proved by using a substitution in the integral associated with the expected value, the population mean, variance, skewness, and kurtosis for $X \sim$ Weibull(λ, κ) are

$$E[X] = \frac{1}{\lambda}\Gamma\left(1 + \frac{1}{\kappa}\right) = \frac{1}{\lambda\kappa}\Gamma\left(\frac{1}{\kappa}\right) \qquad V[X] = \frac{1}{\lambda^2}\left\{\frac{2}{\kappa}\Gamma\left(\frac{2}{\kappa}\right) - \left[\frac{1}{\kappa}\Gamma\left(\frac{1}{\kappa}\right)\right]^2\right\}$$

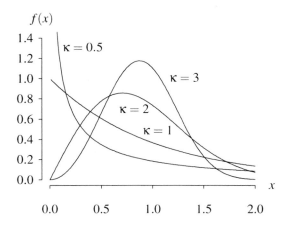

Figure 5.13: Probability density functions for $X \sim \text{Weibull}(1, \kappa)$.

$$E\left[\left(\frac{X-\mu}{\sigma}\right)^3\right] = \left\{\frac{2}{\kappa}\Gamma\left(\frac{2}{\kappa}\right) - \left[\frac{1}{\kappa}\Gamma\left(\frac{1}{\kappa}\right)\right]^2\right\}^{-3/2} \left\{\frac{3}{\kappa}\Gamma\left(\frac{3}{\kappa}\right) - \frac{6}{\kappa^2}\Gamma\left(\frac{1}{\kappa}\right)\Gamma\left(\frac{2}{\kappa}\right) + 2\left[\frac{1}{\kappa}\Gamma\left(\frac{1}{\kappa}\right)\right]^3\right\}$$

$$E\left[\left(\frac{X-\mu}{\sigma}\right)^4\right] = \left\{\frac{2}{\kappa}\Gamma\left(\frac{2}{\kappa}\right) - \left[\frac{1}{\kappa}\Gamma\left(\frac{1}{\kappa}\right)\right]^2\right\}^{-2} \left\{\frac{4}{\kappa}\Gamma\left(\frac{4}{\kappa}\right) - \frac{12}{\kappa^2}\Gamma\left(\frac{1}{\kappa}\right)\Gamma\left(\frac{3}{\kappa}\right)\right.$$

$$\left. + \frac{12}{\kappa^3}\left[\Gamma\left(\frac{1}{\kappa}\right)\right]^2\Gamma\left(\frac{2}{\kappa}\right) - \frac{3}{\kappa^4}\left[\Gamma\left(\frac{1}{\kappa}\right)\right]^4\right\}.$$

Example 5.19 The lifetime X, measured in hours, of a spring that is operated continuously has cumulative distribution function

$$F(x) = \begin{cases} 0 & x \leq 0 \\ 1 - e^{-(\lambda x)^{\kappa}} & x > 0, \end{cases}$$

where $\lambda = 0.001$ and $\kappa = 2$. Find

(a) the probability that a new spring survives 30 hours,

(b) the probability that a spring that has been operating without failure for 50 hours survives an additional 30 hours.

The lifetime of the spring is recognized as a Weibull$(0.001, 2)$ random variable. Since the spring is a mechanical item, it is reasonable to assume that it is degrading with time. Thus, one would expect the probability in part (a) to be greater than the probability in part (b).

(a) The probability that a new spring survives 30 hours is equal to the probability that the random variable X exceeds 30, that is

$$\begin{aligned} P(X > 30) &= 1 - P(X \leq 30) \\ &= 1 - F(30) \\ &= e^{-(0.001 \cdot 30)^2} \\ &= e^{-0.0009} \\ &\cong 0.9991. \end{aligned}$$

(b) The probability that a spring that has been operating without failure for 50 hours survives an additional 30 hours is $P(X > 80 | X > 50)$, which is found using the definition of conditional probability:

$$
\begin{aligned}
P(X > 80 | X > 50) &= \frac{P(X > 80 \text{ and } X > 50)}{P(X > 50)} \\
&= \frac{P(X > 80)}{P(X > 50)} \\
&= \frac{1 - P(X \leq 80)}{1 - P(X \leq 50)} \\
&= \frac{1 - F(80)}{1 - F(50)} \\
&= \frac{e^{-(0.001 \cdot 80)^2}}{e^{-(0.001 \cdot 50)^2}} \\
&= \frac{e^{-0.0064}}{e^{-0.0025}} \\
&= e^{-0.0039} \\
&\cong 0.9961.
\end{aligned}
$$

The spring is degrading with time, so, as expected, the used spring has a lower probability of surviving an additional 30 hours than a new spring.

This chapter has introduced continuous distributions that often arise in probability applications. The key continuous distributions presented in this chapter are summarized in Table 5.1.

Distribution	$f(x)$	Support	Mean	Variance
$U(a, b)$	$\dfrac{1}{b - a}$	$a < x < b$	$\dfrac{a + b}{2}$	$\dfrac{(b - a)^2}{12}$
exponential(λ)	$\lambda e^{-\lambda x}$	$x > 0$	$\dfrac{1}{\lambda}$	$\dfrac{1}{\lambda^2}$
gamma(λ, κ)	$\dfrac{\lambda^\kappa x^{\kappa-1} e^{-\lambda x}}{\Gamma(\kappa)}$	$x > 0$	$\dfrac{\kappa}{\lambda}$	$\dfrac{\kappa}{\lambda^2}$
$N(\mu, \sigma^2)$	$\dfrac{1}{\sigma\sqrt{2\pi}} e^{-(x-\mu)^2/2\sigma^2}$	$-\infty < x < \infty$	μ	σ^2

Table 5.1: Common continuous distributions.

5.6   Exercises

5.1   If $X \sim U(-5, 5)$ find $E\big[\,\big|\,|X| - 2\,\big|\,\big]$.

5.2   A random variable X assumes the value of a Bernoulli$(1/3)$ random variable with probability $1/4$ and assumes the value of a $U(0, 1)$ random variable with probability $3/4$. Find the moment generating function of X.

5.3   A deck is made up of slats that are 6 inches wide with negligible distance between the slats. A frisbee with a 8 inch diameter lands at a random position on the deck. Find the probability that the frisbee covers a portion of three slats.

5.4 Consider the continuous random variable X with probability density function

$$f(x) = \frac{1}{b-a} \qquad a < x < b$$

for real constants $a < b$. Write this probability density function parameterized by its population mean μ and its population standard deviation σ.

5.5 Let $X \sim U(0, 5280)$. Find the probability mass function of $Y = \gcd\left(\lceil X \rceil, 10\right)$, where $\lceil \cdot \rceil$ is the ceiling function and $\gcd(\cdot, \cdot)$ is the greatest common divisor function (for example, $\gcd(9, 12) = 3$).

5.6 A stick of length a is broken at a random point that is equally likely along the length of the stick. What is the probability that the longer piece is more than twice as long as the shorter piece?

5.7 Locating pleasant facilities, such as parks, hospitals, and fire stations, is often done in a manner that minimizes the expected traveling distance for patrons. Likewise, locating unpleasant facilities, such as sewage treatment plants or high power cables is often done in a manner to maximize the expected distance.

    (a) Consider locating a pleasant facility on a road of length a where demand positions X that occur are located as $U(0, a)$ random variables. What is the position of the facility on the road x_0, such that $E[|X - x_0|]$ is minimized?

    (b) Where should the facility be located for any arbitrary demand distribution with probability density function $f(x)$ defined on the length of the road?

    (c) Where should the facility be located for any arbitrary demand distribution with probability mass function $f(x)$ defined on the length of the road?

5.8 Let $X \sim U(2, 9)$. Find $E\left[X - \lfloor X \rfloor\right]$.

5.9 Matrix theory traditionally emphasizes matrices whose elements are real or complex constants. But what if the elements of a matrix are random variables? Such matrices are referred to as "stochastic" or "random" matrices. Although a myriad of questions can be asked concerning random matrices, our emphasis here will be limited to the following question: if the elements of a 3×3 matrix are independent $U(0, 1)$ random variables with positive diagonal elements and negative off-diagonal elements, what is the probability that the matrix has a positive determinant? That is, what is the probability that

$$\begin{vmatrix} +u_{11} & -u_{12} & -u_{13} \\ -u_{21} & +u_{22} & -u_{23} \\ -u_{31} & -u_{32} & +u_{33} \end{vmatrix} > 0,$$

where the u_{ij}'s are mutually independent $U(0, 1)$ random variables? This question is rather vexing using probability theory due to the appearance of some of the random numbers multiple times in the expression for the determinant. Use Monte Carlo simulation in R to estimate the probability of interest.

5.10 Consider a circle with a random radius $R \sim U(0, 5)$. What is the probability that the area of the circle is greater than the circumference?

5.11 What is the probability mass function of the value printed by the following R code?

```
count = 0
for (i in 1:10000)
   if (sum(runif(2)) > 1.5) count = count + 1
print(count)
```

5.12   Let $X \sim U(0, \theta)$. Find the probability density function of

$$Y = \frac{X - \mu}{\sigma},$$

where $\mu = E[X]$ and $\sigma^2 = V[X]$.

5.13   The lifetime X of a certain brand of sixty-watt light bulb is exponentially distributed with population mean 1000 hours, that is,

$$f(x) = \frac{1}{1000}e^{-x/1000} \qquad\qquad x > 0.$$

If 30 such light bulbs are placed on test, find the probability that 10 or fewer of these light bulbs survive to time 1200 by

(a) writing a mathematical expression,

(b) giving an R statement that will compute the probability.

5.14   A box contains the following six flashlights.

- There is one Type 1 flashlight with an exponential(1) lifetime.
- There are two Type 2 flashlights with exponential($1/2$) lifetimes.
- There are three Type 3 flashlight with exponential($1/3$) lifetimes.

You select a flashlight at random from the box.

(a) What is the probability that the flashlight selected lasts more than 4 hours?

(b) Given that the flashlight lasts more than 4 hours, what is the probability that it is a Type 3 flashlight?

5.15   The mean residual life function for a positive random variable X is defined by

$$E(X - x \,|\, X \geq x)$$

when the expectation exists. Find the mean residual life function for a random variable X having an exponential distribution with population mean $1/\lambda$.

5.16   If $E[X^n] = n!$ for $n = 1, 2, \ldots$, find the probability density function of the random variable X.

5.17   In a *time-truncated life test*, there are n light bulbs simultaneously placed on test at time 0. The test is concluded at time $c > 0$. Assuming that the light bulbs lifetimes are drawn from an exponential population with mean $1/\lambda$, where $\lambda > 0$, find the distribution of the *number* of failures that occur by time c.

5.18   Let the random variable X have the chi-square distribution with 14 degrees of freedom. Write a single R statement to calculate $P(5.629 < X < 26.119)$.

5.19   What value is returned by the following APPL statements?

```
X := ChiSquareRV(3);
Y := ChiSquareRV(4);
W := Convolution(X, Y);
Mean(W);
```

5.20   Let $Z \sim N(0, 1)$. Find $z_{0.96}$ using R.

5.21   Find the 99th percentile of a normal random variable with $\mu = 8$ and $\sigma^2 = 9$.

5.22   Find the 95th percentile of the *square* of a standard normal random variable.

5.23   Find the population skewness and kurtosis of a $N(\mu, \sigma^2)$ random variable.

5.24   If $X \sim N(\mu, 100)$, find μ such that $P(X < 22.3) = 0.8790$.

5.25   Write the R statements and associated output to compute the following quantities. Draw a graph of the probability mass function or probability density function illustrating the quantity being computed.

   (a)   The 90th percentile of a $U(0, 20)$ random variable.

   (b)   The 20th percentile of a $N(0, 1)$ random variable.

   (c)   The interquartile range of a $N(5, 9)$ random variable.

   (d)   The interquartile range of a binomial$(10, 0.4)$ random variable.

5.26   The random variable $X \sim N\left(\mu, \sigma^2\right)$ has moment generating function

$$M_X(t) = e^{\mu t + \frac{1}{2}\sigma^2 t^2} \qquad -\infty < t < \infty.$$

What is the moment generating function and distribution of $Y = 3X + 4$?

5.27   Assume that men's heights are $N(70, 16)$ and women's heights are $N(67, 9)$. Let X_1 be the height of a man and X_2 be the height of a woman selected at random. Use Monte Carlo simulation in R to estimate the 1st and 99th percentile of the taller of the man and woman, that is, estimate $y_{0.01}$ and $y_{0.99}$ associated with $Y = \max\{X_1, X_2\}$. Use 1000 simulated pairs.

5.28   Let X be normally distributed with population mean $\mu = 2$ and population variance $\sigma^2 = 9$.

   (a)   What is $P(X < 8)$?

   (b)   What is $P(4 < X < 8)$?

   (c)   What is $P(X > 10)$?

   (d)   Find a constant a such that $P(X < a) = 0.95$.

   (e)   Find a constant b such that $P(-b < X < b) = 0.95$.

   (f)   If six random variables with this probability distribution are observed, what is the probability that exactly four of them are less than eight?

5.29   A fair die is tossed 1000 times.

   (a)   Find the expected number of threes.

   (b)   Find the probability of 200 or fewer threes exactly and by using the normal approximation to the binomial distribution.

   (c)   Find the probability of 200 or fewer threes exactly and by using the normal approximation to the binomial distribution given that there were exactly 500 even numbers that appeared in the 1000 tosses.

5.30   Adult men's heights (in inches) are normally distributed with $\mu = 70$ and $\sigma^2 = 9$. Adult women's heights (in inches) are normally distributed with $\mu = 67$ and $\sigma^2 = 4$. Members of the "Beanstalk Club" must be at least six feet tall.

   (a)   If a man and a woman are selected at random from the respective populations, what is the probability that both are eligible for the Beanstalk Club?

   (b)   If a woman who is eligible for the Beanstalk Club is selected at random, what is her expected height?

5.31   IQ scores are normally distributed with a population mean of 100 and a standard deviation of 10. If six independent IQ scores are collected, what is the probability that three are less than 90, two are between 90 and 120, and one is greater than 120?

5.32   Show that the probability density function for $X \sim N\left(\mu, \sigma^2\right)$ has inflection points at the x-values $\mu \pm \sigma$.

5.33   Show that the standard double-exponential (Laplace) random variable X with probability density function

$$f(x) = \frac{1}{2} e^{-|x|} \qquad -\infty < x < \infty$$

has moment generating function $M(t) = 1/(1 - t^2)$, for $|t| < 1$.

5.34   What is the name of the distribution of an exponential(1) random variable truncated on the left at 3, then shifted 3 units to the left?

5.35   What is the name of the distribution which is a special case of a Weibull random variable with a shape parameter equal to 1?

5.36   What is the name of the distribution of a constant μ added to the product of a standard normal random variable and a constant σ?

5.37   Use each of the following distributions once (and only once) to respond to the following statements: beta, Cauchy, triangular, Weibull, Zipf.

(a) It has undefined moments.

(b) It could be used for modeling the *lifetime* of a cat.

(c) It could be used for modeling the *fraction* of consumers that prefer one soft drink over another.

(d) It could be used for modeling the *number* of puppies in a litter.

(e) It is the distribution of the sum of two independent $U(0, 1)$ random variables.

5.38   For each probability below, give an expression and *carefully* plot (that is, use a computer or carefully plot by hand) on a single set of axes.

(a) $P(\mu - k\sigma < X < \mu + k\sigma)$ when $X \sim N(\mu, \sigma^2)$ for $0 \le k \le 4$.

(b) $P(\mu - k\sigma < X < \mu + k\sigma)$ when X is exponentially distributed with population mean 2 for $0 \le k \le 4$.

(c) $P(\mu - k\sigma < X < \mu + k\sigma)$ when X has a Poisson distribution with population mean 4 for $0 \le k \le 4$.

(d) The lower bound on $P(\mu - k\sigma < X < \mu + k\sigma)$ provided by Chebyshev's inequality for $1 \le k \le 4$.

5.39   Let the random variable X have a special case of the *extreme value distribution* with cumulative distribution function

$$F(x) = e^{-e^{-x}} \qquad -\infty < x < \infty.$$

For real constants a and b satisfying $a < b$, find $P(a < X < b)$.

5.40   Consider the random variable X with the *double exponential* distribution with probability density function

$$f(x) = \frac{1}{2} e^{-|x|} \qquad -\infty < x < \infty.$$

Find $E\left[X^n\right]$ for any positive integer n.

Chapter 6

Joint Distributions

Thus far, we have only considered single random variables, also known as univariate random variables. This chapter introduces the notion of jointly distributed random variables, where the interest is in the probabilistic behavior of two or more random variables simultaneously. In order to ease the transition from one to several random variables, the first four sections consider just two random variables at a time. The final section extends the definitions and results for two random variables to the more general case of n random variables. The first section generalizes probability mass functions, probability density functions, and cumulative distribution functions from univariate distributions to bivariate distributions. The new concepts of marginal and conditional distributions are also introduced in the first section. The second section defines independent random variables and their application. The third section defines the expected value operator in two dimensions, with particular emphasis on the population covariance and correlation between two random variables. The fourth section defines the bivariate normal distribution, which is the two-dimensional analog of the normal distribution. Finally, the last section generalizes these concepts to n dimensions.

There are many other settings where there is an interest in the joint distribution of several random variables simultaneously. For example

- an economist might be interested in the joint distribution of GDP and unemployment,

- a sociologist might be interested in the joint distribution of a wife's height and husband's height,

- a vending operator at a college football stadium might be interested in the joint distribution of soft drink sales and hot dog sales,

- a physician might be interested in the joint distribution of cholesterol level, triglyceride level, and blood pressure.

The purpose of this chapter is to extend the probability models for random variables developed so far to two or more random variables.

6.1   Bivariate Distributions

A U.S. government website lists the number of fatal automobile crashes in 2008 categorized by alcohol-impaired driving status and by time of day, as shown in Table 6.1. There were 261 crashes where the time of the crash was unknown, which have been omitted from the table. Alcohol-impaired driving is when at least one driver or motorcycle rider has a blood alcohol content of 0.08 or higher. The table shows that the percentage of fatal crashes with alcohol impairment varies from a low of 9% during the late morning to a high of 64% in the three-hour period after midnight, so a time-of-day effect seems to be present. The effect of alcohol impairment on the likelihood of a

Time of day	Number of non-alcohol impaired crashes	Number of alcohol impaired crashes	Total	Percent alcohol impaired crashes
Midnight to 2:59 a.m.	1603	2883	4486	64%
3 a.m. to 5:59 a.m.	1413	1361	2774	49%
6 a.m. to 8:59 a.m.	2768	468	3236	14%
9 a.m. to 11:59 a.m.	2992	293	3285	9%
Noon to 2:59 p.m.	3880	476	4356	11%
3 p.m. to 5:59 p.m.	4269	1056	5325	20%
6 p.m. to 8:59 p.m.	3636	1706	5342	32%
9 p.m. to 11:59 p.m.	2640	2312	4952	47%
Total	23,201	10,555	33,756	

Table 6.1: Fatal automobile crash counts in the United States in 2008.

crash is self evident. These statistics should be considered in light of the number of cars on the road and the fact that it is riskier to drive at night.

Table 6.1 shows that two variables, time of day and alcohol-impairment status, seem to influence the probability of a crash. Defining the random variable X to be the various three-hour time segments numbered sequentially beginning at midnight, and the random variable Y to be a binary (Bernoulli) random variable denoting alcohol-impairment status (0 for not impaired and 1 for impaired), a joint probability distribution can be defined as illustrated in Table 6.2. The decimal values in the table are simply the crash counts from Table 6.1 divided by $33,756$, the total number of crashes observed. The $8 \times 2 = 16$ probabilities sum to one, as expected of a probability mass function. The column labeled "Total" reflects what will be defined later in this section as the marginal distribution for X and the row labeled "Total" reflects what will be defined later as the marginal distribution for Y. The marginal distributions are the probability distributions of X and Y ignoring the value of the other random variable.

	$Y = 0$	$Y = 1$	Total
$X = 1$	0.047	0.085	0.132
$X = 2$	0.042	0.041	0.083
$X = 3$	0.082	0.014	0.096
$X = 4$	0.088	0.009	0.097
$X = 5$	0.115	0.014	0.129
$X = 6$	0.126	0.031	0.157
$X = 7$	0.108	0.051	0.159
$X = 8$	0.079	0.068	0.147
Total	0.687	0.313	1.000

Table 6.2: Joint probability mass function of X and Y.

The formal definition of a pair of jointly distributed random variables X and Y is analogous to the definition for a univariate random variable X.

Definition 6.1 Given a random experiment with an associated sample space S, define the *bivariate random variables X and Y* that assign to each element $s \in S$ one and only one pair of real numbers $X(s) = x$ and $Y(s) = y$. The *support* of these random variables is the set of ordered pairs

$$A = \{(x, y) \mid x = X(s), y = Y(s), s \in S\}.$$

As in the univariate case, the $X(s)$ and $Y(s)$ notation is typically shortened to just X and Y for brevity.

Example 6.1 Deal two cards from a well-shuffled deck. Let the random variable X be the number of aces dealt and let the random variable Y be the number of face cards dealt. Find

(a) the support $\mathcal{A}$,

(b) $P(A)$, where $A = \{(x, y) \mid x = y\}$.

(a) In order to find the support of X and Y, first consider the random variables individually. Since two cards are dealt, both X and Y can assume the values 0, 1, and 2. Furthermore, it is not possible for the sum of X and Y to exceed 2, so the support of X and Y can be given by enumerating the six possible (X, Y) pairs:

$$\mathcal{A} = \{(x, y) \mid (x, y) = (0, 0), (0, 1), (0, 2), (1, 0), (2, 0), \text{ or } (1, 1)\}.$$

This would be a tedious way to state the support of X and Y if, for example, 13 cards were dealt, so an alternative, more compact way to state the support is

$$\mathcal{A} = \{(x, y) \mid x = 0, 1, 2; y = 0, 1, 2; x + y \leq 2\}.$$

(b) Since the order that the cards are dealt is not relevant, there are

$$\binom{52}{2} = \frac{52!}{50!\,2!} = \frac{52 \cdot 51}{2 \cdot 1} = 26 \cdot 51 = 1326$$

equally-likely deals. The probability that a deal contains x aces and y face cards is

$$P(X = x, Y = y) = \frac{\binom{4}{x}\binom{12}{y}\binom{36}{2-x-y}}{\binom{52}{2}} \qquad x = 0, 1, 2; y = 0, 1, 2; x + y \leq 2.$$

Calculating these six probabilities results in the entries in Table 6.3. Not surprisingly, the entries in Table 6.3 sum to 1. To find the probability that the same number of aces as face cards are dealt, we need only sum the two appropriate probabilities:

$$P(A) = P(X = Y) = P(X = Y = 0) + P(X = Y = 1) = \frac{630}{1326} + \frac{48}{1326} = \frac{113}{221} \cong 0.5113.$$

x \ y	0	1	2
0	630/1326	432/1326	66/1326
1	144/1326	48/1326	0
2	6/1326	0	0

Table 6.3: Joint probability distribution of X and Y.

Capturing the distribution of probability with probability mass functions and probability density functions is easily extended to two dimensions. For some set $A \subset \mathcal{A}$, if $P(A)$ can be expressed as

$$P(A) = P\big((X, Y) \in A\big) = \sum \sum_A f(x, y)$$

when X and Y are discrete random variables, then $f(x, y)$ is the *joint probability mass function* of X and Y. Likewise, for some set $A \subset \mathcal{A}$, if $P(A)$ can be expressed as

$$P(A) = P((X, Y) \in A) = \int \int_A f(x, y) \, dy \, dx$$

when X and Y are continuous random variables, then $f(x, y)$ is the *joint probability density function* of X and Y.

A joint probability mass function for discrete random variables X and Y must satisfy the *existence conditions*

$$\sum_{\mathcal{A}} \sum f(x, y) = 1 \qquad \text{and} \qquad f(x, y) \geq 0 \text{ for all real } x \text{ and } y.$$

Likewise, a joint probability density function for continuous random variables X and Y must satisfy the *existence conditions*

$$\int \int_{\mathcal{A}} f(x, y) \, dy \, dx = 1 \qquad \text{and} \qquad f(x, y) \geq 0 \text{ for all real } x \text{ and } y.$$

Example 6.2 Deal two cards from a well-shuffled deck. Let the random variable X be the number of aces dealt and let the random variable Y be the number of face cards dealt. Find the joint probability mass function of X and Y and calculate the probability that the hand will contain more aces than face cards.

The joint probability mass function of X and Y was calculated in the previous example as

$$f(x, y) = \frac{\binom{4}{x} \binom{12}{y} \binom{36}{2-x-y}}{\binom{52}{2}} \qquad x = 0, 1, 2; \ y = 0, 1, 2; \ x+y \leq 2.$$

To calculate the probability that the hand will contain more aces than face cards, simply sum over all values of X and Y satisfying $X > Y$, that is

$$
\begin{aligned}
P(X > Y) &= P(X = 1, Y = 0) + P(X = 2, Y = 0) \\
&= f(1, 0) + f(2, 0) \\
&= \frac{144}{1326} + \frac{6}{1326} \\
&= \frac{150}{1326} \\
&= \frac{25}{221} \\
&\cong 0.1131.
\end{aligned}
$$

Example 6.3 A pair of fair dice is tossed. Let X be the value of the smaller number tossed, and Y be the value of the larger number tossed. Find

(a) the joint probability mass function $f(x, y)$,

(b) $P(Y = 2X)$.

(a) The joint probability mass function of X and Y is a folding of the 36 equally-likely outcomes across the $X = Y$ outcome, as shown in Table 6.4. The support of X and Y is

$$\mathcal{A} = \{(x, y) \, | \, x = 1, 2, \ldots, 6; \ y = x, x+1, \ldots, 6\}.$$

x \ y	1	2	3	4	5	6
1	1/36	2/36	2/36	2/36	2/36	2/36
2	0	1/36	2/36	2/36	2/36	2/36
3	0	0	1/36	2/36	2/36	2/36
4	0	0	0	1/36	2/36	2/36
5	0	0	0	0	1/36	2/36
6	0	0	0	0	0	1/36

Table 6.4: Joint probability mass function of X and Y.

(b) To find $P(Y = 2X)$, sum over the three appropriate mass values

$$
\begin{aligned}
P(Y = 2X) &= P(X=1, Y=2) + P(X=2, Y=4) + P(X=3, Y=6) \\
&= \frac{2}{36} + \frac{2}{36} + \frac{2}{36} \\
&= \frac{1}{6}.
\end{aligned}
$$

The last two examples have considered jointly distributed *discrete* random variables X and Y. The next four examples illustrate calculations with jointly distributed *continuous* random variables.

Example 6.4 Xavier and Yolanda agree to meet at the sun dial at 2:00 PM. Neither, however, is particularly punctual, and their actual arrival times to the sun dial are independent and uniformly distributed between 2:00 PM and 3:00 PM. Assuming that each will wait 15 minutes for the other, find the probability that they will actually meet.

Let X be Xavier's arrival time and Y be Yolanda's arrival time, both measured in minutes after the hour. Since X and Y are each $U(0, 60)$ random variables, their individual probability density functions are $f_X(x) = 1/60$ for $0 < x < 60$ and $f_Y(y) = 1/60$ for $0 < y < 60$. It is reasonable to conjecture that the joint probability density function of X and Y is a constant over the square with opposite vertices $(0, 0)$ and $(60, 60)$, that is

$$
f(x, y) = \frac{1}{3600} \qquad 0 < x < 60, 0 < y < 60.
$$

This joint probability density function $f(x, y)$, defined on the 60×60 square region $\mathcal{A}$ shown in Figure 6.1, is constant. This shaded area associated with the arrival times

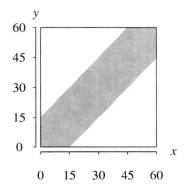

Figure 6.1: Support of X and Y and shaded region of interest.

when Xavier and Yolanda meet is also shown in Figure 6.1. The probability that Xavier and Yolanda will meet is the volume under the joint probability density function above the shaded area. This can be computed via double integration, but it is more easily computed geometrically. Since the area of the two unshaded triangles is half the product of their bases and the heights, the probability that they meet is

$$P\bigl(|X - Y| < 15\bigr) = \frac{1}{3600}\bigl[60^2 - 45^2\bigr] = \frac{1575}{3600} = \frac{7}{16} = 0.4375.$$

This result can be checked via Monte Carlo simulation with the following R statements

```
nrep = 100000
x = runif(nrep, 0, 60)
y = runif(nrep, 0, 60)
sum(abs(x - y) < 15) / nrep
```

which, after a call to set.seed(3), yields

| 0.43489 | 0.43770 | 0.43651 | 0.43520 | 0.43812. |

These values hover about the exact analytic solution $P\bigl(|X - Y| < 15\bigr) = 0.4375$, so the Monte Carlo experiments support the analytic solution. The probability that Xavier and Yolanda meet at the sun dial is easily extended to the probability that two data packets collide in a communications network.

Example 6.5 Let the continuous random variables X and Y have joint probability density function

$$f(x, y) = 1 \qquad\qquad 0 < x < 1, 0 < y < 1.$$

Find $P\bigl(0 < X^2 < Y < 1\bigr)$.

The support $\mathcal{A}$, which is the unit square, and the region of interest are shown in Figure 6.2. The joint probability density function assumes constant value of 1 over the unit square. The desired probability is the volume under $f(x, y)$ above the shaded region in Figure 6.2. This can be calculated with the following double integral:

$$P\bigl(0 < X^2 < Y < 1\bigr) = \int_0^1 \int_{x^2}^1 1\, dy\, dx = \int_0^1 \bigl(1 - x^2\bigr)\, dx = \frac{2}{3}.$$

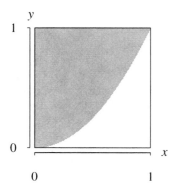

Figure 6.2: Support of X and Y and shaded region of interest.

Example 6.6 Let X and Y have joint probability density function

$$f(x, y) = \frac{1}{2} \qquad 0 < x < y < 2.$$

Find $P(X > a)$, for some real-valued constant a satisfying $0 < a < 2$.

Unlike the previous two examples, the support of X and Y is not a square. The triangular-shaped support and the region of interest (for one particular value of a, namely $a = 0.4$) are shown in Figure 6.3. The probability of interest can be calculated using double integration. Since the triangular region of interest has sides parallel to both the x and y axes, either way of ordering the random variables integration limits requires the same amount of work. We integrate with respect to x first, resulting in

$$
\begin{aligned}
P(X > a) &= \int_a^2 \int_a^y \frac{1}{2} \, dx \, dy \\
&= \int_a^2 \left(\frac{1}{2}y - \frac{1}{2}a \right) dy \\
&= \left[\frac{1}{4}y^2 - \frac{1}{2}ay \right]_a^2 \\
&= 1 - a + \frac{1}{4}a^2 \qquad 0 < a < 2.
\end{aligned}
$$

This probability can also be calculated geometrically. Since both the base and height of the shaded triangle in Figure 6.3 have length $2 - a$, the volume of interest is the product of the area of the shaded region and the height of the joint probability density function

$$P(X > a) = \left(\frac{1}{2} \right) \left(\frac{1}{2} \right)(2 - a)(2 - a) = 1 - a + \frac{1}{4}a^2 \qquad 0 < a < 2,$$

which matches the result obtained by integration.

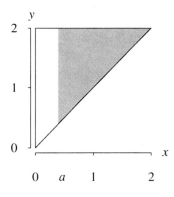

Figure 6.3: Support of X and Y and shaded region of interest.

Example 6.7 Let the random variables X and Y have joint probability density function

$$f(x, y) = e^{-x-y} \qquad x > 0, y > 0.$$

Find the probability that X and Y will lie inside a circle of radius 2 centered at the (x, y) coordinates $(2, 3)$.

In this case, the support $\mathcal{A}$ is the entire first quadrant. The probability that X and Y will lie inside the circle shown in Figure 6.4 is

$$P\big((X-2)^2 + (Y-3)^2 < 4\big) = \int_0^4 \int_{3-\sqrt{4-(x-2)^2}}^{3+\sqrt{4-(x-2)^2}} e^{-x-y} \, dy \, dx \cong 0.2028.$$

This probability is the volume under $f(x, y)$ above the circle. The double integral shown here is not easily computed by hand. The numerical value given here can be computed with the Maple statements

```
int(int(exp(-x - y), y = 3 - sqrt(4 - (x - 2) ^ 2) .. 3 +
    sqrt(4 - (x - 2) ^ 2)), x = 0 .. 4);
evalf(%);
```

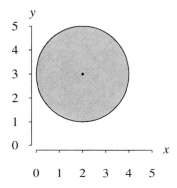

Figure 6.4: Support of X and Y and shaded region of interest.

Two notational dilemmas arise when working with several random variables. First is the choice of letters. Using

$$X \text{ and } Y$$

to denote random variables is best in terms of familiarity with calculus, but you soon run out of letters and the notation does not easily expand for a large number of random variables. On the other hand, using

$$X_1 \text{ and } X_2$$

generalizes well, but involves adding a subscript. We will alternate between these two ways of denoting random variables because each has advantages over the other, depending on the problem at hand. Second, there will be occasions where it is not clear which random variables are associated with a particular joint probability mass function or probability density function. As in the case with univariate random variables, subscripts are added for clarity, for example, $f_{X,Y}(x, y)$ emphasizes that X and Y are the random variables of interest.

Joint cumulative distribution functions

The definition of the cumulative distribution function also extends to two dimensions. The joint cumulative distribution function of two random variables (discrete or continuous) X_1 and X_2 is

$$F(x_1, x_2) = P(X_1 \le x_1, X_2 \le x_2).$$

As was the case with the joint probability density function, subscripts are added to F when it is necessary to emphasize which random variables are under consideration, for example, $F_{X_1, X_2}(x_1, x_2)$.

As in the univariate case, the joint cumulative distribution function $F(x_1, x_2)$ accumulates probability as its arguments increase. So for the extreme cases of the arguments,

$$\lim_{x_1 \to -\infty} \lim_{x_2 \to -\infty} F_{X_1, X_2}(x_1, x_2) = 0$$

and

$$\lim_{x_1 \to \infty} \lim_{x_2 \to \infty} F_{X_1, X_2}(x_1, x_2) = 1.$$

The joint cumulative distribution function is computed in a different manner for discrete and continuous random variables. For discrete random variables, the joint cumulative distribution function of X_1 and X_2 is computed by

$$F(x_1, x_2) = P(X_1 \leq x_1, X_2 \leq x_2) = \sum_{w_1 \leq x_1} \sum_{w_2 \leq x_2} f(w_1, w_2),$$

that is, sum all of the joint probability mass function values that lie to the southwest (with north pointing upward) of the point (x_1, x_2). For continuous random variables, the joint cumulative distribution function of X_1 and X_2 is computed by

$$F(x_1, x_2) = P(X_1 \leq x_1, X_2 \leq x_2) = \int_{-\infty}^{x_1} \int_{-\infty}^{x_2} f(w_1, w_2) \, dw_2 \, dw_1,$$

that is, the area under the joint probability density function above the region in the (x_1, x_2) plane to the southwest of the point (x_1, x_2). When X_1 and X_2 are continuous random variables, the joint probability density function can be found via

$$f(x_1, x_2) = \frac{\partial^2 F(x_1, x_2)}{\partial x_1 \partial x_2}.$$

For discrete random variables, the joint probability mass function can be found by differencing the joint cumulative distribution function.

Example 6.8 Let X_1 and X_2 be continuous random variables with joint probability density function

$$f(x_1, x_2) = x_1 x_2 \qquad 0 < x_1 < 1, 0 < x_2 < 2.$$

Find the joint cumulative distribution function.

The joint cumulative distribution function $F(x_1, x_2)$ needs to be defined over five regions in the (x_1, x_2) plane, as illustrated in Figure 6.5. Region I is the second, third, and fourth quadrants. For any point (x_1, x_2) that falls in region I, the area under the joint probability density function to the southwest of the point is 0. Region II is $\mathcal{A}$, the support of X_1 and X_2. The joint cumulative distribution function on this region is

$$F(x_1, x_2) = \int_0^{x_2} \int_0^{x_1} w_1 w_2 \, dw_1 \, dw_2 = \frac{x_1^2 x_2^2}{4} \qquad 0 < x_1 < 1, \, 0 < x_2 < 2.$$

On region III, the integration limits need to be modified to account for the fact that x_2 is above the support region:

$$F(x_1, x_2) = \int_0^2 \int_0^{x_1} w_1 w_2 \, dw_1 \, dw_2 = x_1^2 \qquad 0 < x_1 < 1, \, x_2 \geq 2.$$

On region IV, the integration limits need to be modified to account for the fact that x_1 is to the right of the support region:

$$F(x_1, x_2) = \int_0^{x_2} \int_0^1 w_1 w_2 \, dw_1 \, dw_2 = \frac{x_2^2}{4} \qquad x_1 \geq 1, \, 0 < x_2 < 2.$$

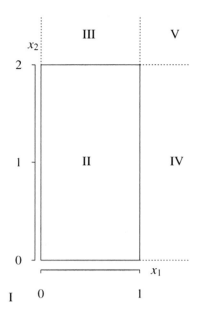

Figure 6.5: Integration regions for determining $F(x_1, x_2)$.

Finally, in region V, the volume under the joint probability density function for every point to the southwest of (x_1, x_2) is 1 because the joint probability density function integrates to one. To summarize, the joint cumulative distribution function of X_1 and X_2 is

$$F(x_1, x_2) = \begin{cases} 0 & x_1 \leq 0 \ \text{or} \ x_2 \leq 0 \\ x_1^2 x_2^2/4 & 0 < x_1 < 1, 0 < x_2 < 2 \\ x_1^2 & 0 < x_1 < 1, x_2 \geq 2 \\ x_2^2/4 & x_1 \geq 1, 0 < x_2 < 2 \\ 1 & x_1 \geq 1, x_2 \geq 2. \end{cases}$$

Now the cumulative probability can be calculated for any (x_1, x_2) pair by plugging the appropriate values into $F(x_1, x_2)$. For example, the probability that both X_1 and X_2 are less than $1/2$, which falls in region II, can be found as

$$P\left(X_1 \leq \frac{1}{2}, X_2 \leq \frac{1}{2}\right) = F\left(\frac{1}{2}, \frac{1}{2}\right) = \frac{(1/2)^2(1/2)^2}{4} = \frac{1}{64}.$$

Example 6.9 A bag contains 6 red balls, 7 white balls, and 8 blue balls. A random sample of 5 balls is drawn without replacement from the bag. If X_1 denotes the number of red balls in the sample and X_2 denotes the number of white balls in the sample, what is $F(2, 3)$?

There are $\binom{21}{5}$ equally likely samples that can be drawn from the bag. Although there are three types of balls being drawn from the bag, this is a *bivariate* distribution because the number of red balls in the sample and the number of white balls in the sample determines the number of blue balls in the sample. The joint probability mass function of X_1 and X_2 is

$$f(x_1, x_2) = \frac{\binom{6}{x_1}\binom{7}{x_2}\binom{8}{5-x_1-x_2}}{\binom{21}{5}} \qquad x_1 = 0, 1, \ldots, 5; x_2 = 0, 1, \ldots, 5; x_1 + x_2 \leq 5,$$

which is defined over a triangular-shaped set of points. This distribution is a generalization of the hypergeometric distribution, and it is a special case of the *bivariate hypergeometric distribution*. The value of $F(2, 3)$ is the sum of all of the joint probability mass function values to the southwest of $(x_1, x_2) = (2, 3)$, which can be found with the double summation

$$F(2, 3) = P(X_1 \leq 2, X_2 \leq 3) = \sum_{w_1=0}^{2} \sum_{w_2=0}^{3} \frac{\binom{6}{w_1}\binom{7}{w_2}\binom{8}{5-w_1-w_2}}{\binom{21}{5}}.$$

This summation is calculated in Maple with the statement

```
sum(sum(binomial(6, w1) * binomial(7, w2) * binomial(8, 5 - w1 - w2) /
        binomial(21, 5), w2 = 0 .. 3), w1 = 0 .. 2);
```

which yields $F(2, 3) = 2501/2907$ or approximately 0.8603.

The univariate cumulative distribution function of one of the variables can be obtained by allowing the argument for the other variable to approach infinity. If X and Y are random variables with joint cumulative distribution function $F(x, y)$, then the cumulative distribution function of X, for example, is

$$
\begin{aligned}
F_X(x) &= P(X \leq x) \\
&= P(X \leq x, Y < \infty) \\
&= \lim_{y \to \infty} P(X \leq x, Y \leq y) \\
&= \lim_{y \to \infty} F(x, y) \\
&= F(x, \infty).
\end{aligned}
$$

So the cumulative distribution of X alone is the joint cumulative distribution function as $y \to \infty$. Similarly, $F_Y(y) = F(\infty, y)$. The distribution of X or Y by themselves is known as a *marginal distribution*, which is the topic of the next subsection.

Marginal distributions

Occasions arise when there is an interest in the distribution of just one of the two random variables. Consider the case of the continuous random variables X and Y with joint probability density function $f(x, y)$. The probability that X lies between two constants a and b is

$$
\begin{aligned}
P(a < X < b) &= P(a < X < b, -\infty < Y < \infty) \\
&= \int_a^b \int_{-\infty}^{\infty} f(x, y)\, dy\, dx \\
&= \int_a^b f_X(x)\, dx.
\end{aligned}
$$

The function

$$f_X(x) = \int_{-\infty}^{\infty} f(x, y)\, dy,$$

which is obtained by integrating y out of the joint probability density function, is known as the *marginal* probability density function of X. In general, the marginal distribution of a random variable considers a random variable by itself. The procedure is analogous for discrete random variables, with integration replaced by summation as given in the following definition.

Definition 6.2 For the discrete random variables X and Y with joint probability mass function $f(x, y)$, the *marginal probability mass functions* for X and Y are

[handwritten: sum out all of the non-necessary variable]

$$f_X(x) = \sum_y f(x, y) \qquad \text{and} \qquad f_Y(y) = \sum_x f(x, y)$$

which are defined over the appropriate supports. For the continuous random variables X and Y with joint probability density function $f(x, y)$, the *marginal probability density functions* for X and Y are

[handwritten: integrate out all un-necessary variables]

$$f_X(x) = \int_{-\infty}^{\infty} f(x, y)\, dy \qquad \text{and} \qquad f_Y(y) = \int_{-\infty}^{\infty} f(x, y)\, dx$$

which are defined over the appropriate supports.

Example 6.10 Let the discrete random variables X and Y have joint probability mass function

$$f(x, y) = \begin{cases} 0.2 & x = 1, y = 1 \\ 0.1 & x = 1, y = 2 \\ 0.3 & x = 1, y = 3 \\ 0.1 & x = 2, y = 1 \\ 0.1 & x = 2, y = 2 \\ 0.2 & x = 2, y = 3. \end{cases}$$

Find the marginal probability mass functions $f_X(x)$ and $f_Y(y)$.

The joint probability mass function is defined on a support $\mathcal{A}$ containing just 6 elements. These 6 values are the entries in Table 6.5. The row sums (calculated by summing over all values of y) are the marginal probability mass function of X; the column sums (calculated by summing over all values of x) are the marginal probability mass function of Y. So the two marginal probability mass functions are

$$f_X(x) = \begin{cases} 0.6 & x = 1 \\ 0.4 & x = 2 \end{cases}$$

and

$$f_Y(y) = \begin{cases} 0.3 & y = 1 \\ 0.2 & y = 2 \\ 0.5 & y = 3. \end{cases}$$

The marginal probability mass function of X gives the probability distribution of X when Y is ignored; likewise, the marginal probability mass function of X gives the probability distribution of X when Y is ignored. These marginal probability mass functions satisfy the same existence conditions as any other probability mass function—that is, they are nonnegative and sum to one.

x \ y	1	2	3	$f_X(x)$
1	0.2	0.1	0.3	0.6
2	0.1	0.1	0.2	0.4
$f_Y(y)$	0.3	0.2	0.5	

Table 6.5: Joint and marginal probability mass functions for X and Y.

Example 6.11 For the continuous random variables X and Y with joint probability density function

$$f(x, y) = \frac{1}{50} \qquad x > 0, y > 0, x + y < 10,$$

find the probability density function of the marginal distribution of Y.

The support of X and Y is the triangular-shaped region shaded in Figure 6.6. The joint probability density function is constant over the triangle. The marginal distribution of Y is found by integrating x out of the joint probability density function

$$
\begin{aligned}
f_Y(y) &= \int_{-\infty}^{\infty} f(x, y)\, dx \\
&= \int_0^{10-y} \frac{1}{50}\, dx \\
&= \frac{10 - y}{50} \qquad 0 < y < 10.
\end{aligned}
$$

The distribution of Y alone has a mode at $y = 0$ and the marginal probability density function decreases linearly on $0 < y < 10$. The geometric interpretation of this marginal distribution is as follows: If one were to place a flashlight at the point $(1000, 5)$ in Figure 6.6 and point it toward the support region, then the beams near $y = 0$ pass through the most density, and the beams near $y = 10$ pass through the least density. This accounts for the fact that the marginal probability density function of Y is highest near $y = 0$ and lowest near $y = 10$.

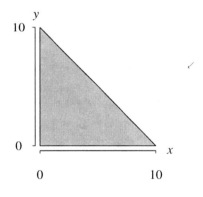

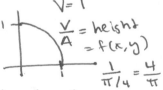

Figure 6.6: Support of X and Y.

Example 6.12 The random variables X and Y are uniformly distributed over the portion of the interior of the unit circle centered at the origin that lies in the first quadrant. Find the probability density function of the marginal distribution of X.

Since the area of the unit circle is π, and the area of the unit circle that falls in the first quadrant is $\pi/4$, the joint probability density function of X and Y is

$$f_{X,Y}(x, y) = \frac{4}{\pi} \qquad x > 0, y > 0, x^2 + y^2 < 1.$$

The marginal probability density function of X is found by integrating y out of the joint probability density function

$$f_X(x) = \int_0^{\sqrt{1-x^2}} \frac{4}{\pi}\, dy = \left[\frac{4y}{\pi}\right]_0^{\sqrt{1-x^2}} = \frac{4}{\pi}\sqrt{1 - x^2} \qquad 0 < x < 1.$$

Example 6.13 Let the continuous random variables X_1 and X_2 have joint probability density function

$$f(x_1, x_2) = x_1 x_2 \qquad 0 < x_1 < 1,\ 0 < x_2 < 2.$$

Find $P(0.2 < X_1 < 0.7)$ by the following three techniques:

(a) computing the value of the double integral over the appropriate limits,

(b) finding the marginal probability density function of X_1 and integrating,

(c) finding the marginal cumulative distribution function of X_1 and using the appropriate arguments.

This is the same joint probability density function that was encountered in Example 6.8, where the joint cumulative distribution function on the rectangular-shaped support was found to be

$$F(x_1, x_2) = x_1^2 x_2^2 / 4 \qquad 0 < x_1 < 1,\ 0 < x_2 < 2.$$

(a) The double integral required to find $P(0.2 < X_1 < 0.7)$ is

$$P(0.2 < X_1 < 0.7) = \int_{0.2}^{0.7} \int_{0}^{2} x_1 x_2\, dx_2\, dx_1 = 0.45.$$

(b) The marginal probability density function of X_1 is

$$f_{X_1}(x_1) = \int_{0}^{2} x_1 x_2\, dx_2 = 2x_1 \qquad 0 < x_1 < 1.$$

The probability that X_1 falls between 0.2 and 0.7 is the area under the marginal probability density function of X_1 between 0.2 and 0.7:

$$P(0.2 < X_1 < 0.7) = \int_{0.2}^{0.7} f_{X_1}(x_1)\, dx_1 = \int_{0.2}^{0.7} 2x_1\, dx_1 = 0.45.$$

(c) The marginal cumulative distribution function of X_1 can be found by using x_1 and 2 (the upper limit of the support for X_2) as arguments in the joint cumulative distribution function of X_1 and X_2:

$$F_{X_1}(x_1) = F(x_1, 2) = x_1^2 \qquad 0 < x_1 < 1.$$

So the probability that X_1 falls between 0.2 and 0.7 is

$$P(0.2 < X_1 < 0.7) = F_{X_1}(0.7) - F_{X_1}(0.2) = 0.49 - 0.04 = 0.45.$$

The last four examples have illustrated the marginal distribution of one of the two random variables in a bivariate distribution, which is the distribution of one of the random variables when the other random variable is not considered. There is a somewhat-similar concept known as a conditional distribution, where the interest is in the distribution of one random variable *given* an observed value of the other random variable.

Conditional distributions

A conditional distribution is the distribution of one random variable given the value of another random variable. When X and Y are discrete random variables, for example, the conditional probability mass function of X given $Y = y$ is

$$
\begin{aligned}
f_{X\,|\,Y=y}(x\,|\,Y=y) &= P(X = x\,|\,Y = y) \\
&= \frac{P(X = x, Y = y)}{P(Y = y)} \\
&= \frac{f(x, y)}{f_Y(y)}
\end{aligned}
$$

by using the definition of conditional probability. So the conditional distribution is the ratio of the joint distribution to the marginal probability density function of the given random variable. This is one of the few cases where the definition of the conditional distributions does not require separate cases for discrete and continuous random variables.

Definition 6.3 Let the (discrete or continuous) random variables X and Y have a joint distribution described by $f(x, y)$ and marginal distributions described by $f_X(x)$ and $f_Y(y)$. The *conditional distribution* of X given $Y = y$ is described by

$$f_{X \mid Y=y}(x \mid Y = y) = \frac{f(x, y)}{f_Y(y)}$$

for $f_Y(y) > 0$, defined over the appropriate support. Likewise, the *conditional distribution* of Y given $X = x$ is described by

$$f_{Y \mid X=x}(y \mid X = x) = \frac{f(x, y)}{f_X(x)}$$

for $f_X(x) > 0$, defined over the appropriate support.

These conditional probability mass functions and probability density functions satisfy the standard existence conditions. For example, when X and Y are discrete, the mathematics required to show that the conditional probability mass function $f_{X \mid Y=y}(x \mid Y = y)$ sums to one is

$$\sum_x f_{X \mid Y=y}(x \mid Y = y) = \sum_x \frac{f(x, y)}{f_Y(y)} = \frac{1}{f_Y(y)} \sum_x f(x, y) = \frac{1}{f_Y(y)} \cdot f_Y(y) = 1.$$

Likewise, when X and Y are continuous, the mathematics required to show that the conditional probability density function $f_{X \mid Y=y}(x \mid Y = y)$ integrates to one is

$$\int_{-\infty}^{\infty} f_{X \mid Y=y}(x \mid Y = y)\, dx = \int_{-\infty}^{\infty} \frac{f(x, y)}{f_Y(y)}\, dx = \frac{1}{f_Y(y)} \int_{-\infty}^{\infty} f(x, y)\, dx = \frac{1}{f_Y(y)} \cdot f_Y(y) = 1.$$

Example 6.14 Consider again the discrete random variables X and Y with joint probability mass function and marginal probability mass functions shown in Table 6.6. Find the probability mass functions following conditional random variables:

$$X \mid Y = 1 \qquad\qquad X \mid Y = 3 \qquad\qquad Y \mid X = 2.$$

The probability mass function of X given $Y = 1$ is

$$f_{X \mid Y=1}(x \mid Y = 1) = \begin{cases} 2/3 & x = 1 \\ 1/3 & x = 2, \end{cases}$$

which is calculated using the $y = 1$ column in Table 6.6. Likewise, the probability mass function of X given $Y = 3$ is

$$f_{X \mid Y=3}(x \mid Y = 3) = \begin{cases} 3/5 & x = 1 \\ 2/5 & x = 2, \end{cases}$$

x \ y	1	2	3	$f_X(x)$
1	0.2	0.1	0.3	0.6
2	0.1	0.1	0.2	0.4
$f_Y(y)$	0.3	0.2	0.5	

Table 6.6: Joint and marginal probability mass functions for X and Y.

which is calculated using the $y = 3$ column in Table 6.6. Finally, the probability mass function of Y given $X = 2$ is

$$f_{Y|X=2}(y|X=2) = \begin{cases} 1/4 & y = 1 \\ 1/4 & y = 2 \\ 1/2 & y = 3, \end{cases}$$

which is calculated using the $x = 2$ row in Table 6.6.

Example 6.15 Deal two cards from a well-shuffled deck. Let the random variable X be the number of aces dealt and let the random variable Y be the number of face cards dealt. Find the conditional distribution of the number of aces in the hand given that there is one face card in the hand.

The joint probability density function was calculated in Example 6.1. The values of $f(x, y)$, along with the associated marginal distributions, are given in Table 6.7.

x \ y	0	1	2	$f_X(x)$
0	630/1326	432/1326	66/1326	1128 / 1326
1	144/1326	48/1326	0	192/1326
2	6/1326	0	0	6/1326
$f_Y(y)$	780/1326	480/1326	66/1326	1

Table 6.7: Joint and marginal probability mass functions of X and Y.

The conditional distribution of X given $Y = 1$ can be found by focusing on the $y = 1$ column of Table 6.7. Using the definition of conditional probability, the distribution is

$$f_{X|Y=1}(x|Y = 1) = \begin{cases} 432/480 & x = 0 \\ 48/480 & x = 1, \end{cases}$$

or

$$f_{X|Y=1}(x|Y = 1) = \begin{cases} 9/10 & x = 0 \\ 1/10 & x = 1, \end{cases}$$

which is a Bernoulli($1/10$) random variable. The interpretation of this conditional distribution is as follows: if it is known that there is a single face card ($Y = 1$) in the hand, then there is a probability of $1/10$ that the other card is an ace ($X = 1$). This particular conclusion could also have been drawn using conditional probability and counting techniques. The probability of having an ace in the hand given that there is one face card is

$$\begin{aligned} P(\text{one ace} \,|\, \text{one face card}) &= \frac{P(\text{one ace and one face card})}{P(\text{one face card})} \\ &= \frac{\binom{4}{1}\binom{12}{1}\binom{36}{0}/\binom{52}{2}}{\binom{12}{1}\binom{40}{1}/\binom{52}{2}} \\ &= \frac{48}{480} \\ &= \frac{1}{10}. \end{aligned}$$

The result can be checked by Monte Carlo simulation with the following R code, which assumes that cards 1–4 are the aces and 41–52 are the face cards.

```
nrep = 500000
nface1 = 0
count = 0
for (i in 1:nrep) {
  aces = 0
  faces = 0
  cards = sample(52, 2)
  if (cards[1] <= 4) aces = aces + 1
  if (cards[1] >= 41) faces = faces + 1
  if (cards[2] <= 4) aces = aces + 1
  if (cards[2] >= 41) faces = faces + 1
  if (faces == 1) {
    nface1 = nface1 + 1
    if (aces == 1) count = count + 1
  }
}
print(count / nface1)
```

Five runs of this simulation, after a call to set.seed(3) yields

| 0.09997 | 0.09976 | 0.10002 | 0.10009 | 0.09959, |

which supports the analytic result $P(\text{one ace} \mid \text{one face card}) = 1/10$.

Example 6.16 Let X and Y be continuous random variables with joint probability density function

$$f_{X,Y}(x, y) = \frac{1}{50} \qquad x > 0, y > 0, x + y < 10.$$

Find

(a) the conditional probability density function of X given that $Y = y$, that is, find $f_{X|Y=y}(x \mid Y = y)$,

(b) $P(3 < X < 5 \mid Y = 2)$.

The joint distribution of X and Y is uniformly distributed over the shaded triangle shown in Figure 6.7.

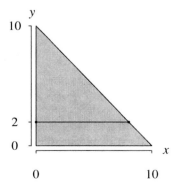

Figure 6.7: Support of X and Y.

(a) As seen earlier, the marginal distribution of Y can be found by integrating x out of the joint probability density function:

$$f_Y(y) = \int_0^{10-y} \frac{1}{50} dx = \frac{10-y}{50} \qquad 0 < y < 10.$$

The conditional distribution of X given $Y = y$ can then be expressed as the conditional probability density function

$$f_{X|Y=y}(x|Y=y) = \frac{f(x,y)}{f_Y(y)} = \frac{1/50}{(10-y)/50} = \frac{1}{10-y} \qquad 0 < x < 10-y; 0 < y < 10.$$

The conditional probability density function of X for *any* value of $Y = y$ between 0 and 10 is constant with respect to x. This can be seen in Figure 6.7 by examining the particular case of $y = 2$. When the joint probability density function, which is constant, is sliced at $y = 2$, the resulting conditional distribution should be $U(0, 8)$. More generally, when the joint probability density function is sliced at y, the resulting conditional distribution should be $U(0, 10-y)$, which is exactly the result achieved above. The two parts of the support of this conditional probability density function should be carefully examined. The first part, $0 < x < 10-y$, gives the allowable values for x (that is, the support), where the upper limit is a function of y. The second part, $0 < y < 10$, gives the allowable values of the variable y.

(b) When $y = 2$, the conditional probability density function becomes

$$f_{X|Y=2}(x|Y=2) = \frac{1}{8} \qquad 0 < x < 8,$$

which is the probability density function of a $U(0, 8)$ random variable. The probability that X falls between 3 and 5 given that $Y = 2$ is found by integrating the conditional probability density function, yielding

$$P(3 < X < 5|Y=2) = \int_3^5 \frac{1}{8} dx = \frac{1}{4}.$$

To summarize this section, the random variables X and Y are defined in an analogous way to their univariate counterparts. The probability distribution of the discrete random variables X and Y on the support $\mathcal{A}$ can be described by the *joint probability mass function*

$$f(x,y) = P(X=x, Y=y).$$

The *joint cumulative distribution function* is defined by

$$F(x,y) = P(X \le x, Y \le y)$$

for both discrete and continuous random variables X and Y. The probability distribution of the continuous random variables X and Y can also be described by the *joint probability density function*

$$f(x,y) = \frac{\partial^2 F(x,y)}{\partial x \partial y}.$$

The *marginal distributions* for X and Y are described by

$$f_X(x) \qquad \text{and} \qquad f_Y(y)$$

and are found by summing out (when X and Y are discrete) or integrating out (when X and Y are continuous) the variable that is not of interest. Finally, the *conditional distributions* for X and Y are described by

$$f_{X|Y=y}(x|Y=y) = \frac{f(x,y)}{f_Y(y)}$$

and

$$f_{Y|X=x}(y|X=x) = \frac{f(x,y)}{f_X(x)}.$$

6.2 Independent Random Variables

Independence was defined for events in Chapter 2. This section extends the definition of independence to random variables.

Recall from Chapter 2 that events A and B are independent if and only if $P(A \cap B) = P(A)P(B)$. Consider the random variables X and Y. Let the event A be $a_1 < X < a_2$ for real constants a_1 and a_2; let the event B be $b_1 < Y < b_2$ for real constants b_1 and b_2. When events A and B are independent, it must be the case that

$$P(a_1 < X < a_2, b_1 < Y < b_2) = P(a_1 < X < a_2)P(b_1 < Y < b_2).$$

So random variables are independent when the joint probability can be factored into the associated marginal probabilities. This is one way of developing the concept of the independence of random variables.

A second way of developing the concept of independence of random variables uses conditional distributions. Without loss of generality, assume that the random variables X and Y are continuous with joint probability density function $f(x, y)$, marginal probability density functions $f_X(x)$ and $f_Y(y)$, and conditional probability density functions $f_{X \mid Y=y}(x \mid Y = y)$ and $f_{Y \mid X=x}(y \mid X = x)$. Using the definition of the conditional probability density function, the joint probability density function can be expressed as

$$f(x, y) = f_{Y \mid X=x}(y \mid X = x)f_X(x).$$

Now if X and Y are independent, it is reasonable to assume that $f_{Y \mid X=x}(y \mid X = x)$ does not depend on the value of x. This means that the marginal probability density function of Y, under the assumption of independence, can be computed as

$$
\begin{aligned}
f_Y(y) &= \int_{-\infty}^{\infty} f(x, y) \, dx \\
&= \int_{-\infty}^{\infty} f_{Y \mid X=x}(y \mid X = x)f_X(x) \, dx \\
&= f_{Y \mid X=x}(y \mid X = x) \int_{-\infty}^{\infty} f_X(x) \, dx \\
&= f_{Y \mid X=x}(y \mid X = x).
\end{aligned}
$$

Substituting into the earlier expression for the joint probability density function reveals that

$$f(x, y) = f_X(x)f_Y(y)$$

when X and Y are independent. This same derivation applies when X and Y are discrete with integrals replaced by summations. This leads to the following definition of independence for the random variables X and Y.

Definition 6.4 Let the random variables X and Y (discrete or continuous) have a joint distribution described by $f(x, y)$ and marginal distributions described by $f_X(x)$ and $f_Y(y)$. The random variables X and Y are *independent* if and only if

$$f(x, y) = f_X(x)f_Y(y)$$

for all real numbers x and y.

Intuitively, if the value of X does not affect the distribution of Y and if the value of Y does not affect the distribution of X, then X and Y are *independent*. There are several comments concerning independence of random variables given below.

- Random variables that are not independent are *dependent*.

- The term *stochastically independent* is used by some authors to avoid confusion with other types of independence (for example, engineering).

- Definition 6.4 is formulated in terms of probability mass functions and probability density functions. An equivalent definition can also be written in terms of cumulative distribution functions. The random variables X and Y are independent if and only if

$$F(x, y) = F_X(x)F_Y(y)$$

for all real numbers x and y.

- For X and Y to be independent, the support of their joint distribution must be a *product space*. If X has support $\mathcal{A}$ and Y has support $\mathcal{B}$, then the product space is

$$\{(x, y) \mid x \in \mathcal{A} \text{ and } y \in \mathcal{B}\}.$$

Geometrically, a product space is a rectangle in the (x, y) coordinate space, with possibly infinite edge lengths. If the support of X and Y is not a product space, one can conclude that X and Y are *dependent* random variables. If the support of X and Y is a product space, one can conclude that X and Y *might* be *independent* random variables.

- This definition extends to more than two random variables (although it is important to differentiate between *pairwise* and *mutually* independent). The n-dimensional definition will be given in the last section of this chapter.

Example 6.17 Consider the random variables X and Y with joint probability mass function given in Table 6.8. Are X and Y independent?

x \ y	1	2	3	$f_X(x)$
1	0.2	0.1	0.3	0.6
2	0.1	0.1	0.2	0.4
$f_Y(y)$	0.3	0.2	0.5	

Table 6.8: Joint and marginal probability mass functions for X and Y.

The random variables X and Y are defined on a product space, so it is *possible* that they are independent. For the random variables X and Y to be independent, the joint distribution must be expressed as the product of the marginal distributions, that is

$$f(x, y) = f_X(x)f_Y(y)$$

for all values of x and y. It is easy to see that the random variables are *dependent* because

$$f(1, 1) = 0.2 \neq 0.18 = (0.6)(0.3) = f_X(1)f_Y(1).$$

Example 6.18 Let X_1 and X_2 be continuous random variables with joint probability density function

$$f(x_1, x_2) = x_1 x_2 \qquad 0 < x_1 < 1, 0 < x_2 < 2.$$

Are X_1 and X_2 independent?

Since X_1 and X_2 are defined on a product space, they might be independent. The marginal distributions are

$$f_{X_1}(x_1) = \int_0^2 x_1 x_2 \, dx_2 = 2x_1 \qquad 0 < x_1 < 1$$

and

$$f_{X_2}(x_2) = \int_0^1 x_1 x_2 \, dx_1 = x_2/2 \qquad 0 < x_2 < 2.$$

Since

$$f(x_1, x_2) = x_1 x_2 = (2x_1)(x_2/2) = f_{X_1}(x_1) f_{X_2}(x_2)$$

on the product space, X_1 and X_2 are *independent*.

There is a useful result that makes it easy to identify whether or not two random variables are independent without calculating the marginal distributions as we did in the previous two examples. The result is stated below, but the proof is left as an exercise.

Theorem 6.1 Let the random variables X and Y have joint distribution described by $f(x, y)$ defined on a product space. Then X and Y are independent if and only if $f(x, y)$ can be written as the product of a nonnegative function of x only and a nonnegative function of y only.

This theorem allows you to quickly determine whether two random variables are independent. First, check to see that the support of their joint distribution is a product space, then check to see if the joint probability mass function or probability density function can be written as the product of two nonnegative functions of each of the variables alone.

Example 6.19 In the previous example, the joint probability density function

$$f(x_1, x_2) = x_1 x_2 \qquad 0 < x_1 < 1, 0 < x_2 < 2$$

was defined on a product space and the joint probability density function could be factored into the product of two nonnegative functions, for example, $f(x_1, x_2) = (x_1)(x_2)$, so X_1 and X_2 are independent random variables.

Example 6.20 Are the random variables X and Y with joint probability density function

$$f(x, y) = 1 \qquad 0 < x < 1, 0 < y < 1$$

independent?

The random variables are defined on a product space, and the joint probability density function can be factored into the product of two nonnegative functions, for example

$$f(x, y) = \left(\frac{11}{17}\right)\left(\frac{17}{11}\right) = 1$$

on the product space, so X and Y are independent. Notice that 11 and 17 were pulled out of thin air. There are obviously lots of ways to factor 1. All that is necessary is to find *one* pair of nonnegative functions in order to establish independence.

Example 6.21 Are the random variables X and Y with joint probability density function

$$f(x, y) = \frac{1}{50} \qquad x > 0, y > 0, x + y < 10$$

independent? *Not product space*

One glance at the support of X and Y tells you that these random variables are *dependent*. The support is triangular, rather than rectangular, so they are not defined on a product space.

Example 6.22 Are the random variables X and Y with joint probability density function

$$f(x, y) = x + y \qquad\qquad 0 < x < 1, 0 < y < 1$$

independent?   *not product of 2 functions*

In this case, the support of X and Y is a product space. The conclusion here, however, is that X and Y are *dependent* because the joint probability density function of the two random variables can't be written as the product of a nonnegative function of x and a nonnegative function of y. Calculating the marginal distributions would confirm that this is the case.

The next example considers two independent discrete random variables X and Y that are defined over a product space.

Example 6.23 Scott, a 70% free throw shooter, attempts n independent free throw shots. Billy, an 80% free throw shooter, attempts m independent free throw shots, which are independent of Scott's shots.

(a) Give an expression for the probability that Scott makes more shots than Billy.

(b) Write a Maple procedure ScottBilly(n, m) that uses the APPL BinomialRV procedure to calculate the required binomial probabilities.

(c) Execute the statement ScottBilly(5, 4) to arrive at an exact fraction for the required probability.

In order for the solution to the problem given below to be valid, each of the free throws taken by Scott and Billy must be independent Bernoulli trials.

(a) Let X be the number of shots made by Scott; let Y be the number of shots made by Billy. Since $X \sim \text{binomial}(n, 7/10)$ and $Y \sim \text{binomial}(m, 8/10)$, and X and Y are independent, the joint probability density function of X and Y is the product of the marginal probability density functions

$$f(x, y) = \binom{n}{x} \left(\frac{7}{10}\right)^x \left(\frac{3}{10}\right)^{n-x} \binom{m}{y} \left(\frac{8}{10}\right)^y \left(\frac{2}{10}\right)^{m-y}$$

for $x = 0, 1, 2, \ldots, n$ and for $y = 0, 1, 2, \ldots, m$. The probability that Scott makes more shots than Billy is

$$P(X > Y) = \sum_{x=1}^{n} \sum_{y=0}^{\min\{x-1, m\}} f(x, y).$$

(b) The most straightforward approach is using loops in the APPL code.

```
ScottBilly := proc(n :: posint, m :: posint)
  local x, y, Scott, Billy, prob:
  Scott := BinomialRV(n, 7 / 10);
  Billy := BinomialRV(m, 8 / 10);
  prob := 0;
  for x from 1 to n do
    for y from 0 to min(x - 1, m) do
      prob := prob + PDF(Scott, x) * PDF(Billy, y);
    od:
  od:
  prob;
end proc;
```

A more succinct procedure without the loops is

```
ScottBilly := proc(n :: posint, m :: posint)
  SF(Difference(BinomialRV(n, 7 / 10), BinomialRV(m, 8 / 10)), 1);
end proc;
```

(c) Using either of the two procedures, the statement

```
ScottBilly(5, 4)
```

yields

$$\frac{5501993}{12500000},$$

which is approximately 0.4402.

The next problem posed is one of the earliest known problems from "geometric probability" known as *Buffon's needle problem*. It was posed by French naturalist and historian Georges–Louis Leclerc, Compte de Buffon (1707–1788). The Wikipedia entry on Buffon shows that his interests were not limited to probability: "Besides his many insights, he also propounded a theory that nature in the New World was inferior to that of Eurasia. He argued that the Americas were lacking in large and powerful creatures, and that even the people were far less virile than their European counterparts. He ascribed this to the marsh odours and dense forests of the continent. These remarks so incensed Thomas Jefferson that he immediately dispatched twenty soldiers to the New Hampshire woods to find a bull moose for Buffon as proof of the 'stature and majesty of American quadrapeds.' It took over two weeks and when shot, the moose lacked imposing horns. Before being shipped back to France, a rack of antlers from a different stag was attached."

Example 6.24 A floor consists of parallel strips of wood, each of width d. A needle of length $l \leq d$ is dropped to the floor. What is the probability that the needle intersects one of the boundaries between the strips of wood?

A graphic of two needle drops onto five planks is shown in Figure 6.8. The needle on the left does not intersect a boundary; The needle on the right intersects a boundary.

Let the random variable X denote the distance between the center of the needle (which divides the needle into two equal pieces of length $l/2$) and the nearest boundary. Figure 6.9 shows a sample needle toss with the dashed projected line of length X.

The next step is to determine what orientations of the needle are associated with a boundary crossing. In order to do so, we must define a second random variable Θ which is the angle between the needle and the projected line from the center of the needle to the nearest boundary. This angle is illustrated in Figure 6.10.

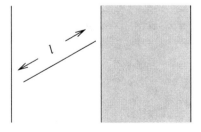
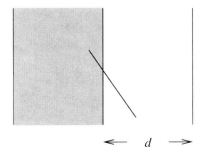

Figure 6.8: Two needle tosses.

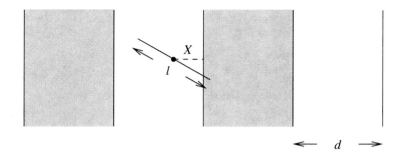

Figure 6.9: One needle toss.

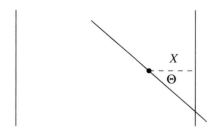

Figure 6.10: The random variables Θ and X.

A boundary crossing occurs whenever $X < \frac{l}{2} \cos \Theta$. The next step is to consider the distributions of X and Θ. It is reasonable to assume that X and Θ are independent random variables because the needle toss is random. Furthermore, since X is the distance to the nearest boundary, its support must be $0 < x < d/2$, and it is furthermore reasonable to assume that $X \sim U(0, d/2)$. The marginal probability density function of X is

$$f_X(x) = \frac{2}{d} \qquad 0 < x < d/2.$$

Likewise, the angle Θ has support $0 < \theta < \pi/2$, and it is also reasonable to assume that the angle is equally likely over its range: $\Theta \sim U(0, \pi/2)$. The marginal probability density function of Θ is

$$f_\Theta(\theta) = \frac{2}{\pi} \qquad 0 < \theta < \pi/2.$$

The rectangular-shaped support of X and Θ and the shaded area associated with a needle crossing a boundary are shown in Figure 6.11. Since the joint probability density function of two independent random variables is the product of the marginal probability density functions, the probability of a crossing can be computed by a double integral:

$$
\begin{aligned}
P\left(X < \frac{l}{2} \cos \theta\right) &= \int_0^{\pi/2} \int_0^{(l/2)\cos\theta} \frac{2}{d} \cdot \frac{2}{\pi} \, dx \, d\theta \\
&= \int_0^{\pi/2} \frac{2l \cos\theta}{\pi d} \, d\theta \\
&= \left[\frac{2l \sin\theta}{\pi d}\right]_0^{\pi/2} \\
&= \frac{2l}{\pi d}.
\end{aligned}
$$

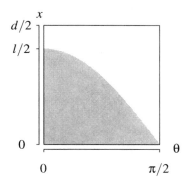

Figure 6.11: The support (rectangle) of Θ and X and the region of interest (shaded).

The fact that π appears on the right-hand side of this equation means that a Monte Carlo simulation can be used to estimate π. Solving the equation for π yields

$$\pi = \frac{2l}{dP\left(X < (l/2)\cos\theta\right)}.$$

Adding hats to the unknown quantities results in

$$\hat{\pi} = \frac{2l}{d\hat{P}\left(X < (l/2)\cos\theta\right)}.$$

So if we drop a needle to the floor repeatedly and count the fraction of crossings, we can arrive at an estimate of π. Arbitrarily setting $d = 1$ and $l = 9/10$, the R code below provides an estimate for π using $100,000$ replications.

```
nrep = 100000
d = 1.0
l = 0.9
x = runif(nrep, 0, d / 2)
theta = runif(nrep, 0, pi / 2)
crosshat = mean(x < (l / 2) * cos(theta))
pihat = 2 * l / (d * crosshat)
```

At this point, we typically run the simulation five times and report the results. Since this is an estimate of π, we are going to simulate the problem to death. When the simulation is run 10,000 times, which is associated with a billion needle tosses, the estimate for π is

$$\hat{\pi} = 3.141612,$$

which differs from $\pi = 3.141592\ldots$ in the fifth digit to the right of the decimal point. With a billion needle tosses, why isn't there more accuracy? As will be seen later, the number of replications must go up by a factor of 100 in order to achieve an additional digit of accuracy. Increased accuracy becomes increasingly more expensive in terms of CPU time for Monte Carlo simulation.

The next example describes an application that is well known to reliability engineers. If a system consists of two components arranged in series and their failure times are independent exponential random variables, then the failure time of the system, which is the minimum of the two component failure times, is also exponentially distributed.

Example 6.25 Let $X_1 \sim$ exponential(λ_1) and $X_2 \sim$ exponential(λ_2) be independent random variables.

(a) Find the probability density function of $X = \min\{X_1, X_2\}$.

(b) Find the probability that $X_1 < X_2$.

Recall from Chapter 5 that the cumulative distribution function of $X_1 \sim$ exponential(λ_1) is

$$F_{X_1}(x_1) = P(X_1 \le x_1) = 1 - e^{-\lambda_1 x_1} \qquad x_1 > 0.$$

Likewise, the cumulative distribution function of $X_2 \sim$ exponential(λ_2) is

$$F_{X_2}(x_2) = P(X_2 \le x_2) = 1 - e^{-\lambda_2 x_2} \qquad x_2 > 0.$$

(a) The probability that the random variable $X = \min\{X_1, X_2\}$ exceeds x is

$$
\begin{aligned}
P(X > x) &= P(\min\{X_1, X_2\} > x) \\
&= P(X_1 > x, X_2 > x) \\
&= P(X_1 > x)\, P(X_2 > x) \\
&= e^{-\lambda_1 x} e^{-\lambda_2 x} \\
&= e^{-(\lambda_1 + \lambda_2)x} \qquad x > 0.
\end{aligned}
$$

Using complementary probability, the cumulative distribution function of X on its support is

$$F_X(x) = P(X \le x) = 1 - e^{-(\lambda_1 + \lambda_2)x} \qquad x > 0.$$

Differentiating with respect to x yields the probability density function of X, which is

$$f_X(x) = (\lambda_1 + \lambda_2)e^{-(\lambda_1 + \lambda_2)x} \qquad x > 0.$$

This can be recognized as the probability density function of an exponential random variable with parameter $\lambda_1 + \lambda_2$, so $X \sim$ exponential$(\lambda_1 + \lambda_2)$. The APPL statement

```
X := Minimum(ExponentialRV(lambda1), ExponentialRV(lambda2));
```

confirms this derivation.

(b) If $X_1 \sim$ exponential(λ_1) and $X_2 \sim$ exponential (λ_2) are independent random variables then the joint probability density function of X_1 and X_2 is the product of the marginal probability density functions

$$f(x_1, x_2) = \lambda_1 e^{-\lambda_1 x_1} \lambda_2 e^{-\lambda_2 x_2} \qquad x_1 > 0, x_2 > 0.$$

The desired probability can be calculated by integrating over the region associated with $x_1 < x_2$

$$
\begin{aligned}
P(X_1 < X_2) &= \int_0^\infty \int_{x_1}^\infty \lambda_1 \lambda_2 e^{-\lambda_1 x_1} e^{-\lambda_2 x_2}\, dx_2\, dx_1 \\
&= \int_0^\infty \lambda_1 e^{-(\lambda_1 + \lambda_2)x_1}\, dx_1 \\
&= \frac{\lambda_1}{\lambda_1 + \lambda_2}.
\end{aligned}
$$

A Monte Carlo simulation can be used to verify the analytic solution. For $\lambda_1 = 1$ and $\lambda_2 = 2$, for instance, five runs of the R code

```
nrep = 100000
lambda1 = 1
lambda2 = 2
x1 = rexp(nrep, lambda1)
x2 = rexp(nrep, lambda2)
mean(x1 < x2)
```

yields

0.3355	0.3339	0.3322	0.3346	0.3322.

These values hover around the analytic solution $P(X_1 < X_2) = \lambda_1/(\lambda_1 + \lambda_2) = 1/(1 + 2) = 1/3$ so this lends credence to the analytic solution. The R function mean averages the number of TRUE values (which are counted as ones), giving an estimate for the desired probability.

The next example considers a parallel system of two components having independent exponential lifetimes.

Example 6.26 Let $X_1 \sim$ exponential(λ_1) and $X_2 \sim$ exponential(λ_2) be the independent lifetimes of two components arranged in a parallel system. The system lifetime is $X = \max\{X_1, X_2\}$, which has probability density function $f_X(x)$, "survivor" function $S_X(x) = P(X \geq x)$ and "hazard" function $h_X(x) = f_X(x)/S_X(x)$. Find $\lim_{x \to \infty} h_X(x)$.

The two components have survivor functions

$$S_{X_1}(x) = e^{-\lambda_1 x} \qquad \text{and} \qquad S_{X_2}(x) = e^{-\lambda_2 x}$$

for $x > 0$. The system survivor function is

$$
\begin{aligned}
S_X(x) &= P(X \geq x) \\
&= P(\max\{X_1, X_2\} \geq x) \\
&= 1 - P(\max\{X_1, X_2\} < x) \\
&= 1 - P(X_1 < x, X_2 < x) \\
&= 1 - P(X_1 < x)P(X_2 < x) \\
&= 1 - (1 - e^{-\lambda_1 x})(1 - e^{-\lambda_2 x}) \\
&= e^{-\lambda_1 x} + e^{-\lambda_2 x} - e^{-(\lambda_1 + \lambda_2)x}
\end{aligned}
$$

for $x > 0$. The probability density function for the system lifetime is

$$f_X(x) = -S_X'(x) = \lambda_1 e^{-\lambda_1 x} + \lambda_2 e^{-\lambda_2 x} - (\lambda_1 + \lambda_2)e^{-(\lambda_1 + \lambda_2)x}$$

for $x > 0$. Thus the hazard function is

$$h_X(x) = \frac{f_X(x)}{S_X(x)} = \frac{\lambda_1 e^{-\lambda_1 x} + \lambda_2 e^{-\lambda_2 x} - (\lambda_1 + \lambda_2)e^{-(\lambda_1 + \lambda_2)x}}{e^{-\lambda_1 x} + e^{-\lambda_2 x} - e^{-(\lambda_1 + \lambda_2)x}}$$

for $x > 0$. To find the limit as $x \to \infty$, consider three cases. First, consider $\lambda_1 < \lambda_2$. In this case, multiply the numerator and denominator of the hazard function by $e^{\lambda_1 x}$ yielding

$$h_X(x) = \frac{\lambda_1 + \lambda_2 e^{-(\lambda_2 - \lambda_1)x} - (\lambda_1 + \lambda_2)e^{-\lambda_2 x}}{1 + e^{-(\lambda_2 - \lambda_1)x} - e^{-\lambda_2 x}}$$

for $x > 0$. In this case, $\lim_{x \to \infty} h(x) = \lambda_1$, which is the failure rate of the stronger component. Similarly, when $\lambda_2 < \lambda_1$, $\lim_{x \to \infty} h(x) = \lambda_2$, which is again the failure rate of the stronger component. Finally, when $\lambda_1 = \lambda_2 = \lambda$, multiply the numerator and denominator of the hazard function by $e^{\lambda x}$ yielding

$$h_X(x) = \frac{\lambda + \lambda - (2\lambda)e^{-\lambda x}}{1 + 1 - e^{-\lambda x}} = \frac{2\lambda\left(1 - e^{-\lambda x}\right)}{2 - e^{-\lambda x}}$$

for $x > 0$. In this case, $\lim_{x \to \infty} h(x) = \lambda$. Combining these three cases into one expression,

$$\lim_{x \to \infty} h_X(x) = \begin{cases} \lambda_1 & \lambda_1 < \lambda_2 \\ \lambda_2 & \lambda_2 < \lambda_1 \\ \lambda & \lambda_1 = \lambda_2 = \lambda, \end{cases}$$

that is, the limiting value of the hazard function is the failure rate of the stronger component:

$$\lim_{x \to \infty} h_X(x) = \min\{\lambda_1, \lambda_2\}.$$

6.3   Expected Values

The definitions of the expectation of a random variable X, namely $E[X]$, and the expectation of a function of a random variable $g(X)$, namely $E[g(X)]$, were presented in Chapter 3. These expected values are generalized here for a pair of random variables X and Y with a joint distribution. The theorem presented below is proved in a similar fashion to its one-dimensional counterpart.

Definition 6.5 Let X and Y be random variables with joint probability mass function $f(x, y)$ if the random variables are discrete or joint probability density function $f(x, y)$ if the random variables are continuous. The *expected value* of $g(X, Y)$ is

$$E[g(X, Y)] = \begin{cases} \displaystyle\sum_x \sum_y g(x, y) f(x, y) & X, Y \text{ discrete} \\ \displaystyle\int_{-\infty}^{\infty} \int_{-\infty}^{\infty} g(x, y) f(x, y)\, dy\, dx & X, Y \text{ continuous} \end{cases}$$

when the sum or integral exist. When the sum or integral diverge, the expected value is undefined.

The expected value of a function of two random variables $g(X, Y)$ is simply the sum or integral of the product of the functions $g(x, y)$ and $f(x, y)$. We illustrate this result with three examples. The first two examples involve discrete random variables and the third example involves continuous random variables.

Example 6.27 Let the discrete random variables X and Y have joint probability mass function

$$f(x, y) = \begin{cases} 0.2 & x = 1, y = 1 \\ 0.1 & x = 1, y = 2 \\ 0.3 & x = 1, y = 3 \\ 0.1 & x = 2, y = 1 \\ 0.1 & x = 2, y = 2 \\ 0.2 & x = 2, y = 3. \end{cases}$$

Find $E[X + Y]$.

The sum of X and Y can assume values between $1 + 1 = 2$ (when $X = 1$ and $Y = 1$) and $2 + 3 = 5$ (when $X = 2$ and $Y = 3$) so the expected value must lie between 2 and 5, and the specific value of the expected value depends on the distribution of probability associated with X and Y. Summing over the six mass values yields

$$
\begin{aligned}
E[X+Y] &= \sum_{x=1}^{2}\sum_{y=1}^{3}(x+y)f(x,y) \\
&= (1+1)(0.2)+(1+2)(0.1)+(1+3)(0.3)+\cdots+(2+3)(0.2) \\
&= 0.4+0.3+1.2+0.3+0.4+1.0 \\
&= 3.6.
\end{aligned}
$$

When a pair of random variables X and Y is drawn from this probability distribution, the expected value of the sum of the two random variables is 3.6.

Example 6.28 A pair of fair n-sided dice are tossed. Find the expected difference between the outcomes.

Let X be the outcome of the first die and Y be the outcome of the second die. Since each die is fair, the marginal distributions of X and Y are each a discrete uniform distribution with lower limit 1 and upper limit n. Since X and Y are independent, the joint probability mass function of X and Y is the product of the marginal probability mass functions:

$$
f(x,y) = \frac{1}{n^2} \qquad x=1,2,\ldots,n;\, y=1,2,\ldots,n.
$$

These random variables have a support over a square array of points. The expected difference of the two outcomes is

$$
\begin{aligned}
E\big[|X-Y|\big] &= \sum_{x=1}^{n}\sum_{y=1}^{n}|x-y|\cdot\frac{1}{n^2} \\
&= \frac{1}{n^2}\sum_{x=1}^{n}\left[\sum_{y=1}^{x}(x-y)+\sum_{y=x+1}^{n}(y-x)\right] \\
&= \frac{1}{n^2}\sum_{x=1}^{n}\left[\frac{(x-1)x}{2}+\frac{(n-x)(n-x+1)}{2}\right] \\
&= \frac{1}{n^2}\sum_{x=1}^{n}\left[\frac{n^2}{2}+\frac{n}{2}+x^2-x(n+1)\right] \\
&= \frac{1}{n^2}\left[\frac{n^3}{2}+\frac{n^2}{2}+\frac{n(n+1)(2n+1)}{6}-\frac{n(n+1)^2}{2}\right] \\
&= \frac{n^2-1}{3n}
\end{aligned}
$$

by using the identities

$$
1+2+\cdots+n=\frac{n(n+1)}{2} \qquad \text{and} \qquad 1^2+2^2+\cdots+n^2=\frac{n(n+1)(2n+1)}{6}.
$$

The analytic result can be checked for a particular value of n using Monte Carlo simulation. For $n=6$ (a pair of standard fair dice), the R statements

```
nrep = 500000
n = 6
mean(abs(ceiling(runif(nrep, 0, n)) - ceiling(runif(nrep, 0, n))))
```

after a call to `set.seed(3)`, yield

1.9435	1.9422	1.9443	1.9467	1.9445,

which hover around the analytic result $E\big[|X-Y|\big] = 35/18 \cong 1.9444$.

Example 6.29 The probability distribution of the continuous random variables X and Y is described by the joint probability density function

$$f(x, y) = \frac{1}{50} \qquad x > 0, y > 0, x + y < 10.$$

Find the expected value of $X^2 Y$.

The expected value can be calculated with a double integral over the support of X and Y of the product of $g(x, y) = x^2 y$ and $f(x, y)$:

$$E\left[X^2 Y\right] = \int_0^{10} \int_0^{10-x} x^2 y \cdot \frac{1}{50} \, dy \, dx = \cdots = \frac{100}{3} \cong 33.3333.$$

The interpretation of this expected value is as follows: The continuous random variables X and Y are uniformly distributed over the triangle shown in Figure 6.12 because $f(x, y)$ is constant. If many (x, y) pairs are selected at random from the triangular region, then one would expect that the average of the $x^2 y$ values will converge to $100/3$.

Monte Carlo simulation can be used to check the analytic result. The acceptance–rejection technique is used to generate uniformly distributed random variates in the triangle. First, random variate pairs (x, y) are generated in the 10×10 square with opposite vertices $(0, 0)$ and $(10, 10)$. If they fall in the triangular-shaped support of X and Y, the pairs are *accepted*. If the pairs fall outside of the support, they are *rejected*. The R code to estimate $E\left[X^2 Y\right]$ is given below.

```
nrep = 100000
x = runif(nrep, 0, 10)
y = runif(nrep, 0, 10)
index = which(x + y < 10)
sum(x[index] ^ 2 * y[index]) / length(index)
```

The which function in R returns the index positions corresponding to random variate pairs that were accepted. (This will be roughly half of the 100,000 random pairs generated in the experiment.) After a call to set.seed(3), five runs of this code yield

33.4264	33.2827	33.0887	33.2674	33.7391.

These values hover around the value obtained analytically, namely $E\left[X^2 Y\right] = 100/3 \cong 33.3333$, so the analytic solution is confirmed.

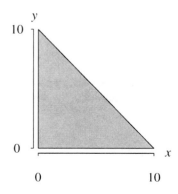

Figure 6.12: Support of X and Y.

There are two results concerning expectations of random variables that often arise in probability problems involving two random variables. The proofs of these results are given for continuous random variables. Analogous proofs for discrete random variables can be constructed by simply replacing integrals with summations. The first result involves sums of functions of random variables.

Theorem 6.2 If X and Y are random variables, then

$$E[g(X) + h(Y)] = E[g(X)] + E[h(Y)]$$

for any functions g and h.

Proof Let $f(x, y)$ denote the joint probability density function of X and Y. The expected value of $g(X) + h(Y)$ is

$$
\begin{aligned}
E[g(X) + h(Y)] &= \int_{-\infty}^{\infty} \int_{-\infty}^{\infty} (g(x) + h(y)) f(x, y) \, dy \, dx \\
&= \int_{-\infty}^{\infty} \int_{-\infty}^{\infty} g(x) f(x, y) \, dy \, dx + \int_{-\infty}^{\infty} \int_{-\infty}^{\infty} h(y) f(x, y) \, dy \, dx \\
&= E[g(X)] + E[h(Y)],
\end{aligned}
$$

which proves the result. $\qquad\square$

The most common application of Theorem 6.2 occurs when $g(X) = X$ and $h(Y) = Y$, which results in the special case

$$E[X + Y] = E[X] + E[Y].$$

In words, this result states that the expected value of the sum of two random variables is the sum of their expected values. Theorem 6.2 generalizes to

$$E[g(X, Y) + h(X, Y)] = E[g(X, Y)] + E[h(X, Y)].$$

Unlike Theorem 6.2, which placed no restrictions on the distribution of X and Y, the next result concerning products of random variables assumes that X and Y are independent random variables.

Theorem 6.3 If X and Y are independent random variables, then

$$E[g(X)h(Y)] = E[g(X)] E[h(Y)]$$

for any functions g and h.

Proof Since X and Y are independent random variables, the joint probability density function can be expressed as the product of the marginal probability density functions $f(x, y) = f_X(x) f_Y(y)$, so

$$
\begin{aligned}
E[g(X)h(Y)] &= \int_{-\infty}^{\infty} \int_{-\infty}^{\infty} g(x) h(y) f(x, y) \, dx \, dy \\
&= \int_{-\infty}^{\infty} \int_{-\infty}^{\infty} g(x) h(y) f_X(x) f_Y(y) \, dx \, dy \\
&= \int_{-\infty}^{\infty} g(x) f_X(x) \, dx \int_{-\infty}^{\infty} h(y) f_Y(y) \, dy \\
&= E[g(X)] E[h(Y)]
\end{aligned}
$$

which proves the result. $\qquad\square$

The most common application of Theorem 6.3 occurs when $g(X) = X$, and $h(Y) = Y$, which results in the special case

$$E[XY] = E[X]E[Y],$$

when X and Y are independent, which will be useful in the next subsection. There are several special expected values for two jointly distributed random variables, namely (a) covariance and correlation, (b) conditional expectation, and (c) moment generating functions. Each of these special expected values will be described in the subsections to follow.

Covariance and correlation

The population mean μ and the population variance σ^2 give us information about the central tendency and dispersion of a univariate random variable, respectively. A related measure for two random variables is the *population covariance*, which measures the strength of the *linear* relationship between X and Y.

Definition 6.6 Let X and Y be random variables with finite population means μ_X and μ_Y, respectively. The *population covariance* between X and Y is

$$\text{Cov}(X, Y) = E\big[(X - \mu_X)(Y - \mu_Y)\big].$$

The notation $\text{Cov}(X, Y)$ for population covariance is not universal; it also goes by σ_{XY} and $C(X, Y)$. The order of the arguments in the population covariance are reversible, that is $\text{Cov}(X, Y) = \text{Cov}(Y, X)$.

Definition 6.6 gives the *defining formula* for the population covariance. Defining formulas are usually helpful in thinking about the meaning of a particular quantity being defined. In this case, the interpretation of the *sign* of the population covariance can be understood from this definition.

- If X and Y tend to be on opposite sides of their population means together (that is, if X tends to be above its population mean when Y tends to be below its population mean, or if X tends to be below its population mean when Y tends to be above its population mean), then the argument of the expected value operator is more often negative, so the population covariance is negative.

- If X and Y tend to be on the same sides of their population means together (that is, if X tends to be above its population mean when Y tends to be above its population mean, or if X tends to be below its population mean when Y tends to be below its population mean), then the argument of the expected value operator is more often positive, so the population covariance is positive.

So the population covariance is a quantity that reflects how the random variables X and Y behave together with respect to linear dependence.

> **Example 6.30** A fair coin is tossed twice. Let X be the number of heads that appear and Y be the number of tails that appear. Find the population covariance between X and Y.
>
> This is a case where $Y = 2 - X$ is a linear function of the value of X. The support of the joint distribution of X and Y is
>
> $$\mathcal{A} = \{(x, y) \,|\, (x, y) = (0, 2), (1, 1), (2, 0)\}.$$
>
> The marginal distributions of X and Y are both binomial$(2, 1/2)$ random variables, each with population mean $E[X] = E[Y] = 1$. Since there are $2^2 = 4$ equally-likely outcomes to the random experiment by the multiplication rule, the joint probability mass function

of X and Y is

$$f(x, y) = \begin{cases} 1/4 & x = 0, y = 2 \\ 1/2 & x = 1, y = 1 \\ 1/4 & x = 2, y = 0. \end{cases}$$

Examining the support of this joint probability mass function, whenever X is *above* its mean, Y must be *below* its mean, and vice-versa. Therefore, we expect the population covariance to be negative. The population covariance between X and Y

$$\begin{aligned}
\text{Cov}(X, Y) &= E\big[(X - \mu_X)(Y - \mu_Y)\big] \\
&= \sum_{\mathcal{A}}\sum (x - 1)(y - 1) f(x, y) \\
&= (0 - 1)(2 - 1)\frac{1}{4} + (1 - 1)(1 - 1)\frac{1}{2} + (2 - 1)(0 - 1)\frac{1}{4} \\
&= -\frac{1}{2}
\end{aligned}$$

is indeed negative.

The *units* of the population covariance are the product of the units of X and the units of Y. It is often the case that X and Y are measured in the same units, in which case the population covariance is measured in those units squared. If (X, Y) are a husband's height and a wife's height measured in feet, then the units on the population covariance are square feet.

Covariance is also defined when more than two variables are involved in a probability model. For example, if X, Y, and Z are random variables that are jointly distributed, then the quantities

$$\text{Cov}(X, Y), \text{Cov}(X, Z), \text{and} \text{Cov}(Y, Z)$$

measure the pairwise population covariance between the three random variables.

The population covariance is related to the population variance in the following sense. For a random variable X, the population variance of X is

$$V[X] = \text{Cov}(X, X),$$

which is to say that the population variance of a random variable X is the population covariance between X and itself. Because of this relationship, population variances and covariances are often presented in what is known as a *variance–covariance* matrix, with variances on the diagonal entries and covariances on the off-diagonal entries. In a regrettable notation choice, Σ is often used to denote the variance–covariance matrix. For two random variables X and Y, the symmetric 2×2 variance–covariance matrix is

$$\Sigma = \begin{bmatrix} V[X] & \text{Cov}(X, Y) \\ \text{Cov}(Y, X) & V[Y] \end{bmatrix},$$

which could also have been written as

$$\Sigma = \begin{bmatrix} \text{Cov}(X, X) & \text{Cov}(X, Y) \\ \text{Cov}(Y, X) & \text{Cov}(Y, Y) \end{bmatrix}.$$

For three random variables X, Y, and Z, the symmetric 3×3 variance–covariance matrix is

$$\Sigma = \begin{bmatrix} \text{Cov}(X, X) & \text{Cov}(X, Y) & \text{Cov}(X, Z) \\ \text{Cov}(Y, X) & \text{Cov}(Y, Y) & \text{Cov}(Y, Z) \\ \text{Cov}(Z, X) & \text{Cov}(Z, Y) & \text{Cov}(Z, Z) \end{bmatrix}.$$

All variance–covariance matrices are positive-semidefinite.

The defining formula given in Definition 6.6 is useful for conceptualizing population covariance. The following shortcut formula is oftentimes quicker for calculating the population covariance.

Theorem 6.4 If X and Y are random variables with finite population means μ_X and μ_Y, respectively, then
$$\text{Cov}(X, Y) = E\left[(X - \mu_X)(Y - \mu_Y)\right] = E[XY] - \mu_X\mu_Y.$$

Proof Since the expected value is a linear operator, the population covariance between X and Y is

$$
\begin{aligned}
E\left[(X - \mu_X)(Y - \mu_Y)\right] &= E[XY - \mu_X X - \mu_Y Y + \mu_X\mu_Y] \\
&= E[XY] - \mu_Y E[X] - \mu_X E[Y] + \mu_X\mu_Y \\
&= E[XY] - \mu_X\mu_Y,
\end{aligned}
$$

which proves the result. □

We now find the population covariance between X and Y from the previous example using the shortcut formula.

Example 6.31 A fair coin is tossed twice. Let X be the number of heads that appear and Y be the number of tails that appear. Find the population covariance between X and Y using the shortcut formula.

As before, the joint probability mass function of X and Y is

$$
f(x, y) = \begin{cases}
1/4 & x = 0, y = 2 \\
1/2 & x = 1, y = 1 \\
1/4 & x = 2, y = 0
\end{cases}
$$

and the marginal distributions of both X and Y are each binomial$(2, 1/2)$. The means of these marginal distributions are $\mu_X = E[X] = 1$ and $\mu_Y = E[Y] = 1$. Using the shortcut formula for computing the population covariance,

$$
\begin{aligned}
\text{Cov}(X, Y) &= E[XY] - \mu_X\mu_Y \\
&= \sum_{\mathcal{A}}\sum xy f(x, y) - 1 \cdot 1 \\
&= 0 \cdot 2 \cdot \frac{1}{4} + 1 \cdot 1 \cdot \frac{1}{2} + 2 \cdot 0 \cdot \frac{1}{2} - 1 \\
&= \frac{1}{2} - 1 \\
&= -\frac{1}{2},
\end{aligned}
$$

which matches the population covariance calculation using the defining formula.

Example 6.32 Deal two cards from a well-shuffled deck. Let the random variable X be the number of aces dealt and let the random variable Y be the number of face cards dealt. Find the population covariance between X and Y using the shortcut formula.

This random experiment and associated pair of random variables was first encountered in Example 6.1. The joint probability mass function of X and Y is

$$
f(x, y) = \frac{\binom{4}{x}\binom{12}{y}\binom{36}{2-x-y}}{\binom{52}{2}} \qquad x = 0, 1, 2; y = 0, 1, 2; x + y \le 2.
$$

Calculating these six probabilities results in the entries in Table 6.9, which also includes the marginal probability mass functions as the row sums and column sums. Using the marginal probability mass functions, the expected number of aces is

$$E[X] = \sum_{x=0}^{2} x f_X(x) = 0 \cdot \frac{1128}{1326} + 1 \cdot \frac{192}{1326} + 2 \cdot \frac{6}{1326} = \frac{204}{1326} = \frac{2}{13}$$

and the expected number of face cards is

$$E[Y] = \sum_{y=0}^{2} y f_Y(y) = 0 \cdot \frac{780}{1326} + 1 \cdot \frac{480}{1326} + 2 \cdot \frac{66}{1326} = \frac{612}{1326} = \frac{6}{13}.$$

The expected value of the product of X and Y is particularly easy to calculate because 0 appears so often in the double summation:

$$E[XY] = \sum_{x=0}^{2} \sum_{y=0}^{2-x} xy f(x, y) = 1 \cdot 1 \cdot \frac{48}{1326} = \frac{8}{221}.$$

So the shortcut formula yields the population covariance between X and Y as

$$\text{Cov}(X, Y) = E[XY] - \mu_X \mu_Y = \frac{8}{221} - \frac{2}{13} \cdot \frac{6}{13} = -\frac{100}{2873} \cong -0.03481.$$

How should the sign of this result be interpreted? The negative population covariance indicates that X and Y tend to be on opposite sides of their means more often than being on the same sides of their means. This makes sense intuitively: when there are more aces dealt, there will be fewer face cards; when there are more face cards dealt, there will be fewer aces. The magnitude of this covariance is diluted relative to the previous example, however, because of the presence of the number cards.

Monte Carlo simulation can be used to illustrate the analytic result. The following R code assumes that cards 1–4 are the aces and cards 41–52 are the face cards.

```
nrep = 100000
x = rep(0, nrep)
y = rep(0, nrep)
for (i in 1:nrep) {
  cards = sample(52, 2)
  if (cards[1] <= 4) x[i] = x[i] + 1
  if (cards[2] <= 4) x[i] = x[i] + 1
  if (cards[1] >= 41) y[i] = y[i] + 1
  if (cards[2] >= 41) y[i] = y[i] + 1
}
mean(x * y) - mean(x) * mean(y)
```

x \ y	0	1	2	$f_X(x)$
0	630/1326	432/1326	66/1326	1128 / 1326
1	144/1326	48/1326	0	192/1326
2	6/1326	0	0	6/1326
$f_Y(y)$	780/1326	480/1326	66/1326	1

Table 6.9: Joint and marginal probability mass functions of X and Y.

Five runs of this Monte Carlo simulation yield

$$-0.03504 \qquad -0.03574 \qquad -0.03452 \qquad -0.03469 \qquad -0.03456.$$

These values hover around the true value $\text{Cov}(X, Y) = -100/2873 \cong -0.03481$, so the Monte Carlo simulation supports the analytic result.

Theorem 6.5 If X and Y are random variables with finite population variances and covariance, then
$$V[X + Y] = V[X] + V[Y] + 2\text{Cov}(X, Y).$$

Proof Using the definition of the population variance, Theorem 6.2, and the fact that expected value is a linear operator,

$$\begin{aligned}
V[X + Y] &= E\left[\left((X + Y) - E[X + Y]\right)^2\right] \\
&= E\left[\left((X - \mu_X) + (Y - \mu_Y)\right)^2\right] \\
&= E\left[(X - \mu_X)^2 + (Y - \mu_Y)^2 + 2(X - \mu_X)(Y - \mu_Y)\right] \\
&= E\left[(X - \mu_X)^2\right] + E\left[(Y - \mu_Y)^2\right] + 2E\left[(X - \mu_X)(Y - \mu_Y)\right] \\
&= V[X] + V[Y] + 2\text{Cov}(X, Y),
\end{aligned}$$

which proves the result. $\qquad\qquad\qquad\qquad\qquad\qquad\qquad\qquad\qquad\qquad\qquad\qquad$ $\square$

Example 6.33 A fair coin is tossed twice. Let X be the number of heads that appear and Y be the number of tails that appear. Find the population variance of $X + Y$.

Since $X \sim \text{binomial}(2, 1/2)$ and $Y \sim \text{binomial}(2, 1/2)$, the population variances of the two random variables are
$$V[X] = 2 \cdot \frac{1}{2} \cdot \frac{1}{2} = \frac{1}{2}$$
and
$$V[Y] = 2 \cdot \frac{1}{2} \cdot \frac{1}{2} = \frac{1}{2}$$

because the population variance of a binomial(n, p) random variable is $np(1 - p)$. Using Theorem 6.5, the population variance of the sum of X and Y is

$$V[X + Y] = V[X] + V[Y] + 2\text{Cov}(X, Y) = \frac{1}{2} + \frac{1}{2} - 2 \cdot \frac{1}{2} = 0,$$

where the population covariance was calculated in Examples 6.30 and 6.31. The population variance of the sum of X and Y must be zero because the sum of the heads and tails tossed in the random experiment is always 2, and the population variance of a constant is zero.

The next result concerns the population covariance between independent random variables.

Theorem 6.6 If X and Y are independent random variables, then $\text{Cov}(X, Y) = 0$.

Proof Since X and Y are independent random variables, $E[XY] = E[X]E[Y]$ by Theorem 6.3. Using the shortcut formula for the population covariance from Theorem 6.4, $\text{Cov}(X, Y) = E[XY] - E[X]E[Y] = 0$, which proves the result. $\qquad\qquad\qquad\qquad\qquad$ $\square$

One question that arises from Theorem 6.6 is whether the converse is true. That is, does a joint distribution of random variables X and Y with a population covariance of 0 imply that the random variables are independent? The converse is not true in general, and this is best established with a counterexample. The particular counterexample given next is not unique. There are many similar examples for both discrete and continuous probability distributions for X and Y.

Example 6.34 Show that the random variables X and Y that are uniformly distributed over the support

$$\mathcal{A} = \{(x, y) \mid 1 < x^2 + y^2 < 4\}$$

have a population covariance 0 and are dependent random variables.

The support region is shown in the donut-shaped region shaded in Figure 6.13 (a dot has been placed at the origin). The random variables are dependent by inspection because the support is not a product space. What is left to show is that the population covariance is 0. Since the area of a circle of radius r is πr^2, the area of the shaded region is the area of the outer circle less the area of the inner circle, which is

$$\pi \cdot 2^2 - \pi \cdot 1^2 = 3\pi.$$

So if the random variables X and Y are uniformly distributed over the donut-shaped support, their joint probability density function must be

$$f(x, y) = \frac{1}{3\pi} \qquad 1 < x^2 + y^2 < 4.$$

Using the symmetry of the support $\mathcal{A}$ and the fact that the joint probability density function is constant on $\mathcal{A}$, it is clear that $E[X] = E[Y] = 0$. What remains to be calculated is $E[XY]$. Since

$$E[XY] = \int \int_{\mathcal{A}} xy f(x, y) \, dy \, dx$$

results in some rather tedious integrals, it is worthwhile converting to polar coordinates.

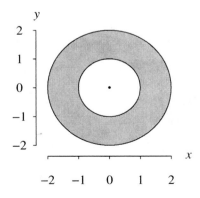

Figure 6.13: Support of X and Y.

Letting $x = r\cos\theta$, $y = r\sin\theta$, and $dy\,dx = r\,dr\,d\theta$, the integral becomes

$$
\begin{aligned}
E[XY] &= \int_0^{2\pi}\int_1^2 (r\cos\theta)(r\sin\theta)\frac{1}{3\pi}r\,dr\,d\theta \\
&= \frac{1}{3\pi}\int_0^{2\pi}\int_1^2 r^3\sin\theta\cos\theta\,dr\,d\theta \\
&= \frac{1}{3\pi}\cdot\frac{15}{4}\int_0^{2\pi}\sin\theta\cos\theta\,dr\,d\theta \\
&= \frac{5}{4\pi}\left[\frac{\sin^2\theta}{2}\right]_0^{2\pi} \\
&= 0.
\end{aligned}
$$

So the population covariance between X and Y using the shortcut formula is

$$\text{Cov}(X,Y) = E[XY] - \mu_X\mu_Y = 0 - 0\cdot 0 = 0.$$

This example provides one instance of a joint distribution of two random variables that have a population covariance 0 yet are dependent random variables. There are many other examples that can be constructed with these properties. This counterexample shows that the converse of Theorem 6.6 is not true in general.

In the case of independent random variables X and Y, the population variance of their sum is the sum of their population variances, as indicated by the next theorem.

Theorem 6.7 If X and Y are independent random variables, then

$$V[X+Y] = V[X] + V[Y].$$

Proof Theorem 6.5 indicates that in general $V[X+Y] = V[X] + V[Y] + 2\text{Cov}(X,Y)$. Since X and Y are assumed to be independent, $\text{Cov}(X,Y) = 0$ by Theorem 6.6, which leads to the result.   □

Theorem 6.7 generalizes to

$$V[aX+bY] = a^2V[X] + b^2V[Y],$$

and the important special case $V[X-Y] = V[X] + V[Y]$, which is important in engineering applications in setting tolerances.

The population covariance suffers from the same weakness as the population variance: if the units of measurement used on the random variable(s) of interest changes, the magnitude of the population variance and covariance changes. This makes the magnitude of the population covariance difficult to interpret. The search for a unitless measure of linear dependence leads to the population correlation coefficient, which we refer to as just the *population correlation*.

Definition 6.7 Let X and Y be random variables with finite population means μ_X and μ_Y, and finite population variances $\sigma_X^2 > 0$ and $\sigma_Y^2 > 0$, respectively. The *population correlation* between X and Y is

$$\rho = \frac{E[(X-\mu_X)(Y-\mu_Y)]}{\sigma_X\sigma_Y}.$$

$$= \frac{\text{Cov}(X,Y)}{\sigma_X\sigma_Y} = \frac{E[XY] - \mu_X\mu_Y}{\sigma_X\sigma_Y}.$$

The choice of notation in Definition 6.7 is not universal. The population correlation also goes by ρ_{XY}, ρ_{12}, $\text{Corr}(X,Y)$. Pulling the constants in the denominator into the expected value operator, the population correlation can be written as

$$\rho = E\left[\left(\frac{X-\mu_X}{\sigma_X}\right)\left(\frac{Y-\mu_Y}{\sigma_Y}\right)\right],$$

which is the expected value of the product of the two standardized random variables.

This seemingly minor tweak to the definition of the population covariance, namely dividing by $\sigma_X \sigma_Y$, results in some rather remarkable properties for ρ. As mentioned earlier, ρ is unitless. If the random variables X and Y are measured in feet, inches, or millimeters, one always arrives at the same value for ρ. It is unaffected by change of scale. We prove three other results concerning ρ below.

Theorem 6.8 If X and Y are independent, then $\rho = 0$.

Proof When the random variables X and Y are independent, $\text{Cov}(X, Y) = 0$ by Theorem 6.6. Since $\text{Cov}(X, Y) = 0$ is the numerator of the definition of the population correlation, $\rho = 0$. $\qquad\square$

Theorem 6.9 If X and Y are random variables with population correlation ρ, then

$$-1 \leq \rho \leq 1.$$

Proof Consider the random variable

$$\frac{X}{\sigma_X} + \frac{Y}{\sigma_Y}.$$

By Theorem 6.5, the population variance of this random variable is

$$
\begin{aligned}
V\left[\frac{X}{\sigma_X} + \frac{Y}{\sigma_Y}\right] &= V\left[\frac{X}{\sigma_X}\right] + V\left[\frac{Y}{\sigma_Y}\right] + 2\text{Cov}\left(\frac{X}{\sigma_X}, \frac{Y}{\sigma_Y}\right) \\
&= \frac{V[X]}{\sigma_X^2} + \frac{V[Y]}{\sigma_Y^2} + 2\frac{\text{Cov}(X, Y)}{\sigma_X \sigma_Y} \\
&= 1 + 1 + 2\rho \\
&= 2(1 + \rho).
\end{aligned}
$$

Since all population variances must be nonnegative,

$$2(1 + \rho) \geq 0,$$

which simplifies to

$$\rho \geq -1.$$

Following this same line of reasoning but working with the random variable

$$\frac{X}{\sigma_X} - \frac{Y}{\sigma_Y}$$

leads to

$$\rho \leq 1.$$

Combining these two inequalities yields the desired result

$$-1 \leq \rho \leq 1. \qquad\square$$

So the population correlation must lie between -1 and 1. This indicates that not only the sign, but also the magnitude of the population correlation is meaningful. The sign of the population correlation indicates the direction of the linear relationship between X and Y; the magnitude of the population correlation measures the strength of the linear relationship between X and Y.

Example 6.35 Let the discrete random variables X and Y have joint probability mass function $f(x, y)$ given by the entries in Table 6.10. Find the population correlation between X and Y.

x \ y	1	2	3	$f_X(x)$
1	0.2	0.1	0.3	0.6
2	0.1	0.1	0.2	0.4
$f_Y(y)$	0.3	0.2	0.5	

Table 6.10: Joint and marginal probability mass functions for X and Y.

The calculations that are required to find the population correlation are given below.

$$E[XY] = \sum_x \sum_y xyf(x, y) = (1)(1)(0.2) + (1)(2)(0.1) + \cdots + (2)(3)(0.2) = 3.1,$$

$$E[X] = \sum_x xf_X(x) = (1)(0.6) + (2)(0.4) = 1.4,$$

$$E[Y] = \sum_y yf_Y(y) = (1)(0.3) + (2)(0.2) + (3)(0.5) = 2.2,$$

$$\text{Cov}(X, Y) = E[XY] - \mu_X\mu_Y = 3.1 - (1.4)(2.2) = 0.02,$$

$$V[X] = E[X^2] - \mu_X^2 = (1)^2(0.6) + (2)^2(0.4) - (1.4)^2 = 0.24,$$

$$V[Y] = E[Y^2] - \mu_Y^2 = (1)^2(0.3) + (2)^2(0.2) + (3)^2(0.5) - (2.2)^2 = 0.76,$$

$$\rho = \frac{0.02}{\sqrt{0.24}\sqrt{0.76}} \cong 0.0468.$$

The fact that this population correlation is positive indicates that X and Y tend to be on the same side of their means together. The fact that the population is much closer to 0 than to 1 indicates that the positive association between X and Y is rather weak.

One final result explores the meaning of the population correlation at its extremes: $\rho = \pm 1$.

Theorem 6.10 Let X and Y be random variables with population correlation ρ. The population correlation ρ equals -1 if and only if the support of X and Y lies on a line with negative slope. The population correlation ρ equals 1 if and only if the support of X and Y lies on a line with positive slope.

Proof Suppose that $\rho = -1$. Consider again the random variable from the previous proof

$$\frac{X}{\sigma_X} + \frac{Y}{\sigma_Y}$$

with population variance

$$V\left[\frac{X}{\sigma_X} + \frac{Y}{\sigma_Y}\right] = 2(1 + \rho).$$

Since $\rho = -1$

$$V\left[\frac{X}{\sigma_X} + \frac{Y}{\sigma_Y}\right] = 0.$$

A random variable with population variance 0 is degenerate, which means that for some real-valued constant c,

$$P\left(\frac{X}{\sigma_X} + \frac{Y}{\sigma_Y} = c\right) = 1.$$

Since $\sigma_X > 0$ and $\sigma_Y > 0$ because the population correlation exists, the event inside of the probability statement is equivalent to X and Y falling on a line with negative slope. The converse and the case of $\rho = 1$ are shown in a similar fashion. □

We have encountered one example where the support fell on a line with a negative slope. In this case we expect that the population correlation will be -1.

Example 6.36 A fair coin is tossed twice. Let X be the number of heads that appear and Y be the number of tails that appear. Find the population correlation between X and Y.

As before, the marginal distributions of X and Y are both binomial$(2, 1/2)$ random variables, each with population mean $E[X] = E[Y] = 1$ and population variance $V[X] = V[Y] = 2 \cdot (1/2) \cdot (1/2) = 1/2$. Since there are $2^2 = 4$ equally-likely outcomes to the random experiment by the multiplication rule, the joint probability mass function of X and Y is

$$f(x, y) = \begin{cases} 1/4 & x = 0, y = 2 \\ 1/2 & x = 1, y = 1 \\ 1/4 & x = 2, y = 0. \end{cases}$$

The three values in the support set fall on a line with a negative slope, so we could simply invoke Theorem 6.10 and conclude that the population correlation is $\rho = -1$. We instead opt to work through the details. The population covariance between X and Y is

$$
\begin{aligned}
\text{Cov}(X, Y) &= E\left[(X - \mu_X)(Y - \mu_Y)\right] \\
&= \sum_\mathcal{A}\sum (x - 1)(y - 1) f(x, y) \\
&= (0 - 1)(2 - 1)\frac{1}{4} + (1 - 1)(1 - 1)\frac{1}{2} + (2 - 1)(0 - 1)\frac{1}{4} \\
&= -\frac{1}{2}.
\end{aligned}
$$

(handwritten): $= E[xy] - \mu_x\mu_y$
$= \sum\sum xy\, f(x,y) - 1\cdot 1 = (0)(2)\frac{1}{4}$
$+ (1)(1)\frac{1}{2}$
$+ (2)(0)\frac{1}{4}$
$= \frac{1}{2} - 1$
$= \boxed{-\frac{1}{2}}$

The population correlation is

$$\rho = \frac{\text{Cov}(X, Y)}{\sigma_X \sigma_Y} = \frac{-1/2}{1/2} \boxed{= -1}$$

(handwritten): flipping coin twice is binomial $(2, \frac{1}{2})$. thus $V[X] = V[Y] = np(1-p) = \frac{1}{2}$

as predicted by Theorem 6.10.

More generally, the population correlation ρ can be interpreted as a measure of the intensity of the concentration of probability about a line with positive or negative slope. This concludes the introduction to covariance and correlation. Another topic that falls under the general heading of expected values is conditional expectation, which follows naturally from conditional distributions.

Conditional expected values

Conditional expected values, or conditional expectations, are an extension of conditional distributions. We define conditional expected values in the general case.

(handwritten): $f_{Y|X=x}(y|X=x) = \dfrac{f(x,y)}{f_X(x)}$ $\quad f_{X|Y=y}(x|Y=y) = \dfrac{f(x,y)}{f_Y(y)}$

Definition 6.8 Let X and Y be discrete random variables, then the conditional expected value of $g(Y)$ given $X = x$ is

$$E[g(Y)\,|\,X = x] = \sum_y g(y) f_{Y|X=x}(y\,|\,X = x).$$

Let X and Y be continuous random variables, then the conditional expected value of $g(Y)$ given $X = x$ is

$$E[g(Y)\,|\,X = x] = \int_{-\infty}^{\infty} g(y) f_{Y|X=x}(y\,|\,X = x)\,dy.$$

The conditional expected value of $g(X)$ given $Y = y$ is defined in a similar fashion. We begin with a simple example of finding conditional expectations.

Example 6.37 Let the discrete random variables X and Y have joint probability mass function $f(x, y)$ given by the entries in Table 6.11. Find the conditional expected value of Y given $X = 2$.

x \ y	1	2	3	$f_X(x)$
1	0.2	0.1	0.3	0.6
2	0.1	0.1	0.2	0.4
$f_Y(y)$	0.3	0.2	0.5	

Table 6.11: Joint and marginal probability mass functions for X and Y.

The conditional probability mass function of Y given $X = 2$ is the ratio of the joint probability mass function to $f_X(2)$ which yields

$$f_{Y|X=2}(y\,|\,X = 2) = \begin{cases} 1/4 & y = 1 \\ 1/4 & y = 2 \\ 1/2 & y = 3. \end{cases}$$

Using Definition 6.8, the conditional expected value of Y given $X = 2$ is

$$E[Y\,|\,X = 2] = 1 \cdot \frac{1}{4} + 2 \cdot \frac{1}{4} + 3 \cdot \frac{1}{2} = \frac{9}{4}.$$

There are some expected values that are difficult to calculate by using the definition of the expected value directly, but by conditioning on a second random variable, can be calculated easily. Here are three such examples, beginning with the famous "thief of Baghdad" problem.

Example 6.38 A thief is in a fiendish, dark, circular dungeon with three identical doors. Once the thief chooses a door and passes through it, the door locks behind him. The three doors lead to

- a 6 hour tunnel that leads to freedom,
- a 3 hour tunnel that returns him to the dungeon,
- a 9 hour tunnel that returns him to the dungeon.

Each door is chosen with equal probability. When the thief is dropped back into the dungeon after choosing the second or third door, he is a "Markov" thief in the sense that there is a memoryless subsequent choice of doors. He isn't able to mark the doors in any way. What is his expected time to escape?

Let the random variable X be the time to escape. Writing the probability density function of X is tedious, particularly so because of the denumerable support—there is no upper bound on the time to escape. Try a few values and you will see the problems that arise. So finding $E[X]$ by its definition is a non-starter in this case. It is much easier to use conditional expected values. Let Y be the door number chosen initially by the thief. The expected time to escape is easily calculated by conditioning on the thief's first door choice:

$$
\begin{aligned}
E[X] &= E[X \mid Y = 1] P(Y = 1) + E[X \mid Y = 2] P(Y = 2) + E[X \mid Y = 3] P(Y = 3) \\
&= (6) \cdot \left(\frac{1}{3}\right) + (3 + E[X]) \cdot \left(\frac{1}{3}\right) + (9 + E[X]) \cdot \left(\frac{1}{3}\right) \\
&= 2 + 1 + \frac{E[X]}{3} + 3 + \frac{E[X]}{3} \\
&= 6 + \frac{2E[X]}{3}.
\end{aligned}
$$

The memoryless aspect of his door choice was used to find the expected values associated with $Y = 1$ and $Y = 2$. Solving this equation for $E[X]$ yields

$$
E[X] = 18.
$$

It will take the thief an average of 18 hours to escape from the dungeon.

The analytic solution above can be checked in R using Monte Carlo simulation. The R code below simulates the time to escape of 100,000 thieves, printing the average of their escape times.

```
nrep = 100000
time = 0
for (i in 1:nrep) {
  escape = 0
  while (escape == 0) {
    door = ceiling(runif(1, 0, 3))
    if (door == 1) {
      escape = 1
      time = time + 6
    }
    if (door == 2) time = time + 3
    if (door == 3) time = time + 9
  }
}
print(time / nrep)
```

The code is run five times after an initial call to set.seed(3) to initialize the random number stream. The average time to escape for the five runs is

17.9445   17.9757   18.0908   18.0489   18.0314.

Since these values hover around 18 hours, the Monte Carlo simulation supports of the analytic result.

The thief of Baghdad problem illustrated the case where $f(x)$ was difficult to write. There are cases where $f(x)$ can be written, but the summations or integrals associated with the calculation of an expected value are difficult. The next example involves an alternative calculation of the population mean of the geometric distribution, which was introduced in Section 4.3.

Example 6.39 Let the random variable X be the number of tails tossed prior to the first head in repeated tosses of a biased coin, where $p = P(\text{tossing a head})$ on an individual toss. Find $E[X]$.

The random variable X has probability mass function

$$f(x) = p(1-p)^x \qquad\qquad x = 0, 1, 2, \ldots,$$

which is recognized as the geometric(p) distribution. There were two derivations of the population mean given in Section 4.3, and this example contains a third derivation that uses conditional expectation. Let Y be the result of the first coin toss ($Y = 0$ for a tail and $Y = 1$ for a head). Conditioning on the results of the first toss and using the *memoryless* property of the geometric distribution,

$$E[X] = E[X \,|\, Y = 0]\,P(Y = 0) + E[X \,|\, Y = 1]\,P(Y = 1).$$

Since the coin does not remember the results of the previous flip,

$$\begin{aligned}
E[X] &= (1 + E[X]) \cdot (1 - p) + 0 \cdot p \\
&= 1 - p + E[X] - pE[X].
\end{aligned}$$

Solving for $E[X]$ yields

$$E[X] = \frac{1-p}{p},$$

which matches the result derived in Section 4.3.

Example 6.40 A biased coin is flipped repeatedly until both a head and tail have appeared. What is the expected number of tosses required?

Let X be the number of flips required to get both a head and a tail. Let p be the probability of flipping a head on a single flip. As in the previous example, let Y be the results of the first coin flip ($Y = 0$ for a tail and $Y = 1$ for a head). Conditioning on the outcome of the first flip, and using the population mean of a Geometric(p) random variable whose support begins at 1,

$$\begin{aligned}
E[X] &= E[X \,|\, Y = 0]P(Y = 0) + E[X \,|\, Y = 1]P(Y = 1) \\
&= \left[1 + \frac{1}{p}\right] \cdot (1 - p) + \left[1 + \frac{1}{1-p}\right] \cdot p \\
&= 1 + \frac{p}{1-p} + \frac{1-p}{p}.
\end{aligned}$$

The conditional expected values given above are associated with the Geometric(p) distribution with support beginning at 1. When the coin is fair ($p = 1/2$), for example,

$$E[X] = 1 + \frac{1/2}{1/2} + \frac{1/2}{1/2} = 3,$$

which is the smallest value of $E[X]$ for $0 < p < 1$, as illustrated in Figure 6.14. This particular result is also apparent via a thought experiment. It takes exactly one toss to get the first unique outcome, then a Geometric($1/2$) number of tosses, to get the other outcome. Since the population mean of a Geometric($1/2$) random variable is 2, the expected number of tosses required is 3.

Monte Carlo simulation can be used to check the analytic result for a specific value of p. The R code for estimating the expected number of tosses required based on 100,000

$E[X]$

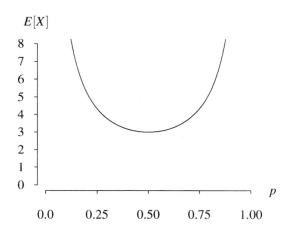

Figure 6.14: Expected number of tosses required as a function of p.

replications follows. The code uses $p = 1/2$, which is associated with the tossing of a fair coin. The call to `floor(runif(1, 0, 2))` simulates the flip of a coin, returning the floor of a random variable that is uniformly distributed between 0 and 2, that is a 0 or a 1 with equal probability.

```
nrep  = 100000
total = 0
for (i in 1:nrep) {
  x = 2
  firstflip = floor(runif(1, 0, 2))
  while (floor(runif(1, 0, 2)) == firstflip) x = x + 1
  total = total + x
}
print(total / nrep)
```

Five runs of this simulation yield

2.99823	2.99809	2.99019	3.00055	2.99825,

which supports the analytic result $E[X] = 3$ in the case of a fair coin.

The conditional expectation in this example was computed for a *specific* value that the random variable X assumes. Expectations can also be computed by conditioning on a *general* value, as illustrated in the next example.

Example 6.41 For the continuous random variables X and Y with joint probability density function

$$f(x, y) = \frac{1}{50} \qquad x > 0, y > 0, x + y < 10,$$

find $f_{X|Y=y}(x \mid Y = y)$ and $E\left[X^k \mid Y = y\right]$ for some real constant $k > 0$.

The support of X and Y is given in Figure 6.15. The marginal probability density function of Y is

$$f_Y(y) = \int_0^{10-y} \frac{1}{50}\,dx = \frac{10-y}{50} \qquad 0 < y < 10,$$

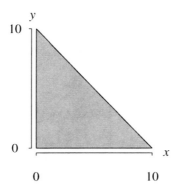

Figure 6.15: Support of X and Y.

so the conditional distribution of X given $Y = y$ is

$$f_{X|Y=y}(x|Y=y) = \frac{f(x,y)}{f_Y(y)} = \frac{1/50}{(10-y)/50} = \frac{1}{10-y} \qquad 0 < x < 10-y; 0 < y < 10.$$

To consider a specific instance, the conditional distribution of X given $Y = 2$ is

$$f_{X|Y=2}(x|Y=2) = \frac{1}{8} \qquad 0 < x < 8,$$

which is the probability density function for a $U(0, 8)$ random variable. Finally, the conditional expectation is computed as

$$
\begin{aligned}
E\left[X^k | Y = y\right] &= \int_0^{10-y} x^k \frac{1}{10-y}\, dx \\
&= \frac{1}{10-y}\left[\frac{x^{k+1}}{k+1}\right]_0^{10-y} \\
&= \frac{(10-y)^k}{k+1} \qquad 0 < y < 10, k > 0.
\end{aligned}
$$

Consider one specific instance of this conditional expectation with $k = 1$ and $Y = 2$,

$$E[X|Y=2] = \frac{8}{2} = 4,$$

as expected because the population mean of a $U(0, 8)$ random variable is 4.

The previous two examples have considered conditional expectations in the discrete and continuous cases. We now shift the emphasis to a new concept related to conditional expectations. Consider the previous example with $k = 1$, where the conditional expectation of X given $Y = y$ was found to be
$$E[X|Y=y] = \frac{10-y}{2}$$
for any $0 < y < 10$. What if, instead of a specific value of y, we considered this conditional expected value with y replaced with Y as $E[X|Y]$? In this case, the expected value, which is usually a constant, is
$$Z = E[X|Y] = \frac{10-Y}{2},$$

which is a random variable. But what is its distribution? The support of the distribution is straightforward. Since the random variable Y has support on $0 < y < 10$, the random variable Z has support on $0 < z < 5$. The distribution of Z can be found using the cumulative distribution function technique. First, the cumulative distribution function of Y is

$$F_Y(y) = \int_0^y f_Y(w)\,dw = \int_0^y \frac{10 - w}{50}\,dw = \left[\frac{10w - w^2/2}{50}\right]_0^y = \frac{10y - y^2/2}{50} \qquad 0 < y < 10.$$

Next, the cumulative distribution function of Z on its support is

$$
\begin{aligned}
F_Z(z) &= P(Z \le z) \\
&= P\left(\frac{10 - Y}{2} \le z\right) \\
&= P(10 - Y \le 2z) \\
&= P(-Y \le 2z - 10) \\
&= P(Y \ge 10 - 2z) \\
&= 1 - P(Y \le 10 - 2z) \\
&= 1 - F_Y(10 - 2z) \\
&= 1 - \frac{10(10 - 2z) - (10 - 2z)^2/2}{50} \\
&= \frac{z^2}{25} \qquad 0 < z < 5.
\end{aligned}
$$

Differentiating with respect to z gives the probability density function

$$f_Z(z) = \frac{2z}{25} \qquad 0 < z < 5,$$

which is the distribution of the random variable

$$Z = E[X \,|\, Y] = \frac{10 - Y}{2}.$$

This notion of a conditional expected value is illustrated in the next two examples.

Example 6.42 Let X and Y be continuous random variables with support on the unit square with opposite corners at $(0, 0)$ and $(1, 1)$. Find $V[V[E[X \,|\, Y]]]$.

This alphabet soup is best attacked from the inside out. From the previous discussion, we know that $E[X \,|\, Y]$ is a random variable. Furthermore, $V[E[X \,|\, Y]]$ is the population variance of that random variable, which is a constant. Since the population variance of a constant is zero, $V[V[E[X \,|\, Y]]] = 0$.

Example 6.43 For the random variables X and Y with joint probability density function

$$f(x, y) = 2 \qquad x > 0, y > 0, x + y < 1,$$

find $V\big[E[Y \,|\, X]\big]$.

The marginal probability density function of X is

$$f_X(x) = \int_0^{1-x} 2\,dy = 2 - 2x \qquad 0 < x < 1.$$

The conditional probability density function of Y given X is

$$f_{Y|X=x}(y \,|\, X = x) = \frac{f(x, y)}{f_X(x)} = \frac{2}{2 - 2x} = \frac{1}{1 - x} \qquad 0 < y < 1 - x;\ 0 < x < 1.$$

Thus the expected value of Y given $X = x$ is

$$E[Y \mid X = x] = \int_0^{1-x} y \cdot \frac{1}{1-x} \, dy = \frac{1-x}{2} \qquad\qquad 0 < x < 1.$$

The random variable

$$E[Y \mid X] = \frac{1-X}{2}$$

has population variance

$$V\big[E[Y \mid X]\big] = V\left[\frac{1-X}{2}\right] = \frac{1}{4} V[X] = \frac{1}{4} \cdot \frac{1}{18} = \frac{1}{72}$$

because

$$E[X] = \int_0^1 x \cdot (2 - 2x) \, dx = \frac{1}{3},$$

$$E[X^2] = \int_0^1 x^2 \cdot (2 - 2x) \, dx = \frac{1}{6},$$

$$V[X] = E[X^2] - \left(E[X]\right)^2$$

and, by the shortcut formula for variance,

$$V[X] = \frac{1}{6} - \left(\frac{1}{3}\right)^2 = \frac{1}{18}.$$

We can derive several results associated with this type of conditional expectation. We prove two such results here.

Theorem 6.11 If X and Y are random variables, then

$$E[X] = E\big[E[X \mid Y]\big]$$

when the expectations exist.

Proof The proof is given for the case when X and Y are continuous random variables. An analogous proof for discrete random variables replaces integrals by summations. The expected value of X is

$$
\begin{aligned}
E[X] &= \int_{-\infty}^{\infty} \int_{-\infty}^{\infty} x f(x, y) \, dx \, dy \\
&= \int_{-\infty}^{\infty} \int_{-\infty}^{\infty} x f_{X \mid Y = y}(x \mid Y = y) \, f_Y(y) \, dx \\
&= \int_{-\infty}^{\infty} E[X \mid Y = y] \, f_Y(y) \, dx \\
&= E\big[E[X \mid Y]\big],
\end{aligned}
$$

which proves the result.   □

In Section 4.3, we showed that the population mean of a Geometric(p) random variable whose support begins at 1 is $1/p$. We now use Theorem 6.11 to calculate the population variance of a Geometric(p) random variable.

Example 6.44 Let X be the number of independent Bernoulli trials, each with probability of success p, required to produce the first success. Thus, $X \sim$ Geometric(p), where the geometric distribution is parameterized from 1. Find $V[X]$ using conditional expectation.

This problem can be solved by conditioning on the outcome of the first Bernoulli trial. Let $Y = 0$ if the first Bernoulli trial is a failure and $Y = 1$ if the first Bernoulli trial is a success. Using the shortcut formula for variance,

$$V[X] = E\left[X^2\right] - E[X]^2,$$

so the problem boils down to calculating $E\left[X^2\right]$. Conditioning on the value of Y, using the fact that $E[X] = 1/p$, and invoking Theorem 6.11 yields

$$
\begin{aligned}
E\left[X^2\right] &= E\left[E\left[X^2 \mid Y\right]\right] \\
&= E\left[X^2 \mid Y = 0\right] P(Y = 0) + E\left[X^2 \mid Y = 1\right] P(Y = 1) \\
&= (1-p)E\left[(1+X)^2\right] + p \\
&= (1-p)\left(1 + 2E[X] + E\left[X^2\right]\right) + p \\
&= 1 + \frac{2(1-p)}{p} + (1-p)E\left[X^2\right].
\end{aligned}
$$

There was just one bit of finesse in this sequence of steps, and that was replacing $E\left[X^2 \mid Y = 0\right]$ with $E\left[(1+X)^2\right]$. This step is legitimate by the memoryless property of the geometric distribution. If the first Bernoulli trial results in failure, then the random trial number of the first success is just one more than it was originally. Equating the left and right sides of this string of equalities gives

$$E\left[X^2\right] = 1 + \frac{2(1-p)}{p} + (1-p)E\left[X^2\right],$$

which can be solved for $E\left[X^2\right]$ resulting in

$$E\left[X^2\right] = \frac{2-p}{p^2}.$$

Finally, returning to the shortcut formula for the population variance,

$$V[X] = E\left[X^2\right] - E[X]^2 = \frac{2-p}{p^2} - \left(\frac{1}{p}\right)^2 = \frac{1-p}{p^2}$$

is the population variance of a geometric(p) random variable parameterized from 1.

Theorem 6.12 If X and Y are random variables, then

$$V[X] = E\left[V[X \mid Y]\right] + V\left[E[X \mid Y]\right]$$

when the expectations exist.

Proof Using the shortcut formula for computing the population variance, the first term on the right-hand side of the equation can be written as

$$
\begin{aligned}
E\left[V[X \mid Y]\right] &= E\left[E\left[X^2 \mid Y\right] - E[X \mid Y]^2\right] \\
&= E\left[E\left[X^2 \mid Y\right]\right] - E\left[E[X \mid Y]^2\right] \\
&= E\left[X^2\right] - E[X]^2 - E\left[E[X \mid Y]^2\right] + E[X]^2 \\
&= V[X] - E\left[E[X \mid Y]^2\right] + E\left[E[X \mid Y]^2\right] \\
&= V[X] - V\left[E[X \mid Y]\right].
\end{aligned}
$$

Solving for $V[X]$ proves the result. $\qquad \square$

A nice result proceeds directly from Theorem 6.12. Examining the first term on the right-hand side of the result, namely, $E[V[X\,|\,Y]]$, it can be concluded that this expected value of a conditional variance must be nonnegative because population variances are always nonnegative. This results in the inequality

$$V[E[X\,|\,Y]] \le V[X],$$

which is attributed to Rao and Blackwell.

Example 6.45 The binomial distribution was introduced in Chapter 4 as the number of successes in n independent Bernoulli trials, each with a probability of success p. The population mean and variance of $X \sim \text{binomial}(n, p)$ are

$$E[X] = np \qquad \text{and} \qquad V[X] = np(1-p).$$

In particular, if $p = 1/2$ (for example, if the Bernoulli trials are flips of a fair coin and X counts the number of heads that appear in n flips), then the population mean and variance are

$$E[X] = n/2 \qquad \text{and} \qquad V[X] = n/4.$$

But what if an application arises where n is fixed but p is unknown? Such an application might arise in an area of quality control known as *acceptance sampling*, where a fixed number of items are sampled from a large lot. To be more specific, assume that p is no longer a fixed probability, but is itself a random variable. Furthermore, assume that you have so little information about the probability of success on each Bernoulli trial that you can only state that the probability of success P is a random variable that is equally likely to fall between 0 and 1. That is to say, $P \sim U(0, 1)$. The problem here is to find $E[X]$ and $V[X]$ when $X \sim \text{binomial}(n, P)$, where $P \sim U(0, 1)$.

To begin with, our earlier discussion about population means and variances indicates that

$$E[X\,|\,P = p] = np,$$

which defines the random variable

$$E[X\,|\,P] = nP.$$

Likewise,

$$V[X\,|\,P = p] = np(1-p),$$

which defines the random variable

$$V[X\,|\,P] = nP(1-P).$$

Using Theorem 6.11, the expected value of X is

$$\begin{aligned} E[X] &= E\big[E[X\,|\,P]\big] \\ &= E[nP] \\ &= nE[P] \\ &= \frac{n}{2} \end{aligned}$$

because $E[P] = 1/2$. Using Theorem 6.12, the population variance of X is

$$\begin{aligned} V[X] &= E\big[V[X\,|\,P]\big] + V\big[E[X\,|\,P]\big] \\ &= E\big[nP(1-P)\big] + V[nP] \\ &= nE[P] - nE\big[P^2\big] + n^2V[P] \\ &= \frac{n}{2} - \frac{n}{3} + \frac{n^2}{12} \\ &= \frac{n^2 + 2n}{12} \end{aligned}$$

because $E[P] = 1/2$, $E[P^2] = 1/3$, and $V[P] = 1/12$. Table 6.12 shows this quadratic growth of $V[X]$ in n when $P \sim U(0, 1)$ versus the linear growth of $V[X]$ in n when $p = 1/2$. The influence of p being random is to spread out the distribution of X.

	$n = 1$	$n = 2$	$n = 3$	$n = 4$	$n = 5$	$n = 6$
$V[X]$ when $p = 1/2$	$1/4$	$1/2$	$3/4$	1	$5/4$	$3/2$
$V[X]$ when $P \sim U(0, 1)$	$1/4$	$2/3$	$5/4$	2	$35/12$	4

Table 6.12: $V[X]$ for $p = 1/2$ and $P \sim U(0, 1)$.

Monte Carlo simulation can be used to support the expression for $V[X]$ for a specific value of n for $P \sim U(0, 1)$. For $n = 6$, five runs of the R code

```
nrep = 500000
n = 6
p = runif(nrep)
x = rbinom(nrep, n, p)
var(x)
```

results in

    3.998727   4.003016   4.001058   4.006192   3.997157,

which supports the $V[X] = 4$ entry in Table 6.12 associated with $n = 6$.

Joint moment generating functions

The definition of the moment generating function extends to the case of two random variables X_1 and X_2.

> **Definition 6.9** Let (X_1, X_2) be a random vector. The *joint moment generating function* of (X_1, X_2) is
>
> $$M(t_1, t_2) = E\left[e^{t_1 X_1 + t_2 X_2}\right]$$
>
> provided that the expected value exists on $-h_1 < t_1 < h_1$ and $-h_2 < t_2 < h_2$ for some positive constants h_1 and h_2.

As was the case with the joint probability mass function, probability density function, and cumulative distribution function, subscripts are added for clarity; for example, $M_{X_1, X_2}(t_1, t_2)$. Joint moment generating functions can be used to generate moments as they did in the univariate case. For the random variables X_1 and X_2, the following five moments are examples of moments that can be calculated by taking partial derivatives of the joint moment generating function, then using $t_1 = 0$ and $t_2 = 0$ as arguments:

$$E[X_1] = \left[\frac{\partial M(t_1, t_2)}{\partial t_1}\right]_{(t_1, t_2) = (0, 0)},$$

$$E[X_2] = \left[\frac{\partial M(t_1, t_2)}{\partial t_2}\right]_{(t_1, t_2) = (0, 0)},$$

$$E[X_1^2] = \left[\frac{\partial^2 M(t_1, t_2)}{\partial t_1^2}\right]_{(t_1, t_2) = (0, 0)},$$

$$E[X_2^2] = \left[\frac{\partial^2 M(t_1, t_2)}{\partial t_2^2}\right]_{(t_1, t_2) = (0, 0)},$$

$$E[X_1 X_2] = \left[\frac{\partial^2 M(t_1, t_2)}{\partial t_1 \partial t_2}\right]_{(t_1, t_2) = (0, 0)}.$$

In addition, the marginal moment generating functions can be found as

$$M_{X_1}(t_1) = E\left[e^{t_1 X_1}\right] = M(t_1, 0),$$

$$M_{X_2}(t_2) = E\left[e^{t_2 X_2}\right] = M(0, t_2).$$

Finally, the random variables X_1 and X_2 are independent if and only if

$$M(t_1, t_2) = M_{X_1}(t_1) M_{X_2}(t_2)$$

for all t_1 and t_2 where the joint moment generating function exists.

Example 6.46 Let X_1 and X_2 be continuous random variables whose distribution is described by the joint probability density function

$$f(x_1, x_2) = \lambda_1 e^{-\lambda_1 x_1} \lambda_2 e^{-\lambda_2 x_2} \qquad\qquad x_1 > 0, \ x_2 > 0$$

for real positive parameters λ_1 and λ_2. Find the joint moment generating function for X_1 and X_2 and use it to find $E[X_1 X_2]$.

The joint moment generating function is

$$
\begin{aligned}
M(t_1, t_2) &= E\left[e^{t_1 X_1 + t_2 X_2}\right] \\
&= \int_0^\infty \int_0^\infty e^{t_1 x_1 + t_2 x_2} f(x_1, x_2)\, dx_1\, dx_2 \\
&= \int_0^\infty \int_0^\infty e^{t_1 x_1 + t_2 x_2} \lambda_1 e^{-\lambda_1 x_1} \lambda_2 e^{-\lambda_2 x_2}\, dx_1\, dx_2 \\
&= \int_0^\infty \int_0^\infty \lambda_1 e^{-(\lambda_1 - t_1)x_1} \lambda_2 e^{-(\lambda_2 - t_2)x_2}\, dx_1\, dx_2 \\
&= \int_0^\infty \lambda_1 e^{-(\lambda_1 - t_1)x_1}\, dx_1 \int_0^\infty \lambda_2 e^{-(\lambda_2 - t_2)x_2}\, dx_2 \\
&= \frac{\lambda_1}{\lambda_1 - t_1} \cdot \frac{\lambda_2}{\lambda_2 - t_2} \qquad\qquad t_1 < \lambda_1, t_2 < \lambda_2.
\end{aligned}
$$

The joint moment generating function does not exist for all t_1 and t_2 because the integrals only converge for $t_1 < \lambda_1$ and $t_2 < \lambda_2$. Returning to Definition 6.9, this joint moment generating function exists in a rectangle about the origin by letting $h_1 = \lambda_1/2$ and $h_2 = \lambda_2/2$. In order to find $E[X_1 X_2]$, first take partial derivatives of the joint moment generating function with respect to t_1 and t_2:

$$\frac{\partial^2 M(t_1, t_2)}{\partial t_1 \partial t_2} = \frac{\lambda_1}{(\lambda_1 - t_1)^2} \cdot \frac{\lambda_2}{(\lambda_2 - t_2)^2}$$

for $t_1 < \lambda_1$ and $t_2 < \lambda_2$. Next, compute the expected value of the product of X_1 and X_2 by using $t_1 = 0$ and $t_2 = 0$ as arguments in the partial derivative expression yielding

$$E[X_1 X_2] = \left[\frac{\partial^2 M(t_1, t_2)}{\partial t_1 \partial t_2}\right]_{(t_1, t_2) = (0,0)} = \frac{1}{\lambda_1 \lambda_2}.$$

These quantities can be calculated by hand as shown above or with the Maple statements given below.

```
assume(lambda1 > 0);
assume(lambda2 > 0);
f := lambda1 * exp(-lambda1 * x1) * lambda2 * exp(-lambda2 * x2);
m := int(int(exp(t1 * x1 + t2 * x2) * f, x1 = 0 .. infinity),
                                        x2 = 0 .. infinity);
e := diff(diff(m, t1), t2);
subs({t1 = 0, t2 = 0}, e);
```

The derivation of $E[X_1X_2]$ given above has been done generally, but for this specific problem the rather tedious calculations could have been avoided by using some of the results presented earlier in this chapter. The value of $E[X_1X_2]$ could have been determined by using the following series of fortunate events: (a) the joint probability density function is defined over a product space and can be factored into nonnegative functions that depend on x_1 alone and x_2 alone, so X_1 and X_2 are independent by Theorem 6.1, (b) the joint probability density function can be factored into the product of

$$f_{X_1}(x_1) = \lambda_1 e^{-\lambda_1 x_1} \qquad\qquad x_1 > 0$$

and

$$f_{X_2}(x_2) = \lambda_2 e^{-\lambda_2 x_2} \qquad\qquad x_2 > 0,$$

which are both functions that integrate to one, so these can be recognized as the marginal probability density functions of X_1 and X_2; (c) the two marginal probability density functions can be recognized as those of $X_1 \sim$ exponential(λ_1) and $X_2 \sim$ exponential(λ_2); (d) the population means of the two exponential distributions are $E[X_1] = 1/\lambda_1$ and $E[X_2] = 1/\lambda_2$; (e) by Theorem 6.3, the expected value of the product is the product of the expected values for independent random variables, so

$$E[X_1X_2] = E[X_1]E[X_2] = \frac{1}{\lambda_1\lambda_2},$$

which matches the result obtained using the joint moment generating function.

One of the important discrete univariate distributions presented in Chapter 4 was the binomial distribution, which models the number of successes in n independent Bernoulli trials, each with probability of success p. An important assumption associated with a Bernoulli trial is that there are just two possible outcomes: success and failure. But what if there are three possible outcomes for each trial? Examples include political affiliation (Republican, Democrat, Independent) and the outcomes to certain sporting events (win, lose, tie). The next example introduces the *trinomial distribution* which models the counts of the outcomes to n independent trials (no longer called Bernoulli trials) of an experiment with three outcomes. The reason that this is a bivariate distribution is that knowledge of the number of Type 1 outcomes, denoted by X_1 and the number of Type 2 outcomes, denoted by X_2, determines the number of Type 3 outcomes as $n - X_1 - X_2$.

Example 6.47 Consider n independent trials, each with three possible outcomes. Let p_1, p_2, and p_3 denote the probability of each of the three outcomes on each trial, where $p_1 + p_2 + p_3 = 1$. Let X_1 be the number of Type 1 outcomes. Let X_2 be the number of Type 2 outcomes. There are $n - X_1 - X_2$ Type 3 outcomes. Determine (a) the joint probability mass function of X_1 and X_2, (b) the joint moment generating function of X_1 and X_2, (c) the marginal probability mass functions of X_1 and X_2, (d) the conditional distributions of $X_1 | X_2 = x_2$ and $X_2 | X_1 = x_1$, and (e) the population correlation between X_1 and X_2.

(a) The probability of x_1 Type 1 outcomes, x_2 Type 2 outcomes, and $n - x_1 - x_2$ Type 3 outcomes *in a specific order* is

$$p_1^{x_1} p_2^{x_2} p_3^{n-x_1-x_2}.$$

The number of ways of ordering the n outcomes is

$$\binom{n}{x_1, x_2, n-x_1-x_2} = \frac{n!}{x_1!x_2!(n-x_1-x_2)!}.$$

Therefore, the joint probability mass function of X_1 and X_2 is

$$f(x_1, x_2) = \frac{n!}{x_1!x_2!(n-x_1-x_2)!} p_1^{x_1} p_2^{x_2} p_3^{n-x_1-x_2}$$

for (x_1, x_2) in the support set $\mathcal{A} = \{x_1 = 0, 1, 2, \ldots; x_2 = 0, 1, 2, \ldots; x_1 + x_2 \le n\}$ and the parameter space $\Omega = \{p_1 > 0; p_2 > 0; p_3 > 0; p_1 + p_2 + p_3 = 1; n = 1, 2, \ldots\}$. The joint probability mass function values correspond to successive terms in the expansion of $(p_1 + p_2 + p_3)^n$. The support set $\mathcal{A}$ is not a product space, so X_1 and X_2 are dependent random variables.

(*b*) Using the binomial theorem (twice), the joint moment generating function of X_1 and X_2 is

$$
\begin{aligned}
M(t_1, t_2) &= E\left[e^{t_1 X_1 + t_2 X_2}\right] \\
&= \sum_{x_1=0}^{n} \sum_{x_2=0}^{n-x_1} e^{t_1 x_1 + t_2 x_2} \frac{n!}{x_1! x_2! (n - x_1 - x_2)!} p_1^{x_1} p_2^{x_2} p_3^{n - x_1 - x_2} \\
&= \sum_{x_1=0}^{n} \sum_{x_2=0}^{n-x_1} \frac{n!}{x_1! x_2! (n - x_1 - x_2)!} \left(p_1 e^{t_1}\right)^{x_1} \left(p_2 e^{t_2}\right)^{x_2} p_3^{n - x_1 - x_2} \\
&= \sum_{x_1=0}^{n} \frac{n!}{x_1! (n - x_1)!} \left(p_1 e^{t_1}\right)^{x_1} \sum_{x_2=0}^{n-x_1} \frac{(n - x_1)!}{x_2! (n - x_1 - x_2)!} \left(p_2 e^{t_2}\right)^{x_2} p_3^{n - x_1 - x_2} \\
&= \sum_{x_1=0}^{n} \frac{n!}{x_1! (n - x_1)!} \left(p_1 e^{t_1}\right)^{x_1} \left(p_2 e^{t_2} + p_3\right)^{n - x_1} \\
&= \left(p_1 e^{t_1} + p_2 e^{t_2} + p_3\right)^n \qquad -\infty < t_1 < \infty, \ -\infty < t_2 < \infty.
\end{aligned}
$$

(*c*) The marginal moment generating functions for X_1 is

$$
M_{X_1}(t_1) = M(t_1, 0) = \left[(1 - p_1) + p_1 e^{t_1}\right]^n \qquad -\infty < t_1 < \infty.
$$

This is recognized as the moment generating function of a (univariate) binomial random variable with parameters n and p_1, so the marginal distribution of X_1 is $X_1 \sim$ binomial(n, p_1). This marginal distribution can also be deduced logically by considering all Type 1 outcomes to be "successes" and all Type 2 and Type 3 outcomes to be lumped together as "failures." Using this classification, the distribution of X_1 must be $X_1 \sim$ binomial(n, p_1). Likewise,

$$
M_{X_2}(t_2) = M(0, t_2) = \left[(1 - p_2) + p_2 e^{t_2}\right]^n \qquad -\infty < t_2 < \infty,
$$

which implies that $X_2 \sim$ binomial(n, p_2). The fact that $M_{X_1}(t_1) M_{X_2}(t_2) \ne M(t_1, t_2)$ is further evidence that X_1 and X_2 are dependent random variables, as concluded earlier.

(*d*) The conditional probability mass function of X_1 given $X_2 = x_2$ can be found by taking the ratio of the joint probability mass function to the marginal probability mass function

$$
\begin{aligned}
f_{X_1 | X_2 = x_2}(x_1 | X_2 = x_2) &= \frac{f(x_1, x_2)}{f_{X_2}(x_2)} \\
&= \frac{(n - x_2)!}{x_1! (n - x_1 - x_2)!} \cdot \frac{p_1^{x_1} p_3^{n - x_1 - x_2}}{(1 - p_2)^{n - x_2}} \\
&= \frac{(n - x_2)!}{x_1! (n - x_1 - x_2)!} \cdot \left(\frac{p_1}{1 - p_2}\right)^{x_1} \cdot \left(\frac{p_3}{1 - p_2}\right)^{n - x_1 - x_2}
\end{aligned}
$$

which is defined on the support $x_1 = 0, 1, 2, \ldots, n - x_2$ and is valid for any for any $x_2 = 0, 1, 2, \ldots, n$. This probability mass function is recognized as a binomial$\left(n - x_2, \frac{p_1}{1 - p_2}\right)$ random variable. This can also be deduced logically. Given that the random variable X_2 assumes the value x_2, there are $n - x_2$ other trials that must be allocated to the other two types of outcomes. Designating Type 1 outcomes as "successes" and Type 3 outcomes as "failures," the probability of a success on each trial is $\frac{p_1}{1-p_2}$, so the distribution of the number of successes is binomial$\left(n - x_2, \frac{p_1}{1-p_2}\right)$. Using a similar line of reasoning, it can be concluded that the conditional distribution of X_2 given $X_1 = x_1$ is binomial$\left(n - x_1, \frac{p_2}{1-p_1}\right)$.

(e) Finally, to find the population correlation, five expected values are required: $E[X_1]$, $E[X_2]$, $V[X_1]$, $V[X_2]$, and $E[X_1 X_2]$. Fortunately, the first four can be easily calculated because the marginal distributions of X_1 and X_2 are binomial. Since $X_1 \sim \text{binomial}(n, p_1)$,

$$E[X_1] = np_1 \qquad\qquad V[X_1] = np_1(1 - p_1).$$

Likewise, since $X_2 \sim \text{binomial}(n, p_2)$,

$$E[X_2] = np_2 \qquad\qquad V[X_2] = np_2(1 - p_2).$$

In order to calculate $E[X_1 X_2]$, take the mixed partial derivative of the joint moment generating function

$$\frac{\partial^2 M(t_1, t_2)}{\partial t_1 \partial t_2} = n(n - 1)\left(p_1 e^{t_1} + p_2 e^{t_2} + p_3\right)^{n-2} p_1 e^{t_1} p_2 e^{t_2}$$

for $-\infty < t_1 < \infty$ and $-\infty < t_2 < \infty$. Next, compute the expected value of the product of X_1 and X_2 by using $t_1 = 0$ and $t_2 = 0$ as arguments in the partial derivative expression:

$$E[X_1 X_2] = \left[\frac{\partial^2 M(t_1, t_2)}{\partial t_1 \partial t_2}\right]_{t_1 = t_2 = 0} = n(n - 1)p_1 p_2.$$

Finally, the population correlation is

$$
\begin{aligned}
\rho &= \frac{E[X_1 X_2] - E[X_1]E[X_2]}{\sqrt{V[X_1]V[X_2]}} \\[2mm]
&= \frac{n(n - 1)p_1 p_2 - (np_1)(np_2)}{\sqrt{np_1(1 - p_1) \cdot np_2(1 - p_2)}} \\[2mm]
&= -\sqrt{\frac{p_1 p_2}{(1 - p_1)(1 - p_2)}}.
\end{aligned}
$$

Why is the population correlation negative? There are only n available slots for Type 1 and Type 2 outcomes, so the more of the slots that are taken by Type 1 outcomes, the fewer available slots remain for Type 2 outcomes. The population correlation is not -1, however, because of the diluting effect of the Type 3 outcomes. This diluting effect be seen graphically in Figure 6.16, which plots ρ as a function of $p_1 = p_2 = p$. When the Type 1 and Type 2 outcomes are equally likely,

$$\rho = -\frac{p}{1 - p}.$$

Near $p = 0$, when Type 1 and Type 2 outcomes are unlikely, Type 3 outcomes dominate, so the population correlation between X_1 and X_2 is nearly 0. At the other extreme near $p = 1/2$, Type 3 outcomes are rare, so Type 1 and Type 2 outcomes dominate, and X_2 is approximately $n - X_1$, so the outcomes lie on a line with negative slope, resulting in a population correlation that approaches -1. In the case when p_1 and p_2 are not equal, Figure 6.17 shows level surfaces of the population correlation. The ten level surfaces plotted are for $\rho = -0.1, -0.2, \ldots, -0.9, -1$, displayed beginning in the southwest

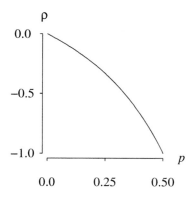

Figure 6.16: Population correlation ρ as a function of $p = p_1 = p_2$.

corner of the plot. Consider the two limiting extreme cases as ρ approaches $\rho = 0$ and $\rho = -1$. Based on the plot, $\rho = 0$ is associated with $p_1 = 0$ or $p_2 = 0$. When either Type 1 or Type 2 outcomes are impossible, then there is no correlation between Type 1 and Type 2 outcomes. The level surface associated with the population correlation approaching $\rho = -1$ is the line $p_2 = 1 - p_1$. When Type 3 outcomes are impossible, the number of Type 1 outcomes X_1 and the number of Type 2 outcomes $X_2 = n - X_1$ must lie on a line with negative slope. This implies that the population correlation must approach $\rho = -1$. All of these limiting extreme cases correspond to a reduction of the trinomial distribution to the binomial distribution because one of the three outcomes is effectively impossible.

This concludes the discussion of expected values for bivariate distributions. There are fewer well-known bivariate distributions than univariate distributions due to mathematical tractability issues. In addition to the trinomial distribution investigated in the previous example, the bivariate normal distribution is a well-known bivariate distribution. It is used so often that it is given its own section.

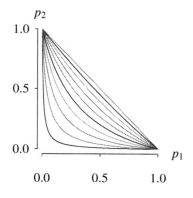

Figure 6.17: Level surfaces of ρ as a function of p_1 and p_2.

6.4 Bivariate Normal Distribution

Just as the normal distribution plays a central role as a univariate distribution, the bivariate normal distribution is the fundamental bivariate distribution. This distribution has several mathematically and geometrically elegant properties and also proves to be quite useful in applications. Some of the early development associated with a distribution like the bivariate normal was done by Sir Francis Galton in the late 19th century concerning data pairs (X, Y) consisting of the average heights of parents, adjusted for gender differences, X, versus the adult heights of their offspring, adjusted for gender differences, Y.

Definition 6.10 The continuous random variables X and Y with joint probability density function

$$f(x, y) = \frac{1}{2\pi\sigma_X\sigma_Y\sqrt{1-\rho^2}}\, e^{-\frac{1}{2(1-\rho^2)}\left[\left(\frac{x-\mu_X}{\sigma_X}\right)^2 - 2\rho\left(\frac{x-\mu_X}{\sigma_X}\right)\left(\frac{y-\mu_Y}{\sigma_Y}\right) + \left(\frac{y-\mu_Y}{\sigma_Y}\right)^2\right]},$$

which is defined on the support $\mathcal{A} = \{(x, y) \mid -\infty < x < \infty, -\infty < y < \infty\}$ with the associated parameter space

$$\Omega = \{(\mu_X, \mu_Y, \sigma_X, \sigma_Y, \rho) \mid -\infty < \mu_X < \infty, -\infty < \mu_Y < \infty, \sigma_X > 0, \sigma_Y > 0, -1 < \rho < 1\}$$

are *bivariate normal random variables* with parameters μ_X, μ_Y, σ_X, σ_Y, and ρ.

The choice of symbols used for the five parameters will come as no surprise in that they also happen to be the following expected values:

$$E[X] = \mu_X \qquad E[Y] = \mu_Y \qquad V[X] = \sigma_X^2 \qquad V[Y] = \sigma_Y^2$$

and ρ is the population correlation. A plot of the joint probability density function for one particular choice of the five parameters is shown in Figure 6.18. Although the support for X and Y covers all of $\mathcal{R}^2$, only a square region is shown in the figure. For all values of the parameters, the bivariate normal probability density function is unimodal with the mode at (μ_X, μ_Y). The height of the joint probability density function at the mode is

$$f(\mu_X, \mu_Y) = \frac{1}{2\pi\sigma_X\sigma_Y\sqrt{1-\rho^2}}.$$

It is hard to distinguish one bivariate normal distribution from another based on three-dimensional graphs of the joint probability density function like the one in Figure 6.18. They all look like mountains. Level surfaces of the joint probability density function tend to be more visually distinct.

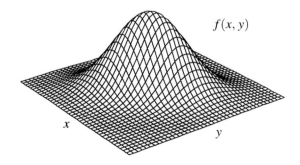

Figure 6.18: The joint probability density function of a bivariate normal distribution.

The rather complicated joint probability density function $f(x, y)$ has level surfaces that are concentric ellipses. Level surfaces are also known as contours and they are the set of points at which $f(x, y)$ assumes a constant value. Consider first the simple case when $\mu_X = \mu_Y = \rho = 0$. In this case, equating the joint probability density function to a constant and performing some algebra gives the usual form of an ellipse

$$\frac{x^2}{\sigma_X^2} + \frac{y^2}{\sigma_Y^2} = c$$

for some constant c. In a more general case where ρ is nonzero, but the population means continue to be 0, the equation for the ellipse becomes a bit more complicated:

$$\frac{x^2}{\sigma_X^2} - \frac{2\rho xy}{\sigma_X \sigma_Y} + \frac{y^2}{\sigma_Y^2} = c$$

for some constant c. Finally, in the most general case, the ellipse has the form

$$\frac{(x - \mu_X)^2}{\sigma_X^2} - \frac{2\rho(x - \mu_X)(y - \mu_Y)}{\sigma_X \sigma_Y} + \frac{(y - \mu_Y)^2}{\sigma_Y^2} = c$$

for some constant c. One particular ellipse gets its own name. The *population concentration ellipse* is the level surface containing the ordered pairs

$$(\mu_X - \sigma_X, \mu_Y - \sigma_Y), (\mu_X - \sigma_X, \mu_Y + \sigma_Y), (\mu_X + \sigma_X, \mu_Y - \sigma_Y), (\mu_X + \sigma_X, \mu_Y + \sigma_Y).$$

Figure 6.19 displays four population concentration ellipses for four different sets of parameters for the bivariate normal distribution. The ellipse in the upper-left plot is a circle (a special case of an ellipse), so one can conclude that the variances are equal, that is $\sigma_X^2 = \sigma_Y^2$, and $\rho = 0$. The ellipse

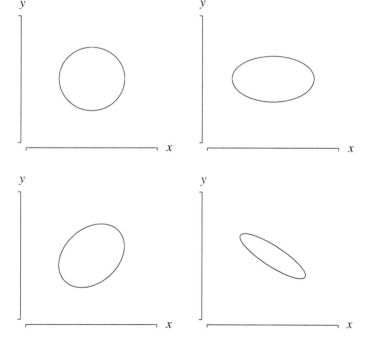

Figure 6.19: Level surfaces of the joint probability density function.

in the upper-right plot has a major axis that is parallel to the x-axis, so one can conclude that there is more spread to the distribution in the x direction than in the y direction. Therefore, $\sigma_X^2 > \sigma_Y^2$ and $\rho = 0$. The ellipse in the lower-left plot has a major axis that has a positive slope, so one can conclude that $\rho > 0$. Finally, ellipse in the lower-right plot has a major axis that has a negative slope, so one can conclude that $\rho < 0$. In addition, because the ellipse in the lower-right plot has a greater eccentricity than the ellipse in the lower-left plot, one can conclude that the magnitude of the population correlation $|\rho|$ is larger for the bivariate normal distribution associated with the lower-right plot. In the limit as $|\rho| \to 1$, the eccentricity of the concentric ellipses increases, and the ellipses converge to lines.

A special case of the bivariate normal distribution occurs when $\rho = 0$. In this case the joint probability density function reduces to

$$
\begin{aligned}
f(x, y) &= \frac{1}{2\pi\sigma_X\sigma_Y} e^{-\frac{1}{2}\left[\left(\frac{x-\mu_X}{\sigma_X}\right)^2 + \left(\frac{y-\mu_Y}{\sigma_Y}\right)^2\right]} \\
&= \frac{1}{\sqrt{2\pi}\sigma_X} e^{-\frac{1}{2}\left(\frac{x-\mu_X}{\sigma_X}\right)^2} \cdot \frac{1}{\sqrt{2\pi}\sigma_Y} e^{-\frac{1}{2}\left(\frac{y-\mu_Y}{\sigma_Y}\right)^2} \\
&= f_X(x) f_Y(y) \qquad -\infty < x < \infty, \; -\infty < y < \infty.
\end{aligned}
$$

So when $\rho = 0$, the joint probability density function can be factored into the product of the marginal probability density functions over the product space, which results in the following theorem.

Theorem 6.13 Let the random variables X and Y have the bivariate normal distribution. The random variables are independent if and only if $\rho = 0$.

It is important to note the difference between this theorem for bivariate normal random variables and the statement for any bivariate distribution given in Theorem 6.8.

Returning to the general case, we now determine the marginal distributions for X and Y. To find $f_X(x)$, (a) integrate y out of the joint probability density function, (b) make the substitutions $r = (x - \mu_X)/\sigma_X$ and $s = (y - \mu_Y)/\sigma_Y$, (c) complete the square of the quantity in the exponent, and (d) use the fact that probability density functions of normal random variables integrate to one as follows.

$$
\begin{aligned}
f_X(x) &= \int_{-\infty}^{\infty} f(x, y)\, dy \\
&= \int_{-\infty}^{\infty} \frac{1}{2\pi\sigma_X\sigma_Y\sqrt{1-\rho^2}} e^{-\frac{1}{2(1-\rho^2)}\left[\left(\frac{x-\mu_X}{\sigma_X}\right)^2 - 2\rho\left(\frac{x-\mu_X}{\sigma_X}\right)\left(\frac{y-\mu_Y}{\sigma_Y}\right) + \left(\frac{y-\mu_Y}{\sigma_Y}\right)^2\right]} dy \\
&= \frac{1}{2\pi\sigma_X\sqrt{1-\rho^2}} \int_{-\infty}^{\infty} e^{-\frac{1}{2(1-\rho^2)}\left(r^2 - 2\rho rs + s^2\right)} ds \\
&= \frac{1}{2\pi\sigma_X\sqrt{1-\rho^2}} \int_{-\infty}^{\infty} e^{-\frac{1}{2(1-\rho^2)}\left(s^2 - 2\rho rs + \rho^2 r^2 + r^2 - \rho^2 r^2\right)} ds \\
&= \frac{1}{2\pi\sigma_X\sqrt{1-\rho^2}} e^{-r^2/2} \int_{-\infty}^{\infty} e^{-\frac{1}{2(1-\rho^2)}(s-\rho r)^2} ds \\
&= \frac{1}{2\pi\sigma_X\sqrt{1-\rho^2}} e^{-r^2/2} \sqrt{2\pi(1-\rho^2)} \cdot \frac{1}{\sqrt{2\pi(1-\rho^2)}} \int_{-\infty}^{\infty} e^{-\frac{1}{2}\left((s-\rho r)/\sqrt{1-\rho^2}\right)^2} ds \\
&= \frac{1}{\sqrt{2\pi}\sigma_X} e^{-r^2/2} \\
&= \frac{1}{\sqrt{2\pi}\sigma_X} e^{-\frac{1}{2}\left(\frac{x-\mu_X}{\sigma_X}\right)^2} \qquad -\infty < x < \infty.
\end{aligned}
$$

This can be recognized as the probability density function of a normally distributed random variable with population mean μ_X and population variance σ_X^2. Similarly, the marginal probability density function of Y is normally distributed with population mean μ_Y and population variance σ_Y^2. These results are summarized in Theorem 6.14

Theorem 6.14 Let the random variables X and Y have the bivariate normal distribution. The marginal distributions of X and Y are

$$X \sim N(\mu_X, \sigma_X^2) \qquad \text{and} \qquad Y \sim N(\mu_Y, \sigma_Y^2).$$

The fact that the marginal distributions are normally distributed implies that the population means and population variances of the two random variables are

$$E[X] = \mu_X \qquad E[Y] = \mu_Y \qquad V[X] = \sigma_X^2 \qquad V[Y] = \sigma_Y^2$$

as alluded to earlier.

The joint moment generating function for the bivariate normal distribution also assumes a mathematically tractable form:

$$M(t_1, t_2) = e^{\mu_1 t_1 + \mu_2 t_2 + \frac{1}{2}(\sigma_X^2 t_1^2 + 2\rho\sigma_X\sigma_Y t_1 t_2 + \sigma_Y^2 t_2^2)} \qquad -\infty < t_1 < \infty, \; -\infty < t_2 < \infty.$$

The derivation of $M(t_1, t_2)$ is left as an exercise.

The conditional distributions can be determined in the usual fashion, for example,

$$f_{Y|X=x}(y|X=x) = \frac{f(x, y)}{f_X(x)} \qquad -\infty < y < \infty.$$

After some tedious algebra, the conditional probability density function simplifies to

$$f_{Y|X=x}(y|X=x) = \frac{1}{\sqrt{2\pi(1-\rho^2)}\sigma_Y} \, e^{-\frac{1}{2(1-\rho^2)\sigma_Y^2}\left(y - \mu_Y - \frac{\rho\sigma_Y(x-\mu_X)}{\sigma_X}\right)^2} \qquad -\infty < y < \infty.$$

This conditional probability density function can be recognized as a univariate normal probability density function. A similar result exists for the distribution of X given $Y = y$. These results are summarized in Theorem 6.15

Theorem 6.15 Let the random variables X and Y have the bivariate normal distribution. The conditional distributions of X and Y are

$$Y|X = x \sim N\left(\mu_Y + \frac{\rho\sigma_Y(x - \mu_X)}{\sigma_X}, \; \left(1 - \rho^2\right)\sigma_Y^2\right)$$

and

$$X|Y = y \sim N\left(\mu_X + \frac{\rho\sigma_X(y - \mu_Y)}{\sigma_Y}, \; \left(1 - \rho^2\right)\sigma_X^2\right).$$

It is interesting to note that the variances of the conditional distributions are independent of the values of x and y, but are reduced by the factor $1 - \rho^2$ by knowing that $X = x$ or $Y = y$. This property is known as *homoscedasticity* (or homogeneity of variance) and can simplify the mathematics for a particular problem. It implies that no matter where the joint probability density function of the bivariate distribution is cut and rescaled to form the conditional distribution, the population variance is always the same. It is worthwhile thinking about the values of the conditional population variances in the extreme cases ($\rho = 0$ and $\rho \to \pm 1$) to see if these population variances match what you expect for the bivariate normal model. The conditional population means and variances follow directly from Theorem 6.15:

$$
\begin{aligned}
E[Y \,|\, X = x] &= \mu_Y + \frac{\rho \sigma_Y (x - \mu_X)}{\sigma_X} \\
V[Y \,|\, X = x] &= (1 - \rho^2) \sigma_Y^2 \\
E[X \,|\, Y = y] &= \mu_X + \frac{\rho \sigma_X (y - \mu_Y)}{\sigma_Y} \\
V[X \,|\, Y = y] &= (1 - \rho^2) \sigma_X^2.
\end{aligned}
$$

Example 6.48 The pioneering statistician Karl Pearson analyzed 1078 pairs of heights of fathers, X, and the adult heights of their sons, Y, each measured in inches. Using a statistical technique known as the *method of moments*, we assume that the population parameters are equal to the associated sample values. Although subsequent work in mathematical statistics will show that it is not proper to equate sample statistics to population parameters and treat them as constants, we will violate that rule in this case. The numbers here are more meaningful than pulling values out of thin air. The parameters in a bivariate normal model are

$$
\mu_X = 67.7 \qquad \mu_Y = 68.7 \qquad \sigma_X = 2.74 \qquad \sigma_Y = 2.81 \qquad \rho = 0.501.
$$

If the bivariate normal distribution is an appropriate model for the joint distribution of the heights of fathers and sons, for a father who is six feet tall ($x = 72$), find

(a) the expected height of his son,

(b) the standard deviation of the height of his son,

(c) the probability that the son is over six feet tall.

Keeping the units in inches, we use Theorem 6.15 to answer the three questions based on the fact that

$$
Y \,|\, X = x \sim N\left(\mu_Y + \frac{\rho \sigma_Y (x - \mu_X)}{\sigma_X}, \; (1 - \rho^2) \sigma_Y^2 \right)
$$

or

$$
Y \,|\, X = 72 \sim N\left(68.7 + \frac{0.501 \cdot 2.81(72 - 67.7)}{2.74}, \; (1 - 0.501^2) \, 2.81^2 \right)
$$

or

$$
Y \,|\, X = 72 \sim N(70.91, 5.91).
$$

(a) The conditional expected height of the son is $E[Y \,|\, X = 72] = 70.91$ inches.

(b) The conditional standard deviation of the height of the son is $\sqrt{V[Y \,|\, X = 72]} = \sqrt{5.91} = 2.43$ inches.

(c) The probability that the son is over six feet tall is found by standardizing the conditional random variable:

$$
\begin{aligned}
P(Y > 72 \,|\, X = 72) &= P\left(\frac{Y - 70.91}{2.43} > \frac{72 - 70.91}{2.43} \right) \\
&= P(Z > 0.4486),
\end{aligned}
$$

where $Z \sim N(0, 1)$. Using the R statement `1 - pnorm(0.4486)`, or, without standardizing `1 - pnorm(72, 70.91, 2.43)`, results in

$$
P(Y > 72 \,|\, X = 72) = 0.3269.
$$

So the probability that a son will be over six feet tall given that his father is six feet tall is 0.3269. This probability is only meaningful if the bivariate normal distribution is an appropriate probability model.

In Pearson's data set, why did mom get left out of the picture? Clearly her height is just as important as the father's in terms of determining the height of the son. Galton performed an analysis of a data set that incorporated the heights of both parents. He let X be the average of the father's height and 1.08 times the mother's height (the 1.08 adjusts for gender). Then Y was the offspring's adult height (with female offspring heights multiplied by 1.08). In this case, one would expect that the data would look roughly the same as the father-son data, but now because both parents are involved, the population correlation ρ should be somewhat higher.

Generating random variates for Monte Carlo simulation is more complicated for bivariate distributions than it was for univariate distributions. One simple algorithm that works for any bivariate distribution is to generate the random variate $X = x$ from the marginal distribution for X, then generate Y from the conditional distribution of Y given $X = x$, where x is the value generated from the marginal distribution. This results in a random variate pair (X, Y) from the bivariate distribution. Obviously, the role of X and Y can be swapped in this scenario—Y can be generated first, then X second.

Returning to the bivariate normal distribution, the algorithm described in the previous paragraph can be implemented in the following fashion.

$$\text{generate } X \sim N(\mu_X, \sigma_X^2)$$
$$\text{generate } Y \sim N\left(\mu_Y + \frac{\rho \sigma_Y (X - \mu_X)}{\sigma_X}, (1 - \rho^2) \sigma_Y^2\right)$$

This algorithm requires that we have a random variate generator for normal variates. In R, the function `rnorm` generates normally distributed random variates using an algorithm that will be discussed in the next chapter.

Example 6.49 Generate 1078 random (X, Y) pairs for the bivariate normal model with parameters

$$\mu_X = 67.7 \qquad \mu_Y = 68.7 \qquad \sigma_X = 2.74 \qquad \sigma_Y = 2.81 \qquad \rho = 0.501.$$

The R code given below generates the random pairs.

```
mx = 67.7
my = 68.7
sx = 2.74
sy = 2.81
rho = 0.501
n = 1078
x = rnorm(n, mx, sx)
y = rnorm(n, my + rho * sy * (x - mx) / sx, sy * sqrt((1 - rho ^ 2)))
```

The (X, Y) pairs that are generated by the R code and stored in x and y are shown in the scatterplot in Figure 6.20. The scatterplot is consistent with intuition. Taller fathers tend to produce taller sons, but it is not a deterministic relationship, otherwise all of the points would fall on a line, for example. The variability associated with the heights of the sons is apparent from the scatterplot. Consistent with the bivariate normal distribution assumption, the generated random variates fall in a pattern consistent with the ellipse-shaped level surfaces.

The variates generated in Figure 6.20 were for a positive population correlation. What would the scatterplot look like if the population correlation were negative? Also, how should this be interpreted for heights? Figure 6.21 shows 1078 random (X, Y) pairs when $\rho = -0.9$. The negative population correlation implies that the slope of the major axis of a level surface of the joint probability density

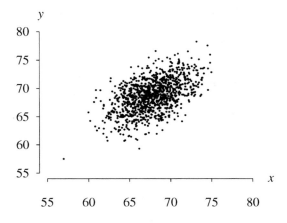

Figure 6.20: Scatterplot of 1078 bivariate normal random variates ($\rho = 0.505$).

function has a negative slope. Also, because the *magnitude* of the population correlation is higher, we expect the data pairs to cluster more closely to the line, which they do. Finally, the interpretation here is completely *inconsistent* with reality. Taller-than-average fathers tend to have shorter-than-average sons, who in turn have taller-than-average sons. This generation-skipping phenomenon is inconsistent with what we see in nature.

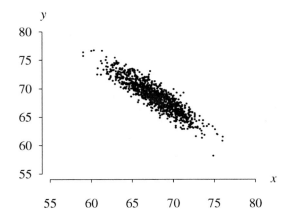

Figure 6.21: Scatterplot of 1078 bivariate normal random variates ($\rho = -0.9$).

Matrix theory has been hiding in the bivariate normal distribution, and this section concludes by exposing the latent linear algebra. Formulating the bivariate normal distribution in terms of matrices provides a natural bridge to the *n*-dimensional *multivariate normal distribution*. The notation is helped along if the bivariate normal random variables X and Y are replaced by X_1 and X_2. The corresponding five parameters in the bivariate normal distribution are μ_{X_1}, μ_{X_2}, $\sigma_{X_1}^2$, $\sigma_{X_2}^2$, and ρ. The random variables can be thought of as a random 2×1 column vector X defined by

$$X = \begin{pmatrix} X_1 \\ X_2 \end{pmatrix}.$$

This random vector has a population mean vector

$$\boldsymbol{\mu} = \left(\begin{array}{c} \mu_{X_1} \\ \mu_{X_2} \end{array} \right)$$

and symmetric, positive semi-definite 2×2 variance–covariance matrix

$$\boldsymbol{\Sigma} = \left(\begin{array}{cc} V[X_1] & \mathrm{Cov}(X_1, X_2) \\ \mathrm{Cov}(X_1, X_2) & V[X_2] \end{array} \right).$$

This variance–covariance matrix can be written in terms of the parameters of the distribution as

$$\boldsymbol{\Sigma} = \left(\begin{array}{cc} \sigma_{X_1}^2 & \rho\sigma_{X_1}\sigma_{X_2} \\ \rho\sigma_{X_1}\sigma_{X_2} & \sigma_{X_2}^2 \end{array} \right).$$

Using this matrix formulation, the joint probability density function can be written in matrix form

$$f(x_1, x_2) = \left(2\pi\sqrt{|\boldsymbol{\Sigma}|} \right)^{-1} e^{-\frac{1}{2}\left[(\boldsymbol{x}-\boldsymbol{\mu})'\boldsymbol{\Sigma}^{-1}(\boldsymbol{x}-\boldsymbol{\mu})\right]} \qquad -\infty < x_1 < \infty, \; -\infty < x_2 < \infty,$$

where $\boldsymbol{\Sigma}^{-1}$ is the inverse of the variance–covariance matrix, $|\boldsymbol{\Sigma}|$ is the determinant of the variance–covariance matrix, and the $'$ denotes transpose. The reader is encouraged to work through the mathematics required to see that this joint probability density function is equivalent to the one given in Definition 6.10. The paragraphs below draw analogies between the univariate normal and bivariate normal distributions.

In the case of a univariate normally distributed random variable X, we use the shorthand

$$X \sim N(\mu, \sigma^2)$$

to indicate that X has the normal distribution with population mean μ and population variance σ^2. The analogous shorthand for the bivariate random vector $\boldsymbol{X}$ is

$$\boldsymbol{X} \sim N(\boldsymbol{\mu}, \boldsymbol{\Sigma})$$

which indicates that the bivariate random vector $\boldsymbol{X} = (X_1, X_2)'$ has the bivariate normal distribution with population mean vector $\boldsymbol{\mu} = (\mu_{X_1}, \mu_{X_2})'$ and population variance–covariance matrix $\boldsymbol{\Sigma}$.

The univariate normally distributed random variable X with population mean μ and population variance σ^2 can be standardized to a standard normal random variable Z by the transformation

$$Z = \frac{X - \mu}{\sigma}.$$

The inverse of this operation is performed via

$$X = \mu + \sigma Z$$

which converts $Z \sim N(0, 1)$ to $X \sim N(\mu, \sigma^2)$. This is useful in random variate generation. If an algorithm can be determined for generating a standard normal random variate Z, then this is the proper way to convert it to a $N(\mu, \sigma^2)$ random variate. Is there an equivalent statement for the bivariate normal distribution?

In order to address this question, some additional notation is necessary. The symmetric, positive semi-definite variance–covariance matrix $\boldsymbol{\Sigma}$ can always be written as

$$\boldsymbol{\Sigma} = \boldsymbol{\Gamma}'\boldsymbol{\Lambda}\boldsymbol{\Gamma},$$

where $\boldsymbol{\Lambda}$ is a 2×2 diagonal matrix with the eigenvalues of $\boldsymbol{\Sigma}$ in decreasing order of magnitude on the diagonal, and $\boldsymbol{\Gamma}'$ is a 2×2 matrix with the associated normalized eigenvectors of $\boldsymbol{\Sigma}$ as columns.

This decomposition of the variance–covariance matrix is known as *spectral decomposition*. Define the square root of this matrix to be

$$\Sigma^{1/2} = \Gamma' \Lambda^{1/2} \Gamma,$$

where $\Lambda^{1/2}$ is a diagonal matrix with the square roots of the eigenvalues in decreasing order of magnitude on the diagonal. Finally, the analogous statement from the previous paragraph is

$$X = \mu + \Sigma^{1/2} Z,$$

where Z is a 2×1 vector of independent standard normal random variables. As in the one-dimensional case, this equation can be used to generate bivariate normal random variate pairs. The random variate generation algorithm for the bivariate normal distribution using matrix notation is illustrated next for a simple case with integer parameters.

Example 6.50 Let the random vector $X = (X_1, X_2)'$ have the bivariate normal distribution with population mean vector $\mu = (\mu_{X_1}, \mu_{X_2})' = (7, 6)'$ and population variance–covariance matrix

$$\Sigma = \begin{pmatrix} 8 & 2 \\ 2 & 5 \end{pmatrix}.$$

Use the matrix equation

$$X = \mu + \Sigma^{1/2} Z$$

to generate and plot random variate pairs from this bivariate normal distribution.

The population correlation is

$$\rho = \frac{\text{Cov}(X_1, X_2)}{\sigma_{X_1} \sigma_{X_2}} = \frac{2}{\sqrt{8}\sqrt{5}} = \frac{1}{\sqrt{10}} \cong 0.3162,$$

so we expect to see a positive population correlation between the random variates generated. The eigenvalues of the variance–covariance matrix are $\lambda_1 = 9$ and $\lambda_2 = 4$. These eigenvalues have corresponding eigenvectors

$$\begin{pmatrix} 2 \\ 1 \end{pmatrix} \qquad \text{and} \qquad \begin{pmatrix} 1 \\ -2 \end{pmatrix}.$$

These eigenvectors can be normalized so that their norm is 1 by dividing by $\sqrt{5}$. The spectral decomposition of the variance–covariance matrix is

$$\Sigma = \Gamma' \Lambda \Gamma = \begin{pmatrix} 2/\sqrt{5} & 1/\sqrt{5} \\ 1/\sqrt{5} & -2/\sqrt{5} \end{pmatrix} \begin{pmatrix} 9 & 0 \\ 0 & 4 \end{pmatrix} \begin{pmatrix} 2/\sqrt{5} & 1/\sqrt{5} \\ 1/\sqrt{5} & -2/\sqrt{5} \end{pmatrix}.$$

The square root of the variance–covariance matrix (the two-dimensional analogue of the standard deviation) is

$$\Sigma^{1/2} = \Gamma' \Lambda^{1/2} \Gamma = \begin{pmatrix} 2/\sqrt{5} & 1/\sqrt{5} \\ 1/\sqrt{5} & -2/\sqrt{5} \end{pmatrix} \begin{pmatrix} 3 & 0 \\ 0 & 2 \end{pmatrix} \begin{pmatrix} 2/\sqrt{5} & 1/\sqrt{5} \\ 1/\sqrt{5} & -2/\sqrt{5} \end{pmatrix}$$

which simplifies to

$$\Sigma^{1/2} = \begin{pmatrix} 14/5 & 2/5 \\ 2/5 & 11/5 \end{pmatrix}.$$

Therefore, an algorithm for generating a random pair (X_1, X_2) from this bivariate normal distribution is

$$X = \mu + \Sigma^{1/2} Z$$

which translates to

$$X = \begin{pmatrix} X_1 \\ X_2 \end{pmatrix} = \begin{pmatrix} 7 \\ 6 \end{pmatrix} + \begin{pmatrix} 14/5 & 2/5 \\ 2/5 & 11/5 \end{pmatrix} \begin{pmatrix} Z_1 \\ Z_2 \end{pmatrix},$$

where Z_1 and Z_2 are independent standard normal random variates. The R code below assigns the population mean vector μ and the variance–covariance matrix Σ to mu and sigma, calculates the eigenvectors and eigenvalues of the variance–covariance matrix Σ using the eigen function, calculates Γ and Λ, generates $n = 200$ random vectors X, and plots them in Figure 6.22.

```
mu = matrix(c(7, 6), 2, 1)
sigma = matrix(c(8, 2, 2, 5), 2, 2, byrow = T)
lam = diag(eigen(sigma)$values)
gam = t(eigen(sigma)$vectors)
sig = t(gam) %*% sqrt(lam) %*% gam
n   = 200
z1 = rnorm(n)
z2 = rnorm(n)
x1  = mu[1] + sig[1, 1] * z1 + sig[1, 2] * z2
x2  = mu[2] + sig[2, 1] * z1 + sig[2, 2] * z2
plot(x1, x2)
```

The generated pairs are centered about the population mean vector $\mu = (7, 6)$ and display a weak positive population correlation as expected.

To complete the brief overview of using matrices to define the bivariate normal distribution, the joint moment generating function can be written as

$$M(t_1, t_2) = e^{t'\mu + t'\Sigma t/2},$$

for all $t \in \mathcal{R}^2$, where $t = (t_1, t_2)'$.

To summarize this section, the bivariate normal distribution

• has level surfaces associated with its joint probability density function that are ellipses,

• has normally distributed marginal distributions,

• has normally distributed conditional distributions with constant variances, and

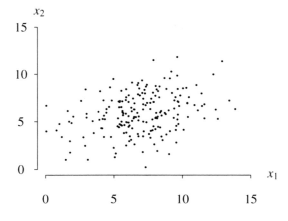

Figure 6.22: Scatterplot of 200 bivariate normal random variates ($\rho = 0.3162$).

- can be formulated using vectors and matrices.

The bivariate normal distribution is a special case of the multivariate normal distribution, which is introduced in the next section. The first four sections of this chapter have focused on joint distributions of just *two* random variables simultaneously. The final section generalizes all of these notions to n random variables that are jointly distributed.

6.5 Multivariate Distributions

The transition from the joint distribution of two random variables to the joint distribution of n random variables results in definitions that look quite similar to those given earlier in the chapter. We again have the choice between using X, Y, and Z or $X_1, X_2, \ldots, X_n$ to denote random variables. In the definitions given here we choose the latter set of symbols because of their generality. The formal definition of a vector of n jointly distributed random variables follows.

Definition 6.11 Given a random experiment with an associated sample space S, define the *multivariate random variables* $X_1, X_2, \ldots, X_n$ that assign to each element $s \in S$ one and only one vector of real numbers $X_1(s) = x_1, X_2(s) = x_2, \ldots, X_n(s) = x_n$. The *support* of these random variables is the set of ordered values

$$\mathcal{A} = \{(x_1, x_2, \ldots, x_n) \,|\, x_1 = X_1(s), x_2 = X_2(s), \ldots, x_n = X_n(s), s \in S\}.$$

The terms *multivariate random variables* to describe $X_1, X_2, \ldots, X_n$ and *random vectors* to describe $\boldsymbol{X} = (X_1, X_2, \ldots, X_n)'$ are equivalent terms that are used interchangeably in this section. If the n random variables are all discrete, then their probability distribution is described by a *joint probability mass function* $f(x_1, x_2, \ldots, x_n)$. For any set A defined on the support of $X_1, X_2, \ldots, X_n$, if $P(A)$ can be expressed as

$$P(A) = P\big((X_1, X_2, \ldots, X_n) \in A\big) = \sum\sum\cdots\sum_A f(x_1, x_2, \ldots, x_n),$$

then $f(x_1, x_2, \ldots, x_n)$ is the joint probability mass function of $X_1, X_2, \ldots, X_n$.

Likewise, if $X_1, X_2, \ldots, X_n$ are all continuous random variables and for any set $A \subset \mathcal{A}$, if $P(A)$ can be expressed as

$$P(A) = P\big((X_1, X_2, \ldots, X_n) \in A\big) = \int \int \cdots \int_A f(x_1, x_2, \ldots, x_n)\, dx_1\, dx_2 \ldots dx_n,$$

then $f(x_1, x_2, \ldots, x_n)$ is the *joint probability density function* of $X_1, X_2, \ldots, X_n$.

Joint probability mass functions and joint probability density functions must satisfy the usual existence conditions. When the random variables $X_1, X_2, \ldots, X_n$ are discrete and are defined over the support $\mathcal{A}$, the joint probability mass function must satisfy the two existence conditions

$$\sum\sum\cdots\sum_{\mathcal{A}} f(x_1, x_2, \ldots, x_n) = 1 \qquad \text{and} \qquad f(x_1, x_2, \ldots, x_n) \geq 0 \text{ for all real } x_1, x_2, \ldots, x_n.$$

Likewise, a joint probability density function for continuous random variables $X_1, X_2, \ldots, X_n$ defined over the support $\mathcal{A}$ must satisfy the two existence conditions

$$\int \int \cdots \int_{\mathcal{A}} f(x_1, x_2, \ldots, x_n) dx_1\, dx_2 \ldots dx_n = 1$$

and

$$f(x_1, x_2, \ldots, x_n) \geq 0 \text{ for all real } x_1, x_2, \ldots, x_n.$$

The examples that follow illustrate probability distributions for multivariate distributions in the discrete and continuous cases.

Example 6.51 Let X_1, X_2, X_3, and X_4 be the number of spades, hearts, diamonds, and clubs, respectively, in a five-card poker hand dealt from a well-shuffled deck. Find the joint probability mass function of X_1, X_2, X_3, and X_4.

Since the sampling of the five cards is performed without replacement from the deck, the joint probability mass function of of X_1, X_2, X_3, and X_4 is

$$f(x_1, x_2, x_3, x_4) = \frac{\binom{13}{x_1}\binom{13}{x_2}\binom{13}{x_3}\binom{13}{x_4}}{\binom{52}{5}} \qquad (x_1, x_2, x_3, x_4) \in \mathcal{A}$$

where $\mathcal{A} = \{(x_1, x_2, x_3, x_4) \mid x_i = 0, 1, 2, 3, 4, 5; \ i = 1, 2, 3, 4; \ \sum_{i=1}^{4} x_i = 5\}$. The distribution of X_1, X_2, X_3, and X_4 is a special case of the *multivariate hypergeometric distribution*. The probability mass function could also be written in terms of just x_1, x_2, and x_3 because x_4 is determined by the other three variables. In this particular problem, $x_4 = 5 - x_1 - x_2 - x_3$.

Example 6.52 Let the continuous random variables X_1, X_2, and X_3 have joint probability density function

$$f(x_1, x_2, x_3) = 1 \qquad 0 < x_1 < 1, 0 < x_2 < 1, 0 < x_3 < 1.$$

Find $P(X_1 < X_2 < X_3)$.

The probability that $X_1 < X_2 < X_3$ can be determined by inspection. Since the support is the interior of the unit cube and the probability density function is constant over the support, the random variables X_1, X_2, and X_3 play symmetric roles. Furthermore, since there are 3! different orderings of the random variables by the multiplication rule, the desired probability is

$$P(X_1 < X_2 < X_3) = \frac{1}{3!} = \frac{1}{6}.$$

The more tedious solution via integration is presented here nonetheless, in order to show how appropriate integration limits are determined for the more general case. The appropriate triple integral to compute the desired probability is

$$P(X_1 < X_2 < X_3) = \int_0^1 \int_{x_1}^1 \int_{x_2}^1 1 \, dx_3 \, dx_2 \, dx_1 = \frac{1}{6},$$

which can be computed by the Maple statement

```
int(int(int(1, x3 = x2 .. 1), x2 = x1 .. 1), x1 = 0 .. 1);
```

Example 6.53 Let the continuous random variables X_1, X_2, and X_3 have joint probability density function

$$f(x_1, x_2, x_3) = cx_1 x_2^2 x_3^3 \qquad 0 < x_1 < 1, 0 < x_2 < 2, 0 < x_3 < 3.$$

Find the constant c so that $f(x_1, x_2, x_3)$ is a legitimate joint probability density function and find $P(X_1 < X_2 < X_3)$.

The support for this trivariate distribution is the interior of a rectangular solid with three different side lengths similar to the shape of a cereal box. Since the product of positive real numbers is positive, it is clear that the joint probability density function is nonnegative on the support as long as $c > 0$. The constant c must ensure that the triple integral over the support is one, that is

$$\int_0^1 \int_0^2 \int_0^3 cx_1 x_2^2 x_3^3 \, dx_3 \, dx_2 \, dx_1 = 1.$$

Using the Maple statement

```
int(int(int(c * x1 * x2 ^ 2 * x3 ^ 3, x3 = 0 .. 3),
                                      x2 = 0 .. 2),
                                      x1 = 0 .. 1);
```

the left-hand side of this equation is $27c$, so the constant c is $c = 1/27$. The joint probability density function can then be written as

$$f(x_1, x_2, x_3) = \frac{x_1 x_2^2 x_3^3}{27} \qquad 0 < x_1 < 1, 0 < x_2 < 2, 0 < x_3 < 3.$$

The probability that $X_1 < X_2 < X_3$ can be determined by integration:

$$P(X_1 < X_2 < X_3) = \int_0^1 \int_{x_1}^2 \int_{x_2}^3 \frac{x_1 x_2^2 x_3^3}{27} \, dx_3 \, dx_2 \, dx_1 = \frac{7361}{8505} \cong 0.8655$$

which can be computed by the Maple statement

```
int(int(int(x1 * x2 ^ 2 * x3 ^ 3 / 27, x3 = x2 .. 3),
                                        x2 = x1 .. 2),
                                        x1 = 0 .. 1);
```

The definition of the joint cumulative distribution function also extends to n-dimensional jointly distributed random variables. For both discrete and continuous random variables, the joint cumulative distribution function is defined to be

$$F(x_1, x_2, \ldots, x_n) = P(X_1 \le x_1, X_2 \le x_2, \ldots, X_n \le x_n).$$

As usual, the commas in the probability function can be interpreted as "and." The joint cumulative distribution function is computed in a different manner for discrete and continuous random variables. For discrete random variables, the joint cumulative distribution function of $X_1, X_2, \ldots, X_n$ is computed by summing:

$$F(x_1, x_2, \ldots, x_n) = \sum_{w_1 \le x_1} \sum_{w_2 \le x_2} \cdots \sum_{w_n \le x_n} f(w_1, w_2, \ldots, w_n).$$

For continuous random variables, the joint cumulative distribution function of $X_1, X_2, \ldots, X_n$ is computed by integrating:

$$F(x_1, x_2, \ldots, x_n) = \int_{-\infty}^{x_1} \int_{-\infty}^{x_2} \cdots \int_{-\infty}^{x_n} f(w_1, w_2, \ldots, w_n) \, dw_n \ldots dw_2 \, dw_1.$$

Also, when $X_1, X_2, \ldots, X_n$ are continuous random variables, the joint probability density function can be found by taking partial derivatives of the joint cumulative distribution function with respect to each of the variables:

$$f(x_1, x_2, \ldots, x_n) = \frac{\partial^n F(x_1, x_2, \ldots, x_n)}{\partial x_1 \partial x_2 \ldots \partial x_n}.$$

Example 6.54 Let X_1, X_2, X_3 have joint probability density function

$$f(x_1, x_2, x_3) = 6e^{-x_1 - 2x_2 - 3x_3} \qquad x_1 > 0, x_2 > 0, x_3 > 0.$$

Find the joint cumulative distribution function.

Since the support of the random variables X_1, X_2, X_3 is the interior of the octant where all three of the variables are positive, the triple integrals required to find the cumulative distribution function can be placed in any order, for example:

$$F(x_1, x_2, x_3) = \int_0^{x_1} \int_0^{x_2} \int_0^{x_3} 6e^{-w_1 - 2w_2 - 3w_3} \, dw_3 \, dw_2 \, dw_1$$

for any (x_1, x_2, x_3) on the support $\mathcal{A}$. This triple integral is easily computed by hand or with the Maple statement

```
int(int(int(6 * exp(-w1 - 2 * w2 - 3 * w3), w3 = 0 .. x3),
                                            w2 = 0 .. x2),
                                            w1 = 0 .. x1);
```

which simplifies to

$$F(x_1, x_2, x_3) = \begin{cases} 0 & x_1 \leq 0 \text{ or } x_2 \leq 0 \text{ or } x_3 \leq 0 \\ \left(1 - e^{-x_1}\right)\left(1 - e^{-2x_2}\right)\left(1 - e^{-3x_3}\right) & x_1 > 0, x_2 > 0, x_3 > 0. \end{cases}$$

The first piece of this piecewise function is defined on seven of the eight octants; the second piece of this piecewise function is defined on the support $\mathcal{A}$.

Example 6.55 Let X_1, X_2, X_3 have joint probability density function

$$f(x_1, x_2, x_3) = \frac{1}{1000} \qquad 0 < x_1 < 10, 0 < x_2 < 10, 0 < x_3 < 10.$$

What is the minimum number of regions required to define the joint cumulative distribution function?

The joint probability density function assumes the value $1/1000$ on the $10 \times 10 \times 10$ cube in the first octant, and is zero elsewhere. The cumulative distribution function is $F(x_1, x_2, x_3) = 0$ on the region

$$\{(x_1, x_2, x_3) \,|\, x_1 \leq 0 \text{ or } x_2 \leq 0 \text{ or } x_3 \leq 0\},$$

which encompasses seven of the eight octants in $x_1 x_2 x_3$ space. The cumulative distribution function must also be defined separately on the support

$$\mathcal{A} = \{(x_1, x_2, x_3) \,|\, 0 < x_1 < 10, 0 < x_2 < 10, 0 < x_3 < 10\}.$$

There are three regions where one of the three variable has achieved its upper limit and the other two variables lie between 0 and 10, for example

$$\{(x_1, x_2, x_3) \,|\, 0 < x_1 < 10, 0 < x_2 < 10, x_3 \geq 10\}.$$

Also, there are three regions where two of the variables have achieved their upper limit but the other variable lies between 0 and 10, for example,

$$\{(x_1, x_2, x_3) \,|\, x_1 \geq 10, x_2 \geq 10, 0 < x_3 < 10\}.$$

Finally, the cumulative distribution function is $F(x_1, x_2, x_3) = 1$ on the region

$$\{(x_1, x_2, x_3) \,|\, x_1 \geq 10, x_2 \geq 10, x_3 \geq 10\}.$$

So there are a total of $1 + 1 + 3 + 3 + 1 = 9$ different regions where a different set of triple integrals is required to compute the cumulative distribution function. Setting up these integration limits is left as an exercise. The alert reader might have noticed a row of Pascal's triangle in the 1–3–3–1 sequence in the previous summation. If the joint probability density function had been defined in an n-dimensional hypersquare, there would have been $2^n + 1$ different regions required to define the joint cumulative distribution function, and all but the first and last are computed with an n-fold integral.

For jointly distributed random variables $X_1, X_2, \ldots, X_n$, the *marginal distribution* of one of the random variables is found by summing (in the discrete case) or integrating (in the continuous case) out all variables that are not of interest. For example, if $X_1, X_2, \ldots, X_n$ are continuous random variables, then the marginal probability density function of X_1 is found by integrating $x_2, x_3, \ldots, x_n$ out of the joint probability density function as follows:

$$f_{X_1}(x_1) = \int_{-\infty}^{\infty} \int_{-\infty}^{\infty} \cdots \int_{-\infty}^{\infty} f(x_1, x_2, \ldots, x_n) \, dx_2 \, dx_3 \ldots dx_n.$$

As a more concrete illustration, consider the random vector of $n = 5$ economic measures associated with a particular country:

X_1: gross domestic product $[0, \infty)$,
X_2: unemployment as a percentage $[0, 100]$,
X_3: national debt $(-\infty, \infty)$,
X_4: annual energy consumption $(0, \infty)$,
X_5: trade deficit $(-\infty, \infty)$.

The support of each of these economic measures follows the words describing the quantity of interest. If we know the joint probability density function of X_1, X_2, X_3, X_4, X_5, how do we find the joint probability density function of X_1, X_4, X_5? In this case, we integrate x_2 and x_3 over their ranges out of the joint probability density function, that is

$$f_{X_1.X_4.X_5}(x_1, x_4, x_5) = \int_{-\infty}^{\infty} \int_{0}^{100} f(x_1, x_2, x_3, x_4, x_5) \, dx_2 \, dx_3.$$

Example 6.56 Consider three random variables X_1, X_2, X_3 with joint probability density function

$$f(x_1, x_2, x_3) = \lambda_1 \lambda_2 \lambda_3 e^{-\lambda_1 x_1 - \lambda_2 x_2 - \lambda_3 x_3} \qquad (x_1, x_2, x_3) \in \mathcal{A},$$

where the support is $\mathcal{A} = \{(x_1, x_2, x_3) \mid x_1 > 0, x_2 > 0, x_3 > 0\}$ and the *parameter space* is $\Omega = \{(\lambda_1, \lambda_2, \lambda_3) \mid \lambda_1 > 0, \lambda_2 > 0, \lambda_3 > 0\}$. Find the marginal probability density function for X_1.

The marginal probability density function for X_1 is found by integrating x_2 and x_3 out of the joint probability density function:

$$
\begin{aligned}
f_{X_1}(x_1) &= \int_0^\infty \int_0^\infty \lambda_1 \lambda_2 \lambda_3 e^{-\lambda_1 x_1} e^{-\lambda_2 x_2} e^{-\lambda_3 x_3} \, dx_2 \, dx_3 \\
&= \int_0^\infty \lambda_1 \lambda_3 e^{-\lambda_1 x_1} e^{-\lambda_3 x_3} \left[-e^{-\lambda_2 x_2} \right]_0^\infty dx_3 \\
&= \lambda_1 e^{-\lambda_1 x_1} \left[-e^{-\lambda_3 x_3} \right]_0^\infty \\
&= \lambda_1 e^{-\lambda_1 x_1} \qquad x_1 > 0.
\end{aligned}
$$

This marginal probability density function can be recognized as an exponential(λ_1) probability density function.

The conditional probability density function of one or more continuous random variables given the value(s) of the other random variables is found by computing the ratio of the joint probability density function to the joint or marginal probability density function of the random variable(s) whose values are known. The process is the same for discrete random variables. As a simple illustration, consider the continuous random variables X_1, X_2, X_3 with a joint probability density function

$$f(x_1, x_2, x_3).$$

In order to find the conditional bivariate probability density function of X_2 and X_3 given $X_1 = x_1$, take the ratio of the joint probability density function to the marginal probability density function:

$$f(x_2, x_3 \mid X_1 = x_1) = \frac{f(x_1, x_2, x_3)}{f_{X_1}(x_1)}$$

over the support of X_2 and X_3 whenever $f_{X_1}(x_1) \neq 0$. Returning to the earlier illustration involving the five economic measures

X_1: gross domestic product,
X_2: unemployment as a percentage,
X_3: national debt,
X_4: annual energy consumption,
X_5: trade deficit,

the joint probability density function of unemployment (X_2) and national debt (X_3) given specific values for the gross domestic product $(X_1 = x_1)$, annual energy consumption $(X_4 = x_4)$, and trade deficit $(X_5 = x_5)$ is

$$f_{X_2,X_3|X_1=x_1,X_4=x_4,X_5=x_5}(x_2, x_3 \,|\, X_1 = x_1, X_4 = x_4, X_5 = x_5) = \frac{f(x_1, x_2, x_3, x_4, x_5)}{f_{X_1,X_4,X_5}(x_1, x_4, x_5)}$$

over the support of X_2 and X_3 wherever $f_{X_1,X_4,X_5}(x_1, x_4, x_5) \neq 0$.

Example 6.57 A *copula* is a joint distribution whose marginal distributions are each uniformly distributed. Copulas have a wide variety of application areas, including random variate generation, financial risk assessment, actuarial science, and engineering. One simple one-parameter copula is the trivariate distribution of the random variables X_1, X_2, X_3 with joint probability density function

$$f(x_1, x_2, x_3) = 1 + \alpha(1 - 2x_1)(1 - 2x_2)(1 - 2x_3) \qquad (x_1, x_2, x_3) \in \mathcal{A},$$

where $\mathcal{A} = \{(x_1, x_2, x_3) \,|\, 0 < x_1 < 1, 0 < x_2 < 1, 0 < x_3 < 1\}$ and the single parameter α satisfies $-1 \leq \alpha \leq 1$. For this joint distribution, find

(a) the marginal distribution of X_3,

(b) the joint marginal distribution of X_1 and X_2,

(d) the conditional distribution of X_3 given $X_1 = x_1$ and $X_2 = x_2$,

(d) the conditional population correlation between X_1 and X_2 given $X_3 = x_3$.

(a) In order to find the marginal distribution of X_3, integrate x_1 and x_2 out of the joint probability density function, which yields

$$f_{X_3}(x_3) = \int_0^1 \int_0^1 \left(1 + \alpha(1 - 2x_1)(1 - 2x_2)(1 - 2x_3)\right) dx_1 \, dx_2 = 1 \qquad 0 < x_3 < 1.$$

The marginal distribution of X_3 is $U(0, 1)$. By the symmetry of the functional form of the joint probability density function, this means that $X_1 \sim U(0, 1)$ and $X_2 \sim U(0, 1)$. So the joint distribution of X_1, X_2, X_3 satisfies the definition of a copula in the sense that each of the marginals is uniform.

(b) In order to find the marginal distribution of X_1 and X_2, integrate x_3 out of the joint probability density function, which yields

$$f_{X_1,X_2}(x_1, x_2) = \int_0^1 \left(1 + \alpha(1 - 2x_1)(1 - 2x_2)(1 - 2x_3)\right) dx_3 = 1$$

on the support $0 < x_1 < 1, 0 < x_2 < 1$. The bivariate marginal distribution for X_1 and X_2 is uniformly distributed over the unit square in the x_1x_2 plane. Again by the symmetry of the functional form of the joint probability density function, the distribution of the other two pairs of random variables (that is, (X_1, X_3) and (X_2, X_3)) is also uniformly distributed on the unit square.

(c) The conditional distribution of X_3 given $X_1 = x_1$ and $X_2 = x_2$ is the ratio of the joint probability density function of X_1, X_2, X_3 to the marginal joint probability density function of X_1 and X_2:

$$
\begin{aligned}
f_{X_3 \mid X_1=x_1, X_2=x_2}(x_3 \mid X_1 = x_1, X_2 = x_2) &= \frac{f(x_1, x_2, x_3)}{f(x_1, x_2)} \\
&= 1 + \alpha(1 - 2x_1)(1 - 2x_2)(1 - 2x_3)
\end{aligned}
$$

on the support $0 < x_3 < 1$, for any values of x_1 and x_2 satisfying $0 < x_1 < 1$ and $0 < x_2 < 1$.

(d) We now compute the conditional population correlation between X_1 and X_2 given $X_3 = x_3$. The conditional distribution of X_1 and X_2 given $X_3 = x_3$ is

$$
\begin{aligned}
f_{X_1, X_2 \mid X_3=x_3}(x_1, x_2 \mid X_3 = x_3) &= \frac{f(x_1, x_2, x_3)}{f_{X_3}(x_3)} \\
&= 1 + \alpha(1 - 2x_1)(1 - 2x_2)(1 - 2x_3)
\end{aligned}
$$

on the support $0 < x_1 < 1, 0 < x_2 < 1$ for any x_3 satisfying $0 < x_3 < 1$. The conditional expected value of $X_1 X_2$ given $X_3 = x_3$ is

$$
E[X_1 X_2 \mid X_3 = x_3] = \int_0^1 \int_0^1 x_1 x_2 \left(1 + \alpha(1 - 2x_1)(1 - 2x_2)(1 - 2x_3)\right) dx_2 \, dx_1,
$$

which simplifies to

$$
E[X_1 X_2 \mid X_3 = x_3] = \frac{9 + \alpha(1 - 2x_3)}{36}.
$$

Since X_1 and X_2 are independent $U(0, 1)$ random variables, their population means and variances are

$$
E[X_1] = \frac{1}{2} \qquad V[X_1] = \frac{1}{12} \qquad E[X_2] = \frac{1}{2} \qquad V[X_2] = \frac{1}{12}.
$$

Finally, using Theorem 6.4 and Definition 6.7, the conditional population correlation between X_1 and X_2 given $X_3 = x_3$ is

$$
\rho = \frac{E[X_1 X_2] - \mu_{X_1} \mu_{X_2}}{\sigma_{X_1} \sigma_{X_2}} = \frac{\frac{9 + \alpha(1 - 2x_3)}{36} - (1/2)(1/2)}{1/12} = \frac{\alpha(1 - 2x_3)}{3}
$$

for $0 < x_3 < 1$. Because the product of the two factors in the numerator of this expression lie between -1 and 1, the conditional population correlation can only assume values in the limited range

$$
-\frac{1}{3} < \rho < \frac{1}{3}.
$$

The same difficulty encountered in Chapter 2 concerning the definition of independence of more than two events also arises when defining independence of more than two random variables. We use the term *mutually independent* in the same fashion as in Chapter 2.

Definition 6.12 Let the random variables $X_1, X_2, \ldots, X_n$ (discrete or continuous) have a joint distribution described by $f(x_1, x_2, \ldots, x_n)$ and marginal distributions described by $f_{X_1}(x_1), f_{X_2}(x_2), \ldots, f_{X_n}(x_n)$. The random variables $X_1, X_2, \ldots, X_n$ are *mutually independent* if and only if

$$
f(x_1, x_2, \ldots, x_n) = f_{X_1}(x_1) f_{X_2}(x_2) \ldots f_{X_n}(x_n)
$$

for all real numbers $x_1, x_2, \ldots, x_n$.

Some authors prefer to drop the term "mutually" when the context is clear and simply state that $X_1, X_2, \ldots, X_n$ are independent random variables. Definition 6.12 was formulated using probability mass functions (when the random variables are discrete) and probability density functions (when the random variables are continuous). It can also be formulated using cumulative distribution functions, as indicated in the following definition.

Definition 6.13 Let the random variables $X_1, X_2, \ldots, X_n$ (discrete or continuous) have a joint cumulative distribution function $F(x_1, x_2, \ldots, x_n)$ and marginal cumulative distribution functions $F_{X_1}(x_1), F_{X_2}(x_2), \ldots, F_{X_n}(x_n)$. The random variables $X_1, X_2, \ldots, X_n$ are *mutually independent* if and only if

$$F(x_1, x_2, \ldots, x_n) = F_{X_1}(x_1)F_{X_2}(x_2)\ldots F_{X_n}(x_n)$$

for all real numbers $x_1, x_2, \ldots, x_n$.

Still another way of viewing independence is to consider the probability that the random variables $X_1, X_2, \ldots, X_n$ fall in an n-dimensional rectangle. If $X_1, X_2, \ldots, X_n$ are mutually independent random variables then

$$P(a_1 < X_1 < b_1, a_2 < X_2 < b_2, \ldots, a_n < X_n < b_n) = \prod_{i=1}^{n} P(a_i < x_i < b_i)$$

for real constants $a_1, a_2, \ldots, a_n$ and $b_1, b_2, \ldots, b_n$.

Any of these three perspectives on mutually independent random variables implies the following: if the random variables $X_1, X_2, \ldots, X_n$ are *mutually independent* then they are *pairwise independent*. This is illustrated in the following example.

Example 6.58 Let X_1, X_2, X_3 be mutually independent Bernoulli($1/2$) random variables. Show that they are pairwise independent.

The points in Figure 6.23 show the support of X_1, X_2, and X_3. The marginal distributions of X_1, X_2, and X_3 are identical, for example

$$f_{X_1}(x_1) = \frac{1}{2} \qquad x_1 = 0, 1.$$

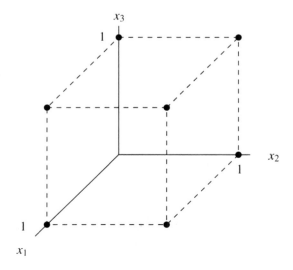

Figure 6.23: The support of X_1, X_2, and X_3.

Using Definition 6.12, the joint probability mass function of X_1, X_2, X_3 is uniformly distributed over the eight vertices of the unit cube in $x_1 x_2 x_3$ space:

$$f(x_1, x_2, x_3) = f_{X_1}(x_1) f_{X_2}(x_2) f_{X_3}(x_3) = \frac{1}{8} \qquad x_1 = 0, 1; \; x_2 = 0, 1; \; x_3 = 0, 1.$$

The marginal probability mass function of X_1 and X_2 is

$$f(x_1, x_2) = f_{X_1}(x_1) f_{X_2}(x_2) = \frac{1}{4} \qquad x_1 = 0, 1; \; x_2 = 0, 1.$$

Invoking Theorem 6.1, since X_1 and X_2 are defined on a product space and their joint probability mass function can be factored into functions of x_1 and x_2 alone, the random variables are pairwise independent. The two other pairs are also pairwise independent by symmetry.

The converse of the result that mutual independence implies pairwise independence is not true in general. Pairwise independence does not necessarily imply mutual independence, as illustrated by the following counterexample.

Example 6.59 Let X_1, X_2, and X_3 have joint probability mass function

$$f(x_1, x_2, x_3) = \frac{1}{4} \qquad (x_1, x_2, x_3) \in \{(1, 0, 0), (0, 1, 0), (0, 0, 1), (1, 1, 1)\}.$$

Show that the random variables are pairwise independent, but not mutually independent.

The support of X_1, X_2, and X_3 is shown in Figure 6.24. The joint probability mass function of X_1 and X_2 is

$$f_{X_1, X_2}(x_1, x_2) = \frac{1}{4} \qquad (x_1, x_2) \in \{(0, 0), (0, 1), (1, 0), (1, 1)\}.$$

The other two pairs of bivariate probability mass functions are identical by symmetry. Furthermore, the marginal distributions of X_1, X_2, and X_3 are each Bernoulli($1/2$) with probability mass function

$$f_{X_i}(x_i) = \frac{1}{2} \qquad x_i = 0, 1$$

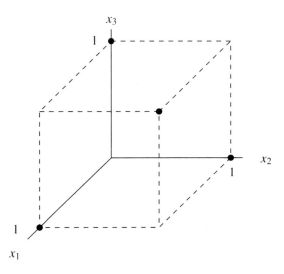

Figure 6.24: The support of X_1, X_2, and X_3.

for $i = 1, 2, 3$. This means that the random variables are *pairwise independent* because, for example,

$$f_{X_1, X_2}(x_1, x_2) = f_{X_1}(x_1) f_{X_2}(x_2),$$

but they are not *mutually independent* because

$$f(x_1, x_2, x_3) \neq f_{X_1}(x_1) f_{X_2}(x_2) f_{X_3}(x_3).$$

Example 6.60 Find $P(X_1 < X_2 < X_3 < X_4)$ when X_1, X_2, X_3, X_4 are mutually independent and identically distributed

(a) continuous random variables,

(b) discrete random variables, each with probability mass function

$$f(x) = \frac{x}{15} \qquad\qquad x = 1, 2, 3, 4, 5.$$

(a) There are $4! = 24$ possible orderings of X_1, X_2, X_3, and X_4 by the multiplication rule. Therefore,

$$P(X_1 < X_2 < X_3 < X_4) = \frac{1}{24},$$

since all of the orderings are equally likely due to the fact that the four random variables are mutually independent and identically distributed.

(b) Ties must be considered because the random variables are discrete. There are five different ways to achieve $X_1 < X_2 < X_3 < X_4$:

$$X_1 = 1, \ X_2 = 2, \ X_3 = 3, \ X_4 = 4$$
$$X_1 = 1, \ X_2 = 2, \ X_3 = 3, \ X_4 = 5$$
$$X_1 = 1, \ X_2 = 2, \ X_3 = 4, \ X_4 = 5$$
$$X_1 = 1, \ X_2 = 3, \ X_3 = 4, \ X_4 = 5$$
$$X_1 = 2, \ X_2 = 3, \ X_3 = 4, \ X_4 = 5.$$

Summing the probabilities associated with these five outcomes yields

$$
\begin{aligned}
P(X_1 < X_2 < X_3 < X_4) \ &= \ P(X_1 = 1, X_2 = 2, X_3 = 3, X_4 = 4) \\
&\quad + \cdots + P(X_1 = 2, X_2 = 3, X_3 = 4, X_4 = 5) \\
&= \ \frac{1}{15} \cdot \frac{2}{15} \cdot \frac{3}{15} \cdot \frac{4}{15} + \cdots + \frac{2}{15} \cdot \frac{3}{15} \cdot \frac{4}{15} \cdot \frac{5}{15} \\
&= \ \frac{24 + 30 + 40 + 60 + 120}{15^4} \\
&= \ \frac{274}{50,625} \\
&\cong \ 0.005412.
\end{aligned}
$$

The next example addresses a probability question associated with a *random polynomial*, one with random, rather than constant, coefficients.

Example 6.61 Let the continuous random variables A, B, and C be mutually independent $U(0, 1)$ random variables. Find the probability that the random quadratic equation

$$Ax^2 + Bx + C = 0$$

has real roots.

The joint probability density function $f(a, b, c)$ is defined on the interior of the unit cube $\mathcal{A} = \{(a, b, c) \mid 0 < a < 1, 0 < b < 1, 0 < c < 1\}$. Using Definition 6.12, the trivariate probability density function of A, B, and C is the product of the marginal probability density functions:

$$f(a, b, c) = 1 \qquad 0 < a < 1, 0 < b < 1, 0 < c < 1.$$

The quadratic equation has real roots when the discriminant $B^2 - 4AC$ is greater than or equal to zero. The support $\mathcal{A}$, along with the trace of the edge of the region of interest through the boundary of the unit cube when $a = 1$, $b = 1$, and $c = 1$ is shown in Figure 6.25. The portion on the diagram when $b = 1$ is $4ac = 1$, which is important when setting up the limits for the triple integral. Since large values of b correspond to real roots, it is placed as the inside integration limit. One set of appropriate integration limits for computing the probability that the random polynomial has real roots is

$$P\left(B^2 - 4AC \geq 0\right) = \int_0^{1/4} \int_0^1 \int_{\sqrt{4ac}}^1 1 \, db \, dc \, da + \int_{1/4}^1 \int_0^{1/4a} \int_{\sqrt{4ac}}^1 1 \, db \, dc \, da.$$

The limits of integration are appropriate, of course, for any joint probability density function defined on the interior of the unit cube which has a functional form more complicated than $f(a, b, c) = 1$. This integral can be computed with the Maple statement

```
int(int(int(1, b = sqrt(4 * a * c) .. 1),
            c = 0 .. 1),
        a = 0 .. 1 / 4) +
int(int(int(1, b = sqrt(4 * a * c) .. 1),
            c = 0 .. 1 / (4 * a)),
        a = 1 / 4 .. 1);
```

which yields a probability of real roots

$$P\left(B^2 - 4AC \geq 0\right) = \frac{5}{36} + \frac{\ln 4}{12} \cong 0.2544.$$

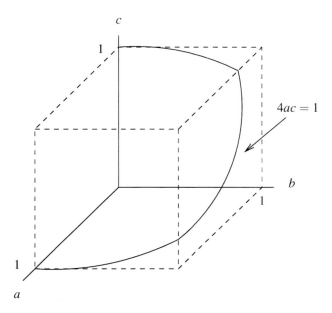

Figure 6.25: The support of A, B, and C and the trace of the region of integration.

This analytic result is supported by Monte Carlo simulation, as illustrated by the R code that follows with 100,000 replications. The code returns the fraction of the random quadratic functions that contain real roots. Since the marginal distributions of A, B, and C are independent $U(0, 1)$ random variables, they can be generated with the `runif` function.

```
nrep = 100000
A = runif(nrep)
B = runif(nrep)
C = runif(nrep)
mean(B ^ 2 - 4 * A * C > 0)
```

The upper-case letters for variable names are used to avoid a conflict with the R *collect* function c that is used to define vectors. Five runs of this simulation yield

| 0.25403 | 0.25422 | 0.25399 | 0.25358 | 0.25499 |

which hover around the analytic solution $P\left(B^2 - 4AC \geq 0\right) \cong 0.2544$.

The *binomial* distribution was introduced in Chapter 4. The *trinomial* distribution was introduced earlier in this chapter. Both of these distributions are special cases of the *multinomial* distribution.

The binomial distribution was based on the notion of n mutually independent Bernoulli trials, each with two possible outcomes: success and failure. The trinomial distribution also had n mutually independent trials, but each trial had three possible outcomes. The multinomial distribution also has n mutually independent trials, but each trial has k possible outcomes, with associated probabilities $p_1, p_2, \ldots, p_k$, where $p_1 + p_2 + \cdots + p_k = 1$. Let the discrete random variable X_i denote a count of the number of type i outcomes in the n trials, for $i = 1, 2, \ldots, k$. We now develop the joint distribution of $X_1, X_2, \ldots, X_k$.

The joint probability mass function of $X_1, X_2, \ldots, X_k$ is

$$f(x_1, x_2, \ldots, x_k) = P(X_1 = x_1, X_2 = x_2, \ldots, X_k = x_k).$$

Because of the mutual independence of the trials, the probability of exactly x_1 Type 1 outcomes, x_2 Type 2 outcomes, $\ldots$, and x_k Type k outcomes *in a specific order* is

$$p_1^{x_1} p_2^{x_2} \cdots p_k^{x_k}.$$

The number of ways of ordering the n outcomes is the multinomial coefficient

$$\binom{n}{x_1, x_2, \ldots, x_k} = \frac{n!}{x_1! x_2! \ldots x_k!}.$$

Therefore, the joint probability mass function of $X_1, X_2, \ldots, X_k$ is

$$f(x_1, x_2, \ldots, x_k) = \frac{n!}{x_1! x_2! \ldots x_k!} p_1^{x_1} p_2^{x_2} \cdots p_k^{x_k} \qquad (x_1, x_2, \ldots, x_k) \in \mathcal{A}.$$

The support set $\mathcal{A}$ is unusually tedious for this distribution. First, because there are n trials, the total of the counting variables $x_1, x_2, \ldots, x_k$ must sum to n, that is

$$x_1 + x_2 + \cdots + x_k = n.$$

Second, each x_i value can assume an integer value between 0 and n. Combining these two facts yields the support

$$\mathcal{A} = \{(x_1, x_2, \ldots, x_k) \mid x_i = 0, 1, \ldots, n, \text{ for } i = 1, 2, \ldots, k; \, x_1 + x_2 + \cdots + x_k = n\}.$$

The parameter space for the multinomial distribution is also tedious. There are $k + 2$ parameters for the multinomial distribution: $n, k, p_1, p_2, \ldots, p_k$. Letting $\mathbf{N}$ denote the set of positive integers, the parameter space is

$$\Omega = \{(n, k, p_1, p_2, \ldots, p_k) \,|\, n \in \mathbf{N}; \; k = 2, 3, \ldots; \; 0 < p_i < 1 \text{ for } i = 1, 2, \ldots, k; \; p_1 + \cdots + p_k = 1\}.$$

The marginal distributions of the multinomial are each binomial, that is $X_i \sim \text{binomial}(n, p_i)$, for $i = 1, 2, \ldots, k$. The multinomial distribution includes the trinomial and binomial distributions as special cases when $k = 3$ and $k = 2$. The multinomial distribution can be applied to an instructor assigning one of $k = 5$ grades (A, B, C, D, F) to each of n students in a class. Another example includes a shoe store that carries an inventory of n pairs of shoes with k different shoe sizes. The multinomial distribution is applied below to a problem involving n rolls of a die.

Example 6.62 A die is weighted, or "loaded," so that the number of spots X that appear on the up face when the die is rolled has probability mass function

$$f(x) = \frac{x}{21} \qquad x = 1, 2, 3, 4, 5, 6.$$

If this loaded die is rolled 21 times, find the probability of rolling one one, two twos, three threes, four fours, five fives, and six sixes.

This problem fits into the multinomial distribution framework because the $n = 21$ trials (rolls) are mutually independent, and there are $k = 6$ outcomes to each trial. The parameters for the multinomial distribution are

$$n = 21, k = 6, p_1 = \frac{1}{21}, p_2 = \frac{2}{21}, p_3 = \frac{3}{21}, p_4 = \frac{4}{21}, p_5 = \frac{5}{21}, p_6 = \frac{6}{21}.$$

The probability of rolling one one, two twos, three threes, four fours, five fives, and six sixes is

$$f(1, 2, 3, 4, 5, 6) = P(X_1 = 1, X_2 = 2, X_3 = 3, X_4 = 4, X_5 = 5, X_6 = 6)$$

which evaluates to

$$\frac{21!}{1!2!3!4!5!6!} \left(\frac{1}{21}\right)^1 \left(\frac{2}{21}\right)^2 \left(\frac{3}{21}\right)^3 \left(\frac{4}{21}\right)^4 \left(\frac{5}{21}\right)^5 \left(\frac{6}{21}\right)^6 \cong 0.001417.$$

Why is this probability so small? Although the outcome in question is the most likely occurrence for 21 rolls of this loaded die, there are a myriad of outcomes associated with the sample space of 6^{21} sequences of rolls, each with a relatively small probability of occurrence.

The smallest and largest of a group of random variables is oftentimes of interest. Examples include the time to failure of a series system of components (minimum), the time to failure of a parallel system of components (maximum), and the amount of rainfall associated with a 100-year flooding event (maximum). The example that follows considers the minimum of n independent exponential random variables.

Example 6.63 Let $X_1, X_2, \ldots, X_n$ be mutually independent random variables. Furthermore, let

$$X_i \sim \text{exponential}(\lambda_i)$$

for $i = 1, 2, \ldots, n$, where $\lambda_1, \lambda_2, \ldots, \lambda_n$ are positive parameters. Find the distribution of $X = \min\{X_1, X_2, \ldots, X_n\}$.

The cumulative distribution function technique will be used to solve the problem. The cumulative distribution function of X_i is

$$F_{X_i}(x_i) = 1 - e^{-\lambda_i x_i} \qquad x_i > 0,$$

for $i = 1, 2, \ldots, n$. Using complementary probability, equivalent events, and mutual independence, the cumulative function of X is

$$
\begin{aligned}
F_X(x) &= P(X \leq x) \\
&= P(\min\{X_1, X_2, \ldots, X_n\} \leq x) \\
&= 1 - P(\min\{X_1, X_2, \ldots, X_n\} > x) \\
&= 1 - P(X_1 > x, X_2 > x, \ldots, X_n > x) \\
&= 1 - P(X_1 > x)P(X_2 > x)\ldots P(X_n > x) \\
&= 1 - e^{-\lambda_1 x} e^{-\lambda_2 x} \ldots e^{-\lambda_n x} \\
&= 1 - e^{-(\lambda_1 + \lambda_2 + \cdots + \lambda_n)x} \qquad x > 0.
\end{aligned}
$$

This cumulative distribution function can be recognized as the cumulative distribution function of an exponential$(\lambda_1 + \lambda_2 + \cdots + \lambda_n)$ random variable. This result is well known to design engineers. If a series system consists of n electrical components with exponential times to failure with failure rates $\lambda_1, \lambda_2, \ldots, \lambda_n$, then the system time to failure is also exponentially distributed with a failure rate that is the sum of the component failure rates.

The notion of an *expected value* of some function of n random variables follows directly from the two-dimensional case, as shown next.

Definition 6.14 Let X_1, X_2, $\ldots$, X_n be random variables with joint probability mass function $f(x_1, x_2, \ldots, x_n)$ if the random variables are discrete or joint probability density function $f(x_1, x_2, \ldots, x_n)$ if the random variables are continuous. The *expected value* of $g(X_1, X_2, \ldots, X_n)$ is

$$E[g(X_1, X_2, \ldots, X_n)] = \sum_{x_1} \sum_{x_2} \cdots \sum_{x_n} g(x_1, x_2, \ldots, x_n) f(x_1, x_2, \ldots, x_n)$$

when $X_1, X_2, \ldots, X_n$ are discrete and

$$E[g(X_1, X_2, \ldots, X_n)] = \int_{-\infty}^{\infty} \int_{-\infty}^{\infty} \cdots \int_{-\infty}^{\infty} g(x_1, x_2, \ldots, x_n) f(x_1, x_2, \ldots, x_n) \, dx_1 \, dx_2 \ldots dx_n$$

when $X_1, X_2, \ldots, X_n$ are continuous and when the sum or integral exist. When the sum or integral diverge, the expected value is undefined.

Definition 6.14 is illustrated for a trivariate distribution encountered earlier in this section.

Example 6.64 Let the continuous random variables X_1, X_2, and X_3 have joint probability density function

$$f(x_1, x_2, x_3) = \frac{x_1 x_2^2 x_3^3}{27} \qquad 0 < x_1 < 1, 0 < x_2 < 2, 0 < x_3 < 3.$$

Find $E[X_1 + X_2 X_3]$.

Using Theorem 6.14 and leaving the integration details to the interested reader, the expectation is computed by

$$E[X_1 + X_2 X_3] = \int_0^1 \int_0^2 \int_0^3 (x_1 + x_2 x_3) \frac{x_1 x_2^2 x_3^3}{27} \, dx_3 \, dx_2 \, dx_1 = \frac{64}{15} \cong 4.2667.$$

Unlike the random quadratic equation problem given in Example 6.61, which had a difficult analytic solution and a simple Monte Carlo solution, this problem takes a bit of work to simulate the expected value. One piece of luck however, is that since the support is a product space and the joint probability density function can be factored into nonnegative functions of x_1, x_2, and x_3 alone, the random variables are mutually independent by a generalization of Theorem 6.1. Leaving the integration details out, the joint probability density function can be written as a product of the marginal probability density functions as

$$f(x_1, x_2, x_3) = f_{X_1}(x_1) f_{X_2}(x_2) f_{X_3}(x_3) = (2x_1) \left(\frac{3}{8} x_2^2 \right) \left(\frac{4}{81} x_3^3 \right) \qquad (x_1, x_2, x_3) \in \mathcal{A},$$

where $\mathcal{A} = \{ (x_1, x_2, x_3) \, | \, 0 < x_1 < 1, 0 < x_2 < 2, 0 < x_3 < 3 \}$. These marginal probability density functions are converted to marginal cumulative distribution functions and inverted, yielding

$$F_{X_1}^{-1}(u) = u^{1/2} \qquad F_{X_2}^{-1}(u) = 2u^{1/3} \qquad F_{X_3}^{-1}(u) = 3u^{1/4}$$

for $0 < u < 1$. These inverse cumulative distribution functions are used to generate the variates X_1, X_2, and X_3 in the R Monte Carlo simulation code given below.

```
nrep = 100000
x1 = runif(nrep) ^ (1 / 2)
x2 = 2 * runif(nrep) ^ (1 / 3)
x3 = 3 * runif(nrep) ^ (1 / 4)
mean(x1 + x2 * x3)
```

Five runs of this simulation code yields

$$4.2571 \qquad 4.2682 \qquad 4.2697 \qquad 4.2710 \qquad 4.2674,$$

which is consistent with the analytic result $E[X_1 + X_2 X_3] = 64/15 \cong 4.2667$.

There are several results concerning expectations of random variables that often arise in probability problems involving several random variables. The proofs of these results are given for continuous random variables. Analogous proofs for discrete random variables can be constructed by simply replacing integrals with summations. The first result involves sums of functions of random variables.

Theorem 6.16 If $X_1, X_2, \ldots, X_n$ are random variables, then

$$E \left[\sum_{i=1}^{n} g_i(X_i) \right] = \sum_{i=1}^{n} E[g_i(X_i)]$$

for any functions $g_1, g_2, \ldots, g_n$.

Proof Let $f(x_1, x_2, \ldots, x_n)$ denote the joint probability density function of the random variables $X_1, X_2, \ldots, X_n$. The expected value of $g_1(X_1) + g_2(X_2) + \cdots + g_n(X_n)$ is

$$
\begin{aligned}
E \left[\sum_{i=1}^{n} g_i(X_i) \right] &= \int_{-\infty}^{\infty} \int_{-\infty}^{\infty} \cdots \int_{-\infty}^{\infty} \left(\sum_{i=1}^{n} g_i(x_i) \right) f(x_1, x_2, \ldots, x_n) \, dx_1 \, dx_2 \ldots dx_n \\
&= \sum_{i=1}^{n} \left(\int_{-\infty}^{\infty} \int_{-\infty}^{\infty} \cdots \int_{-\infty}^{\infty} g_i(x_i) f(x_1, x_2, \ldots, x_n) \, dx_1 \, dx_2 \ldots dx_n \right) \\
&= \sum_{i=1}^{n} E[g_i(X_i)],
\end{aligned}
$$

which proves the result.   □

The most common application of Theorem 6.16 occurs when $g_i(X_i) = X_i$ for $i = 1, 2, \ldots, n$, which results in the special case

$$E[X_1 + X_2 + \cdots + X_n] = E[X_1] + E[X_2] + \cdots + E[X_n].$$

In words, the expected value of the sum of random variables is the sum of their expected values. A second, slightly more general special case, concerns expected values of linear combinations. For real constants $a_1, a_2, \ldots, a_n$,

$$E[a_1X_1 + a_2X_2 + \cdots + a_nX_n] = a_1E[X_1] + a_2E[X_2] + \cdots + a_nE[X_n].$$

These results apply to both mutually independent and dependent random variables.

Example 6.65 Find the population mean of a binomial random variable X with parameters n and p.

The population mean of a binomial random variable was derived in Section 4.2 as a less-than-pleasant summation using the standard definition of the expected value. This problem is reworked here using Theorem 6.16, providing a much cleaner solution. Let $X_1, X_2, \ldots, X_n$ be mutually independent and identically distributed Bernoulli(p) random variables, each with population mean p. Since $X = X_1 + X_2 + \cdots + X_n \sim$ binomial(n, p), its population mean is

$$E[X] = E[X_1 + X_2 + \cdots + X_n] = E[X_1] + E[X_2] + \cdots + E[X_n] = p + p + \cdots + p = np,$$

which matches the result from Section 4.2.

When the X_i's are identically distributed, it is sometimes possible to focus on just one of the X_i values. This is illustrated in the following two examples as arbitrarily choosing X_1.

Example 6.66 If n married couples are sitting at random positions around a round table, what is the expected number of wives sitting next to their husbands?

Let X_i be 0 if couple i is not seated together and let X_i be 1 if couple i is seated together, for $i = 1, 2, \ldots, n$. Since X_i assumes only binary values, it is a Bernoulli random variable and $E[X_i] = P(X_i = 1)$, for $i = 1, 2, \ldots, n$. Fix the position of wife i. The probability that husband i is seated next to her is $\frac{2}{2n-1}$ because the husband occupying two of the equally-likely $2n - 1$ positions (one on her left and one on her right) corresponds to the couple sitting together. Let $X = X_1 + X_2 + \cdots + X_n$ be the number of couples sitting together. The expected number of couples seated together is

$$
\begin{aligned}
E[X] &= E[X_1 + X_2 + \cdots + X_n] \\
&= E[X_1] + E[X_2] + \cdots + E[X_n] \\
&= nE[X_1] \\
&= nP(X_1 = 1) \\
&= n\left[\frac{2}{2n-1}\right] \\
&= \frac{2n}{2n-1}
\end{aligned}
$$

for $n \geq 2$. For $n = 1$ couple, $E[X] = 1$. The reader is encouraged to check this result by hand for the case of $n = 2$ couples by enumerating the $3! = 6$ equally likely circular permutations and computing the expected number of adjacent couples. The expected number of adjacent couples approaches 1 as $n \to \infty$.

Example 6.67 What is the expected number of distinct birthdays for n people?

As usual, assume that all of the people are unrelated, all birthdays are equally likely, and ignore leap years. Let X_i be 0 if day i is not someone's birthday and let X_i be 1 if day i is someone's birthday, for $i = 1, 2, \ldots, 365$. Since X_i assumes only binary values, it is a Bernoulli random variable and $E[X_i] = P(X_i = 1)$, for $i = 1, 2, \ldots, n$. Let the random variable $X = X_1 + X_2 + \cdots + X_{365}$ be the number of distinct birthdays. Thus, the expected number of distinct birthdays is

$$
\begin{aligned}
E[X] &= E[X_1 + X_2 + \cdots + X_{365}] \\
&= E[X_1] + E[X_2] + \cdots + E[X_{365}] \\
&= 365 E[X_1] \\
&= 365 P(X_1 = 1) \\
&= 365 \left[1 - \left(\frac{364}{365} \right)^n \right],
\end{aligned}
$$

where $P(X_1 = 1)$ is the probability that one or more people were born on January 1, which is one minus the probability that none of the n people were born on January 1. Consistent with intuition, $E[X] = 1$ when $n = 1$ and $E[X]$ approaches 365 as $n \to \infty$.

The next result concerns the expected value of products of mutually independent random variables. Unlike Theorem 6.16, which placed no restrictions on the distribution of $X_1, X_2, \ldots, X_n$, the next result requires that $X_1, X_2, \ldots, X_n$ are mutually independent random variables. It is again proven only in the case of continuous random variables.

Theorem 6.17 If $X_1, X_2, \ldots, X_n$ are mutually independent random variables, then

$$
E \left[\prod_{i=1}^{n} g_i(X_i) \right] = \prod_{i=1}^{n} E[g_i(X_i)]
$$

for any functions $g_1, g_2, \ldots, g_n$.

Proof Since $X_1, X_2, \ldots, X_n$ are mutually independent random variables by assumption, their joint probability density function can be expressed as the product of the marginal probability density functions, that is

$$
f(x_1, x_2, \ldots, x_n) = f_{X_1}(x_1) f_{X_2}(x_2) \cdots f_{X_n}(x_n)
$$

over the support $\mathcal{A}$. The expected value of interest is

$$
\begin{aligned}
E \left[\prod_{i=1}^{n} g_i(X_i) \right] &= \int_{-\infty}^{\infty} \int_{-\infty}^{\infty} \cdots \int_{-\infty}^{\infty} \left[\prod_{i=1}^{n} g_i(x_i) \right] f(x_1, x_2, \ldots, x_n) \, dx_1 \, dx_2 \ldots dx_n \\
&= \int_{-\infty}^{\infty} \int_{-\infty}^{\infty} \cdots \int_{-\infty}^{\infty} \left[\prod_{i=1}^{n} g_i(x_i) f_{X_i}(x_i) \right] dx_1 \, dx_2 \ldots dx_n \\
&= \prod_{i=1}^{n} \left[\int_{-\infty}^{\infty} g_i(x_i) f_{X_i}(x_i) dx_i \right] \\
&= \prod_{i=1}^{n} E[g_i(X_i)]
\end{aligned}
$$

which proves the result. $\qquad\square$

The most common application of Theorem 6.17 occurs when $X_1, X_2, \ldots, X_n$ are mutually independent random variables, $g_1(X_1) = X_1, g_2(X_2) = X_2, \ldots, g_n(X_n) = X_n$, which results in the special case

$$E[X_1 X_2 \ldots X_n] = E[X_1] E[X_2] \ldots E[X_n].$$

A second application of Theorem 6.17 occurs when $X_1, X_2, \ldots, X_n$ are mutually independent random variables, $g_1(X_1) = e^{tX_1}, g_2(X_2) = e^{tX_2}, \ldots, g_n(X_n) = e^{tX_n}$, which results in the special case

$$E\left[\prod_{i=1}^{n} e^{tX_i}\right] = \prod_{i=1}^{n} E\left[e^{tX_i}\right]$$

or

$$E\left[e^{t(X_1 + X_2 + \cdots + X_n)}\right] = \prod_{i=1}^{n} E\left[e^{tX_i}\right].$$

Recognizing both sides of this equation as moment generating functions, we conclude that for mutually independent random variables $X_1, X_2, \ldots, X_n$ the moment generating function of the sum of the random variables is the product of the marginal moment generating functions:

$$M_{X_1 + X_2 + \cdots + X_n}(t) = \prod_{i=1}^{n} M_{X_i}(t).$$

This formula will be the topic of the third section of the next chapter.

The results concerning expected values also include variance. One of the most well-known results is given below and illustrated with two examples.

Theorem 6.18 If $X_1, X_2, \ldots, X_n$ are random variables, then

$$V\left[\sum_{i=1}^{n} a_i X_i\right] = \sum_{i=1}^{n} a_i^2 V[X_i] + 2\sum\sum_{i<j} a_i a_j \text{Cov}(X_i, X_j)$$

for any real constants $a_1, a_2, \ldots, a_n$.

Proof Let $\mu_{X_1}, \mu_{X_2}, \ldots, \mu_{X_n}$ denote the population means of $X_1, X_2, \ldots, X_n$. Also, let $X = a_1 X_1 + a_2 X_2 + \cdots + a_n X_n$. To find the population variance of X, begin with the definition of the population variance:

$$
\begin{aligned}
V[X] &= E\left[(X - E[X])^2\right] \\
&= E\left[\left(\sum_{i=1}^{n} a_i X_i - E\left[\sum_{i=1}^{n} a_i X_i\right]\right)^2\right] \\
&= E\left[\left(\sum_{i=1}^{n} a_i(X_i - \mu_{X_i})\right)^2\right] \\
&= E\left[\sum_{i=1}^{n} a_i^2 (X_i - \mu_{X_i})^2 + \sum\sum_{i \neq j} a_i a_j (X_i - \mu_{X_i})(X_j - \mu_{X_j})\right] \\
&= \sum_{i=1}^{n} a_i^2 E\left[(X_i - \mu_{X_i})^2\right] + \sum\sum_{i \neq j} a_i a_j E\left[(X_i - \mu_{X_i})(X_j - \mu_{X_j})\right] \\
&= \sum_{i=1}^{n} a_i^2 V[X_i] + \sum\sum_{i \neq j} a_i a_j \text{Cov}(X_i, X_j) \\
&= \sum_{i=1}^{n} a_i^2 V[X_i] + 2\sum\sum_{i<j} a_i a_j \text{Cov}(X_i, X_j)
\end{aligned}
$$

because E is a linear operator and $\text{Cov}(X_i, X_j) = \text{Cov}(X_j, X_i)$. $\qquad\qquad\square$

There are two special cases of Theorem 6.18 that arise in practice and are useful in problem solving. The first case is when all of the constants $a_1, a_2, \ldots, a_n$ are equal to one. In this case the population variance of the sum $X_1 + X_2 + \cdots + X_n$ is

$$V\left[\sum_{i=1}^{n} X_i\right] = \sum_{i=1}^{n} V[X_i] + 2\sum\sum_{i<j} \text{Cov}(X_i, X_j).$$

This special case also has a nice interpretation in terms of the variance–covariance matrix. Define the variance–covariance matrix of $X_1, X_2, \ldots, X_n$ as the $n \times n$ symmetric, positive semidefinite matrix

$$\Sigma = \begin{pmatrix} V[X_1] & \text{Cov}(X_1, X_2) & \cdots & \text{Cov}(X_1, X_n) \\ \text{Cov}(X_2, X_1) & V[X_2] & \cdots & \text{Cov}(X_2, X_n) \\ \vdots & \vdots & \ddots & \vdots \\ \text{Cov}(X_n, X_1) & \text{Cov}(X_n, X_2) & \cdots & V[X_n] \end{pmatrix}$$

which contains the population variances of the random variables on the diagonal elements and the population covariances between X_i and X_j in the (i, j) position of the matrix. A careful inspection of this first special case of Theorem 6.18 shows that the population variance of the sum equals the sum of the elements in the variance–covariance matrix. .

The second case is when $X_1, X_2, \ldots, X_n$ are mutually independent random variables and the constants $a_1, a_2, \ldots, a_n$ are equal to one. Recall that mutual independence implies pairwise independence, and, from Theorem 6.6, pairwise independence implies that the population covariance is 0. This means that all of the terms in the double summation in Theorem 6.18 vanish, and the expression reduces to

$$V\left[\sum_{i=1}^{n} X_i\right] = \sum_{i=1}^{n} V[X_i].$$

So in the special case of mutual independence, the population variance of a sum is the sum of the population variances.

Example 6.68 Find the population variance of a binomial random variable X with parameters n and p.

The population variance of a binomial random variable was given in Section 4.2 as $\sigma^2 = np(1 - p)$. Theorem 6.18 provides an easy derivation. Let $X_1, X_2, \ldots, X_n$ be mutually independent and identically distributed Bernoulli(p) random variables, each with population variance $p(1 - p)$. Since $X = X_1 + X_2 + \cdots + X_n$ is binomial(n, p), its population variance is

$$V[X] = V[X_1 + X_2 + \cdots + X_n] = V[X_1] + V[X_2] + \cdots + V[X_n] = np(1 - p).$$

Example 6.69 Let $X_1, X_2, \ldots, X_n$ be mutually independent random variables that are identically distributed (statistics textbooks often use the abbreviation *iid* for independent and identically distributed in this case). Assume that $E[X_i] = \mu$ and $V[X_i] = \sigma^2$, for $i = 1, 2, \ldots, n$. Define the *sample mean* of $X_1, X_2, \ldots, X_n$ as

$$\bar{X} = \frac{1}{n}\sum_{i=1}^{n} X_i$$

which is just the arithmetic average of the random variables. Find the population mean and variance of $\bar{X}$.

The sample mean is a random variable because its definition consists of the sum of n random variables in its numerator. The population mean of $\bar{X}$ is

$$
\begin{aligned}
E\left[\bar{X}\right] &= E\left[\frac{1}{n}\sum_{i=1}^{n} X_i\right] \\
&= \frac{1}{n}E\left[\sum_{i=1}^{n} X_i\right] \\
&= \frac{1}{n}\sum_{i=1}^{n} E\left[X_i\right] \\
&= \frac{1}{n}\sum_{i=1}^{n} \mu \\
&= \frac{1}{n}n\mu \\
&= \mu
\end{aligned}
$$

via Theorem 6.16. This is a reassuring result for statisticians. If the sample mean is used to estimate the (typically unknown) population mean μ, then it is "on target" in the sense that its population mean is μ. Statisticians use the term *unbiased* for such an estimator: the sample mean $\bar{X}$ is an unbiased estimator of the population mean μ.

The population variance of $\bar{X}$ is

$$
\begin{aligned}
V\left[\bar{X}\right] &= V\left[\frac{1}{n}\sum_{i=1}^{n} X_i\right] \\
&= \frac{1}{n^2}V\left[\sum_{i=1}^{n} X_i\right] \\
&= \frac{1}{n^2}\sum_{i=1}^{n} V\left[X_i\right] \\
&= \frac{1}{n^2}\sum_{i=1}^{n} \sigma^2 \\
&= \frac{1}{n^2}n\sigma^2 \\
&= \frac{\sigma^2}{n}
\end{aligned}
$$

via the second special case of Theorem 6.18. This is also good news for the statistician. The population variance of $\bar{X}$ goes to zero in the limit as $n \to \infty$. If enough samples are taken, the unbiased estimator $\bar{X}$ has a population variance that approaches zero. This works well, in principle, if sampling is cheap, as it is in Monte Carlo simulation. On the other hand, if each data value collected involves crashing a car into a wall, economics gets in the way of the limit.

Example 6.70 What is the expected number of rolls of a fair die required to see all six outcomes? What is the population variance of the number of rolls of a fair die required to see all six outcomes?

Define the mutually independent discrete random variables $X_1, X_2, \ldots, X_6$ as follows:

- X_1 is the number of rolls to see the first unique number,

- X_2 is the number of additional rolls to see the second unique number,
- X_3 is the number of additional rolls to see the third unique number,
- X_4 is the number of additional rolls to see the fourth unique number,
- X_5 is the number of additional rolls to see the fifth unique number, and
- X_6 is the number of additional rolls to see the sixth unique number.

Based on these definitions, the number of rolls required to see all six outcomes is

$$X = X_1 + X_2 + X_3 + X_4 + X_5 + X_6.$$

Assuming that the geometric distribution is parameterized with a support beginning at 1, the distribution of $X_1, X_2, \ldots, X_6$ is

- X_1 is a degenerate random variable at 1,
- $X_2 \sim \text{Geometric}(5/6)$,
- $X_3 \sim \text{Geometric}(4/6)$,
- $X_4 \sim \text{Geometric}(3/6)$,
- $X_5 \sim \text{Geometric}(2/6)$,
- $X_6 \sim \text{Geometric}(1/6)$.

Since the population mean of a Geometric(p) random variable is $1/p$, the expected number of rolls to see all six outcomes is

$$
\begin{aligned}
E[X] &= E[X_1 + X_2 + X_3 + X_4 + X_5 + X_6] \\
&= E[X_1] + E[X_2] + E[X_3] + E[X_4] + E[X_5] + E[X_6] \\
&= 1 + \frac{6}{5} + \frac{6}{4} + \frac{6}{3} + \frac{6}{2} + 6 \\
&= \frac{147}{10} \\
&= 14.7
\end{aligned}
$$

rolls. Likewise, since the population variance of a Geometric(p) random variable is $(1-p)/p^2$, the population variance of the number of rolls to see all six outcomes is

$$
\begin{aligned}
V[X] &= V[X_1 + X_2 + X_3 + X_4 + X_5 + X_6] \\
&= V[X_1] + V[X_2] + V[X_3] + V[X_4] + V[X_5] + V[X_6] \\
&= 0 + \frac{1/6}{(5/6)^2} + \frac{2/6}{(4/6)^2} + \frac{3/6}{(3/6)^2} + \frac{4/6}{(2/6)^2} + \frac{5/6}{(1/6)^2} \\
&= \frac{3899}{100} \\
&= 38.99
\end{aligned}
$$

squared rolls.

A Monte Carlo simulation in R with 100,000 replications of the experiment is given below.

```
nrep = 100000
x = rep(0, nrep)
for (i in 1:nrep) {
  y = rep(0, 6)
  numrolls = 0
```

```
  while (prod(y) == 0) {
    numrolls = numrolls + 1
    roll = ceiling(runif(1, 0, 6))
    y[roll] = 1
  }
  x[i] = numrolls
}
mean(x)
var(x)
```

The vector x holds the number of rolls necessary to see all six outcomes for the replication. The six elements of the vector y are set to zeros at the beginning of each simulation replication, then switched to ones when the associated number is rolled. Five runs of this simulation yield estimates for the expected number of rolls to see all outcomes of

    14.7172   14.7210   14.7079   14.6661   14.7395

and associated population variances

    39.3205   38.9130   38.9754   38.2586   39.7227

which is consistent with the analytic solution $E[X] = 14.7$ and $V[X] = 38.99$.

The joint moment generating function for n random variables $X_1, X_2, \ldots, X_n$ completely defines the joint probability distribution of the random variables and is defined in a similar fashion to its two-dimensional counterpart.

Definition 6.15 Let $(X_1, X_2, \ldots, X_n)'$ be a random vector. The *joint moment generating function* of $(X_1, X_2, \ldots, X_n)'$ is

$$M(t_1, t_2, \ldots, t_n) = E\left[e^{t_1 X_1 + t_2 X_2 + \cdots + t_n X_n}\right]$$

provided that the expected value exists on $-h_1 < t_1 < h_1, -h_2 < t_2 < h_2, \ldots, -h_n < t_n < h_n$ for some positive constants $h_1, h_2, \ldots, h_n$.

The moment generating function for the marginal distribution of X_i is found by setting the arguments for the variables that are not of interest to 0, that is

$$M(0, 0, \ldots, 0, t_i, 0, \ldots, 0).$$

It is also possible to get the joint moment generating function for any subgroup of random variables in a similar fashion. For example, consider the three random variables X_1, X_2, X_3. The moment generating function for X_1 and X_3 is

$$M(t_1, 0, t_3).$$

Finally, the random variables $X_1, X_2, \ldots, X_n$ are mutually independent if and only if

$$M(t_1, t_2, \ldots, t_n) = \prod_{i=1}^{n} M(0, 0, \ldots, 0, t_i, 0, \ldots, 0).$$

Vectors and matrices can help simplify notation for complex problems involving multidimensional random variables. Let $X = (X_1, X_2, \ldots, X_n)'$ be an $n \times 1$ vector of random variables with finite first and second moments. It is tradition in probability and statistics to write vectors as column vectors, and we keep this tradition here. The population mean of X is the $n \times 1$ vector

$$\mu = E[X]$$

where the E operator is applied component-wise to the elements of X. The variance–covariance matrix of X is the $n \times n$ symmetric, positive semi-definite matrix Σ whose (i, j)th entry is the population covariance between X_i and X_j, for $i = 1, 2, \ldots, n$ and $j = 1, 2, \ldots, n$. So the random variables of interest, their population means, their population variances, and the population covariances between all pairs are described by

$$X = \begin{pmatrix} X_1 \\ X_2 \\ \vdots \\ X_n \end{pmatrix}, \quad \mu = \begin{pmatrix} \mu_{X_1} \\ \mu_{X_2} \\ \vdots \\ \mu_{X_n} \end{pmatrix}, \quad \Sigma = \begin{pmatrix} V[X_1] & \mathrm{Cov}(X_1, X_2) & \cdots & \mathrm{Cov}(X_1, X_n) \\ \mathrm{Cov}(X_2, X_1) & V[X_2] & \cdots & \mathrm{Cov}(X_2, X_n) \\ \vdots & \vdots & \ddots & \vdots \\ \mathrm{Cov}(X_n, X_1) & \mathrm{Cov}(X_n, X_2) & \cdots & V[X_n] \end{pmatrix}.$$

We now illustrate the generalization of two results that should be familiar in the one-dimensional case to the n-dimensional case by using vector and matrix notation.

Example 6.71 The defining formula for the population variance of a one-dimensional random variable X is

$$\sigma^2 = E\left[(X - \mu)^2\right].$$

Is there an analogous, matrix-based formula in n-dimensions?

There is an analogous formula in n-dimensions, which is

$$\Sigma = E\left[(X - \mu)(X - \mu)'\right],$$

where the expected value operator E is assumed to apply component-wise on the elements of the $n \times n$ matrix $(X - \mu)(X - \mu)'$. This is most easily seen by multiplying out the matrix expression inside the expected value operator:

$$(X - \mu)(X - \mu)' = \begin{pmatrix} X_1 - \mu_{X_1} \\ X_2 - \mu_{X_2} \\ \vdots \\ X_n - \mu_{X_n} \end{pmatrix} \begin{pmatrix} X_1 - \mu_{X_1} & X_2 - \mu_{X_2} & \cdots & X_n - \mu_{X_n} \end{pmatrix}.$$

The product of these two vectors yields the $n \times n$ matrix

$$\begin{pmatrix} (X_1 - \mu_{X_1})(X_1 - \mu_{X_1}) & (X_1 - \mu_{X_1})(X_2 - \mu_{X_2}) & \cdots & (X_1 - \mu_{X_1})(X_n - \mu_{X_n}) \\ (X_2 - \mu_{X_2})(X_1 - \mu_{X_1}) & (X_2 - \mu_{X_2})(X_2 - \mu_{X_2}) & \cdots & (X_2 - \mu_{X_2})(X_n - \mu_{X_n}) \\ \vdots & \vdots & \ddots & \vdots \\ (X_n - \mu_{X_n})(X_1 - \mu_{X_1}) & (X_n - \mu_{X_n})(X_2 - \mu_{X_2}) & \cdots & (X_n - \mu_{X_n})(X_n - \mu_{X_n}) \end{pmatrix}.$$

Taking the expected value of each element of this matrix gives the variance–covariance matrix, which validates the formula $\Sigma = E\left[(X - \mu)(X - \mu)'\right]$. The formula collapses to $V[X] = \sigma^2 = E\left[(X - \mu)^2\right]$, as it should when $n = 1$.

Example 6.72 Consider a one-dimensional random variable X with population mean μ_X and population variance σ_X^2. Using the properties of the expected value and population variance operators, the random variable

$$Y = aX + b$$

has population mean

$$\mu_Y = E[Y] = a\mu_X + b$$

and population variance

$$\sigma_Y^2 = V[Y] = a^2\sigma_X^2$$

for real constants a and b. Is there a comparable matrix-based result in n-dimensions?

There is an analogous result in n-dimensions: If the $n \times 1$ vector of random variables X has an $n \times 1$ vector of population means μ_X and variance–covariance matrix Σ_X, then the $m \times 1$ vector of random variables

$$Y = AX + B$$

has population mean

$$\mu_Y = E[Y] = A\mu_X + B$$

and variance–covariance matrix

$$\Sigma_Y = A\Sigma_X A'$$

for an $m \times n$ matrix of real constants A and an $m \times 1$ vector of real constants B.

The matrix result collapses to the 1-dimensional result when $n = 1$ and $m = 1$. The matrix result collapses to special cases of Theorem 6.16 and Theorem 6.18 when $m = 1$. The reader is encouraged to work through the details.

Rather than prove this result in general, we illustrate some of the computations in the case of the linear transformation from the trivariate ($n = 3$) distribution of the random vector $X = (X_1, X_2, X_3)'$ with population mean vector $\mu_X = (\mu_{X_1}, \mu_{X_2}, \mu_{X_3})'$ and variance–covariance matrix Σ_X to a bivariate ($m = 2$) distribution of

$$Y = AX + B$$

via the transformation matrices

$$A = \begin{pmatrix} a_{11} & a_{12} & a_{13} \\ a_{21} & a_{22} & a_{23} \end{pmatrix} \qquad B = \begin{pmatrix} b_1 \\ b_2 \end{pmatrix}.$$

The transformed random variables are

$$Y = AX + B = \begin{pmatrix} a_{11} & a_{12} & a_{13} \\ a_{21} & a_{22} & a_{23} \end{pmatrix} \begin{pmatrix} X_1 \\ X_2 \\ X_3 \end{pmatrix} + \begin{pmatrix} b_1 \\ b_2 \end{pmatrix}$$

or

$$Y = AX + B = \begin{pmatrix} a_{11}X_1 + a_{12}X_2 + a_{13}X_3 + b_1 \\ a_{21}X_1 + a_{22}X_2 + a_{23}X_3 + b_2 \end{pmatrix}.$$

By the result given above, the population mean of Y is

$$\mu_Y = E[Y] = A\mu_X + B = \begin{pmatrix} a_{11} & a_{12} & a_{13} \\ a_{21} & a_{22} & a_{23} \end{pmatrix} \begin{pmatrix} \mu_{X_1} \\ \mu_{X_2} \\ \mu_{X_3} \end{pmatrix} + \begin{pmatrix} b_1 \\ b_2 \end{pmatrix}$$

or

$$\mu_Y = E[Y] = A\mu_X + B = \begin{pmatrix} a_{11}\mu_{X_1} + a_{12}\mu_{X_2} + a_{13}\mu_{X_3} + b_1 \\ a_{21}\mu_{X_1} + a_{22}\mu_{X_2} + a_{23}\mu_{X_3} + b_2 \end{pmatrix}.$$

Using the less-than-perfect notation σ_{ij} to denote the (i, j)th entry of Σ_X, the variance–covariance matrix of X, the variance–covariance matrix of Y is

$$\Sigma_Y = A\Sigma_X A' = \begin{pmatrix} a_{11} & a_{12} & a_{13} \\ a_{21} & a_{22} & a_{23} \end{pmatrix} \begin{pmatrix} \sigma_{11} & \sigma_{12} & \sigma_{13} \\ \sigma_{21} & \sigma_{22} & \sigma_{23} \\ \sigma_{31} & \sigma_{32} & \sigma_{33} \end{pmatrix} \begin{pmatrix} a_{11} & a_{21} \\ a_{12} & a_{22} \\ a_{13} & a_{23} \end{pmatrix}.$$

Multiplying these matrices results in a 2×2 matrix with each element consisting of a sum of nine terms

$$\Sigma_Y = A\Sigma_X A' = \begin{pmatrix} \sum_{i=1}^{3}\sum_{j=1}^{3} a_{1i}a_{1j}\sigma_{ij} & \sum_{i=1}^{3}\sum_{j=1}^{3} a_{1i}a_{2j}\sigma_{ij} \\ \sum_{i=1}^{3}\sum_{j=1}^{3} a_{2i}a_{1j}\sigma_{ij} & \sum_{i=1}^{3}\sum_{j=1}^{3} a_{2i}a_{2j}\sigma_{ij} \end{pmatrix}.$$

The diagonal elements of Σ_Y are correct by Theorem 6.18. The correctness of the off-diagonal elements is left as an exercise.

No discussion of multivariate distributions would be complete without an introduction to the *multivariate normal distribution*. The multivariate normal distribution is typically the first distribution of choice for modeling n correlated continuous random variables $X_1, X_2, \ldots, X_n$ with bell-shaped marginal probability density functions. As was the case with the bivariate normal distribution, the multivariate normal distribution has elegant mathematical properties and practical applications that make it the distribution of choice for many modeling situations. Matrix theory will again be used to shorten the expressions used to describe the multivariate normal distribution. Consider an $n \times 1$ vector of random variables $X = (X_1, X_2, \ldots, X_n)'$, an associated $n \times 1$ vector of population means $\mu = (\mu_1, \mu_2, \ldots, \mu_n)'$, and an associated variance–covariance matrix Σ. The random vector X has the multivariate normal distribution if its joint probability density function has the form

$$f(x_1, x_2, \ldots, x_n) = \frac{1}{(2\pi)^{n/2}|\Sigma|^{1/2}}\, e^{-\frac{1}{2}(\boldsymbol{x}-\boldsymbol{\mu})'\Sigma^{-1}(\boldsymbol{x}-\boldsymbol{\mu})} \qquad (x_1, x_2, \ldots, x_n) \in \mathcal{A},$$

where

- $\boldsymbol{x} = (x_1, x_2, \ldots, x_n)'$,

- Σ^{-1} is the inverse of the variance–covariance matrix,

- $|\Sigma|$ is the determinant of the variance–covariance matrix,

- the support is

$$\mathcal{A} = \{(x_1, x_2, \ldots, x_n) \,|\, -\infty < x_i < \infty,\ \text{for } i = 1, 2, \ldots, n\},$$

- the parameter space is

$$\Omega = \{(\mu, \Sigma) \,|\, \mu \in \mathcal{R}^n, \Sigma \text{ is an } n \times n \text{ symmetric, positive semi-definite matrix}\}.$$

Just as in the univariate case when

$$X \sim N(\mu, \sigma^2)$$

was used as shorthand to indicate that X was normally distributed with population mean μ and population variance σ^2, the shorthand

$$X \sim N(\mu, \Sigma)$$

is used to indicate that the vector X is normally distributed with population mean vector μ and variance–covariance matrix Σ. A multivariate normal distribution is defined by n population means in the vector μ, n population variances on the diagonal elements of Σ, and $(n-1)n/2$ population covariances on the off-diagonal diagonal of Σ. Dividing by two accounts for the fact that Σ is symmetric. So the multivariate normal distribution is defined by

$$n + n + \frac{(n-1)n}{2} = \frac{n^2 + 3n}{2}$$

parameters. In the univariate normal case ($n = 1$), the two parameters are μ and σ^2. In the bivariate normal case ($n = 2$), the five parameters are μ_{X_1}, μ_{X_2}, $\sigma_{X_1}^2$, $\sigma_{X_2}^2$, and ρ. The reader is encouraged to check and see that the joint probability density function reduces to the univariate normal probability density function

$$f(x) = \frac{1}{\sqrt{2\pi}\sigma} e^{-\frac{1}{2}\left(\frac{x-\mu}{\sigma}\right)^2} \qquad -\infty < x < \infty$$

when $n = 1$ and the bivariate normal probability density function

$$f(x_1, x_2) = \frac{1}{2\pi\sigma_{X_1}\sigma_{X_2}\sqrt{1-\rho^2}} e^{-\frac{1}{2(1-\rho^2)}\left[\left(\frac{x_1-\mu_{X_1}}{\sigma_{X_1}}\right)^2 - 2\rho\left(\frac{x_1-\mu_{X_1}}{\sigma_{X_1}}\right)\left(\frac{x_2-\mu_{X_2}}{\sigma_{X_2}}\right) + \left(\frac{x_2-\mu_{X_2}}{\sigma_{X_2}}\right)^2\right]}$$

on the support $\mathcal{A} = \{(x_1, x_2) \mid -\infty < x_1 < \infty, -\infty < x_2 < \infty\}$ when $n = 2$.

The joint moment generating function for $\boldsymbol{X} \sim N(\boldsymbol{\mu}, \boldsymbol{\Sigma})$ is

$$M(t_1, t_2, \ldots, t_n) = E\left[e^{t'X}\right] = e^{t'\mu} e^{t'\Sigma t/2}$$

for $-\infty < t_i < \infty$, $i = 1, 2, \ldots, n$, where $\boldsymbol{t} = (t_1, t_2, \ldots, t_n)'$. Equating all of the arguments except t_i to 0 in the joint moment generating function yields

$$M(0, 0, \ldots, t_i, \ldots, 0) = e^{t_i\mu_i} e^{t_i^2\sigma_i^2/2}$$

for $-\infty < t_i < \infty$, $i = 1, 2, \ldots, n$. Using the moment generating function for a univariate normal random variable derived in Section 5.4, this particular moment generating function can be recognized as that of a $N(\mu_i, \sigma_i^2)$ random variable. This implies that all marginal distributions of the multivariate normal distribution are univariate normal, that is

$$X_i \sim N(\mu_i, \sigma_i^2)$$

for $i = 1, 2, \ldots, n$.

In the univariate case, the notion of a "standard" normal random variable Z with population mean 0 and population variance 1 was introduced. Is there a similar notion for the multivariate normal distribution? If the population mean vector $\boldsymbol{\mu}$ consists entirely of zeros and the variance–covariance matrix $\boldsymbol{\Sigma}$ is the identity matrix $\boldsymbol{I}$ (all ones on the diagonal elements and all zeros on the off-diagonal elements so that all marginal distributions have population variance 1 and all population covariances between pairs of random variables are 0), that is

$$\boldsymbol{Z} \sim N(\boldsymbol{0}, \boldsymbol{I})$$

then $\boldsymbol{Z}$ has the *standard multivariate normal distribution*. Consistent with the univariate notation, define the random vector $\boldsymbol{Z} = (Z_1, Z_2, \ldots, Z_n)'$, where $Z_1, Z_2, \ldots, Z_n$ are mutually independent $N(0, 1)$ random variables. The joint probability density function of $\boldsymbol{Z}$ is

$$f_{Z_1, Z_2, \ldots, Z_n}(z_1, z_2, \ldots, z_n) = \prod_{i=1}^{n} \frac{1}{\sqrt{2\pi}} e^{-\frac{1}{2}z_i^2} = \left(\frac{1}{2\pi}\right)^{n/2} e^{-\frac{1}{2}\sum_{i=1}^{n} z_i^2} = \left(\frac{1}{2\pi}\right)^{n/2} e^{-\frac{1}{2}z'z}$$

on $\mathcal{A} = \{(z_1, z_2, \ldots, z_n) \mid (z_1, z_2, \ldots, z_n) \in \mathcal{R}^n\}$ where $\boldsymbol{z} = (z_1, z_2, \ldots, z_n)'$. Using mutual independence, the definition of the joint moment generating function, and Theorem 6.16, the joint moment generating function of $Z_1, Z_2, \ldots, Z_n$ is

$$
\begin{aligned}
M_{Z_1, Z_2, \ldots, Z_n}(t_1, t_2, \ldots, t_n) &= E\left[e^{t_1 Z_1 + t_2 Z_2 + \cdots + t_n Z_n}\right] \\
&= E\left[e^{t_1 Z_1}\right] E\left[e^{t_2 Z_2}\right] \ldots E\left[e^{t_n Z_n}\right] \\
&= e^{t_1^2/2} e^{t_2^2/2} \ldots e^{t_n^2/2} \\
&= e^{t't/2}
\end{aligned}
$$

for $-\infty < t_i < \infty$, $i = 1, 2, \ldots, n$. This is the same joint moment generating function that would be obtained by using $\mu = 0$ and $\Sigma = I$ in the general expression for the joint moment generating function.

There are several results for the normal distribution in the univariate case that generalize nicely to the multivariate case. This paragraph outlines one such instance. In the univariate case, if X is a normal random variable with population mean μ and population variance σ^2, that is

$$X \sim N(\mu, \sigma^2),$$

then

$$Y = aX + b \sim N(a\mu + b, a^2\sigma^2),$$

where a and b are real constants, as shown in Theorem 5.3. In the multivariate case, if $X = (X_1, X_2, \ldots, X_n)'$ is an $n \times 1$ vector of random variables having the multivariate normal distribution with population mean $\mu = (\mu_1, \mu_2, \ldots, \mu_n)'$ and variance–covariance matrix Σ, that is

$$X \sim N(\mu, \Sigma),$$

then

$$Y = AX + B \sim N(A\mu + B, A\Sigma A')$$

where $Y = (Y_1, Y_2, \ldots, Y_m)'$ is an $m \times 1$ vector of random variables, A is an $m \times n$ matrix of constants, and B is an $m \times 1$ vector of constants. This linear transformation reduces to the univariate case when $n = 1$ and $m = 1$.

As was the case with the bivariate normal distribution, the symmetric, positive semi-definite variance–covariance matrix Σ can always be written as

$$\Sigma = \Gamma'\Lambda\Gamma,$$

where Λ is an $n \times n$ diagonal matrix with the eigenvalues of Σ in decreasing order of magnitude on the diagonal, and Γ' is an $n \times n$ matrix with the associated normalized eigenvectors of Σ as columns. In symbols, $\Lambda = \text{diag}(\lambda_1, \lambda_2, \ldots, \lambda_n)$, where $\lambda_1 \geq \lambda_2 \geq \cdots \geq \lambda_n \geq 0$ are the eigenvalues of Σ. This decomposition of the variance–covariance matrix is known as *spectral decomposition*. The matrix Γ is an *orthonormal* matrix, that is, $\Gamma^{-1} = \Gamma'$, which implies that $\Gamma\Gamma' = I$. Define the square root of the variance–covariance matrix to be

$$\Sigma^{1/2} = \Gamma'\Lambda^{1/2}\Gamma,$$

where $\Lambda^{1/2}$ is a diagonal matrix with the square roots of the eigenvalues in decreasing order of magnitude on the diagonal. Finally, define

$$\left(\Sigma^{1/2}\right)^{-1} = \Sigma^{-1/2} = \Gamma'\Lambda^{-1/2}\Gamma.$$

These matrix manipulations are illustrated for a trivariate normal distribution in the next example.

Example 6.73 Consider the trivariate normal distribution ($n = 3$) with arbitrary population mean vector $\mu = (\mu_1, \mu_2, \mu_3)'$ and variance–covariance matrix

$$\Sigma = \begin{bmatrix} 6 & -2 & 0 \\ -2 & 4 & 2 \\ 0 & 2 & 6 \end{bmatrix}.$$

This variance–covariance matrix can be decomposed in Maple via the statements (lists read row-wise)

```
with(linalg);
Sigma := matrix(3, 3, [6, -2, 0, -2, 4, 2, 0, 2, 6]);
eigenvalues(Sigma);
eigenvectors(Sigma);
```

yielding the eigenvalues $\lambda_1 = 8$, $\lambda_2 = 6$, $\lambda_3 = 2$ and associated eigenvectors and normalized eigenvectors:

$$\begin{pmatrix} -1 \\ 1 \\ 1 \end{pmatrix}, \begin{pmatrix} 1 \\ 0 \\ 1 \end{pmatrix}, \begin{pmatrix} -1 \\ -2 \\ 1 \end{pmatrix} \text{ and } \begin{pmatrix} -1/\sqrt{3} \\ 1/\sqrt{3} \\ 1/\sqrt{3} \end{pmatrix}, \begin{pmatrix} 1/\sqrt{2} \\ 0 \\ 1/\sqrt{2} \end{pmatrix}, \begin{pmatrix} -1/\sqrt{6} \\ -2/\sqrt{6} \\ 1/\sqrt{6} \end{pmatrix}.$$

So the decomposition is

$$\Sigma = \Gamma'\Lambda\Gamma = \begin{bmatrix} -1/\sqrt{3} & 1/\sqrt{2} & -1/\sqrt{6} \\ 1/\sqrt{3} & 0 & -2/\sqrt{6} \\ 1/\sqrt{3} & 1/\sqrt{2} & 1/\sqrt{6} \end{bmatrix} \begin{bmatrix} 8 & 0 & 0 \\ 0 & 6 & 0 \\ 0 & 0 & 2 \end{bmatrix} \begin{bmatrix} -1/\sqrt{3} & 1/\sqrt{3} & 1/\sqrt{3} \\ 1/\sqrt{2} & 0 & 1/\sqrt{2} \\ -1/\sqrt{6} & -2/\sqrt{6} & 1/\sqrt{6} \end{bmatrix}.$$

The purpose of the previous discussion concerning the spectral decomposition of the variance–covariance matrix Σ is to provide an analogy from an important transformation in the univariate normal case. If $X \sim N(\mu, \sigma^2)$, the transformation

$$Z = \frac{X - \mu}{\sigma}$$

converts a normal random variable X to a standard normal random variable Z. Likewise, if $Z \sim N(0, 1)$, the transformation

$$X = \mu + \sigma Z$$

converts a standard normal random variable Z to a normal random variable X. These transformations can be generalized to the multivariate normal distribution. If $X \sim N(\mu, \Sigma)$, the transformation

$$Z = \Sigma^{-1/2}(X - \mu)$$

converts a normal random vector X to a standard normal random vector Z. Likewise, if $Z \sim N(0, I)$, the transformation

$$X = \mu + \Sigma^{1/2} Z$$

converts a standard normal random vector Z to a normal random vector X. These results have practical applications in random variate generation for Monte Carlo simulation.

When the $n \times 1$ random vector $X = (X_1, X_2, \ldots, X_n)'$, which has the multivariate normal distribution with population mean μ and variance–covariance matrix Σ has its elements partitioned into the first m elements and the other $n - m$ elements, then each portion of the vector is also multivariate normal. More specifically, if X, μ, and Σ can be partitioned as:

$$X = \begin{bmatrix} X_1 \\ X_2 \end{bmatrix}, \mu = \begin{bmatrix} \mu_1 \\ \mu_2 \end{bmatrix}, \Sigma = \begin{bmatrix} \Sigma_{11} & \Sigma_{12} \\ \Sigma_{21} & \Sigma_{22} \end{bmatrix}.$$

then

$$X_1 \sim N(\mu_1, \Sigma_{11})$$

and

$$X_2 \sim N(\mu_2, \Sigma_{22}).$$

Furthermore, X_1 and X_2 are independent if and only if $\Sigma_{12} = 0$.

Entire books have been devoted to the properties of the multivariate normal distribution. Some of these properties link the multivariate normal distribution to distributions encountered earlier in the text. Here is one final example. Let $X = (X_1, X_2, \ldots, X_n)' \sim N(\boldsymbol{\mu}, \boldsymbol{\Sigma})$, where $\boldsymbol{\Sigma}$ is positive definite. The random variable

$$(X - \boldsymbol{\mu})' \boldsymbol{\Sigma}^{-1} (X - \boldsymbol{\mu}) \sim \chi^2(n),$$

where $\chi^2(n)$ is the chi-square distribution with n degrees of freedom. .

We end the chapter with a longer example from a field of study known as *queueing theory* which analyzes the dynamics of waiting lines. More specifically, queueing theory uses probability to model the dynamics of customers arriving to a system with one or more waiting lines in order to receive service from one or more servers.

One of the most fundamental queue models that arises is known in queueing theory as the $M/M/1$ queue. Each letter or number in this sequence has meaning to researchers and practitioners in queueing theory. The first field defines the stochastic mechanism associated with the arrival stream of customers to the queue. In the case of an $M/M/1$ queue, the initial M, which stands for "Markov," indicates that the times between arrivals are a sequence of independent exponential(λ) random variables. In other words, the arrival process is a Poisson process with arrival rate λ. The second field defines the distribution of the service times. In the case of an $M/M/1$ queue, the second M also stands for "Markov," and indicates that the service times are a sequence of independent exponential(μ) random variables. (The unfortunate notational choice of μ to denote the service rate conflicts, of course, with the symbol that is universally used for the population mean.) The third field defines the number of parallel servers at the queue. In the case of an $M/M/1$ queue, the 1 implies that there is a single server. To summarize, the three symbols that describe the $M/M/1$ queue have the following meanings:

- the times between customer arrivals are mutually independent exponential(λ) random variables,

- the service times are mutually independent exponential(μ) random variables, and

- there is a single server.

There are other implicit assumptions associated with an $M/M/1$ queue, for example,

- the *queue discipline* which defines the order in which customers wait in line is first come, first served (also known as FIFO, which stands for first in, first out),

- the *queue capacity* is infinite; the line lengthens to whatever length is needed,

- there is no reneging, that is, no customer leaves the line after entering it,

- there is no passing in the queue,

- the delay time between processing customers is negligible, and

- the server never takes any breaks.

Queueing theorists use the term *sojourn time* to describe the time in the system for a customer (the time lapse between the arrival time of a customer and the departure time after receiving service). The sojourn time is the sum of the time spent waiting in the queue (if any) plus the time in service. Our interest here is focused on the sojourn times of the first two customers, denoted by T_1 and T_2. In particular, we would like to derive the joint probability density function of T_1 and T_2, then calculate the population covariance between T_1 and T_2. To begin this derivation, we first prove a general theorem involving mutually independent exponential random variables which will be useful in deriving the joint probability density function of the first two sojourn times. Assume that the system begins with the queue empty and the server idle.

Theorem 6.19 Let $X_1 \sim$ exponential(λ_1), $X_2 \sim$ exponential(λ_2), and $X_3 \sim$ exponential(λ_3) be mutually independent random variables. The joint probability density function of

$$(T_1, T_2) = (X_1 + X_2, X_1 + X_3)$$

is

$$f_{T_1,T_2}(t_1, t_2) = \begin{cases} \dfrac{\lambda_1\lambda_2\lambda_3 \left(e^{\lambda_1 t_1} - e^{(\lambda_2+\lambda_3)t_1}\right) e^{-\lambda_1 t_1 - \lambda_2 t_1 - \lambda_3 t_2}}{\lambda_1 - \lambda_2 - \lambda_3} & 0 < t_1 < t_2 \\[4mm] \dfrac{\lambda_1\lambda_2\lambda_3 \left(e^{\lambda_1 t_2} - e^{(\lambda_2+\lambda_3)t_2}\right) e^{-\lambda_2 t_1 - \lambda_1 t_2 - \lambda_3 t_2}}{\lambda_1 - \lambda_2 - \lambda_3} & 0 < t_2 < t_1. \end{cases}$$

Proof By conditioning, the joint cumulative distribution function of T_1 and T_2 is

$$
\begin{aligned}
F_{T_1,T_2}(t_1, t_2) &= P(T_1 \le t_1, T_2 \le t_2) \\
&= P(X_1 + X_2 \le t_1, X_1 + X_3 \le t_2) \\
&= P(X_2 \le t_1 - X_1, X_3 \le t_2 - X_1) \\
&= \int_0^{\min\{t_1,\, t_2\}} P(X_2 \le t_1 - x_1, X_3 \le t_2 - x_1 \mid X_1 = x_1) f_{X_1}(x_1) dx_1 \\
&= \int_0^{\min\{t_1,\, t_2\}} P(X_2 \le t_1 - x_1 \mid X_1 = x_1) P(X_3 \le t_2 - x_1 \mid X_1 = x_1) \cdot f_{X_1}(x_1) dx_1 \\
&= \int_0^{\min\{t_1,\, t_2\}} \left(1 - e^{-\lambda_2(t_1 - x_1)}\right)\left(1 - e^{-\lambda_3(t_2 - x_1)}\right) \lambda_1 e^{-\lambda_1 x_1} dx_1 \\
&= \begin{cases} \displaystyle\int_0^{t_1} \left(1 - e^{-\lambda_2(t_1 - x_1)}\right)\left(1 - e^{-\lambda_3(t_2 - x_1)}\right) \lambda_1 e^{-\lambda_1 x_1} dx_1 & 0 < t_1 < t_2 \\[4mm] \displaystyle\int_0^{t_2} \left(1 - e^{-\lambda_2(t_1 - x_1)}\right)\left(1 - e^{-\lambda_3(t_2 - x_1)}\right) \lambda_1 e^{-\lambda_1 x_1} dx_1 & 0 < t_2 < t_1. \end{cases}
\end{aligned}
$$

After evaluating the integrals and differentiating, the joint probability density function of T_1 and T_2 is

$$f_{T_1,T_2}(t_1, t_2) = \begin{cases} \dfrac{\lambda_1\lambda_2\lambda_3 \left(e^{\lambda_1 t_1} - e^{(\lambda_2+\lambda_3)t_1}\right) e^{-\lambda_1 t_1 - \lambda_2 t_1 - \lambda_3 t_2}}{\lambda_1 - \lambda_2 - \lambda_3} & 0 < t_1 < t_2 \\[4mm] \dfrac{\lambda_1\lambda_2\lambda_3 \left(e^{\lambda_1 t_2} - e^{(\lambda_2+\lambda_3)t_2}\right) e^{-\lambda_2 t_1 - \lambda_1 t_2 - \lambda_3 t_2}}{\lambda_1 - \lambda_2 - \lambda_3} & 0 < t_2 < t_1, \end{cases}$$

which proves the theorem.   □

We now return to the queueing problem.

Example 6.74 Find the joint probability density function of T_1 and T_2, the sojourn times of the first two customers in an $M/M/1$ queue with arrival rate λ and service rate μ. Use this joint probability density function to determine the population covariance between T_1 and T_2.

When considering the first two customer's sojourn times, the experience of all customers that arrive to the system after the first two customers is irrelevant. In order to determine the joint distribution of the first two sojourn times T_1 and T_2, we partition the outcomes into the following two distinct scenarios.

- **Scenario 1**. *The first customer departs before the second customer arrives.* This is the easier of the two scenarios to handle mathematically. Both customers spend no time waiting in the queue. This means that T_1 and T_2 are independent exponential(μ) random variables. Their joint probability density function is simply the product of their marginal probability density functions.

- **Scenario 2**. *The first customer departs after the second customer arrives.* This is the more difficult of the two scenarios to handle mathematically because part of the first customer's service time is the second customer's queueing time. As in Scenario 1, $T_1 \sim$ exponential(μ), but T_2 is now the sum of customer 2's wait and service times.

The moment that the first customer arrives to the system, there is a competition for which event will occur next. Either the first customer will depart next (which is Scenario 1), which occurs with probability

$$p_1 = \frac{\mu}{\lambda + \mu}$$

or the second customer will arrive next (which is Scenario 2), which occurs with probability

$$p_2 = \frac{\lambda}{\lambda + \mu}.$$

The two probabilities were found using the result derived in part (b) of Example 6.25. This means that the joint probability density function of T_1 and T_2 is a *mixture* of the joint probability density functions associated with Scenarios 1 and 2 using the mixing probabilities p_1 and p_2.

As alluded to earlier, the joint probability density function of T_1 and T_2 under Scenario 1 is the product of the marginal probability density functions:

$$f_{T_1, T_2}(t_1, t_2) = f_{T_1}(t_1) \cdot f_{T_2}(t_2) = \mu e^{-\mu t_1} \cdot \mu e^{-\mu t_2} \qquad t_1 > 0, \, t_2 > 0.$$

To find the joint probability density function of T_1 and T_2 under Scenario 2 requires the use of some properties of the exponential distribution.

As shown by the hash marks on the time axis in Figure 6.26, there are four distinct events that occur over time in Scenario 2:

- customer 1 arrives,
- customer 2 arrives,
- customer 1 departs, and
- customer 2 departs.

The sojourn times T_1 and T_2 and the X_i values from Theorem 6.19 are labeled in Figure 6.26. The three time segments have the following probability distributions.

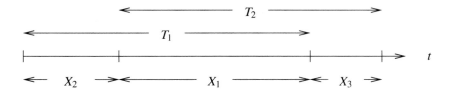

Figure 6.26: $M/M/1$ event times under Scenario 2.

- $X_2 \sim$ exponential$(\lambda + \mu)$ by part (a) of Example 6.25.

- $X_1 \sim$ exponential(μ) by the memoryless property of the exponential distribution given in Theorem 5.1. The remaining service time of the first customer, which is also the waiting time in the queue of the second customer, is exponential(μ).

- $X_3 \sim$ exponential(μ) because X_3 is the service time for customer 3.

Theorem 6.19 can now be invoked to determine the joint probability density function of T_1 and T_2 in Scenario 2. Substituting $\lambda_1 = \mu$, $\lambda_2 = \lambda + \mu$, and $\lambda_3 = \mu$ into the mixture of Scenarios 1 and 2 yields the joint probability density function of T_1 and T_2 as

$$
f_{T_1, T_2}(t_1, t_2) = \begin{cases} \dfrac{\mu^2 \left(\lambda e^{-\mu t_2} + \mu e^{-\lambda t_1 - \mu t_1 - \mu t_2} \right)}{\lambda + \mu} & 0 < t_1 \leq t_2 \\[4mm] \dfrac{\mu^2 \left(\lambda e^{-\lambda t_1 - \mu t_1 + \lambda t_2} + \mu e^{-\lambda t_1 - \mu t_1 - \mu t_2} \right)}{\lambda + \mu} & 0 < t_2 < t_1. \end{cases}
$$

Using this joint probability density function, the population covariance between the sojourn times of customers 1 and 2 is

$$
\mathrm{Cov}(T_1, T_2) = \frac{\lambda(\lambda + 2\mu)}{(\lambda + \mu)^2 \mu^2}.
$$

6.6   Exercises

6.1   A bag contains three balls numbered 1, 2, and 3. Two balls are sampled *with replacement*. Let X_1 be the number on the first ball sampled and let X_2 be the number on the second ball sampled. Find the probability mass function of the sample mean

$$
\frac{X_1 + X_2}{2}.
$$

6.2   A bag contains three balls numbered 1, 2, and 3. Two balls are sampled *without replacement*. Let X_1 be the number on the first ball sampled and let X_2 be the number on the second ball sampled. Find the probability mass function of the sample mean

$$
\frac{X_1 + X_2}{2}.
$$

6.3   The continuous random variables X and Y have joint probability density function

$$
f(x, y) = \frac{1}{100} \qquad 0 < x < 10, 0 < y < 10.
$$

Find $P(2X + Y < 12)$.

6.4   The continuous random variables X and Y have joint probability density function

$$
f(x, y) = \frac{1}{2} \qquad |x| + |y| < 1.
$$

Find $P(X > 1/2)$.

6.5   The random variables X and Y are uniformly distributed over the *interior* of a circle of radius 1 centered at the origin. Find the marginal probability density function $f_X(x)$.

6.6 A fair coin is flipped three times. Let X be the number of heads observed and Y be the number of tails observed. Find the joint probability mass function of X and Y.

6.7 For random variables X and Y with joint cumulative distribution function $F(x, y)$, write $P(a < X \le b, c < Y \le d)$ in terms of the joint cumulative distribution function.

6.8 Consider three points chosen randomly and independently on the circumference of a circle. What is the probability that the center of the circle is in the interior of the triangle obtained by connecting the three points?

6.9 Mama Mia's Memoryless Market has two types of checkouts: regular and self-serve. The checkout time (in minutes) for the regular checkout is exponential(1). The checkout time (in minutes) for the self-serve checkout is exponential(2). What is the probability that a regular checkout time is more than 3 minutes longer than a self-serve checkout time?

6.10 The bivariate Loehr distribution is uniformly distributed over the L-shaped polygon with vertices $(0, 0)$, $(0, 1)$, $(1/k, 1)$, $(1/k, 1/k)$, $(1, 1/k)$, and $(1, 0)$, in the xy plane, where k is a large positive integer.

  (a) Find $f(x, y)$.
  (b) Find $f_X(x)$.

6.11 Consider the R code below which generates random variates x and y.

```
x = runif(1, 0, 10)
y = runif(1, x, 10)
```

The R variable y is a realization of a continuous random variable Y. Find $f_Y(y)$.

6.12 Consider the random variables X and Y with joint probability density function

$$f(x, y) = 2 \qquad 0 < x < y < 1.$$

Find $F(x, y)$ for any point (x, y) in the support set $\mathcal{A}$. Use a calculus and a non-calculus (geometrical) approach to solve this problem.

6.13 Let X denote the number of tosses of a fair coin required to obtain the first occurrence of a "head." Let Y denote the number of rolls of a fair die required to obtain the first occurrence of a "2." Find $P(X \le Y)$.

6.14 Let the ordered pair (X, Y) be a randomly selected point on the *circumference* of a unit circle centered at the origin. Find the joint probability density function $f_{X,Y}(x, y)$.

6.15 Find $F(x, y)$ for the random variables X and Y with joint probability density function

$$f(x, y) = 2 \qquad x > 0, y > 0, x + y < 1.$$

6.16 Let X_1 be uniformly distributed between zero and one. Let x_1 be the realization of X_1 (that is, x_1 is a sample value of X_1). Let X_2 be uniformly distributed between zero and x_1.

  (a) Find the marginal probability density function of X_2.
  (b) Find the population mean of X_2 and execute a Monte Carlo simulation that supports the value that you derive analytically.

6.17 Let X and Y be continuous random variables with joint probability density function $f(x, y)$ defined on $\mathcal{A} = \{(x, y) \mid -\infty < x < \infty, y > |x|\}$. Set up expressions (with appropriate integration limits and support) for the marginal distributions of X and Y.

6.18   What is the probability that there are more spades than hearts in a five-card poker hand?

6.19   Kent leaves for work every day between 5:00 AM and 5:10 AM. He needs to be at work by 6:00 AM. He can take the highway or cut through town. The travel time to work via the highway is between 45 and 55 minutes. The travel time to work by cutting through town is between 40 and 60 minutes. Assuming that his departure and travel times are uniform on the given ranges, find the probability he will be on time by the two routes.

6.20   Let X be a Poisson random variable with a random parameter Λ, which has probability density function

$$f_\Lambda(\lambda) = \lambda e^{-\lambda} \qquad \lambda > 0.$$

Find the probability mass function of X. *Hint*: The conditional probability mass function of X given λ is $f_{X\,|\,\Lambda=\lambda}(x\,|\,\Lambda=\lambda) = \lambda^x e^{-\lambda}/x!$ for $x = 0, 1, 2, \ldots$, and $\lambda > 0$.

6.21   Let X_1 and X_2 be the numbers on two balls drawn randomly from a bag of billiard balls numbered $1, 2, \ldots, 15$. Find the joint probability mass function of $Y_1 = \min\{X_1, X_2\}$ and $Y_2 = \max\{X_1, X_2\}$ and the marginal probability mass function of Y_1 when

   (a) sampling is without replacement,

   (b) sampling is with replacement.

6.22   For the random variables X and Y with joint probability density function

$$f(x, y) = 2 \qquad\qquad 0 < x < y < 1,$$

find the joint cumulative distribution function for all real values of x and y.

6.23   Let X and Y be uniformly distributed on the unit disk with joint probability density function

$$f(x, y) = \frac{1}{\pi} \qquad\qquad x^2 + y^2 < 1.$$

   (a) Find $P(|X| + |Y| > 1)$ without using calculus.

   (b) How many separate regions is the joint cumulative distribution function $F(x, y)$ defined on? *Hint* for part (b): Do not find the joint cumulative distribution function here—just find the number of regions it is defined on. Here is an example of the number of regions where the cumulative distribution function is defined: the joint cumulative distribution function of the product of two independent, identically distributed exponential(λ) random variables, for example, is defined on *two* separate regions:

$$F(x, y) = \begin{cases} 0 & x \le 0 \text{ or } y \le 0 \\ 1 - e^{-\lambda(x+y)} & x > 0, y > 0. \end{cases}$$

6.24   Let the random variables X and Y have joint probability density function

$$f(x, y) = \frac{1}{\pi} e^{-\frac{1}{2}(x^2 + y^2)} \qquad\qquad x > 0, y < 0 \text{ or } x < 0, y > 0.$$

Find the marginal probability density functions. *Hint*: consider $\mathcal{A}$ carefully.

6.25   Consider the random variables X and Y with joint probability density function

$$f(x, y) = \frac{1}{\pi} \qquad\qquad x^2 + y^2 < 1.$$

   (a) Find $P\left(X^2 + Y^2 \le r^2\right)$ for any real $r \ge 0$.

   (b) Set up, but do not evaluate, expressions for $P(X + Y \le r)$ for any real r.

6.26 Let X_1 and X_2 be continuous random variables with joint probability density function

$$f(x_1, x_2) = \frac{1}{9} \qquad 0 < x_2 < x_1^2 < 9, x_1 > 0.$$

Find the marginal distribution for X_1.

6.27 Consider the Cartesian coordinate system. Li-Hsing selects a point that is uniformly distributed between the origin and $(1, 0)$. Huarng selects a point that is uniformly distributed between the origin and $(0, 1)$, independently of Li-Hsing's choice. Find the population median of the distribution of the distance between the two points.

6.28 Consider the random variables X and Y with joint probability density function

$$f(x, y) = cxy \qquad 0 < y < x < 1.$$

(a) What is the value of the constant c?

(b) What is $P(X > 1/2)$?

(c) What is $P(Y > X^2)$?

6.29 There are five balls in a bag numbered 1, 2, 3, 4, and 5. Two balls are drawn at random without replacement. Let X be the smaller number chosen and let Y be the larger number chosen.

(a) What is the joint probability mass function of X and Y?

(b) What is the probability mass function of $Z = Y - X$?

6.30 Let (x_i, y_i), for $i = 1, 2, 3$, denote the vertices of a triangle on the Cartesian coordinate system. Let the random ordered pair (X, Y) be uniformly distributed over the triangle (that is, the joint probability density function of X and Y is the reciprocal of the area of the triangle). Find the marginal distribution of X, $f_X(x)$. *Hint 1*: You may assume, without loss of generality, that $x_1 \leq x_2 \leq x_3$, although you must explicitly consider the cases $x_1 = x_2$ and $x_2 = x_3$. *Hint 2*: For a triangle with side lengths a, b, and c, and semi-perimeter $s = (a+b+c)/2$, Heron's formula gives the area as $\sqrt{s(s-a)(s-b)(s-c)}$.

6.31 Let the random variables X and Y be uniformly distributed over the interior of the unit circle centered at $(0, 0)$.

(a) Write the joint probability density function $f(x, y)$.

(b) Are X and Y independent?

(c) Use no calculus to find the exact value of $P(|X| + |Y| < 1)$.

(d) Use no calculus to find the exact value of $P(3X + Y < 0)$.

(e) Use no calculus to find the exact value of $P(X^2 + Y^2 < 1/4)$.

6.32 Let X and Y be continuous random variables with joint probability density function

$$f(x, y) = \frac{1}{3\pi} \qquad 1 < x^2 + y^2 < 4.$$

Find the marginal probability density function of X.

6.33 Let the independent random variables X_1 and X_2 be drawn from a population with probability mass function

$$f(x) = \begin{cases} 1/3 & x = 2 \\ 2/3 & x = 8. \end{cases}$$

Find the probability mass function of the sample mean

$$\frac{X_1 + X_2}{2}.$$

6.34   Let $X \sim$ binomial$(5, 7/10)$ and $Y \sim$ binomial$(4, 4/5)$ be independent random variables. Find $P(\min\{X, Y\} = 3)$.

6.35   Let the independent random variables X and Y have marginal probability mass functions

$$f_X(x) = \frac{1}{2} \qquad x = 1, 2$$

and

$$f_Y(y) = \frac{1}{3} \qquad y = 1, 2, 3.$$

Find $P(X = Y)$.

6.36   Let T_1 and T_2 be independent lifetimes of two light bulbs. Find the probability density function of time of the first failure $X = \min\{T_1, T_2\}$. Also, find the probability density function of the time between failures $Y = |T_1 - T_2|$. Assume the following pairs of distribution assumptions for T_1 and T_2.

     (a) $T_1 \sim$ exponential(λ) and $T_2 \sim$ exponential(λ).

     (b) $T_1 \sim$ exponential(λ_1) and $T_2 \sim$ exponential(λ_2).

     (c) $T_1 \sim$ Weibull(λ, κ) and $T_2 \sim$ Weibull(λ, κ).

6.37   Let $X_1 \sim$ geometric(p_1) and $X_2 \sim$ geometric(p_2) using the parameterization of the geometric distribution beginning at 0. If X_1 and X_2 are independent, find the probability mass function of $Y_2 = \max\{X_1, X_2\}$.

6.38   Consider the linear programming problem:

$$\begin{array}{llll} \text{maximize} & Ax_1 + & Bx_2 \\ \text{subject to} & x_1 + & x_2 \leq 1 \\ & x_1 & \geq 0 \\ & & x_2 \geq 0 \end{array}$$

where the random coefficients $A \sim U(0, 1)$ and $B \sim U(0, 2)$ are independent random variables. Find the joint probability mass function of X_1 and X_2, the values of x_1 and x_2 that solve the linear program.

6.39   Let the random variable X have probability density function

$$f_X(x) = 1 \qquad 0 < x < 1.$$

Let the random variable Y have probability density function

$$f_Y(y) = \frac{1}{3} \qquad -1 < y < 2.$$

Assume that X and Y are independent.

     (a) Find the probability density function of $V = \min\{X, Y\}$.

     (b) Find the probability density function of $W = \min\{\lfloor 2X \rfloor, \lfloor Y \rfloor\}$.

6.40   Let X_1 and X_2 be independent and identically distributed random variables with support on the open interval $(0, 1)$ and common probability density function $f(x)$. Write an equation that $f(x)$ must satisfy so that $X_1 X_2 \sim U(0, 1)$.

6.41   Let X and Y be independent and identically distributed random variables, each with common cumulative distribution function $F(\cdot)$. What is the the cumulative distribution function of $Z = \max\{X, Y\}$?

6.42 Determine, by inspection, whether X and Y with distribution described by $f(x, y)$ given below, are independent or dependent.

(a) $f(x, y) = \frac{1}{27}\left(x^3 + y\right)$ $\qquad x = 0, 1, 2; \ y = 1, 2$

(b) $f(x, y) = \frac{1}{40}x^2 y$ $\qquad (x, y) = (1, 1), (3, 1), (1, 3), (3, 3)$

(c) $f(x, y) = e^{-x}$ $\qquad x > 0; \ 0 < y < 1$

(d) $f(x, y) = 1/2$ $\qquad |x| + |y| < 1$

(e) $f(x, y) = 1/4$ $\qquad -1 < x < 1; \ -1 < y < 1$

6.43 Let $X_1 \sim \text{exponential}(\lambda_1)$ and $X_2 \sim \text{exponential}(\lambda_2)$ be independent random variables. Find the cumulative distribution function of $Z = \max\{X_1, X_2\}/\min\{X_1, X_2\}$.

6.44 Let X and Y be independent Geometric(p) random variables.

(a) Find the probability that X equals Y.

(b) Find an expression for the probability that Y is a multiple of X.

6.45 Let X be a discrete random variable with probability mass function $f_X(x)$ and support on the nonnegative integers. Likewise, let Y be a discrete random variable with probability mass function $f_Y(y)$ and support on the nonnegative integers. Assume that X and Y are independent. Give an expression for the probability mass function of $Z = X + Y$.

6.46 If X_1 is a Bernoulli random variable with $p_1 = 1/2$ and X_2 is a Bernoulli random variable with $p_2 = 3/4$, are these two random variables necessarily independent?

6.47 Consider an n-component parallel system consisting of identical components. In a *load-sharing* parallel arrangement, the load experienced is redistributed among the remaining surviving components as each component fails. This would be the case for the lug nuts that attach a wheel to a car. More specifically, assume that there are $n = 5$ identical components arranged in a load-sharing parallel system. When there are i functioning components, each component has an exponential time to failure with rate $(6 - i)!$ failures per year. By the memoryless property of the exponential distribution, each surviving component is as good as new when one of the components fails. Find

(a) the population mean time to system failure,

(b) the 95th percentile of the time to system failure.

6.48 Let p_1, p_2, p_3, p_4 be positive real numbers such that $p_1 + p_2 + p_3 + p_4 = 1$. Let the joint probability density function for X and Y be

$$f(x, y) = \begin{cases} p_1 & x = 0, y = 0 \\ p_2 & x = 0, y = 1 \\ p_3 & x = 1, y = 0 \\ p_4 & x = 1, y = 1. \end{cases}$$

Find the population covariance between X and Y.

6.49 Two cards are selected at random and without replacement from a well-shuffled deck. Let X be the number of jacks and let Y be the number of queens. Find the population covariance between X and Y and write a sentence or two explaining the significance of its *sign*.

6.50 A coin has a probability p of turning up heads, where $0 < p < 1$. The coin is tossed three times. Let X be the total number of heads that appear in the three tosses. Let Y be the total number of heads that appear on the last two tosses. Find the population covariance between X and Y. Write a sentence commenting on the *sign* of the population covariance.

6.51 Let X and Y be random variables with joint probability density function

$$f(x, y) = \frac{8xy}{k^4} \qquad 0 < x < y < k,$$

where k is a positive constant.

(a) Find the population covariance between X and Y.

(b) Find $E[Y \mid X = x]$.

(c) Find $V[Y \mid X = x]$.

6.52 The Internal Revenue Service could classify couples using the following random variables

$$X = \begin{cases} 0 & \text{filing jointly} \\ 1 & \text{filing separately} \end{cases}$$

and

$$Y = \begin{cases} 0 & \text{no dependents} \\ 1 & \text{1 or more dependents} \end{cases}$$

The joint probability mass function of X and Y is given below. Find the population covariance between X and Y.

		Y	
		0	1
X	0	0.34	0.46
	1	0.06	0.14

6.53 Ray and Jane are bridge partners. They each receive 13 cards dealt from a well-shuffled deck. Let X be the number of spades in Ray's hand and Y be the number of spades in Jane's hand.

(a) Find the joint probability mass function of X and Y.

(b) Find the population covariance of X and Y.

6.54 Consider the R code below which prints the sample mean of two random variates.

```
x = runif(2, 4, 8)
print(((x[1] + x[2]) / 2)
```

(a) What is the population mean of the value printed?

(b) What is the population variance of the value printed?

6.55 A coin is tossed n times. The probability of tossing a head on a single toss is p, where $0 < p < 1$.

(a) Find the expected value of the product of the *number* of heads tossed and the *number* of tails tossed.

(b) Find the population correlation between the number of heads tossed and the number of tails minus the number of heads. Note that this can be done with little or no mathematics.

(c) Write R statements to verify your solutions to parts (a) and (b) using a Monte Carlo simulation when $n = 17$ and $p = 1/3$ with 1000 replications.

6.56 For random variables X, Y, and Z, show that

$$\text{Cov}(X, Y + Z) = \text{Cov}(X, Y) + \text{Cov}(X, Z).$$

6.57 Without doing any mathematics, what is the population correlation between X, the number of red cards (that is, hearts and diamonds), and Y, the number of black cards (that is, spades and clubs), in a 5-card poker hand dealt from a standard 52-card deck?

6.58 Let X and Y have a bivariate distribution with variance–covariance matrix

$$\Sigma = \begin{bmatrix} 2 & -1 \\ -1 & 3 \end{bmatrix}.$$

Find $V[X+Y]$.

6.59 Let the continuous random variables X and Y be uniformly distributed over the support

$$\mathcal{A} = \left\{ (x, y) \,\big|\, 0 < y < |x| < 1 \right\}.$$

Find

    (a) $P(X > 2/3)$,

    (b) $P(Y > 2/3)$,

    (c) $E\left[XY^2\right]$,

    (d) $E[\arccos X]$,

    (e) $f_Y(y)$.

6.60 A five-card hand is dealt from a well-shuffled deck of cards. Let

    • X_1 be the number of spades,

    • X_2 be the number of diamonds,

    • X_3 be the number of jacks,

    • X_4 be the number of queens.

The population correlation between X_1 and X_2 is denoted by $\rho_{X_1 X_2}$. The population correlation between X_3 and X_4 is denoted by $\rho_{X_3 X_4}$. Without computing the correlations, choose one of the following statements that you believe to be true and write *two sentences* describing why you think it is true.

    (a) $\rho_{X_1 X_2} < \rho_{X_3 X_4}$,

    (b) $\rho_{X_1 X_2} = \rho_{X_3 X_4}$,

    (c) $\rho_{X_1 X_2} > \rho_{X_3 X_4}$.

6.61 Solve Buffon's needle problem for $l > d$.

6.62 In Buffon's needle problem, a needle of length l is tossed n times, and the number of times that it crosses one of the parallel lines that are a distance d apart is denoted by W. In order to estimate π, the fraction of crossings

$$\hat{p} = \frac{W}{n}$$

is equated to $\frac{2l}{\pi d}$ (in the case of $l \leq d$) and π is solved for, yielding an estimate $\hat{\pi}$. The question here is: *how long should the needle be?* If one uses a long needle (l large), $\hat{p}$ will be close to one. On the other hand, if one uses a short needle (l small), $\hat{p}$ will be close to zero. A reasonable middle ground would be to choose l so as to maximize $V\left(\frac{W}{n}\right)$. Find the value of l that maximizes $V\left(\frac{W}{n}\right)$.

6.63 Let X and Y be random variables with $V[X] = 10$, $V[Y] = 16$, and $V[X + Y] = 24$. Find $\text{Cov}(X, Y)$.

6.64 The table below gives the joint probability mass function of X and Y, where $p_1 + p_2 + p_3 + p_4 = 1$. Find $E[E[Y | X]]$.

$x \diagdown y$	0	1
0	p_1	p_2
1	p_3	p_4

6.65 The table below gives the joint probability mass function of X and Y, where $p_1 + p_2 + p_3 + p_4 = 1$. Find $E\left[X^5 | Y = 1\right]$.

$x \diagdown y$	0	1
0	p_1	p_2
1	p_3	p_4

6.66 Let the random variable X be the number of successes in n mutually independent Bernoulli trials and the random variable $Y = n - X$ be the number of failures in these n mutually independent Bernoulli trials. Find ρ, the population correlation between X and Y.

6.67 For the joint probability density function defined by

$$f(x, y) = \frac{2}{k^2} \qquad 0 < x < y < k,$$

for some positive real constant k, find the conditional expected value of X given $Y = y$ and find the population correlation between X and Y.

6.68 Three cards are drawn without replacement from a well-shuffled deck of cards. Let the random variable X be the number of spades and let the random variable Y be the number of diamonds. Find the population covariance between X and Y.

6.69 The random variables X and Y are uniformly distributed over the *interior* of a circle of radius 2 centered at $(3, 4)$. Find $E[X | Y = 5]$.

6.70 Let X be an exponential random variable with population mean 2. Let $Y = 5X$. Find ρ, the population correlation between X and Y.

6.71 A die is weighted so that the number of spots that appear on the up face has probability mass function
$$f(x) = \frac{x}{21} \qquad x = 1, 2, \ldots, 6.$$
A pair of two such dice are rolled n times. Let the random variable X be the number of double sixes that are tossed and the random variable Y be the number of times that a three appears on one or both of the faces. Find the population correlation between X and Y.

6.72 Consider the square with vertices $(1, 1)$, $(-1, 1)$, $(1, -1)$, and $(-1, -1)$. Find the expected rectilinear (Manhattan) distance between a random point chosen in the interior of the square and the origin.

6.73 The random variables X and Y are uniformly distributed over the support

$$\mathcal{A} = \left\{(x, y) \,|\, 0 < x^2 < y < 4\right\}.$$

Find $E[Y | X = 1]$ and $V[Y | X = 1]$.

6.74 The first significant digit X_1 and the second significant digit X_2 have joint probability mass function

$$f(x_1, x_2) = \log_{10}\left(1 + \frac{1}{10x_1 + x_2}\right) \qquad x_1 = 1, 2, \ldots, 9; \ x_2 = 0, 1, 2, \ldots, 9.$$

(a) Are X_1 and X_2 defined on a product space?

(b) Are X_1 and X_2 independent?

(c) Find the marginal distribution of X_1.

(d) Find the marginal distribution of X_2.

(e) Find the mean of X_2 to four digits.

6.75 Let X_1 and X_2 be independent standard normal random variables. Let $Y_1 = X_1 + X_2$ and $Y_2 = 2X_1 - X_2$.

(a) Find the distribution of Y_1.

(b) Find $\text{Cov}(Y_1, Y_2)$.

6.76 The continuous random variables X and Y are uniformly distributed over the support region $\mathcal{A}$ shown in Figure 6.27 (that is, the joint probability density function of X and Y is $f(x, y) = c$, where c is the reciprocal of the shaded area). Draw the support region $\mathcal{A}$, then draw $E[Y \mid X = x]$ and $E[X \mid Y = y]$ on top of $\mathcal{A}$ for appropriate values of x and y.

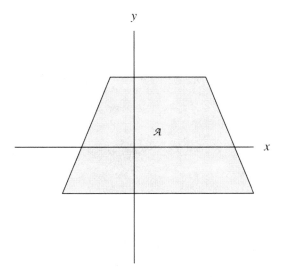

Figure 6.27: The support of X and Y.

6.77 Let the random variables X and Y have joint probability density function

$$f(x, y) = 1 \qquad 0 < x < 1, 0 < y < 1.$$

Find $V[E[Y \mid X]]$.

6.78 Two cards are drawn without replacement from a well-shuffled deck. Let X be the number of aces and Y be the number of face cards (that is, kings, queens, and jacks) drawn. Find $E[V[E[Y \mid X]]]$.

6.79 Consider the random variables X and Y defined on a product space with joint probability density function $f(x, y)$. What is $V[E[V[E[X \mid Y]]]]$?

6.80 Let X and Y have joint probability density function $f(x, y) = kx$, where k is a constant, defined on the support $\mathcal{A} = \{(x, y) \mid 0 < x < y < 1\}$. Find $E[V[E[X \mid Y]]]$.

6.81 A wallet contains ten bills. Two of the bills are $20 bills. Three of the bills are $10 bills. Five of the bills are $5 bills. If three bills are sampled from the wallet at random and without replacement, find the expected monetary value removed from the wallet.

6.82 Let X and Y have joint moment generating function

$$M(t_1, t_2) = p_1 + p_2 e^{t_1} + p_3 e^{t_2} + p_4 e^{t_1 + t_2} \qquad -\infty < t_1 < \infty, \ -\infty < t_2 < \infty,$$

where p_1, p_2, p_3, and p_4 are probabilities that sum to 1. Find the population covariance between X and Y.

6.83 Find the expected number of runs in n flips of a biased coin, where p is the probability of flipping a head for $0 < p < 1$. *Hint*: when $n = 7$, the sequence HHTTTHH has three runs.

6.84 Let the random variables X and Y have joint moment generating function

$$M(t_1, t_2) = \frac{1}{2} e^{t_1 + t_2} + \frac{1}{4} e^{2t_1 + t_2} + \frac{1}{12} e^{t_2} + \frac{1}{6} e^{4t_1 + 3t_2}$$

for all real values of t_1 and t_2.

(a) Find $V[X]$.

(b) Find $P(X < Y)$.

6.85 Two fair dice are tossed. Let the random variable X be the sum of the numbers on the up faces and let the random variable Y be the absolute difference of the numbers on the up faces (for example, tossing a three and a five yields $X = 3 + 5 = 8$ and $Y = |3 - 5| = 2$). Find the population covariance between X and Y.

6.86 Let the random variables X_1 and X_2 have joint moment generating function

$$M(t_1, t_2) = \left(p + q e^{t_1} + r e^{t_2} \right)^m,$$

where $-\infty < t_1 < \infty$ and $-\infty < t_2 < \infty$, $0 < p < 1$, $0 < q < 1$, $0 < r < 1$, $p + q + r = 1$, and m is a positive integer. Find the population covariance between X_1 and X_2.

6.87 A biased coin is flipped repeatedly until both a head and tail have appeared. Let p be the probability that the coin comes up heads on a single toss. What is the expected value and the population variance of the number of tosses required? Check your solutions for a fair coin with a Monte Carlo simulation experiment.

6.88 A candy company produces heart-shaped pieces of candy with brief romantic slogans. You have bag containing 1 heart that reads "LUV YA," 4 hearts that read "HUG ME," and 5 hearts that read "TXT ME." You randomly sample 3 hearts from the bag without replacement. Let the random variable X be the number of "LUV YA" hearts selected and Y be the number of "HUG ME" hearts selected.

(a) Find the joint probability density function of X and Y.

(b) Find the probability that all three types of hearts are represented in the sample.

(c) Find $E[X]$.

6.89 If X denotes the number of 3's and Y denotes the number of even outcomes in n rolls of a fair die, find $\text{Cov}(X, Y)$.

6.90 Let X be the number of kings and let Y be the number of spades in a two-card hand dealt from a well-shuffled deck. Find $E\left[|X - Y|\right]$.

6.91 Let the ordered pair (X, Y) be a randomly selected point on the circumference of a unit circle centered at the origin. Find the population covariance between X and Y.

6.92 Let X and Y have the bivariate normal distribution with parameters μ_X, μ_Y σ_X^2, σ_Y^2, and ρ. Write the variance–covariance matrix for X and Y in terms of these five parameters only.

6.93 Let X and Y have a bivariate normal distribution with variance–covariance matrix

$$\Sigma = \begin{bmatrix} 9 & r \\ r & 1 \end{bmatrix}.$$

Find the allowable values of the constant r.

6.94 Let X and Y have the bivariate normal distribution with parameters μ_X, μ_Y, σ_X, σ_Y, and ρ. Label each of the following statements as true or false.

(a) $\rho = 0 \Rightarrow f_X(x) f_Y(y) = f(x, y)$.

(b) $\rho = 0 \Rightarrow$ level surfaces (contours) of the joint probability density function are circular.

(c) $V[Y \mid X = x]$ is independent of x.

(d) The variance–covariance matrix of (X, Y) is

$$\Sigma = \begin{pmatrix} \sigma_X^2 & \rho \sigma_X \sigma_Y \\ \rho \sigma_X \sigma_Y & \sigma_Y^2 \end{pmatrix}.$$

6.95 Determine whether each statement below is true or false.

(a) The joint probability density function of the random variables X and Y is always the product of their marginal probability density functions.

(b) The binomial, negative binomial, and geometric distributions all have interpretations that can be formulated in terms of Bernoulli trials.

(c) The population variance is the expected distance between a random variable and its population mean.

(d) For any random variable X and any real number x, $f_X(x) = P(X = x)$.

(e) A normal random variable X has the same value for its population mean, median, and mode.

(f) If $\text{Cov}(X, Y) = 0$, then the random variables X and Y are independent.

(g) If X_1 and X_2 are each Bernoulli trials with support values 0 and 1, the Bernoulli trials must be independent.

(h) The exponential distribution is the only probability distribution with the memoryless property.

(i) If X_1 and X_2 have the bivariate normal distribution, then $P(X_1 > 0, X_2 > 0) < 1$.

(j) For any random variable X, the random variable $Y = \lfloor X \rfloor$ is always a discrete random variable.

6.96 Let the random variables X and Y have the bivariate normal distribution with parameters $\mu_X, \mu_Y, \sigma_X, \sigma_Y$, and ρ. Find the value of the constant c that minimizes $V[X + cY]$.

6.97   A fair die is rolled 100 times. Find the population mean and the population variance of the sum of the odd numbers that appear.

6.98   What is the minimum number of times a fair die should be rolled to be at least 90% certain that the fraction of fives rolled is within 0.005 of the true probability of rolling a five, which is $1/6$?

   (a) Give a mathematical statement of this problem.

   (b) Write the R code to calculate the minimum number of rolls required.

6.99   A bag contains four balls numbered 1, 2, 3, and 4. Three balls are drawn at random from the bag. Find the joint probability mass function of X_1, X_2, and X_3, the results of the three draws, if

   (a) sampling is without replacement,

   (b) sampling is with replacement.

6.100 Wes is enrolled in four classes this semester. The final grades in each class are mutually independent random variables with an identical probability distribution described by:

$$P(A) = 0.5 \qquad P(B) = 0.3 \qquad P(C) = 0.2.$$

An A counts for 4 points, a B counts for 3 points, and a C counts for 2 points when calculating a grade point average (GPA).

   (a) Find his expected GPA this semester.

   (b) Find the population variance of his GPA this semester.

   (c) Find the probability that Wes will get two A's, one B, and one C this semester.

6.101 The weather on any day can take on one of two states: sunny and rainy. This Sunday has a 30% chance of rain. On the six days that follow Sunday,

   • the probability that it is rainy is 0.6 if it is rainy on the previous day,

   • the probability that it is rainy is 0.2 if it is sunny on the previous day.

   What is the probability mass function of the number of rainy days during the next week?

6.102 Let X_1, X_2, and X_3 have joint moment generating function

$$M(t_1, t_2, t_3) = E\left[e^{t_1 X_1 + t_2 X_2 + t_3 X_3}\right] = \left(\frac{1}{1 - t_1}\right)\left(\frac{1}{1 - t_2}\right)\left(\frac{1}{1 - t_3}\right)$$

   for $t_1 < 1, t_2 < 1, t_3 < 1$. Are X_1, X_2, and X_3 mutually independent?

6.103 Let X_1, X_2, and X_3 have joint cumulative distribution function

$$F(x_1, x_2, x_3) = \begin{cases} 0 & x_1 < 4 \text{ or } x_2 < 5 \text{ or } x_3 < 6 \\ 1 & x_1 \geq 4 \text{ and } x_2 \geq 5 \text{ and } x_3 \geq 6. \end{cases}$$

   Find $E[X_1 X_2 X_3]$.

6.104 Intelligence quotient (IQ) scores are normally distributed random variables with a mean of 100 and a standard deviation of 10. If nine children are tested, find the probability that two have scores that are less than 90, three have scores between 90 and 110, and four have scores higher than 110.

6.105 Let X_1, X_2, X_3 be random variables with variance–covariance matrix

$$\begin{bmatrix} 8 & 5 & 3 \\ 5 & 7 & 4 \\ 3 & 4 & 6 \end{bmatrix}.$$

Find the variance of $X_1 + X_2 + X_3$.

6.106 Consider three consecutive sojourn times in a single-server queueing node: X_1, X_2, X_3. The population means of these three sojourn times are all equal to 8.1, and their variance–covariance matrix is

$$\Sigma = \begin{bmatrix} 4 & 2 & 1 \\ 2 & 4 & 2 \\ 1 & 2 & 4 \end{bmatrix}.$$

If $\bar{X}$ is the sample mean of the three sojourn times, find $E\left[\bar{X}\right]$ and $V\left[\bar{X}\right]$.

6.107 Let $X_i \sim N(\mu_i, \sigma_i^2)$, $i = 1, 2, 3, 4$. Also, X_1, X_2, X_3, and X_4 are mutually independent random variables. If $F(x, \mu_i, \sigma_i^2)$ is the cumulative distribution function of a normal random variable with population mean μ_i and population variance σ_i^2, for $i = 1, 2, 3, 4$, give an expression for the probability that *exactly one* of the four X_i values is less than k, where k is a real constant.

6.108 Let X_i have a Weibull distribution with $\lambda = 0.05$ and $\kappa = 2.5$ for $i = 1, 2, \ldots, 1000$, which denote the mutually independent lifetimes, in months, of light bulbs lighting the interior of a factory.

    (a) What is the probability that one particular bulb survives six months?

    (b) What is the expected number of bulbs that survive six months?

    (c) What is the time when you would expect that 25% of the bulbs will be burned out?

6.109 Let X_1, X_2, and X_3 be mutually independent and identically distributed random variables each having moment generating function

$$M(t) = \frac{1}{2}e^t + \frac{1}{3}e^{2t} + \frac{1}{6}e^{5t}$$

for all real values of t.

    (a) Find $V[X_1]$.

    (b) Find $P(X_1 + X_2 + X_3 = 4)$.

6.110 Let X, Y, and Z be random variables joint probability density function $f(x, y, z)$ with support on the interior of the unit sphere. Set up an expression (with appropriate integration limits) for $E[X + \cos Y \mid Z = z]$ in terms of the joint probability density function $f(x, y, z)$.

6.111 Let X_1, X_2, X_3, X_4 be mutually independent and identically distributed Bernoulli(p) random variables. Find $P(X_1 = X_2 = X_3 = X_4)$.

6.112 Let X_1, X_2, X_3, X_4 be mutually independent and identically distributed Bernoulli(p) random variables. Find $P(\bar{X} = 1)$, where $\bar{X}$ is the sample mean $\bar{X} = (X_1 + X_2 + X_3 + X_4)/4$.

6.113 Let X_1, X_2, X_3, X_4 be mutually independent and identically distributed Bernoulli(p) random variables. Find $E[X_1 X_2 X_3 X_4]$.

6.114 Let X_1, X_2, X_3, X_4 be mutually independent and identically distributed Bernoulli(p) random variables. Find $V\left[X_1 X_2^2 X_3^3 X_4^4\right]$.

6.115 Let X_1, X_2, X_3, X_4 be mutually independent and identically distributed Bernoulli(p) random variables. Find $P(X_1 \leq X_2 \leq X_3 \leq X_4)$.

6.116 Four cards are drawn with replacement from a well-shuffled deck. Find the probability that all suits are represented.

6.117 Let the joint probability density function for X_1, X_2, X_3 be

$$f(x_1, x_2, x_3) = \lambda_1 \lambda_2 \lambda_3 e^{-\lambda_1 x_1 - \lambda_2 x_2 - \lambda_3 x_3} \qquad x_1 > 0, x_2 > 0, x_3 > 0,$$

where $\lambda_1, \lambda_2, \lambda_3$ are positive real constants.

(a) Find $P(X_1 < X_2)$.

(b) Find $P(X_1 < X_2 < X_3)$.

6.118 Let T_1, T_2, T_3 be the mutually independent component lifetimes associated with the system of components arranged so that the system lifetime T is

$$T = \max\{T_1, \min\{T_2, T_3\}\}.$$

If $T_1 \sim$ exponential(λ_1), $T_2 \sim$ exponential(λ_2), and $T_3 \sim$ exponential(λ_3), where $\lambda_1, \lambda_2,$ and λ_3 are positive rates, find an expression for the population mean time to system failure

(a) by hand,

(b) by APPL.

6.119 If a gambler bets 100 \$1 bills repeatedly at craps, find his expected earnings.

6.120 What is the name of the distribution of the sum of n mutually independent Bernoulli(p) random variables?

6.121 What is the name of the distribution of the sum of the following two independent random variables: $X \sim$ binomial(n, p) and $Y \sim$ binomial(m, p)?

6.122 What is the name of the distribution of the sum of n mutually independent and identically distributed geometric(p) random variables?

6.123 What is the name of the distribution of the sum of the squares of n mutually independent standard normal random variables?

6.124 Let $X_1, X_2, \ldots, X_n$ be mutually independent lifetimes of the components in a series system. Assume each has probability density function

$$f_X(x) = 1 \qquad 0 < x < 1.$$

Let Y be the system lifetime. Find the probability density function of Y.

6.125 Let X_1, X_2, X_3, X_4, X_5 be mutually independent lug nut lifetimes. Assume that each lug nut has probability density function

$$f_X(x) = 2x \qquad 0 < x < 1.$$

The lug nuts are arranged as a five-component parallel system. Find the probability density function of the system lifetime $Y = \max\{X_1, X_2, \ldots, X_5\}$.

6.126 Let $X_1, X_2,$ and X_3 be mutually independent random variables such that $X_1 \sim N(1, 1)$, $X_2 \sim N(4, 4)$, and $X_3 \sim N(9, 9)$. Find the population variance of $X_1 + 5X_2 - 4X_3$.

6.127 Let X_1, X_2, X_3 have the trivariate normal distribution with population mean vector

$$\boldsymbol{\mu} = (4, 5, 6)'$$

and variance–covariance matrix

$$\Sigma = \begin{bmatrix} 7 & 2 & 0 \\ 2 & 8 & 3 \\ 0 & 3 & 9 \end{bmatrix}.$$

Give the name of the distribution and the associated parameter values of the joint marginal distribution of X_2 and X_3.

6.128 Ted has a piece of licorice of unit length. Bill makes sequential mutually independent and identically distributed demands for a length of Ted's licorice X, each with probability density function

$$f(x) = e^{-x} \qquad x > 0.$$

Ted always satisfies Bill's demand if he is able to do so. Bill's demands stop once Ted is not able to satisfy a demand. Let the random variable N be the number of Bill's demands that Ted is able to successfully satisfy. Find $E[N]$.

6.129 Give the APPL statements and associated output to find the probability that the sum of the spots showing in a roll of five fair dice is 17. Check your solution by executing a Monte Carlo simulation.

6.130 Dr. Foote is a podiatrist. He has perused the podiatry literature and found the following population probabilities associated with patients having toe problems.

Toe	Probability
Big toe	0.56
Index toe	0.08
Middle toe	0.07
Ring (?!) toe	0.07
Pinky toe	0.22

If Dr. Foote sees six patients with complaints about one of their toes, find the probability that all of the five different toes are represented in his sample.

6.131 A bag contains 5 balls numbered 1, 2, 3, 4, 5. Three balls are selected at random *with replacement* from the bag. The numbers on the balls selected are X_1, X_2, and X_3. Find the population mean and population variance of

$$Y = X_1 + X_2 + X_3.$$

Check your results by executing a Monte Carlo simulation.

6.132 A bag contains 5 balls numbered 1, 2, 3, 4, 5. Three balls are selected at random *without replacement* from the bag. The numbers on the balls selected are X_1, X_2, and X_3. Find the population mean and population variance of

$$Y = X_1 + X_2 + X_3.$$

Check your results by executing a Monte Carlo simulation.

6.133 Let (X_1, Y_1) and (X_2, Y_2) be two points chosen at random from the interior of a unit square with opposite vertices $(0, 0)$ and $(1, 1)$. Find

$$E\left[(X_1 - X_2)^2 + (Y_1 - Y_2)^2\right].$$

6.134   Professor Monte teaches a graduate course in probability at Riverwater College. He ignores a student's performance throughout the semester and assigns each student an A, B, or C with equal probability. Professor Monte's courses are not very popular. If six students are currently enrolled in his course, find the probability that he will assign two A's, two B's and two C's.

Chapter 7

Functions of Random Variables

This chapter introduces three techniques for determining the probability distribution of a random variable that is a function of one or more other random variables with known probability distribution. In the univariate case, a random variable X has a known probability distribution, and we want to know the distribution of $Y = g(X)$ for some function g. In the bivariate case, the joint probability distribution of X_1 and X_2 is known, and we want to know the distribution of $Y = g(X_1, X_2)$. Finally, in the multivariate case, the joint probability distribution of X_1, X_2, $\ldots$, X_n is known, and we want to know the distribution of $Y = g(X_1, X_2, \ldots, X_n)$. The next three paragraphs give representative applications of these three cases.

A company produces copper wire that has a random radius X whose probability distribution has been established by a long history of sampling. The company would like to know how much copper is required for the wire, so they are interested in the probability distribution of the cross-sectional *area* of the wire $Y = g(X) = \pi X^2$. The techniques presented in this chapter allow them to determine the distribution of the amount of copper consumed in the production of the wire.

The number of daily calls X_1 to a 911 emergency call center that are true emergencies has a Poisson distribution with mean λ_1. The number of daily calls X_2 to the 911 emergency call center that are not true emergencies has a Poisson distribution with mean λ_2. What is the probability distribution of the total number of daily calls to the 911 emergency call center $Y = g(X_1, X_2) = X_1 + X_2$?

A statistician gathers mutually independent and identically distributed data values $X_1, X_2, \ldots, X_n$ from a population with a known probability distribution. He computes the sample mean

$$Y = g(X_1, X_2, \ldots, X_n) = \frac{1}{n} \sum_{i=1}^{n} X_i.$$

What is the probability distribution of the sample mean?

Each of the three sections in this chapter presents a technique for computing the distribution of a function of one or more random variables. The three techniques are listed below.

- The *cumulative distribution function technique* was introduced in the univariate case in Chapter 3. This technique uses standard probability operations to convert the cumulative distribution function of X to the cumulative distribution function of Y. In the univariate case, the cumulative distribution function technique begins with $F_X(x)$ and concludes with $F_Y(y)$.

- The *transformation technique* converts the probability mass (or density) function of X to the probability mass (or density) function of Y. In the univariate case, the transformation technique begins with $f_X(x)$ and concludes with $f_Y(y)$.

- The *moment generating function technique* is particularly useful for finding the probability distribution of sums of mutually independent and identically distributed random variables

$X_1, X_2, \ldots, X_n$. The moment generating function technique begins with the moment generating functions of the individual random variables $M_{X_1}(t)$, $M_{X_2}(t)$, $\ldots$, $M_{X_n}(t)$, and concludes with identifying the moment generating function of the sum of the random variables $M_{X_1 + X_2 + \cdots + X_n}(t)$.

The reason that *three* techniques (rather than a single technique) are introduced here is that the techniques are oftentimes uniquely suited to a particular problem at hand. Some problems can be addressed by all three techniques, but typically one is superior to the others for a specific problem.

Within the three sections in this chapter, the univariate case is typically presented first, followed by the bivariate case, then the multivariate case. Likewise, the discrete case is typically presented before the continuous case because the mathematics tends to be simpler.

Before introducing the three techniques, a brief review of functions, 1–1 functions, and inverses is presented. These concepts come up repeatedly when determining the probability distribution of a function of a random variable.

> **Definition 7.1** A *function* is a relation between a set of elements known as the domain $\mathcal{A}$ and another set of elements known as the range $\mathcal{B}$ that associates each element in $\mathcal{A}$ with exactly one element in $\mathcal{B}$.

Although the sets $\mathcal{A}$ and $\mathcal{B}$ have been defined quite generally in Definition 7.1, they will always be some subset of the real numbers in this chapter. The notation

$$y = g(x)$$

is used to describe the function g, where x is a real number in the domain $\mathcal{A}$ and y is a real number in the range $\mathcal{B}$. The letter g is used for the function that transforms x rather than the more traditional choice of f because probability mass functions and probability density functions have confiscated the letter f. Functions can be described by the equation $y = g(x)$ and also by a graph, as illustrated in the example below.

Example 7.1 Consider the relationships between x and y described by the three equations

$$y = x/3 + 2 \qquad y = x^2 \qquad x = 2(y-2)^2 - 2$$

that are defined on the common domain $\mathcal{A} = \{x \mid -2 < x < 3\}$. The ranges of the equations are

$$\mathcal{B} = \{y \mid 4/3 < y < 3\} \qquad \mathcal{B} = \{y \mid 0 \leq y < 9\} \qquad \mathcal{B} = \left\{y \mid 2 - \sqrt{5/2} < y < 2 + \sqrt{5/2}\right\}$$

and the graphs of these equations are shown in Figure 7.1. The first two equations are functions, that is

$$y = g(x) = x/3 + 2$$

and

$$y = g(x) = x^2$$

both have exactly one element in $\mathcal{B}$ for each element in $\mathcal{A}$. The third equation, however, does *not* represent a function because one or more x-values do not have exactly one element in $\mathcal{B}$. For example, $x = 0$ is associated with $y = 1$ and $y = 3$.

The discussion of the three equations in the previous example leads to the well-known *vertical line test* for functions: a vertical line that is drawn at any x-position on a graph of a function intersects the graph at most once. All vertical lines intersect the graphs of the first two equations at most once. But for the third equation, a vertical line (for example, $x = 1$) intersects the graph of $x = 2(y-2)^2 - 2$ twice, so it is not a function.

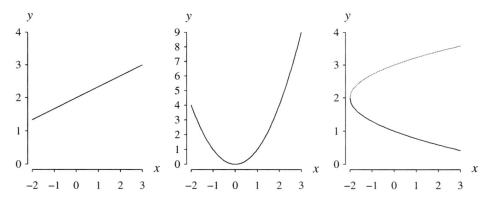

Figure 7.1: Graphs of $y = x/3 + 2$, $y = x^2$, and $x = 2(y-2)^2 - 2$ on $\mathcal{A} = \{x \mid -2 < x < 3\}$.

There are certain functions that are known as 1–1 functions, which are defined next, that play a central role in this chapter.

> **Definition 7.2** Any function $y = g(x)$ that maps a set $\mathcal{A}$ to a set $\mathcal{B}$ such that there is a 1–1 correspondence between the points in the domain $\mathcal{A}$ and the points in the range $\mathcal{B}$ is called a 1–1 function.

The term *bijective* is also used to describe a function that has this 1–1 correspondence between the points in $\mathcal{A}$ and the points in $\mathcal{B}$.

Example 7.2 Consider the function

$$y = g(x) = x^2$$

defined on three different domains:

$$\mathcal{A} = \{x \mid 0 < x < 3\} \qquad \mathcal{A} = \{x \mid x = 0, 1, 2, 3\} \qquad \mathcal{A} = \{x \mid x = -1, 0, 1, 2\}.$$

The graphs of the function on the three domains are shown in Figure 7.2. The first characteristic that stands out is that the first domain, $0 < x < 3$, is continuous on the interval $(0, 3)$, but the other two domains are discrete with four elements in the domain $\mathcal{A}$.

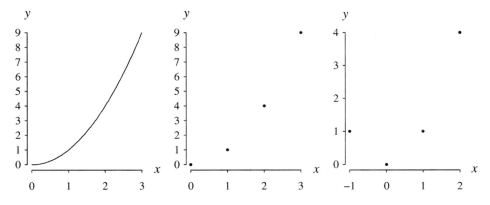

Figure 7.2: Graphs of $y = x^2$ on three domains.

This distinction will translate to discrete and continuous random variables later in the chapter. The function $y = x^2$ is 1–1 on the first two domains because there is a 1–1 correspondence between the points in $\mathcal{A}$ and the points in $\mathcal{B}$. The function is not 1–1, however, on the third domain. The y-value $y = 1$ has not just one, but two corresponding x-values in $\mathcal{A}$, namely $x = -1$ and $x = 1$.

The discussion of 1–1 functions in the previous example leads to the *horizontal line test* for 1–1 functions: a horizontal line that is drawn at any y-position on a graph of a 1–1 function intersects the graph at most once. In many of the problems encountered in this chapter, it is necessary to first determine whether a relationship between x and y is a function (perhaps by applying the vertical line test), then determine if the function is 1–1 (perhaps by applying the horizontal line test) prior to applying one of the results that applies to 1–1 functions.

A 1–1 function $y = g(x)$ has an *inverse function* denoted by $x = g^{-1}(y)$. The inverse function has the property that a round trip from any point x in the domain $\mathcal{A}$ to $\mathcal{B}$ via g, and back from $\mathcal{B}$ to $\mathcal{A}$ via g^{-1} returns back to x. In terms of our notation, a function g is 1–1 if for any $x \in \mathcal{A}$, $g^{-1}(g(x)) = x$. Inverse functions are illustrated for discrete and continuous domains in the next two examples.

Example 7.3 Let the function

$$y = g(x) = 3x + 2$$

be defined on the domain

$$\mathcal{A} = \{x \mid x = 0, 1, 10\}$$

which maps to the range

$$\mathcal{B} = \{y \mid y = 2, 5, 32\}.$$

This linear function passes the vertical and horizontal line tests, so it is a 1–1 function. The inverse function is found by solving $y = g(x)$ for x, yielding

$$x = g^{-1}(y) = \frac{y - 2}{3}.$$

Example 7.4 Let the function

$$y = g(x) = x^2$$

be defined on the domain

$$\mathcal{A} = \{x \mid 0 < x < 3\}$$

which maps to the range

$$\mathcal{B} = \{x \mid 0 < x < 9\}.$$

This segment of the quadratic function passes the vertical and horizontal line tests, so it is a 1–1 function. Being careful to consider the domain of the function, the inverse function is found by solving $y = g(x)$ for x, yielding

$$x = g^{-1}(y) = \sqrt{y}.$$

The round trip relationship associated with the $g(x)$ and $g^{-1}(y)$ can be illustrated for any point in the domain $\mathcal{A}$, say $x = 2$. In this case, $y = g(2) = 2^2 = 4$ is the mapping of the element $x = 2$ in $\mathcal{A}$ to the element $y = 4$ in $\mathcal{B}$. On the return trip, $x = g^{-1}(4) = \sqrt{4} = 2$ is the mapping of the element $y = 4$ in $\mathcal{B}$ back to the element $x = 2$ in $\mathcal{A}$.

Some functions that are not 1–1, might be 2–1, or more generally, many-to-one. For example, the function

$$y = g(x) = |x|$$

on the domain

$$\mathcal{A} = \{x \mid -2 < x < 2\}$$

is a 2–1 function everywhere except $x = 0$. As a second example, the function

$$y = g(x) = \cos x$$

on the domain

$$\mathcal{A} = \{x \mid 0 < x < 4\pi\}$$

is a 4–1 function everywhere except $x = \pi$, $x = 2\pi$, and $x = 3\pi$. Finally, some functions are not simply many-to-one. For example, we refer to the function

$$y = g(x) = x^2$$

on the domain

$$\mathcal{A} = \{x \mid -1 < x < 2\}$$

as a "piecewise many-to-one" function for the purposes of this chapter. The function is 2–1 on $-1 < x < 1$ (except at $x = 0$) and 1–1 on $1 \leq x < 2$.

Using the notation $\mathcal{A}$ and $\mathcal{B}$ for the domain and range of the function $y = g(x)$ was no coincidence. When x and y become the random variables X and Y in the function $Y = g(X)$, the two sets will become the support of X and the support of Y, respectively.

We now begin the discussion of determining the distribution of a function of a random variable, beginning with the cumulative distribution function technique.

7.1   Cumulative Distribution Function Technique

The cumulative distribution function technique in the univariate case was first introduced in Section 3.3. To review, the procedure for finding the distribution of $Y = g(X)$ is to

1. determine the support of X, which is denoted by $\mathcal{A}$,

2. determine the cumulative distribution function of X, which is denoted by $F_X(x)$,

3. determine the support of $Y = g(X)$, which is denoted by $\mathcal{B}$,

4. write the cumulative distribution function of Y as $F_Y(y) = P(Y \leq y) = P\big(g(X) \leq y\big)$,

5. perform algebra on the argument of $P\big(g(X) \leq y\big)$ in order to express $F_Y(y)$ in terms of $F_X(\cdot)$.

The technique begins with the cumulative distribution function of X and ends with the cumulative distribution function of $Y = g(X)$. If the problem requires $f_Y(y)$, it can be found by differentiating the cumulative distribution function when Y is continuous or differencing the cumulative distribution function when Y is discrete. The first two examples illustrate the cumulative distribution function technique for a discrete random variable X.

> **Example 7.5** You roll a fair die and are given \$10 for each spot that appears on the up face. Find the probability mass function of the amount of money that you receive.

Let X be the number of spots on the up face and $Y = g(X) = 10X$ be the number of dollars received. We want to find the probability mass function of Y. Following the steps for the cumulative distribution function technique, the support of X is

$$\mathcal{A} = \{x \mid x = 1, 2, 3, 4, 5, 6\}.$$

Since the die is fair, the probability mass function of X is

$$f_X(x) = \frac{1}{6} \qquad x = 1, 2, 3, 4, 5, 6.$$

A careful and complete version of the cumulative distribution function of X is the step function

$$F_X(x) = \begin{cases} 0 & x < 1 \\ \left\lfloor \frac{x}{6} \right\rfloor & 1 \le x \le 6 \\ 1 & x > 6. \end{cases}$$

For the purposes of the cumulative distribution function technique, however, the cumulative distribution function needs to be defined at only its support values, that is

$$F_X(x) = \frac{x}{6} \qquad x = 1, 2, 3, 4, 5, 6.$$

The function $Y = g(X) = 10X$ maps the support of X to

$$\mathcal{B} = \{y \,|\, y = 10, 20, 30, 40, 50, 60\}.$$

The cumulative distribution function of $Y = g(X) = 10X$ is

$$\begin{aligned} F_Y(y) &= P(Y \le y) \\ &= P(10X \le y) \\ &= P(X \le y/10) \\ &= F_X(y/10) \\ &= y/60 \qquad y = 10, 20, 30, 40, 50, 60. \end{aligned}$$

The question seeks the probability mass function of Y, so the cumulative distribution function of Y must be differenced, for example,

$$f_Y(50) = P(Y = 50) = P(Y \le 50) - P(Y \le 40) = F_Y(50) - F_Y(40) = \frac{50}{60} - \frac{40}{60} = \frac{1}{6}.$$

Continuing this pattern for all values of y in $\mathcal{B}$ results in the probability mass function

$$f_Y(y) = \frac{1}{6} \qquad y = 10, 20, 30, 40, 50, 60.$$

Not surprisingly, each of the dollar values in $\mathcal{B}$ is equally likely.

Example 7.6 Let $X \sim$ Geometric(p), where the geometric distribution is parameterized with support beginning at 1. Find the probability mass function and cumulative distribution function of $Y = X^2$.

From Definition 4.5, the probability mass function of X is

$$f_X(x) = p(1-p)^{x-1} \qquad x = 1, 2, \ldots$$

for $0 < p < 1$, and the associated cumulative distribution function at the mass values is

$$F_X(x) = 1 - (1-p)^x \qquad x = 1, 2, \ldots.$$

The transformation $Y = g(X) = X^2$ maps the support of X, which is

$$\mathcal{A} = \{x \,|\, x = 1, 2, 3, \ldots\}$$

to the support of Y, which is

$$\mathcal{B} = \{y \,|\, y = 1, 4, 9, \ldots\}.$$

The cumulative distribution function of $Y = g(X) = X^2$ is

$$
\begin{aligned}
F_Y(y) &= P(Y \le y) \\
&= P(X^2 \le y) \\
&= P(X \le \sqrt{y}) \\
&= F_X(\sqrt{y}) \\
&= 1 - (1 - p)^{\sqrt{y}} \qquad\qquad y = 1, 4, 9, \ldots.
\end{aligned}
$$

This cumulative distribution function can be differenced to arrive at the probability mass function

$$f_Y(y) = p(1 - p)^{\sqrt{y}-1} \qquad\qquad y = 1, 4, 9, \ldots.$$

This solution can be checked in APPL using the `Transform` procedure. The APPL statements

```
X := GeometricRV(p);
g := [[x -> x * x], [1, infinity]];
Y := Transform(X, g);
```

return the analytic solution. The function $g(X)$ is input as a list of two lists. The first list contains the transformation; the second list gives the range of x-values for which the transformation is valid.

These examples have illustrated the use of the cumulative distribution function technique on the transformation of discrete random variables. The next two examples illustrate the use of the cumulative distribution function technique on the transformation of a univariate continuous random variable.

Example 7.7 Consider the random variable X with probability density function

$$f_X(x) = e^{-x} \qquad\qquad x > 0.$$

Find the probability density function of $Y = 3X$.

The random variable X is recognized as an exponential(1) random variable. The transformation $Y = g(X) = 3X$ is a mapping from the support of X, which is

$$\mathcal{A} = \{x \,|\, x > 0\}$$

to the support of Y, which is

$$\mathcal{B} = \{y \,|\, y > 0\}.$$

The cumulative distribution function of X on its support is

$$F_X(x) = \int_0^x e^{-w} dw = \left[-e^{-w} \right]_0^x = 1 - e^{-x} \qquad\qquad x > 0.$$

The cumulative distribution function of Y is

$$
\begin{aligned}
F_Y(y) &= P(Y \le y) \\
&= P(3X \le y) \\
&= P(X \le y/3) \\
&= F_X(y/3) \\
&= 1 - e^{-y/3} \qquad\qquad y > 0.
\end{aligned}
$$

Finally, the cumulative distribution function of Y is differentiated with respect to y to determine the probability density function of Y:

$$f_Y(y) = \frac{1}{3} e^{-y/3} \qquad y > 0.$$

This can be recognized as the probability density function of an exponential$(1/3)$ random variable.

Example 7.8 Find the probability density function of the square root of a $U(3, 5)$ random variable.

Let $X \sim U(3, 5)$. The transformation $Y = g(X) = \sqrt{X}$ is a mapping from the support of X, which is

$$\mathcal{A} = \{x \,|\, 3 < x < 5\}$$

to the support of Y, which is

$$\mathcal{B} = \left\{ y \,|\, \sqrt{3} < y < \sqrt{5} \right\}.$$

The probability density function of X is

$$f_X(x) = \frac{1}{2} \qquad 3 < x < 5.$$

The cumulative distribution function of X on its support is

$$F_X(x) = \int_3^x \frac{1}{2} \, dw = \frac{x - 3}{2} \qquad 3 < x < 5.$$

The cumulative distribution function of Y is

$$
\begin{aligned}
F_Y(y) &= P(Y \leq y) \\
&= P\left(\sqrt{X} \leq y\right) \\
&= P(X \leq y^2) \\
&= F_X\left(y^2\right) \\
&= \frac{y^2 - 3}{2} \qquad \sqrt{3} < y < \sqrt{5}.
\end{aligned}
$$

The probability density function of Y is found by differentiating the cumulative distribution function of Y with respect to y:

$$f_Y(y) = y \qquad \sqrt{3} < y < \sqrt{5}.$$

This analytic solution can be checked in APPL with the statements

```
X := UniformRV(3, 5);
g := [[x -> sqrt(x)], [3, 5]];
Y := Transform(X, g);
```

which return the analytic solution derived above.

All of the examples considered thus far have been straightforward because the transformation $Y = g(X)$ was a 1–1 function from the support of $\mathcal{A}$ to the support of $\mathcal{B}$ with a closed-form inverse. The next example illustrates the extra layer of complexity that arises when the cumulative distribution function technique is applied to a transformation $Y = g(X)$ that is not a 1–1 function.

Example 7.9 Find the probability density function of the square of a $U(-1, 2)$ random variable.

The function $Y = g(X) = X^2$ transforms the random variable X from its support

$$\mathcal{A} = \{x \mid -1 < x < 2\}$$

to the support of Y, which is

$$\mathcal{B} = \{y \mid 0 \leq y < 4\}.$$

The probability density function of X is

$$f_X(x) = \frac{1}{3} \qquad -1 < x < 2.$$

The cumulative distribution function of X on its support is

$$F_X(x) = \int_{-1}^{x} \frac{1}{3} dw = \left[\frac{w}{3}\right]_{-1}^{x} = \frac{x+1}{3} \qquad -1 < x < 2.$$

So the cumulative distribution function of X is

$$F_X(x) = \begin{cases} 0 & x \leq -1 \\ \dfrac{x+1}{3} & -1 < x < 2 \\ 1 & x \geq 2. \end{cases}$$

Using the cumulative distribution function technique remains straightforward, but the accounting for the various values of y in $\mathcal{B}$ adds an extra layer of complexity. Figure 7.3 contains a graph of the parabola $y = g(x) = x^2$ on the support of X, which is $-1 < x < 2$. For any value y on the interval $0 < y < 1$, the corresponding values of x are $\pm\sqrt{y}$. This is illustrated in Figure 7.3 for $y = 0.36$. The event $Y \leq 0.36$ is equivalent to $-\sqrt{0.36} \leq X \leq \sqrt{0.36}$ or $-0.6 \leq X \leq 0.6$. Similarly, for any value y on the interval $1 < y < 4$, the corresponding value of x is just $\sqrt{y}$. The different situations for these two intervals of y must be accounted for when using the cumulative distribution function

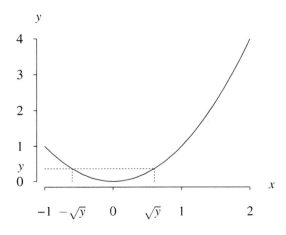

Figure 7.3: Graph of $y = x^2$ on $-1 < x < 2$.

technique. The cumulative distribution function of Y is

$$
\begin{aligned}
F_Y(y) &= P(Y \le y) \\
&= P(X^2 \le y) \\
&= \begin{cases} P\left(-\sqrt{y} \le X \le \sqrt{y}\right) & 0 \le y < 1 \\ P\left(X \le \sqrt{y}\right) & 1 \le y < 4 \end{cases} \\
&= \begin{cases} F_X\left(\sqrt{y}\right) - F_X\left(-\sqrt{y}\right) & 0 \le y < 1 \\ F_X\left(\sqrt{y}\right) & 1 \le y < 4 \end{cases} \\
&= \begin{cases} \dfrac{\sqrt{y}+1}{3} - \dfrac{-\sqrt{y}+1}{3} & 0 \le y < 1 \\ \dfrac{\sqrt{y}+1}{3} & 1 \le y < 4 \end{cases} \\
&= \begin{cases} \dfrac{2\sqrt{y}}{3} & 0 \le y < 1 \\ \dfrac{\sqrt{y}+1}{3} & 1 \le y < 4. \end{cases}
\end{aligned}
$$

Differentiating with respect to y produces the probability density function

$$
f_Y(y) = \begin{cases} \dfrac{1}{3\sqrt{y}} & 0 < y < 1 \\ \dfrac{1}{6\sqrt{y}} & 1 \le y < 4. \end{cases}
$$

This problem can be solved in APPL using the `Transform` function which requires that the transformation $Y = g(X) = X^2$ be divided into monotone segments.

```
X := UniformRV(-1, 2);
g := [[x -> x ^ 2, x -> x ^ 2], [-1, 0, 2]];
Y := Transform(X, g);
```

These statements return the same probability density function for Y that was determined by the analytic solution.

The previous examples have applied the cumulative distribution function technique to transformations of univariate random variables. The technique can also be applied to find the distribution of $Y = g(X_1, X_2)$, where X_1 and X_2 have a joint probability distribution. The first example illustrates the case when X_1 and X_2 are discrete random variables.

Example 7.10 Let $X_1 \sim \text{geometric}(p_1)$ and $X_2 \sim \text{geometric}(p_2)$ be independent random variables with supports beginning at 0. Find the distribution of $Y = \min\{X_1, X_2\}$.

From Definition 4.4, the probability mass function for a geometric random variable X is

$$
f_X(x) = p(1-p)^x \qquad x = 0, 1, 2, \ldots
$$

for $0 < p < 1$, and the associated cumulative distribution function at the mass values is

$$
F_X(x) = 1 - (1-p)^{x+1} \qquad x = 0, 1, 2, \ldots .
$$

The transformation $Y = g(X_1, X_2) = \min\{X_1, X_2\}$ maps the support of X_1 and X_2, which is

$$
\mathcal{A} = \{(x_1, x_2) \,|\, x_1 = 0, 1, 2, \ldots; \, x_2 = 0, 1, 2, \ldots\}
$$

to the support of Y, which is

$$\mathcal{B} = \{y \mid y = 0, 1, 2, \ldots\}.$$

Since X_1 and X_2 are independent, the cumulative distribution function of $Y = g(X) = \min\{X_1, X_2\}$ is

$$
\begin{aligned}
F_Y(y) &= P(Y \le y) \\
&= P(\min\{X_1, X_2\} \le y) \\
&= 1 - P(\min\{X_1, X_2\} > y) \\
&= 1 - P(X_1 > y, X_2 > y) \\
&= 1 - P(X_1 > y)\,P(X_2 > y) \\
&= 1 - (1 - P(X_1 \le y))(1 - P(X_2 \le y)) \\
&= 1 - \left[(1 - p_1)^{y+1}\right]\left[(1 - p_2)^{y+1}\right] \\
&= 1 - \left[(1 - p_1)(1 - p_2)\right]^{y+1} \qquad y = 0, 1, 2, \ldots.
\end{aligned}
$$

This cumulative distribution function is recognized as that of a geometric random variable with parameter $1 - (1 - p_1)(1 - p_2)$. Hence one can conclude that the minimum of two independent geometric random variables is itself geometric.

The next example considers the cumulative distribution function technique for a function of two continuous random variables. The key to solving problems of this type using the cumulative distribution function technique is to determine the appropriate integration limits associated with $g(X_1, X_2) \le y$, where $Y = g(X_1, X_2)$.

Example 7.11 Find the probability density function of the sum of two independent exponential(λ) random variables.

Let X_1 and X_2 denote the independent exponential(λ) random variables. The joint probability density function is the product of the marginal probability density functions:

$$f_{X_1, X_2}(x_1, x_2) = \lambda^2 e^{-\lambda x_1 - \lambda x_2} \qquad x_1 > 0, x_2 > 0.$$

The transformation $Y = g(X_1, X_2) = X_1 + X_2$ maps the support of X_1 and X_2, which is

$$\mathcal{A} = \{(x_1, x_2) \mid x_1 > 0, \ x_2 > 0\}$$

to the support of Y, which is

$$\mathcal{B} = \{y \mid y > 0\}.$$

Using the shaded integration region illustrated for $y = 1.7$ in Figure 7.4, the cumulative distribution function of Y is

$$
\begin{aligned}
F_Y(y) &= P(Y \le y) \\
&= P(X_1 + X_2 \le y) \\
&= \int_0^y \int_0^{y - x_1} \lambda^2 e^{-\lambda x_1 - \lambda x_2}\, dx_2\, dx_1 \\
&= 1 - e^{-\lambda y} - \lambda y e^{-\lambda y} \qquad y > 0.
\end{aligned}
$$

Differentiating with respect to y, the probability density function of Y is

$$f_Y(y) = \lambda^2 y e^{-\lambda y} \qquad y > 0.$$

This probability density function can be recognized as an Erlang(λ, 2) random variable.

This solution can be checked in APPL by using the function `Convolution`, which calculates the probability distribution of the sum of independent random variables.

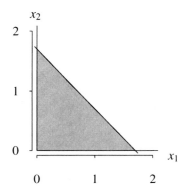

Figure 7.4: Support of X_1 and X_2 (first quadrant) and shaded region $x_1 + x_2 \leq y$.

```
X1 := ExponentialRV(lambda);
X2 := ExponentialRV(lambda);
Y  := Convolution(X1, X2);
PDF(Y);
CDF(Y);
```

These statements return the probability density function and cumulative distribution function for an Erlang(λ, 2) random variable as anticipated.

The result derived in the previous example generalizes from just two independent exponential(λ) random variables to n exponential(λ) random variables. If $X_1, X_2, \ldots, X_n$ are mutually independent exponential(λ) random variables then

$$\sum_{i=1}^{n} X_i \sim \text{Erlang}(\lambda, n).$$

This result will be derived in Section 7.3.

Example 7.12 Find the probability density function of the product of two independent $U(0, 1)$ random variables.

Let X_1 and X_2 denote the independent $U(0, 1)$ random variables. The joint probability density function is the product of the marginal probability density functions:

$$f_{X_1, X_2}(x_1, x_2) = 1 \qquad 0 < x_1 < 1, 0 < x_2 < 1.$$

The transformation $Y = g(X_1, X_2) = X_1 X_2$ maps the support of X_1 and X_2, which is

$$\mathcal{A} = \{(x_1, x_2) \,|\, 0 < x_1 < 1, 0 < x_2 < 1\}$$

to the support of Y, which is

$$\mathcal{B} = \{y \,|\, 0 < y < 1\}.$$

The shaded region illustrated in Figure 7.5 corresponds to $x_1 x_2 \leq y$ for $y = 1/4$ and indicates that the probability that $X_1 X_2 \leq y$ can be computed with two sets of double integrals, but is more easily computed with a single set of double integrals by considering the probability that $X_1 X_2 > y$ and using complimentary probability. The cumulative distribution function of Y is

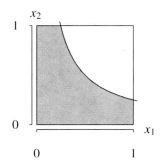

Figure 7.5: Support of X_1 and X_2 (unit square) and shaded region $x_1 x_2 \leq y$.

$$
\begin{aligned}
F_Y(y) &= P(Y \leq y) \\
&= P(X_1 X_2 \leq y) \\
&= 1 - P(X_1 X_2 > y) \\
&= 1 - \int_y^1 \int_{y/x_1}^1 1 \, dx_2 \, dx_1 \\
&= y - y \ln y \qquad 0 < y < 1
\end{aligned}
$$

or

$$
F_Y(y) = \begin{cases} 0 & y \leq 0 \\ y - y \ln y & 0 < y < 1 \\ 1 & y \geq 1. \end{cases}
$$

Differentiating with respect to y,

$$
f_Y(y) = -\ln y \qquad 0 < y < 1.
$$

This result can be verified in APPL using the Product function, which calculates the probability distribution of the product of two independent random variables.

```
X1 := UniformRV(0, 1);
X2 := UniformRV(0, 1);
Y  := Product(X1, X2);
```

A third and final example of applying the cumulative distribution function technique to bivariate continuous random variables illustrates the care that must be taken when the limits of integration change for various values of Y.

Example 7.13 Find the probability density function of the sum of two independent $U(0, 1)$ random variables.

Let X_1 and X_2 denote the independent $U(0, 1)$ random variables. As in the previous example, the joint probability density function is the product of the marginal probability density functions:

$$
f_{X_1, X_2}(x_1, x_2) = 1 \qquad 0 < x_1 < 1, 0 < x_2 < 1.
$$

The transformation $Y = g(X_1, X_2) = X_1 + X_2$ maps the support of X_1 and X_2, which is

$$
\mathcal{A} = \{(x_1, x_2) \mid 0 < x_1 < 1, 0 < x_2 < 1\}
$$

to the support of Y, which is

$$\mathcal{B} = \{y \mid 0 < y < 2\}.$$

Figure 7.6 illustrates the care that must be taken in setting up the integration limits. The shaded region on the left-hand graph corresponds to $x_1 + x_2 \le y$ for $y = 0.5$. For any y value between 0 and 1, only a single set of double integrals is required to calculate the probability that $X_1 + X_2 \le y$. The shaded region on the right-hand graph corresponds to $x_1 + x_2 \le y$ for $y = 1.8$. For any y between 1 and 2, either two sets of double integrals can be used to calculate the probability that $X_1 + X_2 \le y$ or complementary probability can be used with a single set of double integrals. Using the complementary probability approach, the cumulative distribution function for $Y = X_1 + X_2$ is

$$
\begin{aligned}
F_Y(y) \;&=\; P(Y \le y) \\
&=\; P(X_1 + X_2 \le y) \\
&=\; \begin{cases} 0 & y < 0 \\[2mm] \displaystyle\int_0^y \int_0^{y-x_1} 1\,dx_2\,dx_1 & 0 \le y < 1 \\[3mm] \displaystyle 1 - \int_{y-1}^1 \int_{y-x_1}^1 1\,dx_2\,dx_1 & 1 \le y < 2 \\[3mm] 1 & y \ge 2. \end{cases}
\end{aligned}
$$

This cumulative distribution function simplifies to

$$
F_Y(y) = \begin{cases} 0 & y < 0 \\ y^2/2 & 0 \le y < 1 \\ 1 - (2-y)^2/2 & 1 \le y < 2 \\ 1 & y \ge 2. \end{cases}
$$

Differentiating with respect to y to determine the probability density function:

$$
f_Y(y) = \begin{cases} y & 0 \le y < 1 \\ 2 - y & 1 \le y < 2. \end{cases}
$$

This probability density function can be recognized as that of a triangular distribution with minimum 0, mode 1, and maximum 2. The APPL code

```
X1  := UniformRV(0, 1);
X2  := UniformRV(0, 1);
Y   := Convolution(X1, X2);
```

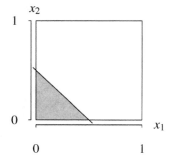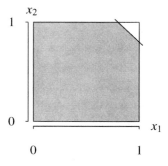

Figure 7.6: Support of X_1 and X_2 (unit square) and shaded regions $x_1 + x_2 \le y$.

confirms that
$$Y = X_1 + X_2 \sim \text{triangular}(0, 1, 2).$$

The fact that the distribution of the sum of two independent $U(0, 1)$ random variables has the triangular distribution should come as no surprise. Just as the total number of spots that appear on the up faces when a pair of fair dice are rolled has a symmetric triangular-shaped probability mass function, the sum of two independent $U(0, 1)$ random variables has a symmetric triangular-shaped probability density function.

You could imagine how tedious it would be to determine the distribution of the sum of four mutually independent $U(0, 1)$ random variables by analytic methods. For cases like this, APPL has a function named `ConvolutionIID` that calls `Convolution` in a loop. The APPL statements

```
X := UniformRV(0, 1);
Y := ConvolutionIID(X, 4);
```

return the random variable Y with probability density function

$$f_Y(y) = \begin{cases} y^3/6 & 0 \le y < 1 \\ -y^3/2 + 2y^2 - 2y + 2/3 & 1 \le y < 2 \\ y^3/2 - 4y^2 + 10y - 22/3 & 2 \le y < 3 \\ -y^3/6 + 2y^2 - 8y + 32/3 & 3 \le y < 4, \end{cases}$$

which is graphed in Figure 7.7. The progression from

- a $U(0, 1)$ distribution for a single $U(0, 1)$ random variable to

- a triangular$(0, 1, 2)$ distribution for the sum of two independent $U(0, 1)$ random variables to

- a somewhat bell-shaped distribution for the sum of four independent $U(0, 1)$ random variables

can be partially explained by the central limit theorem, which will be introduced in the next chapter. The central limit theorem states that the sum of mutually independent and identically distributed random variables converges to the normal distribution as the number of random variables summed approaches infinity. This is the reason that the probability density function becomes more bell-shaped as the number of random variables summed increases.

Implementing the cumulative distribution function technique can be tedious for complex transformations, support regions, and probability distributions. An alternative method that can oftentimes be easier to implement is the transformation technique, which is presented in the next section.

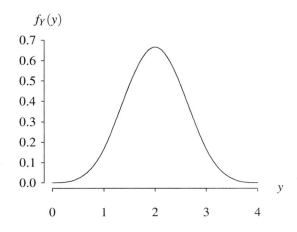

Figure 7.7: Probability density function of the sum of four mutually independent $U(0, 1)$ random variables.

7.2   Transformation Technique

The cumulative distribution function technique is based on the definition of the cumulative distribution function $F_Y(y)$, where Y is a function of one or more random variables. The transformation technique, the topic of this section, could also have been named the probability mass function or probability density function technique because it begins with $f_X(x)$ and ends with $f_Y(y)$, where $Y = g(X)$. We begin describing the technique in the case of univariate discrete random variables.

Theorem 7.1 Let X be a discrete random variable with probability mass function $f_X(x)$ defined on the support $\mathcal{A}$. Let $y = g(x)$ be a 1–1 transformation that maps $\mathcal{A}$ to $\mathcal{B}$. If $x = g^{-1}(y)$ is the inverse transformation for every $y \in \mathcal{B}$, then the probability mass function of $Y = g(X)$ is

$$f_Y(y) = f_X\left(g^{-1}(y)\right) \qquad\qquad y \in \mathcal{B}.$$

Proof The probability mass function of Y is

$$f_Y(y) = P(Y = y) = P\big(g(X) = y\big) = P\left(X = g^{-1}(y)\right) = f_X\left(g^{-1}(y)\right) \qquad y \in \mathcal{B}. \;\square$$

The transformation technique bypasses cumulative distribution functions and works directly with $f_X(x)$ and $f_Y(y)$. The procedure for finding the distribution of $Y = g(X)$ is to

1. determine the support of X, which is denoted by $\mathcal{A}$,

2. determine the probability mass function of X, which is denoted by $f_X(x)$,

3. confirm that $y = g(x)$ is a 1–1 function,

4. if g is a 1–1 function, find its inverse $x = g^{-1}(y)$,

5. determine the support of Y, which is denoted by $\mathcal{B}$,

6. determine the probability mass function of Y, which is $f_Y(y) = f_X\left(g^{-1}(y)\right)$ for every $y \in \mathcal{B}$.

The transformation technique for the discrete random variables X and $Y = g(X)$ begins with the probability mass function of X and ends with the probability mass function of Y. The steps for a continuous random variable X are similar.

Example 7.14 Let X be the daily demand for blueprints at a copy shop with probability mass function

$$f_X(x) = \begin{cases} 0.7 & x = 0 \\ 0.2 & x = 1 \\ 0.1 & x = 2. \end{cases}$$

If blueprints cost \$4 each, find the distribution of daily revenue from blueprints.

The transformation $Y = g(X) = 4X$ is a 1–1 function (because it satisfies the vertical and horizontal line tests) that maps the support of X, which is

$$\mathcal{A} = \{x \,|\, x = 0, 1, 2\}$$

to the support of Y, which is

$$\mathcal{B} = \{y \,|\, y = 0, 4, 8\}.$$

The inverse of $y = g(x) = 4x$ is $x = g^{-1}(y) = y/4$ for all $y \in \mathcal{B}$. Using Theorem 7.1, the probability mass function of Y is

$$f_Y(y) = f_X\left(g^{-1}(y)\right) = \begin{cases} 0.7 & y = 0 \\ 0.2 & y = 4 \\ 0.1 & y = 8, \end{cases}$$

that is, the revenue from blueprints is \$0, \$4, and \$8 with the same probabilities given by the demand.

Example 7.15 Let $X \sim \text{Geometric}(p)$, where the geometric distribution is parameterized from 1. Find the probability mass function of $Y = X^2$.

From Definition 4.5, the probability mass function of X is

$$f_X(x) = p(1-p)^{x-1} \qquad x = 1, 2, \ldots .$$

The transformation $Y = g(X) = X^2$ is a 1–1 function that maps the support of X, which is

$$\mathcal{A} = \{x \,|\, x = 1, 2, 3, \ldots\}$$

to the support of Y, which is

$$\mathcal{B} = \{y \,|\, y = 1, 4, 9, \ldots\}.$$

The inverse of $y = g(x) = x^2$ is $x = g^{-1}(y) = \sqrt{y}$ for all $y \in \mathcal{B}$. Using Theorem 7.1, the probability mass function of Y is

$$f_Y(y) = f_X\left(g^{-1}(y)\right) = p(1-p)^{\sqrt{y}-1} \qquad y = 1, 4, 9, \ldots .$$

This is consistent with the result derived using the cumulative distribution function technique in Example 7.6.

The transformation technique is slightly more complicated when X and $Y = g(X)$ are continuous random variables. The added layer of complication is illustrated by an example, where we find the distribution of $Y = g(X)$ by considering the probability that Y falls in various subsets of $\mathcal{B}$.

Example 7.16 The continuous random variable X has probability density function

$$f(x) = \frac{x}{8} \qquad 0 < x < 4.$$

Find the probability density function of $Y = g(X) = 2X^2$.

The transformation $Y = g(X) = 2X^2$ is 1–1 from

$$\mathcal{A} = \{x \,|\, 0 < x < 4\}$$

to

$$\mathcal{B} = \{y \,|\, 0 < y < 32\}.$$

Consider the open interval (y_1, y_2), where y_1 and y_2 are real constants that satisfy $0 < y_1 < y_2 < 32$. The probability that the random variable Y is in (y_1, y_2) is

$$
\begin{aligned}
P(y_1 < Y < y_2) &= P\left(\sqrt{\frac{y_1}{2}} < \sqrt{\frac{Y}{2}} < \sqrt{\frac{y_2}{2}}\right) \\
&= P\left(\sqrt{\frac{y_1}{2}} < X < \sqrt{\frac{y_2}{2}}\right) \\
&= \int_{\sqrt{y_1/2}}^{\sqrt{y_2/2}} \frac{x}{8}\, dx \\
&= \int_{y_1}^{y_2} \frac{\sqrt{y/2}}{8} \cdot \frac{1}{2\sqrt{2y}}\, dy \\
&= \int_{y_1}^{y_2} \frac{1}{32}\, dy,
\end{aligned}
$$

where the substitution $y = 2x^2$, or, equivalently $x = \sqrt{y/2}$ was used to alter the limits of integration. In performing this substitution, it was necessary to compute

$$\frac{dx}{dy} = \frac{1}{2\sqrt{2y}}.$$

Since

$$P(y_1 < Y < y_2) = \int_{y_1}^{y_2} \frac{1}{32} \, dy,$$

it is concluded that the probability density function for Y is

$$f_Y(y) = \frac{1}{32} \qquad 0 < y < 32.$$

It would be painful to replicate these steps each time we require the distribution of $Y = g(X)$. The theorem that follows focuses on the key step in the derivation, where we switched the integration with respect to x to an integration with respect to y, which can be generically written as

$$P(y_1 < Y < y_2) = \int_{y_1}^{y_2} f_X\left(g^{-1}(y)\right) \cdot \frac{dx}{dy} \cdot dy.$$

The transformation technique for continuous random variables is effectively using the steps illustrated in the previous example to determine the probability density function of Y.

Theorem 7.2 Let X be a continuous random variable with probability density function $f_X(x)$ defined on the support $\mathcal{A}$. Let $y = g(x)$ be a 1–1 transformation that maps $\mathcal{A}$ to $\mathcal{B}$ with inverse $x = g^{-1}(y)$ for every $y \in \mathcal{B}$, and let $\frac{dx}{dy}$ be continuous and nonzero for every $y \in \mathcal{B}$. Then the probability density function of $Y = g(X)$ is

$$f_Y(y) = f_X\left(g^{-1}(y)\right) \left| \frac{dx}{dy} \right| \qquad y \in \mathcal{B}.$$

Proof Consider the case where $y = g(x)$ is a monotone increasing function, which implies that $\frac{dx}{dy} > 0$. The initial string of equalities from the cumulative distribution function technique is

$$F_Y(y) = P(Y \le y) = P\left(g(X) \le y\right) = P\left(X \le g^{-1}(y)\right) = F_X\left(g^{-1}(y)\right)$$

for all $y \in \mathcal{B}$. Differentiating this expression via the chain rule yields the probability density function of Y:

$$f_Y(y) = \frac{d}{dy} F_Y(y) = \frac{d}{dy} F_X\left(g^{-1}(y)\right) = f_X\left(g^{-1}(y)\right) \frac{dx}{dy}$$

for all $y \in \mathcal{B}$. In the case when $y = g(x)$ is a monotone decreasing function, $\frac{dx}{dy} < 0$, so $f_Y(y)$ differs from the monotone increasing case by only a sign change. Since $y = g(x)$ is a 1–1 function, it must be either a monotone increasing function or a monotone decreasing function. The two cases can be combined by taking the absolute value of $\frac{dx}{dy}$, which proves the theorem. □

The next three examples illustrate the use of the transformation technique in the case of a continuous random variable X that is transformed by a 1–1 function $Y = g(X)$. The transformation must be checked to ensure that it is 1–1 in order to use Theorem 7.2. This can be done using the vertical line test (to ensure that $y = g(x)$ is a function) and then the horizontal line test (to ensure that $y = g(x)$ is a 1–1 function).

Example 7.17 Find the probability density function of the square root of a $U(3, 5)$ random variable.

Let $X \sim U(3, 5)$. The probability density function of X is

$$f_X(x) = \frac{1}{2} \qquad 3 < x < 5.$$

The problem asks for the probability density function of $Y = g(X) = \sqrt{X}$. The function $y = g(x)$ is a 1–1 transformation (it passes the vertical and horizontal line tests) from the support of X, which is

$$\mathcal{A} = \{x \,|\, 3 < x < 5\}$$

to the support of Y, which is

$$\mathcal{B} = \left\{ y \,|\, \sqrt{3} < y < \sqrt{5} \right\},$$

with inverse

$$x = g^{-1}(y) = y^2$$

and

$$\frac{dx}{dy} = 2y.$$

Using Theorem 7.2, the probability density function of Y on its support is

$$f_Y(y) = f_X\left(g^{-1}(y)\right) \left| \frac{dx}{dy} \right| = \frac{1}{2} |2y| = y \qquad \sqrt{3} < y < \sqrt{5}.$$

This probability density function is identical the probability density function found using the cumulative distribution function technique in Example 7.8.

Example 7.18 Find the probability density function and expected value of the volume of a sphere with a random radius that has a unit exponential distribution.

It would be perfectly acceptable to let the random variable R denote the random radius of the sphere and the random variable V denote the random volume of the sphere, and use Theorem 7.2 in the following fashion:

$$f_V(v) = f_R\left(g^{-1}(v)\right) \left| \frac{dr}{dv} \right| \qquad v \in \mathcal{B}.$$

To be consistent with the earlier notation, however, we instead let X denote the random radius of the sphere and Y denote the random volume of the sphere. The probability density function of X is that of an exponential random variable with $\lambda = 1$:

$$f_X(x) = e^{-x} \qquad x > 0.$$

The problem asks for the probability density function of the volume of the sphere:

$$Y = g(X) = \frac{4}{3}\pi X^3.$$

The function $y = g(x)$ is a 1–1 transformation from the support of X, which is

$$\mathcal{A} = \{x \,|\, x > 0\}$$

to the support of Y, which is

$$\mathcal{B} = \{y \,|\, y > 0\}$$

with inverse

$$x = g^{-1}(y) = \sqrt[3]{\frac{3y}{4\pi}}$$

and

$$\frac{dx}{dy} = \frac{1}{3}\left(\frac{3y}{4\pi}\right)^{-2/3}\frac{3}{4\pi} = \frac{1}{4\pi}\left(\frac{3y}{4\pi}\right)^{-2/3} = \frac{(3y)^{-2/3}}{(4\pi)^{1/3}} = (36\pi y^2)^{-1/3}.$$

Therefore, by Theorem 7.2, the probability density function of Y is

$$\begin{aligned}
f_Y(y) &= f_X\left(g^{-1}(y)\right)\left|\frac{dx}{dy}\right| \\
&= e^{-(3y/4\pi)^{1/3}}\left|(36\pi y^2)^{-1/3}\right| \\
&= e^{-(3y/4\pi)^{1/3}}(36\pi y^2)^{-1/3} \qquad\qquad y > 0.
\end{aligned}$$

The APPL statements to compute the probability density function of the volume of the sphere are

```
X := ExponentialRV(1);
g := [[x -> 4 * Pi * x ^ 3 / 3], [0, infinity]];
Y := Transform(X, g);
PlotDist(Y);
Mean(Y);
```

which confirm the analytic derivation. The additional call to PlotDist draws a graph of the probability density function of Y, which is shown in Figure 7.8. This probability density function has a vertical asymptote at $y = 0$ and a heavy right-hand tail, which results in $E[Y] = 8\pi \cong 25.1327$.

A Monte Carlo simulation, which generates 100,000 random unit exponential radii and averages the associated volumes, can be used to check the expected volume determined by analytic methods. Five replications of the single R statement

```
mean((4 / 3) * pi * rexp(100000) ^ 3)
```

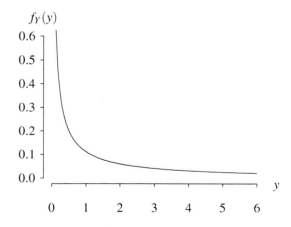

Figure 7.8: Probability density function of Y, the volume of the sphere.

return

$$25.0386 \qquad 25.3754 \qquad 25.6655 \qquad 25.0735 \qquad 24.5519,$$

which is consistent the analytic result $E[Y] = 8\pi \cong 25.1327$.

Example 7.19 A manufacturing facility running at full capacity can produce two tons of a product daily. Due to machine breakdowns, labor shortages, raw material shortages, etc., the actual quantity of product produced daily (in tons) is a random variable X with probability density function

$$f_X(x) = \frac{1}{4}x^3 \qquad 0 < x < 2.$$

Notice that 2 tons is the upper limit of the support of X. This probability distribution has been determined from a large historical data set. The sales price is $900 per ton; raw materials can be purchased for $100 per ton; fixed costs are $1000 per day. Find

(a) the probability density function of the daily profit,

(b) the population mean of the daily profit,

(c) the population standard deviation of the daily profit,

(d) the probability that the facility is profitable on any given day.

Consider first a deterministic analysis with the facility running at full capacity, producing exactly 2 tons of product. The profit in this case is

$$(900 - 100)(2) - 1000 = 600$$

Thus, $600 is the upper limit of the support of the profit. The break-even quantity is found by solving

$$(900 - 100)x - 1000 = 0$$

for x which yields $x = 1.25$ tons produced daily.

(a) Let the random variable Y denote the daily profit, which is given by

$$Y = g(X) = 800X - 1000.$$

The function $y = g(x) = 800x - 1000$ is a 1–1 transformation from the support of X, which is

$$\mathcal{A} = \{x \mid 0 < x < 2\}$$

to the support of Y, which is

$$\mathcal{B} = \{y \mid -1000 < y < 600\}$$

with inverse

$$x = g^{-1}(y) = \frac{y + 1000}{800}$$

and

$$\frac{dx}{dy} = \frac{1}{800}.$$

Therefore, by Theorem 7.2, the probability density function of Y is

$$
\begin{aligned}
f_Y(y) &= f_X\left(g^{-1}(y)\right)\left|\frac{dx}{dy}\right| \\
&= \frac{1}{4}\left(\frac{y + 1000}{800}\right)^3 \left|\frac{1}{800}\right| \\
&= \frac{1}{3200}\left(\frac{y + 1000}{800}\right)^3 \qquad -1000 < y < 600.
\end{aligned}
$$

(b) The population mean profit is

$$\mu_Y = E[Y] = \int_{-1000}^{600} y \cdot \frac{1}{3200} \left(\frac{y + 1000}{800} \right)^3 dy = 280.$$

So the mean profit is $280 daily.

(c) The population variance of the daily profit is

$$\sigma_Y^2 = V[Y] = \int_{-1000}^{600} (y - 280)^2 \frac{1}{3200} \left(\frac{y + 1000}{800} \right)^3 dy = \frac{204,800}{3} \cong 68,266.67.$$

The population standard deviation of the profit is $\sigma_Y = \$261.27$.

(d) The probability that the facility will be profitable on any given day is calculated by finding the probability that Y exceeds 0:

$$P(Y > 0) = \int_0^{600} \frac{1}{3200} \left(\frac{y + 1000}{800} \right)^3 dy = \frac{3471}{4096} \cong 0.8474.$$

The facility will be profitable ($Y > 0$) on about 85% of the days that it operates. The APPL statements to compute the quantities in this problem are given below.

```
X := [[x -> x ^ 3 / 4], [0, 2], ["Continuous", "PDF"]];
g := [[x -> 800 * x - 1000], [0, 2]];
Y := Transform(X, g);
Mean(Y);
sqrt(Variance(Y));
1 - CDF(Y, 0);
```

The previous three examples of the transformation technique for continuous random variables have illustrated the simplest possible application of the technique because it was applied to a 1–1 transformation $Y = g(X)$. The transformation technique can also be applied to 2–1 transformations— with details given in the next result.

Theorem 7.3 Let X be a continuous random variable with probability density function $f_X(x)$ defined on the support $\mathcal{A}$. Let $y = g(x)$ be a 2–1 transformation that maps $\mathcal{A}$ to $\mathcal{B}$. Furthermore, assume that the set $\mathcal{A}$ can be partitioned into two sets $\mathcal{A}_1$ and $\mathcal{A}_2$ that satisfy the following four conditions: $\mathcal{A}_1 \cup \mathcal{A}_2 = \mathcal{A}$, $\mathcal{A}_1 \cap \mathcal{A}_2 = \emptyset$, $y = g(x)$ is a 1–1 transformation that maps $\mathcal{A}_1$ to $\mathcal{B}$ with inverse $x = g_1^{-1}(y)$, and $y = g(x)$ is a 1–1 transformation that maps $\mathcal{A}_2$ to $\mathcal{B}$ with inverse $x = g_2^{-1}(y)$. If the derivatives of the inverse functions are continuous and nonzero for every $y \in \mathcal{B}$, then the probability density function of $Y = g(X)$ is

$$f_Y(y) = f_X \left(g_1^{-1}(y) \right) \left| \frac{dg_1^{-1}(y)}{dy} \right| + f_X \left(g_2^{-1}(y) \right) \left| \frac{dg_2^{-1}(y)}{dy} \right| \qquad y \in \mathcal{B}.$$

Proof Consider the case where $y = g_1(x)$ and $y = g_2(x)$ are continuous monotone increasing functions, which implies that $\frac{dg_1^{-1}(y)}{dy} > 0$ and $\frac{dg_2^{-1}(y)}{dy} > 0$. The cumulative distribution function of Y is

$$
\begin{aligned}
F_Y(y) &= P(Y \le y) \\
&= P\big(g(X) \le y\big) \\
&= P\big(g_1(X) \le y, g_2(X) \le y\big) \\
&= P\big(g_1(X) \le y\big) + P\big(g_2(X) \le y\big) \\
&= P\big(X \le g_1^{-1}(y)\big) + P\big(X \le g_2^{-1}(y)\big) \\
&= F_X \big(g_1^{-1}(y)\big) + F_X \big(g_2^{-1}(y)\big) \qquad y \in \mathcal{B}.
\end{aligned}
$$

Differentiating this expression via the chain rule yields the probability density function of Y:

$$f_Y(y) = \frac{d}{dy} F_Y(y) = f_X\left(g_1^{-1}(y)\right) \frac{dg_1^{-1}(y)}{dy} + f_X\left(g_2^{-1}(y)\right) \frac{dg_2^{-1}(y)}{dy} \qquad y \in \mathcal{B}.$$

In the case when one or both of $y = g_1(x)$ and $y = g_2(x)$ are monotone decreasing functions, the associated derivatives of the inverse functions are negative, so taking the absolute values of the derivatives covers all cases. Since $y = g_1(x)$ and $y = g_2(x)$ are 1–1 functions, they must be either monotone increasing functions or monotone decreasing functions. The two cases can be combined by taking the absolute value of the derivatives of the inverse functions, which proves the theorem. $\qquad\square$

Example 7.20 Find the probability density function of the square of a standard normal random variable.

This problem was previously addressed in Theorem 5.5 using the cumulative distribution function technique. We give an alternative derivation here using the transformation technique. Let $X \sim N(0, 1)$. The probability density function of X is

$$f_X(x) = \frac{1}{\sqrt{2\pi}} e^{-x^2/2} \qquad -\infty < x < \infty.$$

The problem asks for the probability density function of $Y = g(X) = X^2$. The function $y = g(x)$ is a 2–1 transformation from the support of X, which is

$$\mathcal{A} = \{x \mid -\infty < x < \infty\}$$

to the support of Y, which is

$$\mathcal{B} = \{y \mid 0 \le y < \infty\}.$$

The support of X can be partitioned into

$$\mathcal{A}_1 = \{x \mid -\infty < x < 0\} \qquad \text{and} \qquad \mathcal{A}_2 = \{x \mid 0 \le x < \infty\}.$$

The union of $\mathcal{A}_1$ and $\mathcal{A}_2$ is $\mathcal{A}$, the intersection of $\mathcal{A}_1$ and $\mathcal{A}_2$ is $\emptyset$, $y = g_1(x) = x^2$ is a 1–1 mapping from $\mathcal{A}_1$ to $\mathcal{B}$, and $y = g_2(x) = x^2$ is a 1–1 mapping from $\mathcal{A}_2$ to $\mathcal{B}$. Since the assumptions of Theorem 7.3 are satisfied, we compute the two inverse functions

$$x = g_1^{-1}(y) = -\sqrt{y} \qquad \text{and} \qquad x = g_2^{-1}(y) = \sqrt{y}$$

and their derivatives

$$\frac{dg_1^{-1}(y)}{dy} = -\frac{1}{2\sqrt{y}} \qquad \text{and} \qquad \frac{dg_2^{-1}(y)}{dy} = \frac{1}{2\sqrt{y}}.$$

Using Theorem 7.3, the probability density function of Y on its support is

$$f_Y(y) = f_X\left(g_1^{-1}(y)\right) \left| \frac{dg_1^{-1}(y)}{dy} \right| + f_X\left(g_2^{-1}(y)\right) \left| \frac{dg_2^{-1}(y)}{dy} \right| \qquad y \in \mathcal{B}$$

or

$$f_Y(y) = \frac{1}{\sqrt{2\pi}} e^{-y/2} \left| -\frac{1}{2\sqrt{y}} \right| + \frac{1}{\sqrt{2\pi}} e^{-y/2} \left| \frac{1}{2\sqrt{y}} \right| \qquad y \ge 0$$

or

$$f_Y(y) = \sqrt{\frac{1}{2\pi y}} e^{-y/2} \qquad y \ge 0.$$

This is recognized as the probability density function of a chi-square random variable with 1 degree of freedom.

The APPL code to check this result requires just three statements.

```
X := NormalRV(0, 1);
g := [[x -> x ^ 2, x -> x ^ 2], [-infinity, 0, infinity]];
Y := Transform(X, g);
```

As indicated earlier, the function $y = g(x)$ must be broken into piecewise monotone segments so that Transform calculates the distribution of $Y = g(X)$ properly.

The generalization of Theorem 7.3 should now be apparent. If $Y = g(X)$ is a k–1 transformation transformation that maps $\mathcal{A}$ to $\mathcal{B}$, then it can be partitioned into the sets $\mathcal{A}_1$, $\mathcal{A}_2$, ..., $\mathcal{A}_k$ with associated inverse functions $x = g_1^{-1}(y)$, $x = g_2^{-1}(y)$, ..., $x = g_k^{-1}(y)$. If the derivatives with respect to y of the inverse functions are continuous and nonzero for every $y \in \mathcal{B}$, then the probability density function of $Y = g(X)$ is

$$f_Y(y) = \sum_{i=1}^{k} f_X\left(g_i^{-1}(y)\right) \left| \frac{dg_i^{-1}(y)}{dy} \right| \qquad y \in \mathcal{B}.$$

The transformation technique is not limited, however, to just k–1 transformations. The next example shows that the transformation technique can be applied to any transformation $Y = g(X)$ that can be broken into piecewise monotone segments.

Example 7.21 Find the probability density function of the square of a $U(-1, 2)$ random variable.

The probability density function of X is

$$f_X(x) = \frac{1}{3} \qquad -1 < x < 2.$$

The transformation $Y = g(X) = X^2$ is piecewise many-to-1 from the support of X, which is

$$\mathcal{A} = \{x | -1 < x < 2\}$$

to the support of Y, which is

$$\mathcal{B} = \{y | 0 \le y < 4\}.$$

Breaking $\mathcal{A}$ into the sets

$$\mathcal{A}_1 = \{x | -1 < x < 0\} \qquad \mathcal{A}_2 = \{x | 0 \le x < 1\} \qquad \mathcal{A}_3 = \{x | 1 \le x < 2\}$$

yields the following monotonic transformations, their inverses, and their derivatives:

$$\mathcal{A}_1: \quad y = g_1(x) = x^2, \quad x = g_1^{-1}(y) = -\sqrt{y}, \quad \frac{dg_1^{-1}(y)}{dy} = -\frac{1}{2\sqrt{y}},$$

$$\mathcal{A}_2: \quad y = g_2(x) = x^2, \quad x = g_2^{-1}(y) = \sqrt{y}, \quad \frac{dg_2^{-1}(y)}{dy} = \frac{1}{2\sqrt{y}},$$

$$\mathcal{A}_3: \quad y = g_3(x) = x^2, \quad x = g_3^{-1}(y) = \sqrt{y}, \quad \frac{dg_3^{-1}(y)}{dy} = \frac{1}{2\sqrt{y}}.$$

The transformation $Y = g(X) = X^2$ is 2–1 on $\mathcal{A}_1 \cup \mathcal{A}_2$ (with the exception of $x = 0$) and is 1–1 on $\mathcal{A}_3$. Thus, the probability density function of Y is

$$f_Y(y) = \begin{cases} \dfrac{1}{3}\left| -\dfrac{1}{2\sqrt{y}} \right| + \dfrac{1}{3}\left| \dfrac{1}{2\sqrt{y}} \right| & 0 < y < 1 \\[4mm] \dfrac{1}{3}\left| \dfrac{1}{2\sqrt{y}} \right| & 1 \le y < 4 \end{cases}$$

or

$$
f_Y(y) = \begin{cases} \dfrac{1}{3\sqrt{y}} & 0 < y < 1 \\[2mm] \dfrac{1}{6\sqrt{y}} & 1 \le y < 4. \end{cases}
$$

This result is identical to that derived using the cumulative distribution function technique in Example 7.9.

This ends the discussion of the transformation technique in the case of a univariate random variable X defined on the support $\mathcal{A}$ being transformed by $Y = g(X)$. Applications arise when there is an interest in the distribution of some function of two random variables X_1 and X_2. As in the previous cases, we begin with the case when X_1 and X_2 are discrete random variables. Regardless of whether the interest is in a single random variable $Y = g(X_1, X_2)$ or whether the interest is in the joint distribution of

$$ Y_1 = g_1(X_1, X_2) \qquad \text{and} \qquad Y_2 = g_2(X_1, X_2) $$

the following theorem can be applied. Note that a significant abuse of the g^{-1} notation has been allowed in the theorem in order to minimize the number of letters required to denote function names.

Theorem 7.4 Let X_1 and X_2 be discrete random variables with joint probability mass function $f_{X_1, X_2}(x_1, x_2)$ defined on support $\mathcal{A}$. Let the functions

$$ y_1 = g_1(x_1, x_2) $$

$$ y_2 = g_2(x_1, x_2) $$

define a bivariate 1–1 transformation from the two-dimensional set $\mathcal{A}$ to the two-dimensional set $\mathcal{B}$. Assume that the functions $y_1 = g_1(x_1, x_2)$ and $y_2 = g_2(x_1, x_2)$ can be solved uniquely and in closed form for x_1 and x_2 as

$$ x_1 = g_1^{-1}(y_1, y_2) $$

$$ x_2 = g_2^{-1}(y_1, y_2), $$

which maps $\mathcal{B}$ to $\mathcal{A}$. The joint probability mass function of $Y_1 = g_1(X_1, X_2)$ and $Y_2 = g_2(X_1, X_2)$ is

$$ f_{Y_1, Y_2}(y_1, y_2) = f_{X_1, X_2}\big(g_1^{-1}(y_1, y_2), g_2^{-1}(y_1, y_2)\big) \qquad (y_1, y_2) \in \mathcal{B}. $$

Proof The joint probability mass function of Y_1 and Y_2 is

$$
\begin{aligned}
f_{Y_1, Y_2}(y_1, y_2) &= P(Y_1 = y_1, Y_2 = y_2) \\
&= P\big(g_1(X_1, X_2) = y_1, g_2(X_1, X_2) = y_2\big) \\
&= P\big(X_1 = g_1^{-1}(y_1, y_2), X_2 = g_2^{-1}(y_1, y_2)\big) \\
&= f_{X_1, X_2}\big(g_1^{-1}(y_1, y_2), g_2^{-1}(y_1, y_2)\big) \qquad (y_1, y_2) \in \mathcal{B}
\end{aligned}
$$

which proves the result. $\qquad\qquad \square$

There are two settings where Theorem 7.4 can be applied. The first setting is when the joint probability distribution of Y_1 and Y_2 is of interest. In this case, the theorem is applied directly. The second setting is when only Y_1 is of interest. In most cases, Y_2 is not defined, but is required to apply the theorem. The best way to proceed is to define a "dummy" transformation whose only purpose is to enable the use of the theorem, then find the marginal distribution of Y_1 in the usual fashion. In most problems, there are several choices for the dummy transformation that will work successfully. The next example requires the definition of a dummy transformation.

Example 7.22 Let $X_1 \sim \text{Poisson}(\lambda_1)$ and $X_2 \sim \text{Poisson}(\lambda_2)$ be independent random variables with real positive parameters λ_1 and λ_2. Find the probability mass function of the sum of X_1 and X_2.

Recall from Section 4.5 that a $\text{Poisson}(\lambda)$ random variable X has probability mass function

$$f_X(x) = \frac{\lambda^x e^{-\lambda}}{x!} \qquad x = 0, 1, 2, \ldots.$$

Since $X_1 \sim \text{Poisson}(\lambda_1)$ and $X_2 \sim \text{Poisson}(\lambda_2)$ are independent random variables, their joint probability mass function is the product of the marginal probability mass functions:

$$f_{X_1, X_2}(x_1, x_2) = \frac{\lambda_1^{x_1} \lambda_2^{x_2} e^{-\lambda_1 - \lambda_2}}{x_1! x_2!} \qquad x_1 = 0, 1, 2, \ldots; \ x_2 = 0, 1, 2, \ldots.$$

The random variable of interest is

$$Y_1 = g_1(X_1, X_2) = X_1 + X_2.$$

A *dummy* transformation is needed to apply Theorem 7.4. One simple dummy transformation is

$$Y_2 = g_2(X_1, X_2) = X_2.$$

The next step is to check to see if these two functions of X_1 and X_2 result in a 1–1 transformation, which is necessary to apply the theorem. Figure 7.9 shows *part* of the support $\mathcal{A}$ of X_1 and X_2 on the left-hand graph and *part* of the support $\mathcal{B}$ of Y_1 and Y_2 on the right-hand graph. Since $Y_2 = X_2$, rows of points on the left-hand graph map to rows of points on the right-hand graph. Furthermore, each row of points is shifted one to the right of the row of points below it because $Y_1 = X_1 + X_2$. After assessing the transformation in this fashion, we can conclude that

$$y_1 = g_1(x_1, x_2) = x_1 + x_2 \qquad \text{and} \qquad y_2 = g_2(x_1, x_2) = x_2$$

is a 1–1 transformation from

$$\mathcal{A} = \{(x_1, x_2) \,|\, x_1 = 0, 1, 2, \ldots; \ x_2 = 0, 1, 2, \ldots\}$$

to

$$\mathcal{B} = \{(y_1, y_2) \,|\, y_1 = 0, 1, 2, \ldots; \ y_2 = 0, 1, 2, \ldots, y_1\}.$$

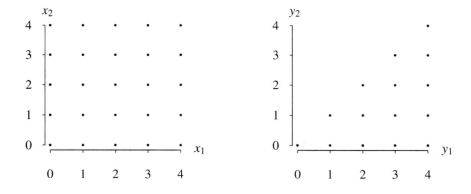

Figure 7.9: Support of X_1 and X_2 and the support of Y_1 and Y_2.

Additionally, the transformation can be solved in closed form for x_1 and x_2:

$$x_1 = g_1^{-1}(y_1, y_2) = y_1 - y_2 \qquad \text{and} \qquad x_2 = g_2^{-1}(y_1, y_2) = y_2.$$

Applying Theorem 7.4, the joint probability mass function of Y_1 and Y_2 is

$$f_{Y_1,Y_2}(y_1, y_2) = \frac{\lambda_1^{y_1-y_2}\lambda_2^{y_2}e^{-\lambda_1-\lambda_2}}{(y_1-y_2)!y_2!} \qquad (y_1, y_2) \in \mathcal{B}.$$

Since the question requires the probability mass function of Y_1 alone, the marginal probability mass function of Y_1 is found by summing out y_2 as follows:

$$
\begin{aligned}
f_{Y_1}(y_1) &= \sum_{y_2=0}^{y_1} \frac{e^{-\lambda_1-\lambda_2}}{y_1!} \cdot \frac{y_1!}{(y_1-y_2)!y_2!}\lambda_1^{y_1-y_2}\lambda_2^{y_2} \\
&= \frac{(\lambda_1+\lambda_2)^{y_1}e^{-\lambda_1-\lambda_2}}{y_1!} \qquad y_1 = 0, 1, \ldots.
\end{aligned}
$$

The derivation of the marginal probability mass function of Y_1 relied on the fact that

$$\frac{y_1!}{(y_1-y_2)!y_2!} = \binom{y_1}{y_2}$$

and the binomial theorem (applied in reverse). The probability mass function of Y_1 is recognized as that of a Poisson random variable with parameter $\lambda_1 + \lambda_2$. So to summarize, if $X_1 \sim \text{Poisson}(\lambda_1)$ and $X_2 \sim \text{Poisson}(\lambda_2)$ are independent random variables, then

$$X_1 + X_2 \sim \text{Poisson}(\lambda_1 + \lambda_2).$$

The APPL code to check this result relies on the `Convolution` function to calculate the distribution of the sum of two independent Poisson random variables.

```
X1 := PoissonRV(lambda1);
X2 := PoissonRV(lambda2);
Y  := Convolution(X1, X2);
```

There are two comments concerning the previous example. First, by induction, this result generalizes to the sum of any number of Poisson random variables. That is, if $X_i \sim \text{Poisson}(\lambda_i)$, for $i = 1, 2, \ldots, n$, and $X_1, X_2, \ldots, X_n$ are mutually independent random variables, then

$$X_1 + X_2 + \cdots + X_n \sim \text{Poisson}(\lambda_1 + \lambda_2 + \cdots + \lambda_n).$$

Second, a *much* easier way of proving this result will be given in the next section.

Example 7.23 Find the population variance of the determinant of a random 2×2 matrix whose entries are mutually independent Bernoulli(p) random variables.

Recall from Section 4.1 that the probability mass function of a Bernoulli(p) random variable X can be written as

$$f(x) = \begin{cases} 1-p & x = 0 \\ p & x = 1, \end{cases}$$

where $0 < p < 1$. The determinant of a 2×2 matrix whose entries are mutually independent Bernoulli(p) random variables X_1, X_2, X_3, X_4 is

$$Y = g(X_1, X_2, X_3, X_4) = \begin{vmatrix} X_1 & X_2 \\ X_3 & X_4 \end{vmatrix} = X_1X_4 - X_2X_3.$$

The problem will be solved in three phases. The first phase determines the distribution of the product of two independent Bernoulli random variables in order to find the distribution of X_1X_4 and X_2X_3. The second phase determines the distribution of the difference of the products, which is the distribution of the random determinant Y. Finally, the variance of Y is computed.

First consider the distribution of X_1X_4. It would be perfectly reasonable to find a dummy transformation and apply Theorem 7.4, but the support is so small that in this case it is easier to just use reason to arrive at the probability mass function. The product X_1X_4 can only assume the values 0 and 1. Furthermore, the only way that X_1X_4 can assume the value 1 is when X_1 and X_4 are both 1, which occurs with probability p^2. Since X_1 and X_4 are independent, the probability mass function of their product is

$$f_{X_1X_4}(x) = \begin{cases} 1 - p^2 & x = 0 \\ p^2 & x = 1. \end{cases}$$

Likewise, the probability mass function of the product of X_2 and X_3 is

$$f_{X_2X_3}(x) = \begin{cases} 1 - p^2 & x = 0 \\ p^2 & x = 1. \end{cases}$$

The determinant of the matrix of mutually independent Bernoulli(p) random variables

$$Y = g(X_1, X_2, X_3, X_4) = \begin{vmatrix} X_1 & X_2 \\ X_3 & X_4 \end{vmatrix} = X_1X_4 - X_2X_3$$

can only assume the values -1, 0, and 1. Because X_1X_4 and X_2X_3 are independent, the probability that $Y = X_1X_4 - X_2X_3 = 0 - 1 = -1$ is

$$P(Y = -1) = P(X_1X_4 = 0)P(X_2X_3 = 1) = \left(1 - p^2\right)p^2.$$

Likewise, the probability that $Y = X_1X_4 - X_2X_3 = 0$ is

$$P(Y = 0) = P(X_1X_4 = 0)P(X_2X_3 = 0) + P(X_1X_4 = 1)P(X_2X_3 = 1) = \left(1 - p^2\right)^2 + p^4.$$

Finally, the probability that $Y = X_1X_4 - X_2X_3 = 1 - 0 = 1$ is

$$P(Y = 1) = P(X_1X_4 = 1)P(X_2X_3 = 0) = p^2\left(1 - p^2\right).$$

Combining these three probabilities produces the probability mass function of the determinant Y, which is symmetric about $y = 0$:

$$f_Y(y) = \begin{cases} p^2\left(1 - p^2\right) & y = -1 \\ p^4 + \left(1 - p^2\right)^2 & y = 0 \\ p^2\left(1 - p^2\right) & y = 1. \end{cases}$$

The last step is to determine the population variance of the determinant. The population mean of the determinant $E[Y]$ must be zero because $f_Y(y)$ is an even function. Using the shortcut formula for the population variance,

$$V[Y] = E\left[Y^2\right] - E[Y]^2 = E\left[Y^2\right] = \sum_{y=-1}^{1} y^2 f_Y(y) = 2p^2\left(1 - p^2\right).$$

The APPL code to solve this problem, which follows that same progression as the analytic solution, relies on the `Product` and `Difference` functions to calculate the distribution of the determinant.

```
X1 := BernoulliRV(p);
X2 := BernoulliRV(p);
X3 := BernoulliRV(p);
X4 := BernoulliRV(p);
T1 := Product(X1, X3);
T2 := Product(X2, X4);
T3 := Difference(T1, T2);
Variance(T3);
```

There is also an APPL procedure named `Determinant`, which can be used to determine the distribution of the determinant of a 2×2 matrix with random elements. This procedure can be used to produce the same result more succinctly with the APPL code given below.

```
X := BernoulliRV(p);
A := [[X, X], [X, X]];
Variance(Determinant(A));
```

This ends the discussion of the use of the transformation technique applied to a bivariate transformation of *discrete* random variables X_1 and X_2. The focus now shifts to the use of the transformation technique applied to a transformation of *continuous* random variables X_1 and X_2. This begins with the analogous result to Theorem 7.4, which again makes an abuse of the inverse notation g^{-1} in order to keep the number of function names to a minimum.

Theorem 7.5 Let X_1 and X_2 be continuous random variables with joint probability density function $f_{X_1, X_2}(x_1, x_2)$ defined on the support $\mathcal{A}$. Let the functions

$$y_1 = g_1(x_1, x_2)$$

$$y_2 = g_2(x_1, x_2)$$

define a bivariate 1–1 transformation from the two-dimensional set $\mathcal{A}$ to the two-dimensional set $\mathcal{B}$. Assume that the functions $y_1 = g_1(x_1, x_2)$ and $y_2 = g_2(x_1, x_2)$ can be solved uniquely and in closed form for x_1 and x_2 as

$$x_1 = g_1^{-1}(y_1, y_2)$$

$$x_2 = g_2^{-1}(y_1, y_2),$$

which maps $\mathcal{B}$ to $\mathcal{A}$. Assume further that the Jacobian of the transformation

$$J = \begin{vmatrix} \dfrac{\partial x_1}{\partial y_1} & \dfrac{\partial x_1}{\partial y_2} \\ \dfrac{\partial x_2}{\partial y_1} & \dfrac{\partial x_2}{\partial y_2} \end{vmatrix},$$

where the bars around the matrix denote the determinant of the matrix, is continuous and nonzero for all $(y_1, y_2) \in \mathcal{B}$. The joint probability density function of $Y_1 = g_1(X_1, X_2)$ and $Y_2 = g_2(X_1, X_2)$ is

$$f_{Y_1, Y_2}(y_1, y_2) = f_{X_1, X_2}\left(g_1^{-1}(y_1, y_2), g_2^{-1}(y_1, y_2)\right) |J| \qquad (y_1, y_2) \in \mathcal{B},$$

where the bars around the Jacobian denote absolute value.

Proof Let the event A be a subset of $\mathcal{A}$. The event B is the mapping of A under the 1–1 transformation $y_1 = g_1(x_1, x_2)$ and $y_2 = g_2(x_1, x_2)$. The probability that Y_1 and Y_2 are in B is

$$
\begin{aligned}
P\big[(Y_1, Y_2) \in B\big] &= P\big[(X_1, X_2) \in A\big] \\
&= \int\!\!\int_A f_{X_1, X_2}(x_1, x_2)\, dx_2\, dx_1 \\
&= \int\!\!\int_B f_{X_1, X_2}\big(g_1^{-1}(y_1, y_2), g_2^{-1}(y_1, y_2)\big)\, |J|\, dy_2\, dy_1
\end{aligned}
$$

by using the substitution $x_1 = g_1^{-1}(y_1, y_2)$ and $x_2 = g_2^{-1}(y_1, y_2)$ and using a result from advanced calculus. This implies that the joint probability density function of $Y_1 = g_1(X_1, X_2)$ and $Y_2 = g_2(X_1, X_2)$ is

$$
f_{Y_1, Y_2}(y_1, y_2) = f_{X_1, X_2}\big(g_1^{-1}(y_1, y_2), g_2^{-1}(y_1, y_2)\big)\, |J| \qquad (y_1, y_2) \in \mathcal{B}. \qquad \square
$$

Example 7.24 Find the probability density function of the product of two independent $U(0, 1)$ random variables.

Let X_1 and X_2 denote the two independent $U(0, 1)$ random variables. We want to find the distribution of $Y_1 = g_1(X_1, X_2) = X_1 X_2$. Since the joint probability density function of X_1 and X_2 is the product of the marginal probability density functions,

$$
f_{X_1, X_2}(x_1, x_2) = 1 \qquad 0 < x_1 < 1, 0 < x_2 < 1.
$$

There are several dummy transformations that would allow us to use Theorem 7.5; we arbitrarily choose the dummy transformation $Y_2 = g_2(X_1, X_2) = X_2$ because the functions can be solved in closed form for X_1 and X_2 and the Jacobian is tractable. A thorough investigation of the transformation

$$
y_1 = g_1(x_1, x_2) = x_1 x_2 \qquad \text{and} \qquad y_2 = g_2(x_1, x_2) = x_2
$$

reveals that it is a bivariate 1–1 transformation from

$$
\mathcal{A} = \{(x_1, x_2)\,|\,0 < x_1 < 1, 0 < x_2 < 1\}
$$

to

$$
\mathcal{B} = \{(y_1, y_2)\,|\,0 < y_1 < 1, 0 < y_1 < y_2 < 1\}.
$$

The support sets $\mathcal{A}$ and $\mathcal{B}$ are illustrated in Figure 7.10. The functions g_1 and g_2 can be solved in closed form for x_1 and x_2 as

$$
x_1 = g_1^{-1}(y_1, y_2) = y_1/y_2 \qquad \text{and} \qquad x_2 = g_2^{-1}(y_1, y_2) = y_2
$$

with associated Jacobian

$$
J = \begin{vmatrix} 1/y_2 & -y_1/y_2^2 \\ 0 & 1 \end{vmatrix} = 1/y_2.
$$

Applying Theorem 7.5, the joint probability density function of $Y_1 = g_1(X_1, X_2) = X_1 X_2$ and $Y_2 = g_2(X_1, X_2) = X_2$ is

$$
f_{Y_1, Y_2}(y_1, y_2) = 1 \cdot \left| \frac{1}{y_2} \right| = \frac{1}{y_2} \qquad (y_1, y_2) \in \mathcal{B}.
$$

Integrating y_2 out of the joint probability density function gives the marginal distribution of Y_1, which is

$$
f_{Y_1}(y_1) = \int_{y_1}^1 \frac{1}{y_2}\, dy_2 = \big[\ln y_2\big]_{y_1}^1 = -\ln y_1 \qquad 0 < y_1 < 1.
$$

This result matches the probability density function determined using the cumulative distribution function technique in Example 7.12.

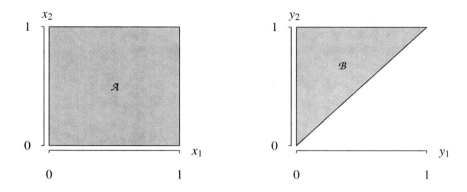

Figure 7.10: The support of X_1 and X_2 and the support of Y_1 and Y_2.

Example 7.25 Find the probability density function of the sum of two independent $U(0, 1)$ random variables.

Let X_1 and X_2 denote the two independent $U(0, 1)$ random variables. We want to find the distribution of $Y_1 = g_1(X_1, X_2) = X_1 + X_2$. As in the previous example, since the joint probability density function of X_1 and X_2 is the product of the marginal probability density functions,

$$f_{X_1, X_2}(x_1, x_2) = 1 \qquad 0 < x_1 < 1, 0 < x_2 < 1.$$

We arbitrarily choose the dummy transformation $Y_2 = g_2(X_1, X_2) = X_1 - X_2$ because the functions can be solved in closed form for X_1 and X_2 and the Jacobian is tractable. The fact that this is a *linear* transformation ensures that the Jacobian is nonzero on $\mathcal{B}$. The transformation

$$y_1 = g_1(x_1, x_2) = x_1 + x_2 \qquad \text{and} \qquad y_2 = g_2(x_1, x_2) = x_1 - x_2$$

is illustrated in Figure 7.11 (the two graphs have different axis scales), and is a bivariate 1–1 transformation from

$$\mathcal{A} = \{(x_1, x_2) \mid 0 < x_1 < 1, 0 < x_2 < 1\}$$

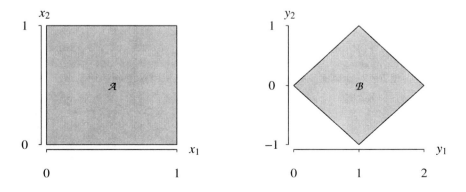

Figure 7.11: The support of X_1 and X_2 and the support of Y_1 and Y_2.

to

$$\mathcal{B} = \{(y_1, y_2) \mid |y_1 - 1| + |y_2| < 1\}.$$

These functions can be solved in closed form for x_1 and x_2 as

$$x_1 = g_1^{-1}(y_1, y_2) = \frac{y_1 + y_2}{2} \qquad \text{and} \qquad x_2 = g_2^{-1}(y_1, y_2) = \frac{y_1 - y_2}{2}$$

with associated Jacobian

$$J = \begin{vmatrix} 1/2 & 1/2 \\ 1/2 & -1/2 \end{vmatrix} = -1/2.$$

Applying Theorem 7.5, the joint probability density function of the random variables $Y_1 = g_1(X_1, X_2) = X_1 + X_2$ and $Y_2 = g_2(X_1, X_2) = X_1 - X_2$ is

$$f_{Y_1, Y_2}(y_1, y_2) = 1 \cdot \left| -\frac{1}{2} \right| = \frac{1}{2} \qquad (y_1, y_2) \in \mathcal{B}.$$

The random variables Y_1 and Y_2 are uniformly distributed over the square support region $\mathcal{B}$. Integrating y_2 out of the joint probability density function gives the marginal distribution of Y_1, which is

$$f_{Y_1}(y_1) = \begin{cases} \displaystyle\int_{-y_1}^{y_1} \frac{1}{2} \, dy_2 & 0 < y_1 < 1 \\[2ex] \displaystyle\int_{y_1-2}^{2-y_1} \frac{1}{2} \, dy_2 & 1 \leq y_1 < 2 \end{cases}$$

or

$$f_{Y_1}(y_1) = \begin{cases} y_1 & 0 < y_1 < 1 \\ 2 - y_1 & 1 \leq y_1 < 2. \end{cases}$$

This result matches the probability density function determined using the cumulative distribution function technique in Example 7.13: $X_1 + X_2 \sim \text{triangular}(0, 1, 2)$.

Example 7.26 Let $X_1 \sim N(0, 1)$ and $X_2 \sim \chi^2(k)$ be independent random variables. Find the distribution of $X_1 / \sqrt{X_2/k}$.

This example differs from the previous two in that the joint probability density function is more complicated. Since the joint probability density function of X_1 and X_2 is the product of the marginal probability density functions,

$$f_{X_1, X_2}(x_1, x_2) = \frac{1}{\sqrt{2\pi}} e^{-\frac{1}{2}x_1^2} \cdot \frac{x_2^{k/2-1} e^{-x_2/2}}{\Gamma(k/2) 2^{k/2}} \qquad -\infty < x_1 < \infty, x_2 > 0.$$

We want to find the distribution of $Y_1 = g_1(X_1, X_2) = X_1 / \sqrt{X_2/k}$. We arbitrarily choose the dummy transformation $Y_2 = g_2(X_1, X_2) = X_2$ because the functions can be solved in closed form for X_1 and X_2 and the Jacobian is tractable. The transformation

$$y_1 = g_1(x_1, x_2) = \frac{x_1}{\sqrt{x_2/k}} \qquad \text{and} \qquad y_2 = g_2(x_1, x_2) = x_2$$

is a bivariate 1–1 transformation from the first two quadrants in the support of X_1 and X_2

$$\mathcal{A} = \{(x_1, x_2) \mid -\infty < x_1 < \infty, x_2 > 0\}$$

to the first two quadrants in the support of Y_1 and Y_2

$$\mathcal{B} = \{(y_1, y_2) \mid -\infty < y_1 < \infty, y_2 > 0\}.$$

These functions can be solved in closed form for x_1 and x_2 as

$$x_1 = g_1^{-1}(y_1, y_2) = y_1 \sqrt{\frac{y_2}{k}} \qquad \text{and} \qquad x_2 = g_2^{-1}(y_1, y_2) = y_2$$

with associated Jacobian

$$J = \begin{vmatrix} \sqrt{\frac{y_2}{k}} & \frac{y_1}{2}\sqrt{\frac{1}{ky_2}} \\ 0 & 1 \end{vmatrix} = \sqrt{\frac{y_2}{k}}.$$

Applying Theorem 7.5, the joint probability density function of $Y_1 = g_1(X_1, X_2) = X_1/\sqrt{X_2/k}$ and $Y_2 = g_2(X_1, X_2) = X_2$ is

$$f_{Y_1, Y_2}(y_1, y_2) = \frac{1}{\sqrt{2\pi}} e^{-\frac{1}{2}y_1^2 y_2/k} \cdot \frac{y_2^{k/2-1} e^{-y_2/2}}{\Gamma(k/2)2^{k/2}} \left| \sqrt{\frac{y_2}{k}} \right| \qquad (y_1, y_2) \in \mathcal{B}$$

which simplifies to

$$f_{Y_1, Y_2}(y_1, y_2) = \frac{1}{\sqrt{2\pi k}\,\Gamma(k/2)\,2^{k/2}} y_2^{(k-1)/2} e^{-(y_2/2)(1+y_1^2/k)} \qquad (y_1, y_2) \in \mathcal{B}.$$

Integrating y_2 out of the joint probability density function gives the marginal distribution of Y_1, which is

$$
\begin{aligned}
f_{Y_1}(y_1) &= \int_0^\infty f_{Y_1, Y_2}(y_1, y_2)\, dy_2 \\
&= \int_0^\infty \frac{1}{\sqrt{2\pi k}\,\Gamma(k/2)\,2^{k/2}} y_2^{(k-1)/2} e^{-(y_2/2)(1+y_1^2/k)}\, dy_2 \\
&= \frac{\Gamma\big((k+1)/2\big)}{\sqrt{\pi k}\,\Gamma(k/2)\,\big(1+y_1^2/k\big)^{(k+1)/2}} \qquad -\infty < y_1 < \infty
\end{aligned}
$$

by using the substitution $u = (y_1/2)\big(1+y_1^2/k\big)$. The distribution of Y_1 is described by the single parameter k. This distribution is well-known in statistics as the t distribution with k degrees of freedom (discovered by William Sealy Gossett under the pseudoname "student"), which is abbreviated with the shorthand $Y_1 \sim t(k)$. This probability density function is a bell-shaped, even function with a mode at $y_1 = 0$ with height

$$f_{Y_1}(0) = \frac{\Gamma\big((k+1)/2\big)}{\sqrt{\pi k}\,\Gamma(k/2)}.$$

The t distribution collapses to the Cauchy distribution (which was introduced in Example 3.45) for $k = 1$ degree of freedom. In addition, the t distribution approaches a standard normal distribution in the limit as $k \to \infty$ (the details are given in Example 8.6). The APPL code

```
X1 := NormalRV(0, 1);
X2 := ChiSquareRV(k);
g1 := [[x -> 1 / sqrt(x / k)], [0, infinity]];
X3 := Transform(X2, g1);
Y1 := Product(X1, X3);
```

returns the probability density function of Y_1 expressed as an integral. The R functions dt, pt, qt, and rt calculate the value of the probability density function of the t distribution, calculate the value of the cumulative distribution function of the t distribution, calculate a percentile of the t distribution, and generate random variates having the t distribution.

Example 7.27 Let $X_1 \sim \chi^2(k_1)$ and $X_2 \sim \chi^2(k_2)$ be independent random variables. Use the transformation technique to find the distribution of $\frac{X_1/k_1}{X_2/k_2}$.

Recall from Section 5.3 that the probability density function of $X \sim \chi^2(k)$ is

$$f_X(x) = \frac{x^{k/2-1}e^{-x/2}}{\Gamma(k/2)2^{k/2}} \qquad\qquad x > 0.$$

Since X_1 and X_2 are independent, their joint probability density function is the product of the marginal probability density functions:

$$f_{X_1,X_2}(x_1, x_2) = \frac{x_1^{k_1/2-1}e^{-x_1/2}}{\Gamma(k_1/2)2^{k_1/2}} \cdot \frac{x_2^{k_2/2-1}e^{-x_2/2}}{\Gamma(k_2/2)2^{k_2/2}} \qquad\qquad x_1 > 0, x_2 > 0$$

or

$$f_{X_1,X_2}(x_1, x_2) = \frac{x_1^{k_1/2-1}x_2^{k_2/2-1}e^{-(x_1+x_2)/2}}{\Gamma(k_1/2)\Gamma(k_2/2)2^{(k_1+k_2)/2}} \qquad\qquad x_1 > 0, x_2 > 0.$$

We want to find the distribution of $Y_1 = g_1(X_1, X_2) = \frac{X_1/k_1}{X_2/k_2}$. We arbitrarily choose the dummy transformation $Y_2 = g_2(X_1, X_2) = X_2$ because the functions can be solved in closed form for X_1 and X_2 and the Jacobian is tractable. The transformation

$$y_1 = g_1(x_1, x_2) = \frac{x_1/k_1}{x_2/k_2} \qquad \text{and} \qquad y_2 = g_2(x_1, x_2) = x_2$$

is a bivariate 1–1 transformation from the support of X_1 and X_2 (the first quadrant)

$$\mathcal{A} = \{(x_1, x_2) \,|\, x_1 > 0, x_2 > 0\}$$

to the support of Y_1 and Y_2 (also the first quadrant)

$$\mathcal{B} = \{(y_1, y_2) \,|\, y_1 > 0, y_2 > 0\}.$$

These functions can be solved in closed form for x_1 and x_2 as

$$x_1 = g_1^{-1}(y_1, y_2) = \frac{k_1}{k_2}y_1y_2 \qquad \text{and} \qquad x_2 = g_2^{-1}(y_1, y_2) = y_2$$

with associated Jacobian

$$J = \begin{vmatrix} k_1y_2/k_2 & k_1y_1/k_2 \\ 0 & 1 \end{vmatrix} = \frac{k_1}{k_2}y_2.$$

Applying the transformation technique, the joint probability density function of $Y_1 = g_1(X_1, X_2) = \frac{X_1/k_1}{X_2/k_2}$ and $Y_2 = g_2(X_1, X_2) = X_2$ is

$$f_{Y_1,Y_2}(y_1, y_2) = \frac{(k_1y_1y_2/k_2)^{k_1/2-1}y_2^{k_2/2-1}e^{-(k_1y_1y_2/k_2+y_2)/2}}{\Gamma(k_1/2)\Gamma(k_2/2)2^{(k_1+k_2)/2}}\left|\frac{k_1}{k_2}y_2\right| \qquad y_1 > 0, y_2 > 0,$$

which, after dropping the absolute value bars, simplifies to

$$f_{Y_1,Y_2}(y_1, y_2) = \frac{k_1(k_1y_1y_2/k_2)^{k_1/2-1}y_2^{k_2/2}e^{-(k_1y_1y_2/k_2+y_2)/2}}{k_2\Gamma(k_1/2)\Gamma(k_2/2)2^{(k_1+k_2)/2}} \qquad y_1 > 0, y_2 > 0.$$

Integrating y_2 out of the joint probability density function gives the marginal distribution of Y_1, which is

$$\begin{aligned} f_{Y_1}(y_1) &= \int_0^\infty f_{Y_1,Y_2}(y_1, y_2)\,dy_2 \\ &= \int_0^\infty \frac{k_1(k_1y_1y_2/k_2)^{k_1/2-1}y_2^{k_2/2}e^{-(k_1y_1y_2/k_2+y_2)/2}}{k_2\Gamma(k_1/2)\Gamma(k_2/2)2^{(k_1+k_2)/2}}\,dy_2 \end{aligned}$$

for $y_1 > 0$. This integral can be worked out by hand beginning with the substitution $w = (k_1 y_1 y_2 / k_2 + y_2)/2$, then using the definition of the gamma function to work the ensuing integral with respect to w. Alternatively, one could use the the following Maple statements to compute the integral.

```
assume(k1, posint);
assume(k2, posint);
f := k1 * (k1 * y1 * y2 / k2) ^ (k1 / 2 - 1) * y2 ^ (k2 / 2) *
     exp(-(k1 * y1 * y2 / k2 + y2) / 2) / (k2 * GAMMA(k1 / 2) *
     GAMMA(k2 / 2) * 2 ^ (k1 / 2 + k2 / 2));
fy1 := int(f, y2 = 0 .. infinity);
```

After simplification, the marginal probability density function of Y_1 is

$$f_{Y_1}(y_1) = \frac{\Gamma((k_1 + k_2)/2) k_1^{k_1/2} k_2^{k_2/2}}{\Gamma(k_1/2)\Gamma(k_2/2)} \cdot \frac{y_1^{k_1/2 - 1}}{(k_1 y_1 + k_2)^{(k_1 + k_2)/2}} \qquad y_1 > 0.$$

This probability density function can be expressed in many different forms, but is often written as

$$f_{Y_1}(y_1) = \frac{1}{y_1 B(k_1/2, k_2/2)} \cdot \sqrt{\frac{(k_1 y_1)^{k_1} k_2^{k_2}}{(k_1 y_1 + k_2)^{k_1 + k_2}}} \qquad y_1 > 0,$$

where B is the beta function (defined in Section 5.5). The distribution of Y_1 is described by the two parameters k_1 and k_2. This distribution is well-known in statistics as the F distribution with k_1 and k_2 degrees of freedom (named after Sir Ronald Fisher for his work on "analysis of variance"), which is abbreviated with the shorthand $Y_1 \sim F(k_1, k_2)$. Like the t distribution, the F distribution arises in statistical applications when it is reasonable to assume that the population from which the data is being drawn has a normal distribution. The population mean (for $k_2 > 2$) and variance (for $k_2 > 4$) of $Y_1 \sim F(k_1, k_2)$ are

$$E[Y_1] = \frac{k_2}{k_2 - 2} \qquad \text{and} \qquad V[Y_1] = \frac{2k_2^2(k_1 + k_2 - 2)}{k_1(k_2 - 2)^2(k_2 - 4)}.$$

The F distribution can be accessed in APPL with the function FRV. The R functions df, pf, qf, and rf calculate the value of the probability density function of the F distribution, calculate the value of the cumulative distribution function of the F distribution, calculate a percentile of the F distribution, and generate random variates having the F distribution.

This transformation can be checked in APPL using calls to the Transform and Product procedures with the following statements.

```
X1 := ChiSquareRV(k1);
X2 := ChiSquareRV(k2);
g1 := [[x -> x / k1], [0, infinity]];
g2 := [[x -> k2 / x], [0, infinity]];
Y1 := Transform(X1, g1);
Y2 := Transform(X2, g2);
F  := Product(Y1, Y2);
```

The probability density function of F matches the probability density function derived analytically above.

The fifth and final example of the application of the transformation technique to find the bivariate distribution of a transformation of continuous random variables X_1 and X_2 with known joint probability density function $f_{X_1, X_2}(x_1, x_2)$ via the 1–1 transformation $Y_1 = g_1(X_1, X_2)$ and $Y_2 = g_2(X_1, X_2)$ results in a well-known random variate generation algorithm called the Box–Muller algorithm.

Example 7.28 Let $X_1 \sim U(0, 1)$ and $X_2 \sim U(0, 1)$ be independent random variables. Find the joint probability density function of

$$Y_1 = \sqrt{-2 \ln X_1} \cos(2\pi X_2) \qquad \text{and} \qquad Y_2 = \sqrt{-2 \ln X_1} \sin(2\pi X_2).$$

Since the joint probability density function of X_1 and X_2 is the product of the marginal probability density functions,

$$f_{X_1, X_2}(x_1, x_2) = 1 \qquad 0 < x_1 < 1, \, 0 < x_2 < 1.$$

To use Theorem 7.5, the transformation achieved via $Y_1 = g_1(X_1, X_2)$ and $Y_2 = g_2(X_1, X_2)$ must be 1–1. Ignoring problems along the x_2 axis, $x_1 = 0$ (that can be ignored because $P(X_1 = 0) = 0$), we see, after thorough investigation of the transformation

$$y_1 = g_1(x_1, x_2) = \sqrt{-2 \ln x_1} \cos(2\pi x_2)$$

and

$$y_2 = g_2(x_1, x_2) = \sqrt{-2 \ln x_1} \sin(2\pi x_2),$$

that this is a bivariate 1–1 transformation from the unit square

$$\mathcal{A} = \{(x_1, x_2) \mid 0 < x_1 < 1, 0 < x_2 < 1\}$$

to the entire plane

$$\mathcal{B} = \{(y_1, y_2) \mid -\infty < y_1 < \infty, -\infty < y_2 < \infty\}.$$

The functions g_1 and g_2 can be solved in closed form for x_1 and x_2 as

$$x_1 = g_1^{-1}(y_1, y_2) = e^{-\frac{1}{2}(y_1^2 + y_2^2)} \qquad \text{and} \qquad x_2 = g_2^{-1}(y_1, y_2) = \frac{1}{2\pi} \arctan\left(\frac{y_2}{y_1}\right)$$

with associated Jacobian

$$J = \begin{vmatrix} -y_1 e^{-\frac{1}{2}(y_1^2 + y_2^2)} & -y_2 e^{-\frac{1}{2}(y_1^2 + y_2^2)} \\ \dfrac{-y_2/y_1^2}{2\pi(1 + (y_2/y_1)^2)} & \dfrac{1/y_1}{2\pi(1 + (y_2/y_1)^2)} \end{vmatrix} = -\frac{1}{2\pi} e^{-\frac{1}{2}(y_1^2 + y_2^2)}.$$

Applying Theorem 7.5, the joint probability density function of $Y_1 = g_1(X_1, X_2) = \sqrt{-2 \ln X_1} \cos(2\pi X_2)$ and $Y_2 = g_2(X_1, X_2) = X_2 = \sqrt{-2 \ln X_1} \sin(2\pi X_2)$ is

$$f_{Y_1, Y_2}(y_1, y_2) = 1 \cdot \left| -\frac{1}{2\pi} e^{-\frac{1}{2}(y_1^2 + y_2^2)} \right| \qquad (y_1, y_2) \in \mathcal{B}$$

or

$$f_{Y_1, Y_2}(y_1, y_2) = \frac{1}{2\pi} e^{-\frac{1}{2}(y_1^2 + y_2^2)} \qquad (y_1, y_2) \in \mathcal{B}.$$

This joint probability density function can be factored into functions of y_1 and y_2 alone as

$$f_{Y_1, Y_2}(y_1, y_2) = \left(\frac{1}{\sqrt{2\pi}} e^{-\frac{1}{2}y_1^2}\right) \left(\frac{1}{\sqrt{2\pi}} e^{-\frac{1}{2}y_2^2}\right) \qquad (y_1, y_2) \in \mathcal{B}.$$

Since $\mathcal{B}$ is a product space and $f_{Y_1, Y_2}(y_1, y_2)$ can be factored into the product of a non-negative function of y_1 and a nonnegative function of y_2, Theorem 6.1 indicates that Y_1 and Y_2 are independent random variables. Furthermore, Y_1 and Y_2 are independent and identically distributed $N(0, 1)$ random variables. This rather insightful observation that the non-trivial transformation of two independent $U(0, 1)$ random variables in the unit square would result in two independent $N(0, 1)$ random variables has resulted in the celebrated *Box–Muller algorithm* in 1958 for generating random standard normal variates. One of the problems with this algorithm is that it requires *two* $U(0, 1)$ random variables and produces *two* $N(0, 1)$ random variates. Most applications require only a single $N(0, 1)$ random variate. The simplest solution is to simply throw away the Y_2 value that is generated. So the Box–Muller algorithm for generating a single standard normal random variate Y_1 can be coded in one line as

$$Y_1 \leftarrow \sqrt{-2 \ln X_1} \cos(2\pi X_2),$$

where X_1 and X_2 are independent random numbers. If a $N(\mu, \sigma^2)$ is required, this $N(0, 1)$ random variate can be multiplied by σ and added to μ using the inverse of the standardizing transformation given in Theorem 5.4. Faster algorithms have been devised subsequently, but the Box–Muller algorithm was the first exact algorithm for generating normal variates.

The one-line algorithm for generating standard normal can be checked in APPL. Due to the principle inverse difficulty with trigonometric functions, however, the random variate generation algorithm must be rewritten as

$$Y_1 \leftarrow \sqrt{-2 \ln X_1} \cos(\pi X_2)$$

before applying `Transform` to the second factor in the expression. The APPL code given below for calculating the distribution of Y_1 uses calls to the `Transform` and `Product` functions.

```
X1 := UniformRV(0, 1);
X2 := UniformRV(0, 1);
g1 := [[x -> sqrt(-2 * log(x))], [0, 1]];
T1 := Transform(X1, g1);
g2 := [[x -> cos(Pi * x)], [0, 1]];
T2 := Transform(X2, g2);
Y1 := Product(T1, T2);
```

This code returns an expression that is mathematically equivalent to

$$f_{Y_1}(y_1) = \frac{1}{\sqrt{2\pi}} e^{-y_1^2/2} \qquad -\infty < y_1 < \infty$$

as expected.

There are two extensions associated with Theorem 7.5 that will be addressed next. The first extends the result from two random variables to n random variables. The proof of the extension is omitted because it follows the same pattern established in the proof of Theorem 7.5. The second extends the result from 1–1 transformations to k–1 transformations. The next result, whose proof is similar to the proof of Theorem 7.5, considers a 1–1 transformation of the random variables $X_1, X_2, \ldots, X_n$ defined on $\mathcal{A}$ to $Y_1, Y_2, \ldots, Y_n$ defined on $\mathcal{B}$ via the 1–1 transformation defined by the n functions

$$Y_i = g_i(X_1, X_2, \ldots, X_n)$$

for $i = 1, 2, \ldots, n$.

> **Theorem 7.6** Let $X_1, X_2, \ldots, X_n$ be continuous random variables with joint probability density function $f_{X_1,X_2,\ldots,X_n}(x_1, x_2, \ldots, x_n)$ defined on the support $\mathcal{A}$. Let the functions
>
> $$y_i = g_i(x_1, x_2, \ldots, x_n)$$
>
> for $i = 1, 2, \ldots, n$, define a 1–1 transformation from the n-dimensional set $\mathcal{A}$ to the n-dimensional set $\mathcal{B}$. Assume that the functions $y_i = g_i(x_1, x_2, \ldots, x_n)$, for $i = 1, 2, \ldots, n$ can be solved uniquely and in closed form for $x_1, x_2, \ldots, x_n$ as
>
> $$x_i = g_i^{-1}(y_1, y_2, \ldots, y_n)$$
>
> for $i = 1, 2, \ldots, n$, which maps $\mathcal{B}$ to $\mathcal{A}$. Assume further that J, the Jacobian of the transformation, which is the determinant of the matrix of partial derivatives of the transformation from $\mathcal{B}$ to $\mathcal{A}$, is continuous and nonzero for all $(y_1, y_2, \ldots, y_n) \in \mathcal{B}$. The joint probability density function of $Y_i = g_i(X_1, X_2, \ldots, X_n)$, for $i = 1, 2, \ldots, n$, is
>
> $$f_{Y_1,Y_2,\ldots,Y_n}(y_1, y_2, \ldots, y_n) = f_{X_1,X_2,\ldots,X_n}\left(g_1^{-1}(y_1, y_2, \ldots, y_n), \ldots, g_n^{-1}(y_1, y_2, \ldots, y_n)\right) |J|$$
>
> for $(y_1, y_2, \ldots, y_n) \in \mathcal{B}$, where the bars around the Jacobian denote absolute value.

Example 7.29 Let X_1, X_2, X_3 be mutually independent exponential(1) random variables. Find the probability density function of $X_1 + X_2 + X_3$.

Recall from Section 5.2 that the probability density function of an exponential(1) random variable X is

$$f_X(x) = e^{-x} \qquad x > 0.$$

Since X_1, X_2, X_3 are mutually independent random variables, their joint probability density function is the product of their marginal probability density functions:

$$f_{X_1,X_2,X_3}(x_1, x_2, x_3) = e^{-x_1-x_2-x_3} \qquad x_1 > 0, x_2 > 0, x_3 > 0.$$

The objective of this example is to find the probability density function of

$$Y_1 = g_1(X_1, X_2, X_3) = X_1 + X_2 + X_3.$$

We arbitrarily choose the dummy transformations $Y_2 = g_2(X_1, X_2, X_3) = X_2 + X_3$ and $Y_3 = g_3(X_1, X_2, X_3) = X_3$ because the functions can be solved in closed form for X_1, X_2, and X_3, and the Jacobian is tractable. The fact that this is a *linear* transformation ensures that the Jacobian is nonzero on $\mathcal{B}$. The transformation

$$
\begin{aligned}
y_1 &= g_1(x_1, x_2, x_3) = x_1 + x_2 + x_3 \\
y_2 &= g_2(x_1, x_2, x_3) = x_2 + x_3 \\
y_3 &= g_3(x_1, x_2, x_3) = x_3
\end{aligned}
$$

is a trivariate 1–1 transformation from

$$\mathcal{A} = \{(x_1, x_2, x_3) \,|\, x_1 > 0, x_2 > 0, x_3 > 0\}$$

to

$$\mathcal{B} = \{(y_1, y_2, y_3) \,|\, 0 < y_3 < y_2 < y_1\}.$$

The functions g_1, g_2, and g_3 can be solved in closed form for x_1, x_2, and x_3 as

$$
\begin{aligned}
x_1 &= g_1^{-1}(y_1, y_2, y_3) = y_1 - y_2 \\
x_2 &= g_2^{-1}(y_1, y_2, y_3) = y_2 - y_3 \\
x_3 &= g_3^{-1}(y_1, y_2, y_3) = \phantom{y_2 - {}} y_3
\end{aligned}
$$

with associated Jacobian

$$J = \begin{vmatrix} 1 & -1 & 0 \\ 0 & 1 & -1 \\ 0 & 0 & 1 \end{vmatrix} = 1.$$

Applying Theorem 7.6, the joint probability density function of $Y_1 = g_1(X_1, X_2, X_3) = X_1 + X_2 + X_3$, $Y_2 = g_2(X_1, X_2, X_3) = X_2 + X_3$, and $Y_3 = g_3(X_1, X_2, X_3) = X_3$ is

$$f_{Y_1, Y_2, Y_3}(y_1, y_2, y_3) = e^{-(y_1 - y_2) - (y_2 - y_3) - y_3} |1| = e^{-y_1} \qquad (y_1, y_2, y_3) \in \mathcal{B}.$$

Integrating y_2 and y_3 out of the joint probability density function gives the marginal distribution of Y_1, which is

$$\begin{aligned} f_{Y_1}(y_1) &= \int_0^{y_1} \int_0^{y_2} e^{-y_1} \, dy_3 \, dy_2 \\ &= \int_0^{y_1} y_2 e^{-y_1} \, dy_2 \\ &= \frac{y_1^2}{2} e^{-y_1} \qquad y_1 > 0. \end{aligned}$$

This probability distribution function can be recognized as an Erlang distribution with $\lambda = 1$ and $k = 3$. The APPL code to confirm the analytic result is given below.

```
X := ExponentialRV(1);
Y := ConvolutionIID(X, 3);
```

The next example illustrates a 1–1 transformation of n random variables that results in a generalization of the beta distribution.

Example 7.30 Let $X_i \sim \text{gamma}(1, \kappa_i)$, for $i = 1, 2, \ldots, n$, be mutually independent random variables. Find the joint probability density function of $Y_1, Y_2, \ldots, Y_n$, where

$$Y_i = \frac{X_i}{X_1 + X_2 + \cdots + X_n}$$

for $i = 1, 2, \ldots, n - 1$, and

$$Y_n = X_1 + X_2 + \cdots + X_n.$$

Recall from Section 5.3 that the probability density function of $X \sim \text{gamma}(1, \kappa)$ is

$$f_X(x) = \frac{x^{\kappa - 1} e^{-x}}{\Gamma(\kappa)} \qquad x > 0.$$

Since $X_1, X_2, \ldots, X_n$ are mutually independent, their joint probability density function is the product of the marginal probability density functions:

$$f_{X_1, X_2, \ldots, X_n}(x_1, x_2, \ldots, x_n) = \prod_{i=1}^{n} \frac{x_i^{\kappa_i - 1} e^{-x_i}}{\Gamma(\kappa_i)} \qquad x_1 > 0, x_2 > 0, \ldots, x_n > 0.$$

The transformation

$$y_1 = g_1(x_1, x_2, \ldots, x_n) = \frac{x_1}{x_1 + x_2 + \cdots + x_n}$$

$$y_2 = g_2(x_1, x_2, \ldots, x_n) = \frac{x_2}{x_1 + x_2 + \cdots + x_n}$$

$$\vdots$$

$$y_{n-1} = g_{n-1}(x_1, x_2, \ldots, x_n) = \frac{x_{n-1}}{x_1 + x_2 + \cdots + x_n}$$

$$y_n = g_n(x_1, x_2, \ldots, x_n) = x_1 + x_2 + \cdots + x_n$$

is an *n*-variate 1–1 transformation from the support of $X_1, X_2, \ldots, X_n$:

$$\mathcal{A} = \{(x_1, x_2, \ldots, x_n) \mid x_1 > 0, x_2 > 0, \ldots, x_n > 0\}$$

to the support of $Y_1, Y_2, \ldots, Y_n$:

$$\mathcal{B} = \{(y_1, y_2, \ldots, y_n) \mid y_1 > 0, y_2 > 0, \ldots, y_n > 0, y_1 + y_2 + \cdots + y_{n-1} < 1\}.$$

These functions can be solved in closed form for $x_1, x_2, \ldots, x_n$ as

$$x_1 = g_1^{-1}(y_1, y_2, \ldots, y_n) = y_1 y_n$$

$$x_2 = g_2^{-1}(y_1, y_2, \ldots, y_n) = y_2 y_n$$

$$\vdots$$

$$x_{n-1} = g_{n-1}^{-1}(y_1, y_2, \ldots, y_n) = y_{n-1} y_n$$

$$x_n = g_n^{-1}(y_1, y_2, \ldots, y_n) = y_n(1 - y_1 - y_2 - \cdots - y_{n-1})$$

with associated Jacobian

$$J = \begin{vmatrix} y_n & 0 & 0 & \cdots & 0 & y_1 \\ 0 & y_n & 0 & \cdots & 0 & y_2 \\ 0 & 0 & y_n & \cdots & 0 & y_3 \\ \vdots & \vdots & \vdots & \ddots & 0 & \vdots \\ 0 & 0 & 0 & \cdots & y_n & y_{n-1} \\ -y_n & -y_n & -y_n & \cdots & -y_n & 1 - y_1 - y_2 - \cdots - y_{n-1} \end{vmatrix}.$$

Using the cofactor expansion across the bottom row of the matrix (and the cofactor expansion of the right-most column within each expansion), the Jacobian is

$$\begin{aligned} J &= y_n \left(y_1 y_n^{n-2} \right) + y_n \left(y_2 y_n^{n-2} \right) + \cdots + y_n \left(y_{n-1} y_n^{n-2} \right) + (1 - y_1 - y_2 - \cdots - y_{n-1}) y_n^{n-1} \\ &= y_1 y_n^{n-1} + y_2 y_n^{n-1} + \cdots + y_{n-1} y_n^{n-1} + (1 - y_1 - y_2 - \cdots - y_{n-1}) y_n^{n-1} \\ &= y_n^{n-1}. \end{aligned}$$

Applying Theorem 7.6, the joint probability density function of $Y_1, Y_2, \ldots, Y_n$ is

$$\prod_{i=1}^{n-1} \frac{(y_i y_n)^{\kappa_i - 1} e^{-y_i y_n}}{\Gamma(\kappa_i)} \cdot \frac{y_n^{\kappa_n - 1}(1 - y_1 - y_2 - \cdots - y_{n-1})^{\kappa_n - 1} e^{-y_n(1 - y_1 - y_2 - \cdots - y_{n-1})}}{\Gamma(\kappa_n)} \cdot \left| y_n^{n-1} \right|$$

for $(y_1, y_2, \ldots, y_n) \in \mathcal{B}$. After dropping the absolute value bars and simplifying,

$$f_{Y_1, Y_2, \ldots, Y_n}(y_1, y_2, \ldots, y_n) = \frac{y_n^{\kappa_1 + \kappa_2 + \cdots + \kappa_n - 1} \prod_{i=1}^{n-1} y_i^{\kappa_i - 1}(1 - y_1 - y_2 - \cdots - y_{n-1})^{\kappa_n - 1} e^{-y_n}}{\Gamma(\kappa_1)\Gamma(\kappa_2)\ldots\Gamma(\kappa_n)}$$

for $(y_1, y_2, \ldots, y_n) \in \mathcal{B}$. This is the solution to the problem as stated. Since Y_n plays a different role than the others in this probability density function, it is worthwhile investigating the distribution of $Y_1, Y_2, \ldots, Y_{n-1}$ alone. Integrating y_n out of the joint probability density function gives the joint distribution for $Y_1, Y_2, \ldots, Y_{n-1}$. Using the definition of the gamma function, this joint probability density function is

$$f_{Y_1, Y_2, \ldots, Y_{n-1}}(y_1, y_2, \ldots, y_{n-1}) = \frac{\Gamma(\kappa_1 + \kappa_2 + \cdots \kappa_n) \prod_{i=1}^{n-1} y_i^{\kappa_i - 1} (1 - y_1 - y_2 - \cdots - y_{n-1})^{\kappa_n - 1}}{\Gamma(\kappa_1)\Gamma(\kappa_2)\ldots\Gamma(\kappa_n)}$$

for $y_1 > 0, y_2 > 0, \ldots, y_{n-1} > 0, y_1 + y_2 + \cdots + y_{n-1} < 1$. This is an $n - 1$ variate distribution with n parameters $\kappa_1, \kappa_2, \ldots, \kappa_n$, which is known as the *Dirichlet distribution*. The Dirichlet distribution is a multivariate generalization of the beta distribution from Section 5.5. Consider the case of $n = 2$. The probability density function of Y_1 is

$$f_{Y_1}(y_1) = \frac{\Gamma(\kappa_1 + \kappa_2)}{\Gamma(\kappa_1)\Gamma(\kappa_2)} y_1^{\kappa_1 - 1} (1 - y_1)^{\kappa_2 - 1} \qquad 0 < y_1 < 1.$$

The case of k–to–1 transformations is handled in an analogous fashion to the one-dimensional case. When the transformation is not 1–1, it must be broken into 1–1 pieces in order to use the transformation technique.

Theorem 7.7 Let $X_1, X_2, \ldots, X_n$ be continuous random variables with joint probability density function $f_{X_1, X_2, \ldots, X_n}(x_1, x_2, \ldots, x_n)$ defined on the support $\mathcal{A}$. Let the functions

$$y_i = g_i(x_1, x_2, \ldots, x_n)$$

for $i = 1, 2, \ldots, n$, define a k-to-1 transformation from the n-dimensional set $\mathcal{A}$ to the n-dimensional set $\mathcal{B}$. Suppose further that $\mathcal{A}$ can be partitioned into the sets $\mathcal{A}_1, \mathcal{A}_2, \ldots, \mathcal{A}_k$ such that $y_i = g_i(x_1, x_2, \ldots, x_n)$, for $i = 1, 2, \ldots, n$, defines a 1–1 transformation from $\mathcal{A}_j$ to $\mathcal{B}$, for $j = 1, 2, \ldots, k$. Assume that the functions $y_i = g_i(x_1, x_2, \ldots, x_n)$, for $i = 1, 2, \ldots, n$ can be solved uniquely and in closed form for $x_1, x_2, \ldots, x_n$ as

$$x_i = g_{ij}^{-1}(y_1, y_2, \ldots, y_n)$$

for $i = 1, 2, \ldots, n$ and $j = 1, 2, \ldots, k$, which maps $\mathcal{B}$ to $\mathcal{A}_j$. Assume further that J_j, the Jacobian of the transformation from $\mathcal{A}_j$ to $\mathcal{B}$, which is the determinant of the matrix of partial derivatives of the transformation from $\mathcal{B}$ to $\mathcal{A}_j$, for $j = 1, 2, \ldots, k$, is continuous and nonzero for all $(y_1, y_2, \ldots, y_n) \in \mathcal{B}$. The joint probability density function of $Y_i = g_i(X_1, X_2, \ldots, X_n)$, for $i = 1, 2, \ldots, n$, is

$$f_{Y_1, Y_2, \ldots, Y_n}(y_1, y_2, \ldots, y_n) = \sum_{j=1}^{k} f_{X_1, X_2, \ldots, X_n}\left(g_{1j}^{-1}(y_1, y_2, \ldots, y_n), \ldots, g_{nj}^{-1}(y_1, y_2, \ldots, y_n)\right) |J_j|$$

for $(y_1, y_2, \ldots, y_n) \in \mathcal{B}$, where the bars around the Jacobians denote absolute value.

Theorem 7.7 will be illustrated in the ensuing discussion on order statistics. This completes the introduction to the transformation technique, which is the most general of the three techniques for finding the distribution of a function of one or more random variables. There is a fundamental transformation that arises so often in statistics that we devote the remainder of this section to investigating its properties. The transformation of interest is just the ordering of a set of data values, resulting in what is known to statisticians as "order statistics."

Order statistics

To motivate the development of the topic of order statistics, consider the following problem that often arises in statistical inference. We know that the sample mean $\bar{X}$ is a reasonable estimator of the population mean μ when data values $X_1, X_2, \ldots, X_n$ constitute mutually independent and identically distributed observations from some unknown population. But what if instead we want to estimate the population median from this random sample? Assuming for simplicity that n is odd, a reasonable way to proceed is to sort the observations into ascending order, then use the middle observation (the sample median) to estimate the population median. In this setting, there is clearly some interest in the probability distribution of the sample median.

Another question that might arise is the estimation of the 95th percentile of the probability distribution associated with some unknown population. If the sample size n is assumed to be large, then a reasonable way to proceed is to sort the observations into ascending order, and use the observation that is 95% of the way into the sorted observations to estimate the 95th percentile. Again, the probability distribution of the sorted values, that is, the order statistics, is of interest.

Order statistics assume a different flavor for discrete and continuous populations. The theory is rather elegant for samples drawn from continuous populations; the theory is tedious for samples drawn from discrete populations due to the possibility of tied observations. We tackle the more difficult case first with three examples that highlight the potential problems that can arise.

Example 7.31 A bag contains 15 balls numbered 1, 2, ..., 15. Five balls are drawn from the bag without replacement. Find the probability mass function of the sample median of the numbers drawn.

Let the random variables $X_1, X_2, \ldots, X_5$ denote the five numbers drawn from the bag. These values are known to statisticians as the *raw data*. Let the random variables $X_{(1)} < X_{(2)} < X_{(3)} < X_{(4)} < X_{(5)}$ denote the original random variables arranged in ascending order, which are known as the *order statistics*. The interest is in the probability mass function of the *sample median*, $X_{(3)}$. Since the sampling of the balls is performed without replacement, the support of $X_{(3)}$ runs from $x = 3$ to $x = 13$:

$$\mathcal{A} = \{x \mid x = 3, 4, \ldots, 13\}.$$

There are $\binom{15}{5}$ equally-likely samples that can be drawn without replacement from the bag. The event $X_{(3)} = x$ is associated with two draws being less than x, one draw being exactly x, and two draws being greater than x. Thus, the probability density function of $X_{(3)}$ is

$$f_{X_{(3)}}(x) = \frac{\binom{x-1}{2}\binom{1}{1}\binom{15-x}{2}}{\binom{15}{5}} \qquad x = 3, 4, \ldots, 13.$$

This probability mass function is plotted in Figure 7.12, which is a symmetric proba-

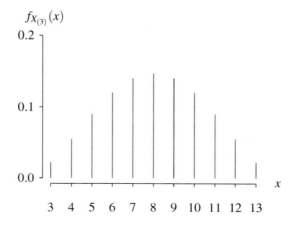

Figure 7.12: Probability mass function $f_{X_{(3)}}(x)$ for the sample median $X_{(3)}$.

bility mass function about $x = 8$. The probability mass function can also be calculated in APPL with the commands

```
X := UniformDiscreteRV(1, 15);
Y := OrderStat(X, 5, 3, "wo");
```

which returns the probability mass function given above. The first three arguments to the `OrderStat` function are the population being sampled, the number of observations being sampled from the population, and the order statistic of interest. The fourth optional argument concerns the type of sampling. In this case, the string `"wo"` tells APPL that sampling is being performed without replacement.

The sampling performed without replacement in the previous example allowed the observations and the associated order statistics to be distinct, which simplifies the problem. The next example uses sampling with replacement, which complicates the analysis because of the possibility of ties.

Example 7.32 A fair die is tossed five times. Find the probability mass function of the sample median of the numbers that appear.

This problem can be recast in terms of balls in a bag as sampling five balls at random and with replacement from a bag containing balls numbered $1, 2, \ldots, 6$. As before, let the random variables $X_1, X_2, \ldots, X_5$ denote the five numbers that appear. Let the random variables $X_{(1)} \leq X_{(2)} \leq X_{(3)} \leq X_{(4)} \leq X_{(5)}$ denote the order statistics. The interest is in the probability mass function of the sample median, $X_{(3)}$. Since the sampling is performed with replacement, the support of $X_{(3)}$ runs from $x = 1$ to $x = 6$:

$$\mathcal{A} = \{x \,|\, x = 1, 2, \ldots, 6\}.$$

By the multiplication rule, there are $6^5 = 7776$ equally-likely outcomes to the random experiment.

The event $X_{(3)} = 1$ is associated with rolling a 1 three or more times. For example, the rolls

$$1, 1, 1, 1, 1$$

and

$$1, 3, 1, 1, 6$$

both correspond to a sample median of 1. So $X_{(3)} = 1$ corresponds to three or more ones appearing in the sample, which can be calculated using the binomial distribution as

$$P(X_{(3)} = 1) = \binom{5}{3}\left(\frac{1}{6}\right)^3\left(\frac{5}{6}\right)^2 + \binom{5}{4}\left(\frac{1}{6}\right)^4\left(\frac{5}{6}\right)^1 + \binom{5}{5}\left(\frac{1}{6}\right)^5\left(\frac{5}{6}\right)^0 = \frac{276}{6^5} = \frac{23}{648}.$$

As an alternative, and more easily generalized solution using the appropriate counting techniques is

$$P(X_{(3)} = 1) = \frac{\binom{5}{2} \cdot 1 \cdot 1 \cdot 1 \cdot 5 \cdot 5 + \binom{5}{1} \cdot 1 \cdot 1 \cdot 1 \cdot 1 \cdot 5 + 1 \cdot 1 \cdot 1 \cdot 1 \cdot 1}{6^5} = \frac{276}{6^5} = \frac{23}{648}.$$

The numerator sums the number of ways of getting exactly three ones, exactly four ones, and exactly five ones in five rolls of a die. Next, determine the probability that the sample median is two, that is $P(X_{(3)} = 2)$. Using a similar approach as finding the probability that the sample median equals 1,

$$P(X_{(3)} = 2) = \frac{\binom{5}{2,1,2} \cdot 1 \cdot 1 \cdot 1 \cdot 4 \cdot 4 + \binom{5}{1,2,2} \cdot 1 \cdot 1 \cdot 1 \cdot 4 \cdot 4 + \cdots + 1 \cdot 1 \cdot 1 \cdot 1 \cdot 1}{6^5} = \frac{1356}{6^5} = \frac{113}{648}.$$

Continuing in this fashion, the probability mass function of $X_{(3)}$ is

$$f_{X_{(3)}}(x) = \begin{cases} 23/648 & x = 1 \\ 113/648 & x = 2 \\ 47/162 & x = 3 \\ 47/162 & x = 4 \\ 113/648 & x = 5 \\ 23/648 & x = 6. \end{cases}$$

This probability mass function is plotted in Figure 7.13, which is a symmetric probability mass function around $x = 7/2$. The probability mass function can be calculated in APPL with the commands

```
X := UniformDiscreteRV(1, 6);
Y := OrderStat(X, 5, 3);
```

which returns the probability mass function given above. The first three arguments to the OrderStat function are the population being sampled, the number of observations being sampled, and the order statistic of interest. The default sampling mechanism is sampling with replacement.

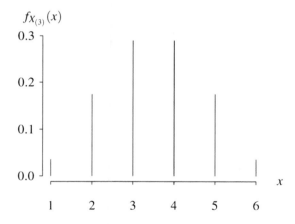

Figure 7.13: Probability mass function $f_{X_{(3)}}(x)$ for the sample median $X_{(3)}$.

The previous two examples have concerned sampling from discrete populations where the outcomes are equally likely. The third and final example of determining the probability mass function of an order statistic when observations are drawn from a discrete population highlights the issues that arise when the observations are not equally likely.

Example 7.33 A spinner has three outcomes, 1, 2, and 3, which occur with probability $1/2$, $1/3$, $1/6$. If the spinner is spun three times, find the distribution of the sample median.

Unlike the previous two examples, the outcomes associated with each observation are not equally likely. Let the random variables X_1, X_2, X_3 denote the results of the three spins. Let the random variables $X_{(1)} \leq X_{(2)} \leq X_{(3)}$ denote the associated order statistics. Sampling is effectively with replacement, so ties must be accounted for. The sample space is small enough for this random experiment, that it can be enumerated. By the multiplication rule, the $3^3 = 27$ outcomes are:

$$
\begin{array}{ccc}
1, 1, 1 & 1, 1, 2 & 1, 1, 3 \\[4pt]
1, 2, 1 & 1, 2, 2 & 1, 2, 3 \\[4pt]
\vdots & \vdots & \vdots \\[4pt]
3, 3, 1 & 3, 3, 2 & 3, 3, 3.
\end{array}
$$

After determining the probabilities associated with the 27 outcomes (for example, all three spins coming up 1 occurs with probability $1/8$), and the order statistics associated with the 27 outcomes, the probability mass function of $X_{(2)}$ is

$$
f_{X_{(2)}}(x) = \begin{cases}
1/2 & x = 1 \\
23/54 & x = 2 \\
2/27 & x = 3.
\end{cases}
$$

Not surprisingly, the spinner's propensity toward a result of 1 on a single spin results in a probability mass function of the sample median that is not symmetric. This probability mass function can be calculated with the APPL commands

```
X := [[1 / 2, 1 / 3, 1 / 6], [1, 2, 3], ["Discrete", "PDF"]];
Y := OrderStat(X, 3, 2);
```

When sampling from discrete populations, APPL can be a useful tool for calculating probability mass functions for order statistics in moderate-sized problems. The APPL `OrderStat` function enumerates all of the possible outcomes to the random experiment and sums the appropriate probabilities associated with the order statistic of interest.

This brief introduction to order statistics now shifts to sampling from continuous populations. Many of the properties of order statistics drawn from continuous populations do not depend on the population probability distribution. Sampling from continuous populations eliminates

- the importance of distinguishing between sampling with and without replacement, and

- the possibility of ties.

Using the same notation as in the discrete population case, the values sampled from the continuous population (that is, the raw data) are denoted by

$$
X_1, X_2, \ldots, X_n.
$$

Since the probability of a tie is zero, the associated order statistics are denoted by the strictly increasing sequence

$$
X_{(1)} < X_{(2)} < \cdots < X_{(n)}.
$$

The focus of the remainder of this section is on the joint and marginal distributions of these order statistics when the observations are realizations of mutually independent and identically distributed random variables. The next example illustrates the derivation of the joint distribution of $n = 2$ order statistics when random sampling is from the uniform distribution using the transformation technique.

Example 7.34 A random sample of $n = 2$ observations is drawn from a $U(0, 1)$ population. Find the joint probability density function of the order statistics.

Let X_1 and X_2 denote the observations and let

$$
X_{(1)} = \min\{X_1, X_2\} \qquad \text{and} \qquad X_{(2)} = \max\{X_1, X_2\}
$$

denote the associated order statistics. Although the ordering of the data might not look like a transformation at first glance, the transformation technique can be applied to this bivariate transformation from X_1 and X_2 to $X_{(1)}$ and $X_{(2)}$. Figure 7.14 shows the support of X_1 and X_2, which is the unit square, on the left-hand graph, and the support of $X_{(1)}$ and $X_{(2)}$ on the right-hand graph. More specifically, consider the mapping of the two points, $(X_1, X_2) = (0.8, 0.9)$ and $(X_1, X_2) = (0.9, 0.8)$, which are plotted on the left-hand graph. The first point, $(X_1, X_2) = (0.8, 0.9)$, which has the random variables drawn in increasing order, is mapped to $(X_{(1)}, X_{(2)}) = (0.8, 0.9)$. In fact, all points that fall above the dashed line in the left-hand graph in Figure 7.14 are mapped directly to identical coordinates on the right-hand graph. The second point, $(X_1, X_2) = (0.9, 0.8)$, which has the random variables drawn in decreasing order, is also mapped to $(X_{(1)}, X_{(2)}) = (0.8, 0.9)$. In fact, all points that fall below the dashed line in the left-hand graph in Figure 7.14 are reflected across the dashed line by the mapping. Using the terminology from earlier in this section, this is a 2–1 bivariate transformation from $\mathcal{A}$ to $\mathcal{B}$, that can be viewed as a folding of $\mathcal{A}$ over the dashed line. We can apply Theorem 7.7 to determine the joint probability density function of the order statistics. The only awkward part of this transformation is the notation; we are working with $X_{(1)}$ and $X_{(2)}$ rather than the usual Y_1 and Y_2.

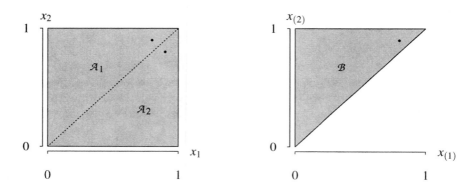

Figure 7.14: The support of X_1 and X_2 and the support of $X_{(1)}$ and $X_{(2)}$.

First, consider the transformation from $\mathcal{A}_1$ to $\mathcal{B}$. The transformation

$$x_{(1)} = g_1(x_1, x_2) = \min\{x_1, x_2\} = x_1 \qquad \text{and} \qquad x_{(2)} = g_2(x_1, x_2) = \max\{x_1, x_2\} = x_2$$

is a bivariate 1–1 transformation from

$$\mathcal{A}_1 = \{(x_1, x_2) \mid 0 < x_1 < x_2 < 1\}$$

to

$$\mathcal{B} = \{(x_{(1)}, x_{(2)}) \mid 0 < x_{(1)} < x_{(2)} < 1\}.$$

These functions can be solved in closed form for x_1 and x_2 as

$$x_1 = g_{11}^{-1}(x_{(1)}, x_{(2)}) = x_{(1)} \qquad \text{and} \qquad x_2 = g_{12}^{-1}(x_{(1)}, x_{(2)}) = x_{(2)}$$

with associated Jacobian

$$J_1 = \begin{vmatrix} 1 & 0 \\ 0 & 1 \end{vmatrix} = 1.$$

Next, consider the transformation from $\mathcal{A}_2$ to $\mathcal{B}$. The transformation

$$x_{(1)} = g_1(x_1, x_2) = \min\{x_1, x_2\} = x_2 \qquad \text{and} \qquad x_{(2)} = g_2(x_1, x_2) = \max\{x_1, x_2\} = x_1$$

is a bivariate 1–1 transformation from

$$\mathcal{A}_2 = \{(x_1, x_2) \,|\, 0 < x_2 < x_1 < 1\}$$

to

$$\mathcal{B} = \{(x_{(1)}, x_{(2)}) \,|\, 0 < x_{(1)} < x_{(2)} < 1\}.$$

These functions can be solved in closed form for x_1 and x_2 as

$$x_1 = g_{21}^{-1}(x_{(1)}, x_{(2)}) = x_{(2)} \qquad \text{and} \qquad x_2 = g_{22}^{-1}(x_{(1)}, x_{(2)}) = x_{(1)}$$

with associated Jacobian

$$J_2 = \begin{vmatrix} 0 & 1 \\ 1 & 0 \end{vmatrix} = -1.$$

Theorem 7.7 with $n = 2$ and $k = 2$ can now be used to find the joint probability density function of $X_{(1)}$ and $X_{(2)}$. First, because X_1 and X_2 are independent, their joint probability density function is

$$f_{X_1, X_2}(x_1, x_2) = 1 \qquad 0 < x_1 < 1, 0 < x_2 < 1.$$

The joint probability density function of $X_{(1)}$ and $X_{(2)}$ is

$$f_{X_{(1)}, X_{(2)}}(x_{(1)}, x_{(2)}) = 1 \cdot |1| + 1 \cdot |-1| = 2 \qquad 0 < x_{(1)} < x_{(2)} < 1.$$

The joint probability density function of $X_{(1)}$ and $X_{(2)}$ is uniformly distributed over the support $\mathcal{B}$ shown in Figure 7.14. So while X_1 and X_2 are independent random variables defined on a product space, the order statistics $X_{(1)}$ and $X_{(2)}$ are dependent random variables. The marginal distributions of $X_{(1)}$ and $X_{(2)}$ are easily calculated as

$$f_{X_{(1)}}(x_{(1)}) = \int_{x_{(1)}}^{1} 2 \, dx_{(2)} = 2(1 - x_{(1)}) \qquad 0 < x_{(1)} < 1$$

and

$$f_{X_{(2)}}(x_{(2)}) = \int_{0}^{x_{(2)}} 2 \, dx_{(1)} = 2x_{(2)} \qquad 0 < x_{(2)} < 1.$$

This example concerning two order statistics drawn from a uniform population is easily generalized to n order statistics drawn from any continuous population in the following theorem.

Theorem 7.8 Let $X_1, X_2, \ldots, X_n$ be mutually independent random variables from a continuous population with probability density function $f(x)$ defined on the support $a < x < b$. The joint probability density function of the associated order statistics $X_{(1)} < X_{(2)} < \cdots < X_{(n)}$ is

$$f(x_{(1)}, x_{(2)}, \ldots, x_{(n)}) = n! f(x_{(1)}) f(x_{(2)}) \ldots f(x_{(n)})$$

for $a < x_{(1)} < x_{(2)} < \cdots < x_{(n)} < b$.

Proof The proof will be constructed for the case of $n = 3$, and it will be apparent that the steps in the proof are completely general for any n. Since X_1, X_2, X_3 are mutually independent random variables, their joint probability density function is the product of their marginal probability density functions:

$$f(x_1, x_2, x_3) = f(x_1) f(x_2) f(x_3) \qquad a < x_1 < b, a < x_2 < b, a < x_3 < b.$$

Theorem 7.7 can be used to determine the joint probability distribution of the order statistics $X_{(1)} < X_{(2)} < X_{(3)}$. By the multiplication rule, there are $3! = 6$ different orderings of the X_1, X_2, X_3 values, so the transformation to the order statistics is a 6–1 transformation from the product space $\mathcal{A}$ to $\mathcal{B}$, with

$$\mathcal{A}_1 = \{(x_1, x_2, x_3) \,|\, a < x_1 < x_2 < x_3 < b\},$$

$$\mathcal{A}_2 = \{(x_1, x_2, x_3) \,|\, a < x_1 < x_3 < x_2 < b\},$$

$$\vdots$$

$$\mathcal{A}_6 = \{(x_1, x_2, x_3) \,|\, a < x_3 < x_2 < x_1 < b\},$$

and

$$\mathcal{B} = \{(x_{(1)}, x_{(2)}, x_{(3)}) \,|\, a < x_{(1)} < x_{(2)} < x_{(3)} < b\}.$$

The six Jacobians associated with the inverse transformations correspond to the six possible ways to permute the elements of the elements in the 3×3 identity matrix such that there is a 1 in every row and every column:

$$J_1 = \begin{vmatrix} 1 & 0 & 0 \\ 0 & 1 & 0 \\ 0 & 0 & 1 \end{vmatrix} = 1, J_2 = \begin{vmatrix} 1 & 0 & 0 \\ 0 & 0 & 1 \\ 0 & 1 & 0 \end{vmatrix} = -1, \ldots, J_6 = \begin{vmatrix} 0 & 0 & 1 \\ 0 & 1 & 0 \\ 1 & 0 & 0 \end{vmatrix} = -1.$$

All of the Jacobians equal ± 1. Invoking Theorem 7.7, the joint probability distribution of the order statistics is

$$f(x_{(1)}, x_{(2)}, x_{(3)}) = 3! f(x_{(1)}) f(x_{(2)}) f(x_{(3)})$$

for $a < x_{(1)} < x_{(2)} < x_{(3)} < b$. In the general case of the order statistics associated with n random variables, sorting the data to arrive at the order statistics constitutes an $n!$–1 transformation with all Jacobians equal to ± 1, which proves the general result.   □

The next two examples illustrate the application of Theorem 7.8.

Example 7.35 Let X_1, X_2, X_3, X_4 be mutually independent random variables from an exponential(λ) population. Find the joint probability density function of the order statistics $X_{(1)}, X_{(2)}, X_{(3)}, X_{(4)}$.

The probability density function of the population distribution is

$$f(x) = \lambda e^{-\lambda x} \qquad x > 0$$

Using Theorem 7.8, the joint probability density function of the order statistics is

$$f_{X_{(1)}, X_{(2)}, X_{(3)}, X_{(4)}}(x_{(1)}, x_{(2)}, x_{(3)}, x_{(4)}) = 24\lambda^4 e^{-\lambda(x_{(1)} + x_{(2)} + x_{(3)} + x_{(4)})}$$

for $0 < x_{(1)} < x_{(2)} < x_{(3)} < x_{(4)}$.

Example 7.36 Let X_1, X_2, X_3, X_4 be a random sample from a population with probability density function $f(x)$ defined on support $a < x < b$. Find the probability density function of $X_{(2)}$.

Using Theorem 7.8 the joint probability density function of $X_{(1)}, X_{(2)}, X_{(3)}, X_{(4)}$ is

$$f_{X_{(1)}, X_{(2)}, X_{(3)}, X_{(4)}}(x_{(1)}, x_{(2)}, x_{(3)}, x_{(4)}) = 24 f(x_{(1)}) f(x_{(2)}) f(x_{(3)}) f(x_{(4)})$$

for $a < x_{(1)} < x_{(2)} < x_{(3)} < x_{(4)} < b$. So the marginal distribution of the second smallest observation is

$$
\begin{aligned}
f_{X_{(2)}}(x_{(2)}) &= 24 f(x_{(2)}) \int_a^{x_{(2)}} \int_{x_{(2)}}^b \int_{x_{(3)}}^b f(x_{(1)}) f(x_{(3)}) f(x_{(4)}) \, dx_{(4)} \, dx_{(3)} \, dx_{(1)} \\
&= 24 f(x_{(2)}) \int_a^{x_{(2)}} \int_{x_{(2)}}^b f(x_{(1)}) f(x_{(3)}) [1 - F(x_{(3)})] \, dx_{(3)} \, dx_{(1)} \\
&= 24 f(x_{(2)}) \int_a^{x_{(2)}} f(x_{(1)}) \frac{[1 - F(x_{(2)})]^2}{2} \, dx_{(1)} \\
&= 24 f(x_{(2)}) F(x_{(2)}) \frac{[1 - F(x_{(2)})]^2}{2} \\
&= 12 F(x_{(2)}) f(x_{(2)}) [1 - F(x_{(2)})]^2 \qquad a < x_{(2)} < b.
\end{aligned}
$$

The expression for the marginal distribution of $X_{(2)}$ is independent of the choice of the population distribution. This is generalized in the next result that concerns the marginal distributions of the order statistics.

Theorem 7.9 Let $X_1, X_2, \ldots, X_n$ be mutually independent random variables from a continuous population with probability density function $f(x)$ and cumulative distribution function $F(x)$ defined on the support $a < x < b$. If $X_{(1)} < X_{(2)} < \cdots < X_{(n)}$ are the associated order statistics, then

$$
f_{X_{(k)}}(x_{(k)}) = \frac{n!}{(k-1)!(n-k)!} [F(x_{(k)})]^{k-1} f(x_{(k)}) [1 - F(x_{(k)})]^{n-k}
$$

for $a < x_{(k)} < b$ and $k = 1, 2, \ldots, n$.

Proof A careful proof of this result would follow along the lines of Example 7.35. We instead give a heuristic argument that generalizes nicely to the joint distribution of selected order statistics. Each of the random variables $X_1, X_2, \ldots, X_n$ will fall in one of the following three disjoint intervals that partition the support $a < x < b$:

- the interval $(a, x_{(k)}]$ with probability $\displaystyle\int_a^{x_{(k)}} f(x) \, dx$,

- the interval $(x_{(k)}, x_{(k)} + h]$ with probability $\displaystyle\int_{x_{(k)}}^{x_{(k)}+h} f(x) \, dx$,

- the interval $(x_{(k)} + h, b)$ with probability $\displaystyle\int_{x_{(k)}+h}^b f(x) \, dx$,

for some fixed real values $x_{(k)}$ and h, where $a < x_{(k)} < b$, $h > 0$, and $x_{(k)} + h < b$. For small values of h, these three probabilities are approximately $F(x_{(k)})$, $h f(x_{(k)})$, and $1 - F(x_{(k)})$. The probability that exactly $k - 1$ of the observations fall in the first interval, one observation falls in the second interval, and $n - k$ of the observations fall in the third interval has the trinomial distribution, with associated approximate probability density function

$$
f_{X_{(k)}}(x_{(k)}) = \frac{n!}{(k-1)!\,1!\,(n-k)!} [F(x_{(k)})]^{k-1} f(x_{(k)})^1 [1 - F(x_{(k)})]^{n-k}
$$

for $a < x_{(k)} < b$ and $k = 1, 2, \ldots, n$. $\qquad\square$

The multinomial approach taken in the proof can be generalized to finding the joint probability density function of two or more order statistics. Consider, for example, the joint distribution of $X_{(i)}$ and $X_{(j)}$, where $1 \le i < j \le n$. In this case, there are five disjoint intervals that partition $a < x < b$ that each of the n random variables can belong to:

- the interval $(a, x_{(i)}]$ with probability $\int_a^{x_{(i)}} f(x)\,dx$,

- the interval $(x_{(i)}, x_{(i)} + h]$ with probability $\int_{x_{(i)}}^{x_{(i)}+h} f(x)\,dx$,

- the interval $(x_{(i)} + h, x_{(j)}]$ with probability $\int_{x_{(i)}+h}^{x_{(j)}} f(x)\,dx$,

- the interval $(x_{(j)}, x_{(j)} + h]$ with probability $\int_{x_{(j)}}^{x_{(j)}+h} f(x)\,dx$,

- the interval $(x_{(j)} + h, b)$ with probability $\int_{x_{(j)}+h}^{b} f(x)\,dx$,

for some fixed real values $x_{(i)}$, $x_{(j)}$, and h, where $a < x_{(i)} < x_{(j)} < b$, $h > 0$, $x_{(i)} + h < x_{(j)}$, and $x_{(j)} + h < b$. Again using the multinomial distribution for small values of h and five outcomes to each of the n trials, the joint probability density function of $X_{(i)}$ and $X_{(j)}$ is

$$f_{X_{(i)},X_{(j)}}(x_{(i)},x_{(j)}) = \frac{n!}{(i-1)!1!(j-i-1)!1!(n-j)!}[F(x_{(i)})]^{i-1}.$$

$$f(x_{(i)})^1[F(x_{(j)}) - F(x_{(i)})]^{j-i-1}f(x_{(j)})^1[1 - F(x_{(j)})]^{n-j}$$

for $a < x_{(i)} < x_{(j)} < b$.

There are three important special cases of Theorem 7.9 that often arise in applications: the probability mass function of the sample minimum, the sample maximum, and the sample median. In the listings below, we assume for simplicity that the sample size is odd for the sample median so that the subscript k can be expressed as $k = (n+1)/2$.

- Sample minimum $(k = 1)$

$$f_{X_{(1)}}(x_{(1)}) = nf(x_{(1)})[1 - F(x_{(1)})]^{n-1} \qquad a < x_{(1)} < b.$$

- Sample maximum $(k = n)$

$$f_{X_{(n)}}(x_{(n)}) = n[F(x_{(n)})]^{n-1}f(x_{(n)}) \qquad a < x_{(n)} < b.$$

- Sample median $(k = (n+1)/2)$

$$f_{X_{(k)}}(x_{(k)}) = \frac{(2k-1)!}{(k-1)!(k-1)!}[F(x_{(k)})]^{k-1}f(x_{(k)})[1 - F(x_{(k)})]^{k-1} \qquad a < x_{(k)} < b.$$

The next example determines the probability density function of the minimum of n exponential random variables.

Example 7.37 A series system consists of n components, each with mutually independent exponential(λ) lifetimes. Find the distribution of the system lifetime.

The probability density function of the lifetime of each component is

$$f(x) = \lambda e^{-\lambda x} \qquad x > 0$$

The cumulative distribution function of the lifetime of each component is

$$F(x) = \begin{cases} 0 & x \leq 0 \\ 1 - e^{-\lambda x} & x > 0. \end{cases}$$

Let $X_1, X_2, \ldots, X_n$ denote the component lifetimes. Since the components are arranged in series, the system fails upon the initial component failure. So we want the probability density function of the first order statistic $X_{(1)} = \min\{X_1, X_2, \ldots, X_n\}$, which is

$$
\begin{aligned}
f_{X_{(1)}}(x_{(1)}) &= nf(x_{(1)})[1 - F(x_{(1)})]^{n-1} \\
&= n\lambda e^{-\lambda x_{(1)}}\left[e^{-\lambda x_{(1)}}\right]^{n-1} \\
&= (n\lambda)\, e^{-(n\lambda)x_{(1)}} \qquad x_{(1)} > 0.
\end{aligned}
$$

The system lifetime is recognized as an exponential($n\lambda$) random variable. The APPL code required to solve this problem can use either MinimumIID:

```
X := ExponentialRV(lambda);
Y := MinimumIID(X, n);
```

or OrderStat:

```
X := ExponentialRV(lambda);
Y := OrderStat(X, n, 1);
```

Example 7.38 Let $X_1, X_2, \ldots, X_n$ be a random sample from a $U(0, 1)$ distribution. Find the distribution of $X_{(n)}$.

The population distribution is described by the probability density function

$$ f(x) = 1 \qquad 0 < x < 1 $$

and cumulative distribution function

$$
F(x) = \begin{cases}
0 & x \le 0 \\
x & 0 < x < 1 \\
1 & x \ge 1.
\end{cases}
$$

Using Theorem 7.9, the largest of the n observations has probability density function

$$ f_{X_{(n)}}(x_{(n)}) = n x_{(n)}^{n-1} \qquad 0 < x_{(n)} < 1. $$

The APPL code required to solve this problem can be done with MaximumIID:

```
X := UniformRV(0, 1);
Y := MaximumIID(X, n);
```

or OrderStat:

```
X := UniformRV(0, 1);
Y := OrderStat(X, n, n);
```

Occasions arise when functions of order statistics are of interest to statisticians. One such function is known as the *sample range*, which is the largest observation minus the smallest observation from a group of random variables $X_1, X_2, \ldots, X_n$.

Example 7.39 Let $X_1, X_2, \ldots, X_n$ mutually independent $U(0, 1)$ random variables. Find the distribution of the sample range $X_{(n)} - X_{(1)}$.

The $U(0, 1)$ population has probability density function

$$f(x) = 1 \qquad 0 < x < 1.$$

The cumulative distribution function of the $U(0, 1)$ population is

$$F(x) = \begin{cases} 0 & x \leq 0 \\ x & 0 < x < 1 \\ 1 & x \geq 1. \end{cases}$$

We start by finding the joint probability density function of $X_{(1)}$ and $X_{(n)}$, which is

$$\begin{aligned} f_{X_{(1)}, X_{(n)}}(x_{(1)}, x_{(n)}) &= \frac{n!}{(n-2)!} f(x_{(1)})[F(x_{(n)}) - F(x_{(1)})]^{n-2} f(x_{(n)}) \\ &= n(n-1)(x_{(n)} - x_{(1)})^{n-2} \qquad 0 < x_{(1)} < x_{(n)} < 1. \end{aligned}$$

We now use Theorem 7.5, to find the distribution of the sample range $Y_1 = X_{(n)} - X_{(1)}$ by using the dummy transformation $Y_2 = X_{(n)}$. The transformation

$$y_1 = g_1(x_{(1)}, x_{(2)}) = x_{(n)} - x_{(1)} \qquad \text{and} \qquad y_2 = g_2(x_{(1)}, x_{(2)}) = x_{(n)}$$

is a bivariate 1–1 transformation from

$$\mathcal{A} = \{(x_{(1)}, x_{(2)}) \mid 0 < x_{(1)} < x_{(n)} < 1\}$$

to

$$\mathcal{B} = \{(y_1, y_2) \mid 0 < y_1 < y_n < 1\}.$$

The functions g_1 and g_2 can be solved in closed form for $x_{(1)}$ and $x_{(n)}$ as

$$x_{(1)} = g_1^{-1}(y_1, y_2) = y_2 - y_1 \qquad \text{and} \qquad x_{(n)} = g_2^{-1}(y_1, y_2) = y_2$$

with associated Jacobian

$$J = \begin{vmatrix} -1 & 1 \\ 0 & 1 \end{vmatrix} = -1.$$

Applying Theorem 7.5, the joint probability density function of $Y_1 = g_1(X_{(1)}, X_{(n)}) = X_{(n)} - X_{(1)}$ and $Y_2 = g_2(X_{(1)}, X_{(n)}) = X_{(n)}$ is

$$f_{Y_1, Y_2}(y_1, y_2) = n(n-1)y_1^{n-2}|-1| = n(n-1)y_1^{n-2} \qquad 0 < y_1 < y_2 < 1.$$

Finally, to find the marginal distribution of Y_1, the sample range,

$$f_{Y_1}(y_1) = \int_{y_1}^{1} n(n-1)y_1^{n-2} dy_2 = n(n-1)y_1^{n-2}(1 - y_1) \qquad 0 < y_1 < 1.$$

The probability density function is shown for $n = 2, 3, 4$ is Figure 7.15. Not surprisingly, the probability distribution is shifted toward 1 as n increases. The more observations that are sampled from the $U(0, 1)$ population, the more likely that $X_{(1)}$ is close to 0 and $X_{(n)}$ is close to 1. APPL has a `RangeStat` function that can be used to find the probability density function of the sample range. For example,

```
X := UniformRV(0, 1);
Y := RangeStat(X, 4);
```

determines the distribution of the sample range $X_{(4)} - X_{(1)}$ when $n = 4$ mutually independent observations are drawn from a $U(0, 1)$ population.

This ends the lengthy introduction to the transformation technique, which is the most general of the three techniques for determining the distribution of the a transformation of one or more random variables. The final technique, known as the *moment generating function technique*, is useful for finding the distribution of sums of mutually independent random variables.

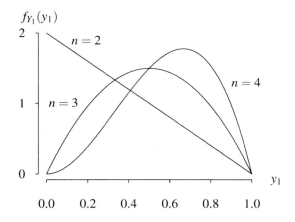

Figure 7.15: Probability density functions of the sample range for $n = 2, 3, 4$.

7.3   Moment Generating Function Technique

The moment generating function technique for determining the probability distribution of the sum of mutually independent random variables is based on the following theorem, which was alluded to in Chapter 6.

Theorem 7.10 Let $X_1, X_2, \ldots, X_n$ be mutually independent random variables with moment generating functions $M_{X_i}(t)$, for $i = 1, 2, \ldots, n$. If $Y = X_1 + X_2 + \cdots + X_n$, then

$$M_Y(t) = M_{X_1}(t)M_{X_2}(t)\ldots M_{X_n}(t).$$

Proof   Since $X_1, X_2, \ldots, X_n$ are mutually independent random variables, the moment generating function of Y is

$$
\begin{aligned}
M_Y(t) &= E\left[e^{tY}\right] \\
&= E\left[e^{t(X_1 + X_2 + \cdots + X_n)}\right] \\
&= E\left[e^{tX_1} e^{tX_2} \ldots e^{tX_n}\right] \\
&= E\left[e^{tX_1}\right] E\left[e^{tX_2}\right] \ldots E\left[e^{tX_n}\right] \\
&= M_{X_1}(t)M_{X_2}(t)\ldots M_{X_n}(t)
\end{aligned}
$$

which proves the result. $\square$

   Using the moment generating function technique involves three steps. The first step is to determine the moment generating functions of the random variables $X_1, X_2, \ldots, X_n$ which are denoted by $M_{X_1}(t), M_{X_2}(t), \ldots, M_{X_n}(t)$. The second step is to apply Theorem 7.10 to determine the moment generating function of $Y = X_1 + X_2 + \cdots + X_n$, that is, $M_Y(t) = M_{X_1}(t)M_{X_2}(t)\ldots M_{X_n}(t)$. The third step is to compare $M_Y(t)$ with the moment generating functions of the univariate probability distributions from Chapters 4 and 5 to determine the distribution of Y.

   The moment generating function technique is limited in the sense that the moment generating function of Y must be one that is easily recognizable from those encountered earlier and it is primarily applied to sums of mutually independent random variables. Oftentimes, as seen in the initial example, the distribution of a sum remains in the same family of probability distributions.

To enhance our ability to recognize the forms of certain moment generating functions, the moment generating functions for several of the popular probability models introduced in Chapters 4 and 5 are listed in Table 7.1. The discrete distributions are listed at the top of the table.

Distribution	$f(x)$	Support	$M(t)$	Domain
Bernoulli(p)	$p^x(1-p)^{1-x}$	$x=0,1$	$1-p+pe^t$	$-\infty < t < \infty$
binomial(n,p)	$\binom{n}{x}p^x(1-p)^{n-x}$	$x=0,1,\ldots,n$	$\left(1-p+pe^t\right)^n$	$-\infty < t < \infty$
geometric(p)	$p(1-p)^x$	$x=0,1,\ldots$	$\dfrac{p}{1-(1-p)e^t}$	$t < -\ln(1-p)$
Geometric(p)	$p(1-p)^{x-1}$	$x=1,2,\ldots$	$\dfrac{pe^t}{1-(1-p)e^t}$	$t < -\ln(1-p)$
negative binomial(r,p)	$\binom{x+r-1}{r-1}p^r(1-p)^x$	$x=0,1,\ldots$	$\left[\dfrac{p}{1-(1-p)e^t}\right]^r$	$t < -\ln(1-p)$
Negative binomial(r,p)	$\binom{x-1}{r-1}p^r(1-p)^{x-r}$	$x=r,r+1,\ldots$	$\left[\dfrac{pe^t}{1-(1-p)e^t}\right]^r$	$t < -\ln(1-p)$
Poisson(λ)	$\dfrac{\lambda^x e^{-\lambda}}{x!}$	$x=0,1,\ldots$	$e^{\lambda(e^t-1)}$	$-\infty < t < \infty$
exponential(λ)	$\lambda e^{-\lambda x}$	$x>0$	$\dfrac{\lambda}{\lambda-t}$	$t<\lambda$
gamma(λ,κ)	$\dfrac{\lambda^\kappa x^{\kappa-1}e^{-\lambda x}}{\Gamma(\kappa)}$	$x>0$	$\left(\dfrac{\lambda}{\lambda-t}\right)^\kappa$	$t<\lambda$
$N(\mu,\sigma^2)$	$\dfrac{1}{\sqrt{2\pi}\sigma}e^{-\frac{1}{2}\left(\frac{x-\mu}{\sigma}\right)^2}$	$-\infty < x < \infty$	$e^{\mu t+\sigma^2 t^2/2}$	$-\infty < t < \infty$

Table 7.1: Moment generating functions for common distributions.

The first example reconsiders the distribution of the sum of n mutually independent Poisson random variables. In the previous section, determining the probability distribution of the sum of just two independent Poisson random variables involved significant tedious algebra. The more general case is easily solved using Theorem 7.10.

Example 7.40 Let $X_1, X_2, \ldots, X_n$ be mutually independent random variables with $X_i \sim$ Poisson(λ_i) for $i = 1, 2, \ldots, n$. Find the probability mass function of

$$Y = X_1 + X_2 + \cdots + X_n.$$

From Table 7.1 the moment generating function of a Poisson(λ) random variable X is

$$M_X(t) = e^{\lambda(e^t-1)} \qquad -\infty < t < \infty.$$

Since $X_i \sim$ Poisson(λ_i), the moment generating function of X_i is

$$M_{X_i}(t) = e^{\lambda_i(e^t-1)} \qquad -\infty < t < \infty,$$

for $i = 1, 2, \ldots, n$. Applying Theorem 7.10, the moment generating function of $Y = X_1 + X_2 + \cdots + X_n$ is

$$
\begin{aligned}
M_Y(t) &= M_{X_1}(t)M_{X_2}(t)\ldots M_{X_n}(t) \\
&= e^{\lambda_1(e^t-1)}e^{\lambda_2(e^t-1)}\ldots e^{\lambda_n(e^t-1)} \\
&= e^{(\lambda_1+\lambda_2+\cdots+\lambda_n)(e^t-1)} \qquad -\infty < t < \infty.
\end{aligned}
$$

This moment generating function is recognized as that of a Poisson($\lambda_1 + \lambda_2 + \cdots + \lambda_n$) random variable, so $Y = X_1 + X_2 + \cdots + X_n$ has probability mass function

$$f_Y(y) = \frac{(\lambda_1 + \lambda_2 + \cdots + \lambda_n)^y e^{-(\lambda_1 + \lambda_2 + \cdots + \lambda_n)}}{y!} \qquad y = 0, 1, 2, \ldots.$$

The result from this example has applications to Poisson processes. If n mutually independent Poisson processes generate a Poisson(λ_1), Poisson(λ_2), ..., Poisson(λ_n) number of events on some time interval, then the total number of events that occur during the time interval (that is, the super-position of the n processes) is Poisson($\lambda_1 + \lambda_2 + \cdots + \lambda_n$).

The next example will show that the moment generating function technique is applied in exactly the same fashion to discrete and continuous random variables.

Example 7.41 Let $X_1, X_2, \ldots, X_n$ be mutually independent random variables with $X_i \sim$ exponential(λ) for $i = 1, 2, \ldots, n$. Find the probability density function of

$$Y = X_1 + X_2 + \cdots + X_n.$$

Since each of the X_i's has the same parameter, statisticians often refer to the random variables $X_1, X_2, \ldots, X_n$ as *independent and identically distributed*, which is abbreviated *iid*. The random variables $X_1, X_2, \ldots, X_n$ are also referred to as a *random sample*. From Table 7.1 the moment generating function of an exponential random variable X is

$$M_X(t) = \frac{\lambda}{\lambda - t} \qquad t < \lambda.$$

Since $X_i \sim$ exponential(λ), the moment generating function of X_i is

$$M_{X_i}(t) = \frac{\lambda}{\lambda - t} \qquad t < \lambda,$$

for $i = 1, 2, \ldots, n$. Applying Theorem 7.10, the moment generating function of $Y = X_1 + X_2 + \cdots + X_n$ is

$$M_Y(t) = M_{X_1}(t) M_{X_2}(t) \ldots M_{X_n}(t) = \frac{\lambda}{\lambda - t} \cdot \frac{\lambda}{\lambda - t} \cdot \ldots \cdot \frac{\lambda}{\lambda - t} = \left(\frac{\lambda}{\lambda - t} \right)^n$$

for $t < \lambda$. Since this moment generating function is recognized as that of an Erlang(λ, n) random variable, $Y = X_1 + X_2 + \cdots + X_n$ has probability density function

$$f_Y(y) = \frac{\lambda^n y^{n-1} e^{-\lambda y}}{(n-1)!} \qquad y > 0.$$

So the sum of n mutually independent and identically distributed exponential(λ) random variables has the Erlang(λ, n) distribution.

Example 7.42 Let $X_1, X_2, \ldots, X_n$ be mutually independent random variables with $X_i \sim \chi^2(k_i)$ for $i = 1, 2, \ldots, n$. Find the probability distribution of

$$Y = X_1 + X_2 + \cdots + X_n.$$

From Table 7.1 the moment generating function of of a chi-square random variable X with k degrees of freedom (which is a gamma random variable with parameters $\kappa = k/2$ and $\lambda = 1/2$) is

$$M_X(t) = (1 - 2t)^{-k/2} \qquad t < 1/2.$$

Since $X_i \sim \chi^2(k_i)$, the moment generating function of X_i is

$$M_{X_i}(t) = (1 - 2t)^{-k_i/2} \qquad t < 1/2,$$

for $i = 1, 2, \ldots, n$. Applying Theorem 7.10, the moment generating function of $Y = X_1 + X_2 + \cdots + X_n$ is

$$
\begin{aligned}
M_Y(t) &= M_{X_1}(t) M_{X_2}(t) \ldots M_{X_n}(t) \\
&= (1 - 2t)^{-k_1/2} \cdot (1 - 2t)^{-k_2/2} \cdot \ldots \cdot (1 - 2t)^{-k_n/2} \\
&= (1 - 2t)^{-(k_1 + k_2 + \cdots + k_n)/2} \qquad t < 1/2.
\end{aligned}
$$

Since this moment generating function is recognized as that of a chi-square random variable with $k_1 + k_2 + \cdots + k_n$ degrees of freedom,

$$Y = X_1 + X_2 + \cdots + X_n \sim \chi^2(k_1 + k_2 + \cdots + k_n).$$

The three examples encountered thus far have involved random variables with familiar distributions. The moment generating function technique can also be applied to sums of random variables that have not yet been encountered, as illustrated in the next example.

Example 7.43 A die is weighted so that the number of spots that appear on the up face has probability mass function

$$f(x) = \frac{x}{21} \qquad x = 1, 2, 3, 4, 5, 6.$$

If this die is tossed 10 times, find the probability that the total number of spots on the up faces is 13.

Let X be the number of spots showing on a single roll of the die. The probability mass function of X,

$$f(x) = \frac{x}{21} \qquad x = 1, 2, 3, 4, 5, 6,$$

has an associated moment generating function

$$M_X(t) = E\left[e^{tX}\right] = \sum_{x=1}^{6} e^{tx} \frac{x}{21} = \frac{1}{21} e^t + \frac{2}{21} e^{2t} + \cdots + \frac{6}{21} e^{6t} \qquad -\infty < t < \infty.$$

Since each of the ten rolls has an identical probability distribution associated with the number of spots showing, the outcomes for each of the ten rolls, $X_1, X_2, \ldots, X_{10}$, each has this moment generating function. Let $Y = X_1 + X_2 + \cdots + X_{10}$ (the support of Y is $10, 11, 12, \ldots, 60$). The moment generating function of Y is

$$M_Y(t) = \left(\frac{1}{21} e^t + \frac{2}{21} e^{2t} + \cdots + \frac{6}{21} e^{6t} \right)^{10} \qquad -\infty < t < \infty.$$

The probability that the total number of spots appearing is 13 is the coefficient of e^{13t} in the expansion of $M_Y(t)$. Consider all possible ways of obtaining a total of 13 spots: 9 ones and a four; 8 ones and a two and a three; and 7 ones and three twos. The associated term in the expansion of $M_Y(t)$ is

$$\binom{10}{1} \left(\frac{1}{21} e^t\right)^9 \left(\frac{4}{21} e^{4t}\right) + \binom{10}{8,1,1} \left(\frac{1}{21} e^t\right)^8 \left(\frac{2}{21} e^{2t}\right) \left(\frac{3}{21} e^{3t}\right) + \binom{10}{3} \left(\frac{1}{21} e^t\right)^7 \left(\frac{2}{21} e^{2t}\right)^3.$$

The associated coefficient is

$$\frac{10 \cdot 4}{21^{10}} + \frac{10 \cdot 9 \cdot 2 \cdot 3}{21^{10}} + \frac{10 \cdot 9 \cdot 8}{3 \cdot 2 \cdot 1} \cdot \frac{8}{21^{10}} = \frac{40 + 540 + 960}{21^{10}} = \frac{1540}{21^{10}} \cong 9.2327 \times 10^{-11},$$

which is not very likely. This solution can be verified in APPL with the the following statements.

```
X := [[x -> x / 21], [1 .. 6], ["Discrete", "PDF"]];
Y := ConvolutionIID(X, 10);
PDF(Y, 13);
```

The moment generating function technique is useful for finding the distribution of sums of mutually independent random variables. It can be adapted to find the distribution of linear combinations of mutually independent random variables. The following result adapts Theorem 7.10 to linear combinations. Its proof is left as an exercise, and is nearly identical to the proof of Theorem 7.10.

Theorem 7.11 Let $X_1, X_2, \ldots, X_n$ be mutually independent random variables with moment generating functions $M_{X_i}(t)$, for $i = 1, 2, \ldots, n$. Let $a_1, a_2, \ldots, a_n$ be nonzero real constants. If $Y = a_1X_1 + a_2X_2 + \cdots + a_nX_n$, then

$$M_Y(t) = M_{a_1X_1}(t)M_{a_2X_2}(t) \ldots M_{a_nX_n}(t).$$

Example 7.44 Let $X_1, X_2, \ldots, X_n$ be mutually independent random variables with $X_i \sim N\left(\mu_i, \sigma_i^2\right)$ for $i = 1, 2, \ldots, n$. Find the probability distribution of

$$Y = a_1X_1 + a_2X_2 + \cdots + a_nX_n.$$

From Table 7.1 the moment generating function of a $N\left(\mu, \sigma^2\right)$ random variable X is

$$M_X(t) = e^{\mu t + \frac{1}{2}\sigma^2 t^2} \qquad -\infty < t < \infty.$$

Since $X_i \sim N\left(\mu_i, \sigma_i^2\right)$, the moment generating function for X_i is

$$M_{X_i}(t) = e^{\mu_i t + \frac{1}{2}\sigma_i^2 t^2} \qquad -\infty < t < \infty,$$

for $i = 1, 2, \ldots, n$. Applying Theorem 7.11, the moment generating function of $Y = a_1X_1 + a_2X_2 + \cdots + a_nX_n$ is

$$
\begin{aligned}
M_Y(t) &= M_{a_1X_1}(t)M_{a_2X_2}(t)\ldots M_{a_nX_n}(t) \\
&= M_{X_1}(a_1t)M_{X_2}(a_2t)\ldots M_{X_n}(a_nt) \\
&= e^{\mu_1 a_1 t + \frac{1}{2}\sigma_1^2 a_1^2 t^2} e^{\mu_2 a_2 t + \frac{1}{2}\sigma_2^2 a_2^2 t^2} \ldots e^{\mu_n a_n t + \frac{1}{2}\sigma_n^2 a_n^2 t^2} \\
&= e^{(a_1\mu_1 + a_2\mu_2 + \cdots + a_n\mu_n)t + \frac{1}{2}\left(a_1^2\sigma_1^2 + a_2^2\sigma_2^2 + \cdots + a_n^2\sigma_n^2\right)t^2} \qquad -\infty < t < \infty.
\end{aligned}
$$

This moment generating function is recognized as that of a normal random variable:

$$Y \sim N\left(a_1\mu_1 + a_2\mu_2 + \cdots + a_n\mu_n, \ a_1^2\sigma_1^2 + a_2^2\sigma_2^2 + \cdots + a_n^2\sigma_n^2\right).$$

To summarize, if $X_i \sim N\left(\mu_i, \sigma_i^2\right)$ for $i = 1, 2, \ldots, n$, and $X_1, X_2, \ldots, X_n$ are mutually independent random variables, then the linear combination $Y = a_1X_1 + a_2X_2 + \cdots + a_nX_n$ is normally distributed with population mean

$$\mu_Y = a_1\mu_1 + a_2\mu_2 + \cdots + a_n\mu_n$$

and population variance

$$\sigma_Y^2 = a_1^2\sigma_1^2 + a_2^2\sigma_2^2 + \cdots + a_n^2\sigma_n^2.$$

7.4 Exercises

7.1 For the table below, place YES, NO, or the appropriate number into each entry.

Mathematical expression	Domain	Function?	1–1 function?	Inverse exists?	$g(9)$	$g^{-1}(9)$
$y = g(x) = 3x - 4$	$\{x \mid x \in \mathcal{R}\}$					
$y = g(x) = x^2$	$\{x \mid x \in \mathcal{R}\}$					
$y = g(x) = \pm\sqrt{x}$	$\{x \mid x \in \mathcal{R}, x \geq 0\}$					

7.2 If X is a strictly continuous random variable with cumulative distribution function $F_X(x)$, write an expression for the cumulative distribution function of $Y = X^4$ in terms of $F_X(x)$.

7.3 Let $X \sim$ exponential(λ_1) and $Y \sim$ exponential(λ_2). Use the convolution formula

$$f_Z(z) = \int_{-\infty}^{\infty} f_X(z-y)f_Y(y)\,dy$$

to determine the probability density function of $Z = X + Y$. Assume that X and Y are independent.

7.4 Assuming that $X_1 \sim$ exponential(λ_1) and $X_2 \sim$ exponential(λ_2) are independent, use the cumulative distribution function technique to find the cumulative distribution function of the sample mean $\bar{X} = (X_1 + X_2)/2$.

7.5 Let X and Y be uniformly distributed over a disk with radius 1 centered at the origin. Find the cumulative distribution function of $Z = \max\{X, Y\}$.

7.6 Consider the pair of random variables (X, Y), which is the solution to the 2×2 set of linear equations

$$Ax + y = 2$$
$$x + y = 3$$

where A is a random variable with probability density function

$$f_A(a) = 2(1 - a) \qquad 0 < a < 1.$$

What is the cumulative distribution function, 99th percentile, and expected value of the rectilinear distance (or Manhattan distance or L_1 norm) from the origin to the random solution to the equation?

7.7 Use APPL to find the population mean of the difference between the two real roots (larger root less the smaller root) of the quadratic equation

$$x^2 + Bx + C = 0,$$

where $B \sim U(2, 3)$ and $C \sim U(0, 1)$ are independent random variables. Verify your result by Monte Carlo simulation.

7.8 Let the random variable X be uniformly distributed between 0 and 1. Find the probability density function of $Y = X^n$, where n is a positive integer.

7.9 Find the population mean, population variance, and 95th percentile of the distance between two random points in the interior of the unit square using APPL. Use Monte Carlo simulation in R to check the results.

7.10 Let X be the number of fours in five tosses of a fair die. Find the probability mass function of $Y = X^2$.

7.11 Let X have probability density function

$$f_X(x) = e^{-x} \qquad x > 0$$

and Y have probability density function

$$f_Y(y) = 2e^{-2y} \qquad y > 0.$$

Assuming that X and Y are independent, use APPL to find $V\big[|X - Y|\big]$, and support your result by Monte Carlo simulation.

7.12 Use APPL to find the population mean area of the triangle created by the origin and the two complex roots of the quadratic equation

$$x^2 + x + C = 0$$

in the complex plane, where $C \sim U(1, 2)$. Support your result by Monte Carlo simulation.

7.13 The continuous random variable X has probability density function

$$f_X(x) = \begin{cases} x & 0 < x < 1 \\ 1/4 & 2 < x < 4. \end{cases}$$

Find the probability density function of $Y = \sqrt{X}$.

7.14 Let X_1 and X_2 be independent random variables, each from a population probability distribution with common probability density function:

$$f(x) = \frac{2}{x^3} \qquad x > 1.$$

Find the joint probability density function of $Y_1 = X_1 X_2$ and $Y_2 = X_2$.

7.15 Let X_1 and X_2 be continuous random variables with support on the entire first quadrant and joint probability density function $f_{X_1, X_2}(x_1, x_2)$. Use the transformation technique to find the probability density function of X_1/X_2.

7.16 Let the continuous random variable X have probability density function $f_X(x)$. What is the probability density function of $Y = mX + b$, where m and b are real constants?

7.17 Let $X \sim U(0, 8)$. Write the function $f_Y(y)$, and its associated support, for the following transformations:

(a) $Y = X/2$,

(b) $Y = \lceil X \rceil$,

(c) $Y = \lceil X \rceil - \lfloor X \rfloor$,

(d) $Y = |X - 2|$,

(e) $Y = \max\{|X - 2|, 1\}$.

Hint: Use reasoning rather than mathematics to write $f_Y(y)$.

7.18 Let $X \sim U(-k, 2k)$ for some positive real constant k. Find the probability density function of $Y = X^2$.

7.19 A single spin of a spinner can result in three equally likely outcomes: 1, 2, and 3. Let X_1 be the result of one spin of the spinner. Let X_2 be the result of a second spin of the spinner. Find the probability mass function of $X_1^2 - X_2^2$.

7.20 Let $X_1 \sim N(0, 1)$, $X_2 \sim \chi^2(1)$, $X_3 \sim \chi^2(n)$ be mutually independent random variables. Find the probability distribution (name and parameter values) of

(a) $-7X_1$,

(b) $X_1^2 + X_3$,

(c) $X_3/(nX_1^2)$,

(d) $X_1/\sqrt{X_2}$,

(e) $\sqrt{n}X_1/\sqrt{X_3}$.

No mathematics is required on this problem; simply write down the solution.

7.21 Two points are chosen at random on the *perimeter* of a unit square.

(a) Find the probability density function, population mean, and population variance of the distance between the two points using APPL. Report the population mean and population variance to ten-digit accuracy.

(b) Use Monte Carlo simulation to check your population mean and population variance from part (a).

7.22 Let X_1, X_2, X_3, be mutually independent exponential(λ) random variables. Find the 96th percentile of the following random variables.

(a) $3\min\{X_1, X_2, X_3\}$,

(b) $X_1 + X_2 - X_3$.

7.23 Consider the linear programming problem:

$$
\begin{array}{lrcl}
\text{maximize} & x_1 + & x_2 & \\
\text{subject to} & Ax_1 + & Bx_2 & \leq 1 \\
& x_1 & & \geq 0 \\
& & x_2 & \geq 0
\end{array}
$$

where the random coefficients A and B are independent $U(0, 1)$ random variables. Find the population median of the value of the objective function, $X_1 + X_2$, at its optimal value, where X_1 and X_2 are the random optimal solution values.

7.24 Let X_1 and X_2 be independent $U(0, \theta)$ random variables.

(a) Find the probability density function of the sample mean

$$\bar{X} = \frac{X_1 + X_2}{2},$$

(b) Find the probability density function of the sample standard deviation

$$S = \sqrt{(X_1 - \bar{X})^2 + (X_2 - \bar{X})^2}.$$

7.25 Let $X_1 \sim U(a, b)$, where $0 < a < b$, and $X_2 \sim U(-\pi, \pi)$ be independent random variables. Find the joint probability density function of

$$Y_1 = X_1 \cos X_2 \qquad \text{and} \qquad Y_2 = X_1 \sin X_2.$$

7.26 Let X_1 and X_2 denote a random sample from a $N(\mu, \sigma^2)$ population. Consider the random variables $(X_1 + X_2)/2$ and $(X_1 - X_2)^2/2$, which happen to be the sample mean and the unbiased version of the sample variance. Graphically and/or analytically examine the properties of this transformation.

7.27 Consider the stochastic project network precedence diagram shown in Figure 7.16. Let X_1, X_2, X_3 be mutually independent $U(0, 1)$ random variables that denote the activity duration times. Find the probability density function of the project completion time, which is $\max\{X_1 + X_2, X_3\}$.

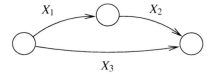

Figure 7.16: Stochastic project network precedence diagram.

7.28 Let $X_1 \sim U(0, 2)$ and X_2 have probability density function

$$f_{X_2}(x_2) = \frac{x_2}{2} \qquad 0 < x_2 < 2.$$

Assuming that X_1 and X_2 are independent random variables, find the joint probability density function of $Y_1 = \min\{X_1, X_2\}$ and $Y_2 = \max\{X_1, X_2\}$, and the marginal distributions of Y_1 and Y_2.

7.29 Two cards are drawn without replacement from a well-shuffled deck of 52 cards. Let X_1 be the number of black cards (that is, clubs and spades) and let X_2 be the number of diamonds. Find the probability mass function of $2X_1^2 + X_2$.

7.30 Find the population median of the volume of a sphere with a random radius X having probability density function

$$f(x) = 2x \qquad 0 < x < 1.$$

7.31 Let the random variable X have probability density function

$$f(x) = \theta x^{\theta - 1} \qquad 0 < x < 1,$$

where θ is a positive parameter. Use the transformation technique to find the probability density function of X^2.

7.32 Poppy Turner is a farmer. He would like to build a rectangular pig pen along a long brick wall using l feet of fencing for the three sides. Let X be the length of the sides of the pen that are perpendicular to the brick wall. If X has probability density function

$$f(x) = \frac{8x}{l^2} \qquad 0 < x < l/2,$$

find the probability density function of the area of the pen.

7.33 Let the random variable X have probability density function

$$f_X(x) = x/8 \qquad 0 < x < 4.$$

Find the probability density function of $Y = g(X) = 2X^2$ via the transformation technique.

7.34 Let $X \sim U(0, \pi)$. Find the probability density function of $Y = \sin X$. *Hint:* the derivative of the arcsin function is $\frac{d}{dy} \arcsin y = 1/\sqrt{1 - y^2}$.

7.35 Let X_1 and X_2 be independent exponential random variables with population means $1/\lambda_1$ and $1/\lambda_2$, respectively. Find the probability density function of X_1/X_2.

7.36   Let the random variable X be uniformly distributed between 0 and 1. Find the probability density function of $Y = \arcsin X$ using

    (a) the cumulative distribution function technique,

    (b) the transformation technique.

7.37   Let X_1 and X_2 be uniformly distributed over the rectangle with opposite corners at the origin and $(c, 2c)$, where c is some real positive constant. Find the distribution of X_1/X_2 using

    (a) the cumulative distribution function technique,

    (b) the transformation technique.

7.38   Find the probability density function of the ratio (quotient) of two independent $U(0, 1)$ random variables using both the cumulative distribution function technique and the transformation technique.

7.39   Let the random variable X have probability density function

$$f(x) = \frac{1}{\pi} \qquad -\pi/2 < x < \pi/2.$$

Find the probability density function of $Y = \sin X$ by the

    (a) cumulative distribution function technique,

    (b) transformation technique.

    Hint: The derivative of $x = \arcsin y$ is $\dfrac{dx}{dy} = \dfrac{1}{\sqrt{1-y^2}}$.

7.40   The random variable X has probability density function

$$f(x) = \lambda e^{-\lambda x} \qquad x > 0,$$

where λ is a positive parameter. Find the probability density function of $Y = |X - k|$, where k is a positive constant.

7.41   If $X_1 \sim$ exponential(λ_1), $X_2 \sim$ exponential(λ_2), and X_1 and X_2 are independent, what is the joint probability density function of $Y_1 = X_1 + X_2$ and $Y_2 = X_2$?

7.42   Let X_1 and X_2 be uniformly distributed over the unit circle, that is,

$$f(x_1, x_2) = \frac{1}{\pi} \qquad x_1^2 + x_2^2 < 1.$$

Find the joint probability density function of $Y_1 = \sqrt{X_1^2 + X_2^2}$ and $Y_2 = \arctan(X_2/X_1)$.

7.43   Isaiah wants to invest $1,000,000 in three consecutive one-year CDs with mutually independent random interest rates:

    • R_1 is exponential with population mean $6/100$,

    • R_2 is exponential with population mean $7/100$,

    • R_3 is exponential with population mean $8/100$.

Assuming that interest is compounded continuously, what is the population mean of the amount in his account at the end of three years rounded to the nearest penny?

7.44 Let the random variables X_1 and X_2 be defined by the joint probability density function

$$f_{X_1, X_2}(x_1, x_2) = \frac{1}{4} \qquad 0 < x_1 < 1, 0 < x_2 < 4.$$

Find the probability density function of $Y_1 = X_1/X_2$ using the dummy transformation $Y_2 = X_2$. Make sure to show the region $\mathcal{B}$, the inverse transformation, and the Jacobian of the inverse transformation.

7.45 Let X_1, X_2, X_3, X_4 be mutually independent and identically distributed $U(0, 1)$ random variables. Use APPL to find the 99th percentile of the sample mean

$$\bar{X} = \frac{X_1 + X_2 + X_3 + X_4}{4}$$

to ten-digit accuracy.

7.46 Write APPL statements to find the expected rectilinear distance (or Manhattan distance or L_1 norm) between two points chosen at random in the interior of the unit square.

7.47 Write APPL statements to find the expected number of real roots of the quadratic equation

$$Ax^2 + Bx + C = 0,$$

where A, B, and C are the outcomes of three rolls of a fair die. Support your solution with a Monte Carlo simulation.

7.48 Let $Z_1, Z_2, \ldots, Z_{20}$ be mutually independent $N(0, 1)$ random variables. Find the 99th percentile of

$$\frac{Z_1}{\sqrt{(Z_2^2 + Z_3^2 + \cdots + Z_{20}^2)/19}}.$$

7.49 Let X_1 be a chi-square random variable with m degrees of freedom, where m is a positive integer. Let X_2 be a chi-square random variable with n degrees of freedom, where n is a positive integer. If X_1 and X_2 are independent, show that

$$E\left[\frac{X_1/m}{X_2/n}\right] = \frac{n}{n-2}.$$

Give any restrictions on m and n that are necessary for your result to hold.

7.50 Consider a circle of random radius $R \sim \text{exponential}(1)$. Two points are selected at random on the circumference of the circle. Let the random variable Y be the length of the line segment (chord) joining the two points. Let Θ be the random angle formed by the line segments from the center of the circle to the two points as measured counterclockwise from the first point to the second point.

(a) Write the APPL statements to find the population mean of Y (*hint:* $Y = 2R\sin(\Theta/2)$).

(b) Write a Monte Carlo simulation to check your answer to part (a).

7.51 Give the APPL statements to find the population variance of the *Euclidean distance* (or L_2 norm) between the x- and y-intercepts of the graph of

$$Ax + By = 1,$$

where A and B are independent $U(0, 1)$ random variables.

7.52 Give the APPL statements to find the population variance of the rectilinear distance (or Manhattan distance or L_1 norm) between the x- and y-intercepts of the graph of

$$Ax + By = 1,$$

where A and B are independent $U(0, 1)$ random variables.

7.53 Scott, a 70% free throw shooter, attempts 5 mutually independent free throw shots. Billy, an 80% free throw shooter, attempts 4 mutually independent free throw shots, which are independent of Scott's shots. They take the product of the number of shots that they each make and donate that number of dollars to their favorite charity.

    (a) Write the APPL code that calculates the population mean amount that the charity will receive from this random experiment as an exact fraction.

    (b) Write a Monte Carlo simulation to support the exact value determined by APPL in part (a).

7.54 Find the 95th percentile of the t distribution with 2 degrees of freedom.

7.55 Let the random variable X have an F distribution with 2 and 7 degrees of freedom. Find $P(0.052 < X < 9.55)$.

7.56 Find the 99th percentile of an F random variable with 6 and 3 degrees of freedom.

7.57 Find the median of a random variable with a t distribution with 94 degrees of freedom.

7.58 Let Z_1, Z_2 and Z_3 be mutually independent $N(0, 1)$ random variables. Determine a constant c such that

$$P\left(\frac{Z_1}{\sqrt{Z_2^2 + Z_3^2}} > c \right) = 0.3.$$

7.59 Let Z_1 and Z_2 be independent $N(0, 1)$ random variables. Determine a constant c such that

$$P\left(\frac{Z_1}{\sqrt{Z_2^2}} < c \right) = 0.99$$

7.60 Let X_1, X_2, X_3, X_4, X_5 be mutually independent random lug nut lifetimes, each with probability density function $f(x)$ and cumulative distribution function $F(x)$. Find the probability density function of the second order statistic, $Y = X_{(2)}$.

7.61 Find the probability density function of the second order statistic, $X_{(2)}$, associated with $n = 5$ observations drawn from a population with probability density function

$$f(x) = \frac{1}{x^2} \qquad x > 1.$$

7.62 Let X_1 and X_2 have joint probability density function

$$f(x_1, x_2) = \frac{x_1 + 5x_2}{3} \qquad 0 < x_1 < 1, 0 < x_2 < 1.$$

    (a) Find $V(X_{(2)})$, where $X_{(1)}$ and $X_{(2)}$ are the order statistics, that is, $X_{(1)} = \min\{X_1, X_2\}$ and $X_{(2)} = \max\{X_1, X_2\}$.

    (b) Support your solution to part (a) using Monte Carlo simulation.

7.63 Write the APPL statements to find the mean of the largest of ten independent and identically distributed $N(2, 1)$ random variables that have been truncated on the left at 0.

7.64 Consider a wire of length 1. Let the continuous random variables X_1 and X_2 denote two random points at which to break the wire.

    (a) If X_1 and X_2 have joint probability density function $f(x_1, x_2)$ defined on the unit square $0 < x_1 < 1$ and $0 < x_2 < 1$, find an expression for the probability that you can form a triangle with the pieces. *Hint*: all three pieces must be shorter than $1/2$ to form a triangle.

    (b) If X_1 and X_2 are independent and uniformly distributed between 0 and 1, find the expected area of the triangle given that you can form a triangle. *Hint*: For a triangle with side lengths a, b, and c, and semi-perimeter $s = (a + b + c)/2$, Heron's formula gives the area as $\sqrt{s(s-a)(s-b)(s-c)}$.

7.65 Let X_1 and X_2 be independent random variables with the following probability density functions:
$$f_{X_1}(x_1) = 1 \qquad 0 < x_1 < 1,$$
$$f_{X_2}(x_2) = 2x_2 \qquad 0 < x_2 < 1.$$
If $X_{(1)} = \min\{X_1, X_2\}$ and $X_{(2)} = \max\{X_1, X_2\}$, find $P(X_{(1)} < 1/3)$.

7.66 Let X_1 and X_2 be uniformly distributed over a disk with radius 1 centered at the origin. Use *no calculus* to find the joint probability density function of $X_{(1)} = \min\{X_1, X_2\}$ and $X_{(2)} = \max\{X_1, X_2\}$. Use a geometric argument.

7.67 Let $X_1, X_2, \ldots, X_{10}$ be mutually independent and identically distributed Weibull(λ, κ) random variables, and let $X_{(1)}, X_{(2)}, \ldots, X_{(10)}$ be the associated order statistics. Give the APPL statement(s) required to find $P(X_{(3)} < 2)$.

7.68 Let $X_1 \sim U(0, 1)$ and $X_2 \sim U(0, 2)$ be independent random variables. Let $X_{(1)} = \min\{X_1, X_2\}$ and $X_{(2)} = \max\{X_1, X_2\}$ be the associated order statistics.

    (a) Find the joint probability density function of $X_{(1)}$ and $X_{(2)}$.

    (b) Find the marginal probability density functions of $X_{(1)}$ and $X_{(2)}$.

7.69 Let X_1 and X_2 be independent $U(0, 1)$ random variables.

    (a) Find the correlation ρ between the order statistics $X_{(1)}$ and $X_{(2)}$.

    (b) Verify your solution to part (a) using Monte Carlo simulation.

7.70 Five billiard balls are drawn without replacement from a bag containing balls numbered $1, 2, \ldots, 15$. Let $X_1, X_2, \ldots, X_5$ be the numbers drawn, and $X_{(1)}, X_{(2)}, \ldots, X_{(5)}$ be the corresponding order statistics.

    (a) Find the probability mass function of the third order statistic, $X_{(3)}$.

    (b) Find the joint probability mass function of $X_{(3)}$ and $X_{(4)}$.

7.71 Let $X_1, X_2, \ldots, X_n$ be a random sample from a $U(0, 1)$ population. Find the median of the distribution of the nth order statistic, $X_{(n)}$.

7.72 Barry and Jeanne independently select points uniformly on the interval $(0, 1)$. Find the expected distance between their two points.

7.73 Let $Z_1, Z_2, \ldots, Z_8$ be mutually independent and identically distributed standard normal random variables. Calculate the probability that the sum of their squares is less than ten.

7.74 Write the APPL statements to find the population variance of the sum of three mutually independent and identically distributed logistic random variables with both of its parameters equal to 1. The probability density function for the logistic distribution is

$$f(x) = \frac{\lambda^\kappa \kappa e^{\kappa x}}{(1 + \lambda^\kappa e^{\kappa x})^2} \qquad -\infty < x < \infty,$$

where λ is a positive scale parameter and κ is a positive shape parameter.

7.75 Let $Z_1, Z_2, \ldots, Z_n$ be mutually independent and identically distributed standard normal random variables and $a_1, a_2, \ldots, a_n$ be nonzero real constants. Find the distribution of

$$Y = \frac{a_1 Z_1 + a_2 Z_2 + \cdots + a_n Z_n}{\sqrt{a_1^2 + a_2^2 + \cdots + a_n^2}}.$$

7.76 Let X_1, X_2, X_3, X_4, and X_5 be mutually independent and identically distributed random variables, each with moment generating function

$$M_{X_1}(t) = M_{X_2}(t) = M_{X_3}(t) = M_{X_4}(t) = M_{X_5}(t) = 0.8 + 0.2e^{3t}$$

for $-\infty < t < \infty$. Find $P(X_1 + X_2 + X_3 + X_4 + X_5 = 6)$.

7.77 Let $X_1, X_2, \ldots, X_n$ be a random sample from a Poisson population with population mean λ. Find an expression for $P(\bar{X} \leq c)$, for any real positive constant c, where $\bar{X}$ is the sample mean.

7.78 Find the probability mass function of the sum of n mutually independent and identically distributed negative binomial random variables. Use the form of the negative binomial distribution whose support begins at 0.

7.79 Let X_1, X_2, X_3, X_4 be mutually independent $\chi^2(n)$ random variables. Find an integer n such that $P(X_1 + X_2 + X_3 + X_4 < 42.98) = 0.99$.

7.80 Let $X_1, X_2, \ldots, X_n$ be mutually independent standard normal random variables.

  (a) Find the 99th percentile of X_1.

  (b) Find the 99th percentile of nX_1.

  (c) Find the 99th percentile of $X_1 + X_2 + \cdots + X_n$.

7.81 The southern border of Virginia is 425 miles long. If 500,000 men, 500,000 women and 500,000 children stand shoulder-to-shoulder along this border, find the probability that their combined length will exceed the length of the border. You may assume that all shoulder widths are independent and that men's, women's, and children's shoulder widths (in inches) are $N(21, 4)$, $N(18, 4)$, and $N(14.85, 9)$ random variables, respectively.

7.82 Boy Scouts often compete in a "Pinewood Derby," where boys race model cars on a two-lane track. Michael's car requires $N(4, 0.04)$ seconds to finish the race and Nozer's car requires $N(4.07, 0.09)$ seconds to finish the race.

  (a) Find the probability that Michael's car wins a single race.

  (b) Find the probability that Michael's car wins the majority of the races in a "best-of-five" series of races.

  (c) Give R statements to compute the answer to part (b).

7.83 Chip is building the foundation for a home. He lays 40 cinder blocks to establish the width of the home. Each cinder block has a width that is normally distributed with a mean of 15.7 inches and a variance of 0.1 square inches. The width of the grout he uses between each block is normally distributed with a mean of 0.3 inches and a variance of 0.09 square inches. Find the probability that the total width of the foundation of his house exceeds 53.5 feet.

7.84 What is the name of the distribution of the reciprocal of an $F(n_1, n_2)$ random variable?

7.85 What is the name of the distribution of the sum of n mutually independent and identically distributed gamma random variables?

7.86 What is the name of the distribution of the minimum of two independent and identically distributed Weibull random variables?

7.87 What is the name of the distribution of the ratio of one chi-square random variable divided by its degrees of freedom, to another independent chi-square random variable divided by its degrees of freedom?

7.88 Consider the continuous random variable X with probability density function

$$f(x) = \begin{cases} x & 0 < x < 1 \\ 1/4 & 1 \leq x < 3. \end{cases}$$

Use APPL to

(a) verify that this is a legitimate probability density function,

(b) find $E[\cos(X)]$,

(c) find the 95th percentile of $X_1 + X_2 + X_3 + X_4$, where X_1, X_2, X_3, X_4 are mutually independent and identically distributed observations drawn from the population described by X above,

(d) find the population median of the product of X and a $U(0, 2)$ random variable.

7.89 Let X and Y be independent standard normal random variables. Find $P(|X - Y| > 3)$.

7.90 Mama Mia's memoryless salon has all n stylists occupied right now, with the patron in chair i having been worked on for X_i minutes, for $i = 1, 2, \ldots, n$. Amazingly (and to your advantage in terms of mathematical tractability), the cutting time at the salon has an exponential distribution with population mean 10 minutes. Assume that there are k patrons presently waiting in line.

(a) Give the name and parameters of the wait time of a customer arriving to Mama's salon right now.

(b) Give the 25th, 50th, and 75th percentile of the wait times for a customer arriving to a salon at this moment when $n = 9$ and $k = 7$.

7.91 Give the APPL statements and output required to find the probability density function of the sample mean of five mutually independent $U(0, 1)$ random variables.

7.92 The thickness of a sheet of paper is normally distributed with population mean $\mu = 0.002$ inches and population standard deviation $\sigma = 0.0001$ inches.

(a) Find the population mean and population variance of the height of a ream (500 sheets) of paper.

(b) Find the 99th percentile of the height of a ream (500 sheets) of paper.

7.93   Let $X_1 \sim \text{gamma}(\lambda_1, \kappa_1)$, $X_2 \sim \text{gamma}(\lambda_2, \kappa_2)$, $X_3 \sim \text{gamma}(\lambda_3, \kappa_3)$, be mutually independent random variables. Find the moment generating function of

$$Y = a_1 X_1 + a_2 X_2 + a_3 X_3$$

for real, positive constants a_1, a_2, a_3.

7.94   Let A and B be 2×2 matrices whose entries are mutually independent $U(0, 1)$ observations. What is the expected value of the trace (the sum of the diagonal elements) of AB?

7.95   Let A be a 2×3 matrix whose entries are mutually independent Bernoulli(p) random variables. Let B be a 3×2 matrix whose entries are mutually independent Bernoulli(p) random variables. What is the population mean and variance of the trace (the sum of the diagonal elements) of AB?

7.96   The *Mellin* transformation of a positive (that is, $f_X(x) = 0$ for all $x \leq 0$) random variable X is

$$L_X(t) = E\left[X^t\right],$$

which exists if $E\left[X^t\right]$ exists for $|t| \leq c$, where c is some positive constant.

   (a) If X has moment generating function $M_X(t)$, show that

$$L_X(t) = M_{\ln X}(t).$$

   (b) If X has Mellin transform $L_X(t)$, Y has Mellin transform $L_Y(t)$, and X and Y are independent, show that XY has Mellin transform

$$L_{XY}(t) = L_X(t) L_Y(t).$$

7.97   Let $X \sim N(2, 4)$ and $Y \sim N(7, 9)$ be independent random variables. Find a constant c such that

$$P(3X + 5Y \geq c) = P(8X - Y + 1 \leq c/4).$$

Support your solution by Monte Carlo simulation.

7.98   The number of calls daily to a call center is $X_1 \sim \text{Poisson}(\lambda_1)$. The number of calls daily to a second call center is $X_2 \sim \text{Poisson}(\lambda_2)$. Assuming that the number of daily calls to the two centers are independent and that there are n total calls to both call centers during a particular day (that is, $X_1 + X_2 = n$), what is the probability distribution of the number of calls to the first call center on that particular day?

Chapter 8

Limiting Distributions

The previous chapter presented three techniques for determining the probability distribution of $Y = g(X)$ given the probability distribution of X, or more generally, determining the probability distribution of $Y = g(X_1, X_2, \ldots, X_n)$ given the joint probability distribution of $X_1, X_2, \ldots, X_n$. The examples and exercises from the previous section were hand picked to result in mathematically tractable probability distributions. Examples of these friendly transformations include:

- the square of a standard normal random variable (which has the chi-square distribution),

- the sum of n mutually independent exponential random variables (which has the Erlang distribution),

- the sum of n mutually independent normal random variables (which has the normal distribution).

There are many transformations in which the mathematics are intractable or are just too tedious to compute by hand. Examples of such hostile transformations include:

- the product of n mutually independent F random variables,

- the sum of n mutually independent uniform random variables,

- the sum of n mutually independent Weibull random variables.

This chapter considers *limiting distributions* which apply to random variables that depend on a parameter n, which is a positive integer and is often referred to in statistics as the *sample size*. Examples of random variables that depend on n include a binomial(n, p) random variable, an Erlang(λ, n) random variable, and a sample mean $\bar{X} = (1/n) \sum_{i=1}^{n} X_i$.

Some of the most useful theorems in probability theory fall in a class known as *limit theorems*, which apply in the limit as the sample size n grows to infinity. These results are also known as *asymptotic* results and they oftentimes exist for all probability distributions rather than just for a single probability distribution. One of the most remarkable and well-known results is the *central limit theorem*, which states that the sum of n mutually independent random variables converges to a normal distribution as $n \to \infty$ under some mild assumptions. We begin the introduction of limiting distributions with a concept known as *convergence in probability*.

8.1   Convergence in Probability

There are instances that arise in probability when the distribution of one or more random variables are a function of the positive integer n. Simple examples include

- the distribution of $X \sim$ binomial(n, p) depends on n,

- the distribution of $X \sim \chi^2(n)$ depends on n,

- if $X_1, X_2, \ldots, X_n$ are mutually independent $N(\mu, \sigma^2)$ random variables, then the distribution of

$$\bar{X} \sim N\left(\mu, \frac{\sigma^2}{n}\right)$$

depends on n.

When the probability distribution of a random variable that is a function of n is intractable, it is often useful to know the limiting probability distribution as $n \to \infty$. We focus first on "convergence in probability."

Definition 8.1 A sequence of random variables $X_1, X_2, \ldots$ *converges in probability* to a random variable X if, for every $\varepsilon > 0$,

$$\lim_{n \to \infty} P(|X_n - X| < \varepsilon) = 1.$$

Three comments concerning Definition 8.1 are germane. First, using complementary probabilities, the limiting probability expression can also be written as

$$\lim_{n \to \infty} P(|X_n - X| \geq \varepsilon) = 0.$$

Second, convergence in probability is indicated by the shorthand $X_n \xrightarrow{P} X$. Finally, an important special case of this definition occurs when the distribution of X is a degenerate distribution at some constant value a. In the case when $P(X = a) = 1$, the sequence of random variables *converges in probability* to the constant a, which is written as $X_n \xrightarrow{P} a$. Convergence in probability to a degenerate distribution is the most common and useful type of convergence in probability. The case of convergence to a random variable X with a degenerate distribution is illustrated in the following example.

Example 8.1 Let $Y_1, Y_2, \ldots, Y_n$ be mutually independent $U(0, \theta)$ random variables, where θ is a positive fixed parameter. Let $X_n = \max\{Y_1, Y_2, \ldots, Y_n\}$. Show that the sequence of random variables $X_1, X_2, \ldots$ converges in probability to a degenerate random variable as $n \to \infty$.

Since each of the Y_i values is a $U(0, \theta)$ random variable, each has probability density function

$$f_Y(y) = \frac{1}{\theta} \qquad 0 < y < \theta,$$

where Y_i is denoted generically by Y here, and has cumulative distribution function on its support

$$F_Y(y) = \int_0^y \frac{1}{\theta} \, dw = \frac{y}{\theta} \qquad 0 < y < \theta.$$

Using the cumulative distribution function technique, the cumulative distribution function of $X_n = \max\{Y_1, Y_2, \ldots, Y_n\}$ on its support is

$$
\begin{aligned}
F_{X_n}(x) &= P(X_n \leq x) \\
&= P\big(\max\{Y_1, Y_2, \ldots, Y_n\} \leq x\big) \\
&= P(Y_1 \leq x, Y_2 \leq x, \ldots, Y_n \leq x) \\
&= P(Y_1 \leq x)P(Y_2 \leq x) \ldots P(Y_n \leq x) \\
&= F_{Y_1}(x)F_{Y_2}(x) \ldots F_{Y_n}(x) \\
&= \left(\frac{x}{\theta}\right)^n \qquad 0 < x < \theta.
\end{aligned}
$$

As n increases, an increasing portion of the probability distribution of X_n shifts to the right toward θ, as illustrated in Figure 8.1 for $n = 1$, $n = 3$, and $n = 20$ when $\theta = 2$. Thus, it is reasonable to conjecture that X_n converges in probability to the constant upper bound of the support of the distributions of the $Y_1, Y_2, \ldots, Y_n$ values, that is

$$X_n \xrightarrow{P} \theta.$$

Computing the probability in Definition 8.1,

$$
\begin{aligned}
P(|X_n - \theta| < \varepsilon) &= P(-\varepsilon < X_n - \theta < \varepsilon) \\
&= P(\theta - \varepsilon < X_n < \theta + \varepsilon) \\
&= P(\theta - \varepsilon < X_n \le \theta) \\
&= P(X_n \le \theta) - P(X_n \le \theta - \varepsilon) \\
&= F_{X_n}(\theta) - F_{X_n}(\theta - \varepsilon) \\
&= 1 - \left(\frac{\theta - \varepsilon}{\theta}\right)^n
\end{aligned}
$$

for any finite sample size n. Taking the limit as $n \to \infty$ of both sides of this equation for some small, fixed $\varepsilon > 0$ yields

$$\lim_{n \to \infty} P(|X_n - \theta| < \varepsilon) = \lim_{n \to \infty} \left[1 - \left(\frac{\theta - \varepsilon}{\theta}\right)^n\right] = 1.$$

Therefore, as conjectured, $X_n \xrightarrow{P} \theta$.

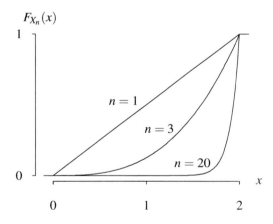

Figure 8.1: Cumulative distribution function $F_{X_n}(x)$ on $0 < x < 2$.

The next example illustrates how Chebyshev's inequality

$$P(|X - \mu| < k\sigma) \ge 1 - \frac{1}{k^2}$$

from Section 3.5 can be used to show convergence in probability.

Example 8.2 Let $X_1, X_2, \ldots, X_n$ be mutually independent observations from a population with finite population mean μ and finite, positive population variance σ^2. Does the sample mean

$$\bar{X}_n = \frac{1}{n} \sum_{i=1}^{n} X_i$$

converge in probability to a constant value?

Recalling from Example 6.69 that μ is the population mean and $\sigma/\sqrt{n}$ is the population standard deviation of $\bar{X}_n$, Chebyshev's inequality implies that

$$P\left(|\bar{X}_n - \mu| < \frac{k\sigma}{\sqrt{n}}\right) \geq 1 - \frac{1}{k^2}.$$

Letting $k = \varepsilon\sqrt{n}/\sigma$ for some small, fixed $\varepsilon > 0$ forces the probability statement to conform to Definition 8.1:

$$P\left(|\bar{X}_n - \mu| < \varepsilon\right) \geq 1 - \frac{\sigma^2}{n\varepsilon^2}.$$

Taking the limit of both sides of the inequality as $n \to \infty$ results in

$$\lim_{n\to\infty} P\left(|\bar{X}_n - \mu| < \varepsilon\right) = 1,$$

which implies that $\bar{X}_n \xrightarrow{P} \mu$. This important result is a one way of stating the *weak law of large numbers*, which is often abbreviated WLLN. This version of the weak law of large numbers states that the sample mean $\bar{X}$ is arbitrarily close to the population mean μ in the limit as $n \to \infty$.

Example 8.3 If $Y_n \sim$ binomial(n, p) then $X_n = Y_n/n$ is the fraction of successes in n mutually independent Bernoulli trials. Show that $X_n = Y_n/n$ converges in probability to p.

This result is known as the "law of averages." The probability integral transformation, which states that $F(X) \sim U(0, 1)$, and the law of averages provide the basis for Monte Carlo simulation. The R code used to estimate probabilities earlier in the text typically concluded by printing count / nrep, where count plays the role of Y_n and nrep plays the role of n. The law of averages states that $X_n = Y_n/n$ is arbitrarily close to p in the limit as $n \to \infty$. The law of averages does not determine how quickly Y_n/n approaches p, but it does indicate that it will get there eventually. The probability integral transformation, on the other hand, is the basis for random variate generation.

Returning to the problem, since the population mean and variance of Y_n are

$$E[Y_n] = np$$

and

$$V[Y_n] = np(1 - p),$$

the population mean and variance of the fraction of successes $X_n = Y_n/n$ are

$$E[X_n] = E\left[\frac{Y_n}{n}\right] = \frac{1}{n}E[Y_n] = p$$

and

$$V[X_n] = V\left[\frac{Y_n}{n}\right] = \frac{1}{n^2}V[Y_n] = \frac{p(1 - p)}{n}.$$

Chebyshev's inequality implies that

$$P\left(|X_n - p| < k\sqrt{\frac{p(1 - p)}{n}}\right) \geq 1 - \frac{1}{k^2}.$$

Letting

$$k = \frac{\varepsilon}{\sqrt{\frac{p(1-p)}{n}}}$$

for some small, fixed $\varepsilon > 0$ forces the probability statement to conform to Definition 8.1, which reduces to

$$P\left(|X_n - p| < \varepsilon\right) \geq 1 - \frac{p(1-p)}{n\varepsilon^2}.$$

Taking the limit of both sides of the inequality as $n \to \infty$ results in

$$\lim_{n \to \infty} P\left(|X_n - p| < \varepsilon\right) = 1,$$

which implies that $X_n \xrightarrow{P} p$. The alert reader will have noticed that this result is nothing more than a special case of the weak law of large numbers applied to Bernoulli trials. It has been treated separately here because of its close association with Monte Carlo simulation.

This ends the introduction to convergence in probability. The next section introduces a slightly weaker type of convergence known as *convergence in distribution*.

8.2 Convergence in Distribution

As was the case with convergence in probability, convergence in distribution also requires a sequence of random variables that depend on n, a positive integer. In this case, the sequence of *cumulative distribution functions* (rather than random variables themselves) converge to another cumulative distribution function. Convergence in distribution is defined below.

Definition 8.2 Let $X_1, X_2, \ldots$ be a sequence of random variables. Let X_n have cumulative distribution function $F_{X_n}(x)$. If $F_X(x)$ is a cumulative distribution function of a random variable X and if

$$\lim_{n \to \infty} F_{X_n}(x) = F_X(x)$$

at every x where $F_X(x)$ is continuous, then X_n is said to *converge in distribution* to the random variable X with cumulative distribution function $F_X(x)$.

Convergence in distribution is indicated by the shorthand $X_n \xrightarrow{D} X$. The rather loose notation $X_n \xrightarrow{D} N(0, 1)$, for instance, will be used to indicate that X_n converges in distribution to a standard normal random variable even though $N(0, 1)$ is a distribution rather than a random variable. The proper way to write this would be to place a random variable on the right-hand side of the expression rather than an abbreviation for a specific distribution. The first example concerns the limiting distribution of a sequence of random variables that are each minimums of random variables with a known distribution. This example illustrates that only considering x values where $F_X(x)$ is continuous is important.

Example 8.4 Let $Y_1, Y_2, \ldots, Y_n$ be mutually independent random variables drawn from a $U(0, 1)$ population. Let the sequence of random variables $X_1, X_2, \ldots$ be defined by $X_n = \min\{Y_1, Y_2, \ldots, Y_n\}$. Show that X_n converges in distribution to a degenerate distribution.

The probability density function of any one of the Y_i, denoted generically here as Y, is

$$f_Y(y) = 1 \qquad 0 < y < 1.$$

The associated cumulative distribution function of Y is

$$F_Y(y) = \begin{cases} 0 & y \leq 0 \\ y & 0 < y < 1 \\ 1 & y \geq 1. \end{cases}$$

The cumulative distribution function of $X_n = \min\{Y_1, Y_2, \ldots, Y_n\}$ on its support is

$$
\begin{aligned}
F_{X_n}(x) &= P(X_n \leq x) \\
&= P\big(\min\{Y_1, Y_2, \ldots, Y_n\} \leq x\big) \\
&= 1 - P\big(\min\{Y_1, Y_2, \ldots, Y_n\} > x\big) \\
&= 1 - P(Y_1 > x, Y_2 > x, \ldots, Y_n > x) \\
&= 1 - P(Y_1 > x)P(Y_2 > x)\ldots P(Y_n > x) \\
&= 1 - \big[1 - P(Y_1 \leq x)\big]\big[1 - P(Y_2 \leq x)\big]\ldots\big[1 - P(Y_n \leq x)\big] \\
&= 1 - \big[1 - F_{Y_1}(x)\big]\big[1 - F_{Y_2}(x)\big]\ldots\big[1 - F_{Y_n}(x)\big] \\
&= 1 - \big[1 - F_Y(x)\big]\big[1 - F_Y(x)\big]\ldots\big[1 - F_Y(x)\big] \\
&= 1 - (1 - x)^n \qquad\qquad 0 < x < 1.
\end{aligned}
$$

This cumulative distribution function is plotted for $n = 1$, $n = 3$, and $n = 20$ in Figure 8.2. As n increases, more of the probability distribution is pushed toward 0 as one would expect because X_n is the minimum of n mutually independent $U(0, 1)$ random variables.

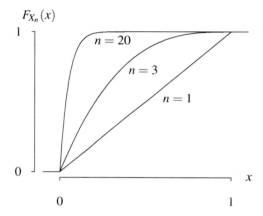

Figure 8.2: Cumulative distribution function $F_{X_n}(x)$ on $0 < x < 1$.

The limiting cumulative distribution function of X_n is

$$
\lim_{n \to \infty} F_{X_n}(x) = \begin{cases} 0 & x \leq 0 \\ 1 & x > 0, \end{cases}
$$

which corresponds to the limiting cumulative distribution function (carefully read the fine print in Definition 8.2 concerning the discontinuity in $F_X(x)$ at $x = 0$ and watch the placement of the strict inequalities in the two cumulative distribution functions)

$$
F_X(x) = \begin{cases} 0 & x < 0 \\ 1 & x \geq 0. \end{cases}
$$

Hence, $X_n \xrightarrow{D} X$, where X is a degenerate random variable at 0. The shorthand to describe this is $X_n \xrightarrow{D} 0$.

    This first example of convergence in distribution considered a limiting distribution that is degenerate. The next example explores a limiting distribution that is not degenerate.

Example 8.5 Let $Y_1, Y_2, \ldots, Y_n$ again be a random sample from a $U(0, 1)$ population. Find the limiting distribution of $X_n = n\left(1 - \max\{Y_1, Y_2, \ldots, Y_n\}\right)$.

This example provides a bit more drama than the previous example, where it was somewhat obvious that the smallest of n mutually independent $U(0, 1)$ random variables converged to a degenerate distribution at 0 in the limit as $n \to \infty$. In this case, X_n is the product of n, which drags the random variable X_n off to infinity in the limit as $n \to \infty$, and $1 - \max\{Y_1, Y_2, \ldots, Y_n\}$, which tends to 0 because the maximum tends to 1 in the limit as $n \to \infty$. Which of the two factors exhibits a stronger pull in the limit as $n \to \infty$? Convergence in distribution provides the answer.

As in the previous example, each of the Y_i values has cumulative distribution function

$$F_Y(y) = \begin{cases} 0 & y \leq 0 \\ y & 0 < y < 1 \\ 1 & y \geq 1. \end{cases}$$

The cumulative distribution function of X_n is

$$
\begin{aligned}
F_{X_n}(x) &= P(X_n \leq x) \\
&= P\left(n(1 - \max\{Y_1, Y_2, \ldots, Y_n\}) \leq x\right) \\
&= P\left(\max\{Y_1, Y_2, \ldots, Y_n\} > 1 - \frac{x}{n}\right) \\
&= 1 - P\left(\max\{Y_1, Y_2, \ldots, Y_n\} \leq 1 - \frac{x}{n}\right) \\
&= 1 - P\left(Y_1 \leq 1 - \frac{x}{n}, Y_2 \leq 1 - \frac{x}{n}, \ldots, Y_n \leq 1 - \frac{x}{n}\right) \\
&= 1 - P\left(Y_1 \leq 1 - \frac{x}{n}\right) P\left(Y_2 \leq 1 - \frac{x}{n}\right) \ldots P\left(Y_n \leq 1 - \frac{x}{n}\right) \\
&= 1 - F_{Y_1}\left(1 - \frac{x}{n}\right) F_{Y_2}\left(1 - \frac{x}{n}\right) \ldots F_{Y_n}\left(1 - \frac{x}{n}\right) \\
&= 1 - \left(1 - \frac{x}{n}\right)^n \qquad 0 < x < n.
\end{aligned}
$$

So taking the limit as n goes to infinity, the limiting cumulative distribution function is

$$\lim_{n \to \infty} F_{X_n}(x) = \begin{cases} 0 & x \leq 0 \\ 1 - e^{-x} & x > 0. \end{cases}$$

This is recognized as the cumulative distribution function of a unit exponential distribution. So $X_n \xrightarrow{D} \text{exponential}(1)$. The tug-of-war between the two factors that define X_n ends in a stalemate: neither n nor $1 - \max\{Y_1, Y_2, \ldots, Y_n\}$ are able to drag X_n off to infinity or 0. The seeds of this stalemate can be seen graphically by plotting the probability density function of X_n for some small values of n. Differentiating the cumulative distribution function with respect to x gives

$$f_{X_n}(x) = \left(1 - \frac{x}{n}\right)^{n-1} \qquad 0 < x < n,$$

for $n = 1, 2, \ldots$. This probability density function is plotted in Figure 8.3 for $n = 1$, $n = 2$, and $n = 3$.

The first two examples illustrated the case of a mathematically tractable cumulative distribution function. The third example illustrates a work around when the cumulative distribution function is not tractable.

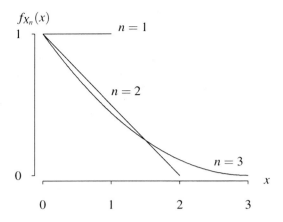

Figure 8.3: Probability density function $f_{X_n}(x)$ on $0 < x < n$.

Example 8.6 Let $X_n \sim t(n)$. Find the limiting distribution of X_n.

Recall from Example 7.26 that the probability density function of a random variable X_n having the t distribution with n degrees of freedom is

$$f_{X_n}(x) = \frac{\Gamma((n+1)/2)}{\sqrt{\pi n}\,\Gamma(n/2)\,(1+x^2/n)^{(n+1)/2}} \qquad -\infty < x < \infty.$$

This is a case of a random variable with an intractable cumulative distribution function, which forces us to work with its probability density function in the following fashion:

$$
\begin{aligned}
\lim_{n\to\infty} F_{X_n}(x) &= \lim_{n\to\infty} \int_{-\infty}^{x} f_{X_n}(w)\,dw \\
&= \lim_{n\to\infty} \int_{-\infty}^{x} \frac{\Gamma((n+1)/2)}{\sqrt{\pi n}\,\Gamma(n/2)} \cdot \frac{1}{(1+w^2/n)^{(n+1)/2}}\,dw \\
&= \int_{-\infty}^{x} \lim_{n\to\infty} \frac{\Gamma((n+1)/2)}{\sqrt{\pi n}\,\Gamma(n/2)} \cdot \frac{1}{(1+w^2/n)^{(n+1)/2}}\,dw \\
&= \int_{-\infty}^{x} \lim_{n\to\infty} \left[\frac{\Gamma((n+1)/2)}{\sqrt{n/2}\,\Gamma(n/2)}\right] \cdot \lim_{n\to\infty}\left[\frac{1}{(1+w^2/n)^{1/2}}\right] \cdot \\
&\qquad \lim_{n\to\infty}\left[\frac{1}{\sqrt{2\pi}} \cdot \frac{1}{(1+w^2/n)^{n/2}}\right] dw \\
&= \int_{-\infty}^{x} 1\cdot 1\cdot \frac{1}{\sqrt{2\pi}} \cdot e^{-w^2/2}\,dw \qquad -\infty < x < \infty.
\end{aligned}
$$

The integrand is recognized as the probability density function of a standard normal random variable. Thus, the limiting distribution of a t random variable with n degrees of freedom in the limit as $n \to \infty$ is a standard normal distribution, that is $X_n \xrightarrow{D} N(0, 1)$. The probability density function of the $t(n)$ distribution is illustrated in Figure 8.4 for $n = 1$ (the Cauchy distribution), $n = 3$, and as $n \to \infty$ (the standard normal distribution). The effect of a small number of degrees of freedom for the t distribution is to have heavier tails and a less area in the body of the distribution relative to the standard normal distribution. As shown in Example 3.45, these heavier tails for the Cauchy distribution (that is, the $t(1)$ distribution) result in undefined moments.

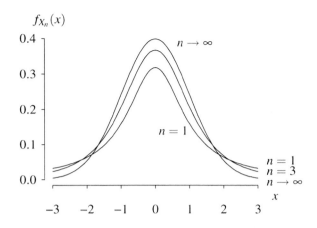

Figure 8.4: Probability density function $f_{X_n}(x)$ for a $t(n)$ random variable on $-\infty < x < \infty$.

There is a second way to show that a sequence of random variables converges in distribution. Since both the cumulative distribution function and the moment generating function completely define the distribution of a random variable, we can also use limiting moment generating functions to show that a sequence of random variable converges in distribution. The examples so far have required us to determine $F_{X_n}(x)$ explicitly for all n, which is oftentimes what we want to avoid. Cases arise occasionally when a limiting moment generating function $M_{X_n}(t)$ is an easier vehicle to arrive at a limiting distribution.

Theorem 8.1 Let $X_1, X_2, \ldots$ be a sequence of random variables. Let X_n have moment generating function $M_{X_n}(t)$ that exists on $-h < t < h$ for all n. If $M_X(t)$ is a moment generating function defined on $|t| \leq h' < h$ and if

$$\lim_{n \to \infty} M_{X_n}(t) = M_X(t)$$

for all $|t| < h'$, then X_n is said to *converge in distribution* to a random variable X with moment generating function $M_X(t)$.

Proving this theorem about convergence in distribution is beyond the scope of this text. So there are now two ways to show convergence in distribution. First, one can show that the sequence of cumulative distribution functions converges, that is $F_{X_n}(x) \to F_X(x)$; second, one can show that the sequence of moment generating functions converges, that is $M_{X_n}(t) \to M_X(t)$. The second approach proves to be particularly useful for sums of mutually independent random variables because

$$M_{X_1 + X_2 + \cdots + X_n}(t) = \prod_{i=1}^{n} M_{X_i}(t)$$

and then the limit can be taken as $n \to \infty$.

An important result from calculus that can be helpful when working with limiting moment generating functions is

$$\lim_{n \to \infty} \left[1 + \frac{a}{n} + \frac{\phi(n)}{n} \right]^{bn} = \lim_{n \to \infty} \left[1 + \frac{a}{n} \right]^{bn} = e^{ab},$$

where ϕ is a function of n, a and b do not depend on n, $\lim_{n \to \infty} \phi(n) = 0$, and $\lim_{n \to \infty} n\big(\phi(n+1) - \phi(n)\big) = 0$. For example,

$$\lim_{n \to \infty} \left[1 - \frac{t^3}{n} + \frac{7t}{n^2} \right]^{-4n} = \lim_{n \to \infty} \left[1 + \frac{-t^3}{n} + \frac{7t/n}{n} \right]^{-4n} = e^{4t^3}.$$

Example 8.7 Let $X_n \sim$ binomial(n, p), where the parameters n and p are chosen to vary so that $\mu = np$ is a constant value for every n, that is $p = \mu/n$, with μ fixed. Find the limiting distribution of X_n.

The moment generating function of X_n is

$$M_{X_n}(t) = E\left[e^{tX_n}\right] = \left[(1-p) + pe^t\right]^n = \left[\left(1 - \frac{\mu}{n}\right) + \frac{\mu}{n}e^t\right]^n = \left[1 + \frac{\mu(e^t - 1)}{n}\right]^n$$

for $-\infty < t < \infty$. Applying the previous result concerning limits, the limiting moment generating function is

$$\lim_{n \to \infty} M_{X_n}(t) = e^{\mu(e^t - 1)}$$

for $-\infty < t < \infty$. The limiting moment generating function of X_n is recognized as that of a Poisson random variable with mean μ from Section 4.5, that is $X_n \xrightarrow{D}$ Poisson(μ). This result forms the basis for the "Poisson approximation to the binomial," which is appropriate when n is large and p is small. When p happens to be large, simply switch the roles of success and failure to make p small (if at first you don't succeed, redefine success). This result was shown using probability mass functions in Section 4.5.

The next example provides a numerical illustration of how well the Poisson approximation to the binomial works for a large n and small p.

Example 8.8 Assume that the probability that a woman will give birth to triplets (without fertility enhancement) is $p = 1/8000$. If a large city has 9000 deliveries in a particular year, find the probability that two or fewer of these deliveries will be triplets.

This problem will be solved in two fashions: exactly using the binomial distribution and approximately using the Poisson approximation to the binomial distribution.

Solution 1. Assuming that each birth is an independent Bernoulli trial with triplets being a success, the probability of two or fewer triplets in the 9000 births is

$$\sum_{x=0}^{2} \binom{9000}{x} \left(\frac{1}{8000}\right)^x \left(\frac{7999}{8000}\right)^{9000-x} \cong 0.89534$$

which can be calculated with the R statement

```
pbinom(2, 9000, 1 / 8000)
```

This solution is difficult to compute numerically because of the large factorials associated with the binomial coefficients.

Solution 2. Since n is large and p is small, the Poisson approximation to the binomial can be used with

$$\mu = np = (9000)(1/8000) = 9/8.$$

The probability of two or fewer triplets in the 9000 births is approximately

$$\sum_{x=0}^{2} \frac{(9/8)^x e^{-9/8}}{x!} = \left(1 + \frac{9}{8} + \frac{81}{128}\right) e^{-9/8} = \frac{353}{128} e^{-9/8} \cong 0.89533$$

which can be calculated with the R statement

```
ppois(2, 9000 / 8000)
```

The probabilities calculated by the two methods differ only in the fifth digit after the decimal point, so they are equivalent for all practical purposes.

The next three examples are a sequence that analyzes the Erlang distribution in the limit as its shape parameter increases using limiting moment generating functions.

Example 8.9 Let $X_n \sim \text{Erlang}(\lambda, n)$, where λ is a constant. Find the limiting distribution of X_n.

Recall from Section 5.3 that the moment generating function of X_n is

$$M_{X_n}(t) = \left(\frac{\lambda}{\lambda - t} \right)^n \qquad t < \lambda.$$

Taking the limit of both sides of this equation as $n \to \infty$ yields the limiting moment generating function

$$\lim_{n \to \infty} M_{X_n}(t) = \lim_{n \to \infty} \left(\frac{\lambda}{\lambda - t} \right)^n \qquad t < \lambda,$$

which approaches 0, 1, or ∞, depending on the value of t. That is, X_n has no limiting distribution. The seeds of this non-convergence are shown in Figure 8.5, which shows the cumulative distribution function of a $\text{Erlang}(\lambda, n)$ random variable for $\lambda = 1$ and $n = 1, 2, 3, 4, 5$. As n increases, the body of the probability distribution moves to the right along the x-axis and does not converge in distribution. The geometry confirms the conclusion that the $\text{Erlang}(\lambda, n)$ distribution does not converge in distribution.

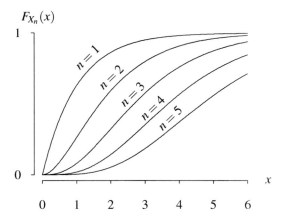

Figure 8.5: Cumulative distribution function $F_{X_n}(x)$ of an $\text{Erlang}(\lambda, n)$ random variable.

The lack of convergence in distribution in the previous example can be overcome in the following fashion. The population mean of an $\text{Erlang}(\lambda, n)$ random variable is

$$E[X_n] = \frac{n}{\lambda}$$

for $n = 1, 2, \ldots$, so dividing X_n by n will result in a new random variable that has the same mean, namely $1/\lambda$, for any value of n. This scheme will prevent the probability distribution from marching off to the right as it did in the previous example. The details are worked out in the next example.

Example 8.10 Let $Y_n \sim \text{Erlang}(\lambda, n)$. Find the limiting distribution of $X_n = Y_n/n$.

Recall from Table 7.1 that the moment generating function of Y_n is

$$M_{Y_n}(t) = \left(\frac{\lambda}{\lambda - t}\right)^n \qquad t < \lambda.$$

So the moment generating function of $X_n = Y_n/n$ is

$$
\begin{aligned}
M_{X_n}(t) &= E\left[e^{tX_n}\right] \\
&= E\left[e^{tY_n/n}\right] \\
&= M_{Y_n}(t/n) \\
&= \left(\frac{\lambda}{\lambda - t/n}\right)^n \\
&= \left(\frac{n\lambda}{n\lambda - t}\right)^n \qquad t < n\lambda.
\end{aligned}
$$

The limiting moment generating function of X_n is

$$
\begin{aligned}
\lim_{n\to\infty} M_{X_n}(t) &= \lim_{n\to\infty} \left(\frac{n\lambda}{n\lambda - t}\right)^n \\
&= \lim_{n\to\infty} \left(\frac{n\lambda - t}{n\lambda}\right)^{-n} \\
&= \lim_{n\to\infty} \left(1 + \frac{-t/\lambda}{n}\right)^{-n} \\
&= e^{t/\lambda} \qquad -\infty < t < \infty.
\end{aligned}
$$

This limiting moment generating function is recognized as that of a degenerate distribution at $1/\lambda$. Hence $X_n \xrightarrow{D} 1/\lambda$. The convergence in distribution toward a degenerate distribution at $1/\lambda$ can be seen for $\lambda = 1$ and $n = 1$, $n = 3$, $n = 20$, and $n = 500$ in the cumulative distribution functions plotted in Figure 8.6.

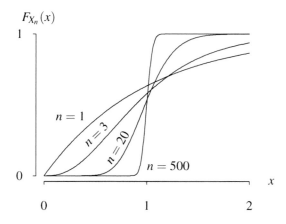

Figure 8.6: Erlang$(n\lambda, n)$ cumulative distribution functions with $\lambda = 1$.

Our strategy of keeping the population mean of X_n constant was effective. The degenerate distribution for the Erlang distribution is certainly an improvement over no limiting distribution at all. But we can do better.

The reason that the limiting distribution degenerated in the previous example has to do with the population variance of X_n. While we kept the population mean of X_n constant at $E[X_n] = E[Y_n/n] = 1/\lambda$ for all n, we let the population variance of X_n,

$$V[X_n] = V\left[\frac{Y_n}{n}\right] = \frac{1}{n^2}V[Y_n] = \frac{1}{n^2} \cdot \frac{n}{\lambda^2} = \frac{1}{n\lambda^2},$$

go to 0 in the limit as $n \to \infty$. The key to correctly finding the limiting distribution of an Erlang random variable lies in keeping both the population mean and the population variance constant as n increases. This can be done by returning to the concept of a *standardized* random variable, which always has population mean 0 and population variance 1. This third and final example concerning the Erlang distribution standardizes Y_n by subtracting its population mean and dividing by its population standard deviation.

Example 8.11 Let $Y_n \sim \text{Erlang}(\lambda, n)$. Find the limiting distribution of

$$X_n = \frac{Y_n - n/\lambda}{\sqrt{n/\lambda^2}}.$$

The random variable Y_n has been transformed by subtracting its population mean, n/λ, and dividing by its population standard deviation, $\sqrt{n/\lambda^2}$, so that X_n has population mean 0 and population variance 1 for all values of n. As in the previous examples, the moment generating function of Y_n is

$$M_{Y_n}(t) = \left(\frac{\lambda}{\lambda - t}\right)^n \qquad t < \lambda.$$

The formula for X_n can be written more compactly as

$$X_n = \frac{Y_n - n/\lambda}{\sqrt{n/\lambda^2}} = \lambda Y_n/\sqrt{n} - \sqrt{n}.$$

So the moment generating function of X_n is

$$
\begin{aligned}
M_{X_n}(t) &= E\left[e^{tX_n}\right] \\
&= E\left[e^{t(\lambda Y_n/\sqrt{n} - \sqrt{n})}\right] \\
&= e^{-t\sqrt{n}}E\left[e^{t\lambda Y_n/\sqrt{n}}\right] \\
&= e^{-t\sqrt{n}}M_{Y_n}\left(t\lambda/\sqrt{n}\right) \\
&= e^{-t\sqrt{n}}\left(\frac{\lambda}{\lambda - t\lambda/\sqrt{n}}\right)^n \\
&= e^{-t\sqrt{n}}\left(\frac{1}{1 - t/\sqrt{n}}\right)^n \\
&= e^{-t\sqrt{n}}\left(1 - t/\sqrt{n}\right)^{-n} \qquad t < \sqrt{n}.
\end{aligned}
$$

The limiting moment generating function of X_n is

$$\lim_{n \to \infty} M_{X_n}(t) = e^{t^2/2}$$

for $-\infty < t < \infty$, which is recognized as the moment generating function of a standard normal distribution. The moment generating function of X_n and the limiting moment generating function can be computed with the Maple statements

```
assume(n, posint);
m := int(y ^ (n - 1) * exp(-y + (y - n) * t / sqrt(n)) / (n - 1)!,
        y = 0 .. infinity);
limit(m, n = infinity);
```

Finally, on the third attempt, we can conclude that $X_n \xrightarrow{D} N(0, 1)$. So an Erlang($\lambda$, n) random variable converges to a normal distribution in the limit as $n \to \infty$. Figure 8.7 illustrates for $\lambda = 1$ that even for n as small as $n = 10$, it is apparent that the Erlang probability density function is indeed approaching a bell-shaped curve.

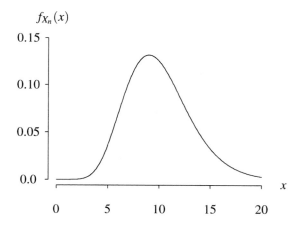

Figure 8.7: Erlang(λ, n) probability density function with $\lambda = 1$ and $n = 10$.

So we have concluded that an Erlang(λ, n) random variable converges in distribution to a normal random variable as $n \to \infty$. Do all probability distributions converge to a normal distributions? Certainly not, with the $U(0, n)$ distribution providing an easy counterexample. What is it about the Erlang distribution that makes it converge to a normal distribution? The answer lies in a relatively obscure example from the previous chapter. In Example 7.41, the moment generating function technique was used to show that the sum of n mutually independent exponential(λ) random variables had the Erlang(λ, n) distribution. This result is the key. Any distribution that can be expressed as the sum of n mutually independent and identically distributed random variables with finite population variances converges in distribution to the normal distribution in the limit as $n \to \infty$. For example, the binomial(n, p) and $\chi^2(n)$ distributions both converge to the normal distribution as $n \to \infty$ because they can be written as sums. The reason for this convergence is one of the most useful results in statistics which is known as the *central limit theorem*, which is introduced in the next section.

This concludes the discussion of convergence in distribution, which can be shown by investigating the cumulative distribution functions or the moment generating functions of a sequence of random variables of interest. The concepts of convergence in probability and convergence in distribution are related. Convergence in probability implies convergence in distribution. Convergence in distribution implies convergence in probability only when convergence is to a degenerate distribution.

8.3   Central Limit Theorem

The sample mean is an important statistic that statisticians frequently use to summarize a data set that consists of the observed values of the random variables $X_1, X_2, \dots, X_n$. The weak law of large numbers, which was proved in Example 8.2, showed that the sample mean $\bar{X}_n$ converges to the population mean μ in the limit as $n \to \infty$. The weak law of large numbers, however, leaves two important questions unanswered:

1. How rapidly does $\bar{X}_n$ tend to μ?

2. How does the shape of the probability density function of $\bar{X}_n$ change as n increases?

We address these two questions in this section. The second question is answered by the *central limit theorem*, which is a cornerstone result in statistical theory.

The first question addresses how rapidly $\bar{X}_n$ tends to μ. The weak law of large numbers states that

$$\bar{X}_n \xrightarrow{P} \mu,$$

or, equivalently,

$$\bar{X}_n - \mu \xrightarrow{P} 0.$$

In order to avoid having $\bar{X}_n - \mu$ degenerate to 0, we can multiply it by a "magnification factor" that keeps the variance of the magnification factor times $X_n - \mu$ constant as n increases. After some experimentation, we find that the factor $\sqrt{n}$ keeps the variance of the random variable

$$\sqrt{n}(\bar{X}_n - \mu)$$

constant for all values of n. This random variable has population mean 0 and population variance

$$V\left[\sqrt{n}(\bar{X}_n - \mu)\right] = n V\left[\bar{X}_n - \mu\right] = n V\left[\bar{X}_n\right] = n \cdot \frac{\sigma^2}{n} = \sigma^2.$$

This choice of $\sqrt{n}$ as a magnification factor implies that $1/\sqrt{n}$ is a measure of how rapidly the sample mean tends toward the population mean. This result highlights one of the limitations of Monte Carlo simulation. Getting one more digit of accuracy in Monte Carlo simulation requires 100 times as many replications because $1/\sqrt{100} = 1/10$. The first few digits of accuracy in a Monte Carlo simulation estimate are obtained quickly; obtaining subsequent digits becomes increasingly expensive in terms of CPU cycles.

We now shift to the second question: how does the shape of the probability density function of $\bar{X}_n$ change as n increases? In the case of sampling drawn from a normally distributed population, this problem can be answered with precision for all values of n. If $X_1, X_2, \ldots, X_n$ are mutually independent $N(\mu, \sigma^2)$ random variables, then $\bar{X}_n \sim N(\mu, \sigma^2/n)$ for any value of n using a minor extension of the result established in Example 7.44. Therefore,

$$\frac{\bar{X}_n - \mu}{\sigma/\sqrt{n}} \sim N(0, 1)$$

for all values of n. What if there is no information about the probability distribution of the population? If only μ and σ are known, then a remarkable and important fact concerning the distribution of the sample mean is that

$$\frac{\bar{X}_n - \mu}{\sigma/\sqrt{n}} \xrightarrow{D} N(0, 1).$$

This remarkable fact is true for observations drawn from *any* distribution with finite population mean μ and finite, positive population variance σ^2. Known as the *central limit theorem*, which is often abbreviated CLT, it allows for approximate statistical inference for large sample sizes when the analyst is unsure of the population distribution.

Theorem 8.2 (Central limit theorem) Let $X_1, X_2, \ldots, X_n$ be mutually independent and identically distributed random variables from a population with finite population mean μ and finite, positive population variance σ^2. Then

$$\frac{\bar{X}_n - \mu}{\sigma/\sqrt{n}} \xrightarrow{D} N(0, 1),$$

where $\bar{X}_n$ is the sample mean.

Proof The central limit theorem will be proved by showing that the limiting moment generating function of the random variable

$$\frac{\bar{X}_n - \mu}{\sigma/\sqrt{n}}$$

approaches $e^{t^2/2}$, which is the moment generating function of a standard normal random variable, as $n \to \infty$. Assume that each X_i has moment generating function $M_{X_i}(t) = M_X(t)$, for $i = 1, 2, \ldots, n$. Without loss of generality, assume further that $\mu_X = 0$ and $\sigma_X = 1$. The reason that this simplifying assumption is valid is that this proof could be applied to the standardized random variable $(X_i - \mu)/\sigma$ which we know from Example 3.54 has population mean and variance

$$E\left[\frac{X_i - \mu}{\sigma}\right] = 0 \qquad \text{and} \qquad V\left[\frac{X_i - \mu}{\sigma}\right] = 1$$

for $i = 1, 2, \ldots, n$. The assumption that $\mu_X = 0$ and $\sigma_X = 1$ reduces the problem to proving that the moment generating function of

$$\frac{\bar{X}_n - \mu}{\sigma/\sqrt{n}} = \frac{\bar{X}_n}{1/\sqrt{n}} = \sum_{i=1}^{n} X_i/\sqrt{n}$$

approaches $e^{t^2/2}$ as $n \to \infty$. Using Theorem 7.10, the moment generating function of $(X_1 + X_2 + \cdots + X_n)/\sqrt{n}$ is

$$\begin{aligned}
M_{(X_1+X_2+\cdots+X_n)/\sqrt{n}}(t) &= \prod_{i=1}^{n} M_{X_i/\sqrt{n}}(t) \\
&= \prod_{i=1}^{n} M_{X_i}(t/\sqrt{n}) \\
&= \left(M_X(t/\sqrt{n})\right)^n.
\end{aligned}$$

Define the log moment generating function as $L_X(t) = \ln M_X(t)$. Since $M_X(0) = 1$, $L_X(0) = 0$. Using the chain rule, the derivative of the log moment generating function evaluated at $t = 0$ is

$$L_X'(0) = \frac{M_X'(0)}{M_X(0)} = \frac{\mu}{1} = 0.$$

Using the quotient rule, the second derivative of the log moment generating function evaluated at $t = 0$ is

$$L_X''(0) = \frac{M_X(0)M_X''(0) - M_X'(0)^2}{M_X(0)^2} = \frac{1 \cdot E\left[X^2\right] - \mu^2}{1^2} = E\left[X^2\right] = 1.$$

Proving that

$$\lim_{n \to \infty} \left(M_X(t/\sqrt{n})\right)^n = e^{t^2/2}$$

is equivalent to proving that

$$\lim_{n \to \infty} n L_X(t/\sqrt{n}) = t^2/2$$

for $-\infty < t < \infty$. Using L'Hospital's rule twice on the two indeterminate forms,

$$
\begin{aligned}
\lim_{n \to \infty} \frac{L_X(t/\sqrt{n})}{1/n} &= \lim_{n \to \infty} \frac{L_X'(t/\sqrt{n})n^{-3/2}t}{2/n^2} \\
&= \lim_{n \to \infty} \frac{L_X'(t/\sqrt{n})t}{2/\sqrt{n}} \\
&= \lim_{n \to \infty} \frac{L_X''(t/\sqrt{n})n^{-3/2}t^2}{2/n^{3/2}} \\
&= \lim_{n \to \infty} \frac{L_X''(t/\sqrt{n})t^2}{2} \\
&= t^2/2
\end{aligned}
$$

for $-\infty < t < \infty$. So by Theorem 8.1,

$$
\frac{\bar{X}_n - \mu}{\sigma/\sqrt{n}} \xrightarrow{D} N(0, 1),
$$

which proves the central limit theorem for any random variable with a moment generating function that exists. $\square$

Almost as remarkable as the central limit theorem itself is the number of different ways that statisticians have devised to express it. For example, the central limit theorem can be written as

$$
\sum_{i=1}^{n} X_i \xrightarrow{D} N\left(n\mu, n\sigma^2\right).
$$

This version of the central limit theorem emphasizes that sums of random variables (regardless of the population or base distribution) tend to be normally distributed in the limit as $n \to \infty$. The two examples that follow highlight this version of the central limit theorem. A second way of expressing the central limit theorem is

$$
\lim_{n \to \infty} P\left(\frac{X_1 + X_2 + \cdots + X_n - n\mu}{\sqrt{n}\sigma} < c\right) = \frac{1}{\sqrt{2\pi}} \int_{-\infty}^{c} e^{-\frac{1}{2}x^2} \, dx
$$

for any real-valued constant c. Obviously, there are dozens of other ways to permute the expression associated with the central limit theorem. Although we do not use the notation here, some statisticians use the notation

$$
\frac{\bar{X}_n - \mu}{\sigma/\sqrt{n}} \overset{a}{\sim} N(0, 1),
$$

to express the central limit theorem. The a above the $\sim$ symbol stands for "asymptotically."

Example 8.12 Let $X_1, X_2, \ldots, X_{12}$ be mutually independent $U(0, 1)$ random variables. Find

$$
P\left(5 < \sum_{i=1}^{12} X_i < 7\right)
$$

exactly using APPL and approximately using the central limit theorem.

The exact solution via APPL is determined using the code

```
X := UniformRV(0, 1);
Y := ConvolutionIID(X, 12);
CDF(Y, 7) - CDF(Y, 5);
```

which yields

$$P\left(5 < \sum_{i=1}^{12} X_i < 7\right) = \frac{27,085,381}{39,916,800} \cong 0.6785.$$

The central limit theorem can be used here even though the sample size $n = 12$ is moderate. Using results from Section 5.1, the population mean and population variance of each of the $n = 12\ U(0, 1)$ observations are

$$\mu = \frac{1}{2} \qquad \sigma^2 = \frac{1}{12}.$$

By the central limit theorem, $\sum_{i=1}^{12} X_i$ is approximately normally distributed with population mean $12 \cdot \frac{1}{2} = 6$ and population variance $12 \cdot \frac{1}{12} = 1$. So the desired probability is approximately

$$
\begin{aligned}
P\left(5 < \sum_{i=1}^{12} X_i < 7\right) &= P\left(\frac{5-6}{\sqrt{1}} < \frac{\sum_{i=1}^{12} X_i - 6}{\sqrt{1}} < \frac{7-6}{\sqrt{1}}\right) \\
&\cong P(-1 < Z < 1) \\
&\cong 0.6827.
\end{aligned}
$$

which is calculated with the R statement

```
pnorm(1) - pnorm(-1)
```

The central limit theorem (barely) provides just two digits of accuracy for computing this particular probability.

The choice of $n = 12$ in the previous example provides an interesting historical sidelight. Prior to the 1958 advent of the Box–Muller algorithm introduced in Example 7.28, there was no exact algorithm for generating normally distributed random variates. An approximate algorithm that was commonly used relied on the central limit theorem. If $X_1, X_2, \ldots, X_{12}$ are mutually independent $U(0, 1)$ random variables, then the random variable

$$X_1 + X_2 + \cdots + X_{12} - 6$$

has population mean 0, population standard deviation 1, and is approximately normally distributed by the central limit theorem. It also happens to be symmetric (just like the standard normal distribution) and therefore has population skewness 0; this results in a perfect match between the first three moments. The APPL statements

```
X := UniformRV(0, 1);
Y := ConvolutionIID(X, 12);
g := [[x -> x - 6], [0, 12]];
Z := Transform(Y, g);
Kurtosis(Z);
```

calculate the population kurtosis of $X_1 + X_2 + \cdots + X_{12} - 6$ as $29/10 = 2.9$, which is very close to 3, the kurtosis of a standard normal random variable. So summing 12 mutually independent $U(0, 1)$ random variates and subtracting 6 results in a probability distribution that is quite close to the standard normal distribution.

APPL's abilities work for small n and mathematically tractable probability distributions, but, as the next example shows, the central limit must be relied on when n is large.

Example 8.13 Let $X_1, X_2, \ldots, X_{100}$ be mutually independent $U(0, 1)$ random variables. Find

$$P\left(45 < \sum_{i=1}^{100} X_i < 55\right)$$

using the central limit theorem.

This is a similar setting to the previous question, but this time APPL is unable to compute the exact solution because it requires too many CPU cycles. Furthermore, since $n = 100$ is much larger than $n = 12$, we expect the central limit theorem to provide more accuracy. Again using results from Section 5.1, the population mean and population variance of each of the $n = 100$ observations are

$$\mu = \frac{1}{2} \qquad \sigma^2 = \frac{1}{12}.$$

This means that by the central limit theorem that $\sum_{i=1}^{100} X_i$ is approximately normally distributed with population mean $100 \cdot \frac{1}{2} = 50$ and population variance $100 \cdot \frac{1}{12} = \frac{25}{3}$. So the desired probability is approximately

$$
\begin{aligned}
P\left(45 < \sum_{i=1}^{100} X_i < 55\right) &= P\left(\frac{45 - 50}{\sqrt{25/3}} < \frac{\sum_{i=1}^{100} X_i - 50}{\sqrt{25/3}} < \frac{55 - 50}{\sqrt{25/3}}\right) \\
&\cong P\left(-\sqrt{3} < Z < \sqrt{3}\right) \\
&\cong 0.9167.
\end{aligned}
$$

which is calculated with the R statement

```
pnorm(sqrt(3)) - pnorm(-sqrt(3))
```

The previous two examples have applied the central limit theorem to a continuous random variable, namely the $U(0, 1)$ distribution. The central limit theorem also applies to discrete random variables, although some extra care must be taken to apply a "continuity correction" to account for the fact that a continuous distribution is being used to approximate a discrete distribution. The continuity correction is illustrated in the next example.

Example 8.14 A fair die is rolled 16 times. Find the probability that the sum of the spots showing is between 40 and 60 inclusive.

Let $X_1, X_2, \ldots, X_{16}$ be the outcomes of the 16 rolls. The APPL code to calculate this probability is given below. The 16 rolls of the fair die is nearing the outer limits of what APPL can handle in terms of exact calculations.

```
X := UniformDiscreteRV(1, 6);
Y := ConvolutionIID(X, 16);
CDF(Y, 60) - CDF(Y, 39);
```

This calculation returns

$$P\left(40 \le \sum_{i=1}^{16} X_i \le 60\right) = \frac{230,715,131,047}{313,456,656,384} \cong 0.7360.$$

The central limit theorem can be applied in this case with $n = 16$ observations from a discrete uniform distribution with a lower limit 1 and an upper limit 6. Recall from

Section 4.7 that the population mean and variance of a discrete uniform distribution with minimum support value a and maximum support value b are

$$\mu = \frac{a+b}{2} = \frac{1+6}{2} = \frac{7}{2} \qquad \text{and} \qquad \sigma^2 = \frac{(b-a+1)^2 - 1}{12} = \frac{35}{12}.$$

So the population mean and variance of the total number of spots showing, $\sum_{i=1}^{16} X_i$, is $16 \cdot \frac{7}{2} = 56$ and $16 \cdot \frac{35}{12} = \frac{140}{3}$. A naive application of the central limit theorem to calculate the desired probability begins in the following fashion:

$$P\left(40 \le \sum_{i=1}^{16} X_i \le 60\right) = P\left(\frac{40-56}{\sqrt{140/3}} \le \frac{\sum_{i=1}^{16} X_i - 56}{\sqrt{140/3}} \le \frac{60-56}{\sqrt{140/3}}\right)$$

$$\cong P\left(\frac{-16}{\sqrt{140/3}} \le Z \le \frac{4}{\sqrt{140/3}}\right)$$

$$\cong 0.7113.$$

This quantity is computed with the R statement

```
pnorm(4 / sqrt(140 / 3)) - pnorm(-16 / sqrt(140 / 3))
```

So why the abysmal one-digit accuracy from the central limit theorem in this case? The problem lies in the integration limits under the normal curve. Figure 8.8 is a graph of the probability mass function of $X = X_1 + X_2 + \cdots + X_{16}$, which has support from $x = 16$ (rolling 16 ones) to $x = 96$ (rolling 16 sixes). Its bell shape indicates that the central limit theorem should provide a reasonably good approximation. The problem that limited us to just one digit of accuracy is that a continuous distribution is being used to approximate a discrete distribution. The appropriate integration limits using what is known as the "continuity correction" is to include that appropriate area under the approximating normal probability density function associated with the two endpoints of interest: $x = 40$ and $x = 60$. The left-hand endpoint of the range, for example, $x = 40$, is best approximated by the area under the approximating normal curve between $x = 39.5$ and $x = 40.5$. If the previous calculation is repeated with $x = 40$ replaced by $x = 39.5$, and $x = 60$ replaced by $x = 60.5$, the new calculation approximates the probability by

$$P\left(\frac{-16.5}{\sqrt{140/3}} \le Z \le \frac{4.5}{\sqrt{140/3}}\right) \cong 0.7371.$$

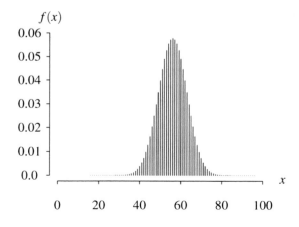

Figure 8.8: Probability mass function of the total number of spots showing.

This quantity is computed with the R statement

```
pnorm(4.5 / sqrt(140 / 3)) - pnorm(-16.5 / sqrt(140 / 3))
```

The use of the continuity correction results in two digits (and almost three digits) of accuracy—a significant improvement over the naive application of the central limit theorem.

Statisticians are often interested in determining the appropriate sample size n in the presence of limited information about a population distribution. The central limit theorem can be used to determine how many observations should be collected.

Example 8.15 Cassie is interested in determining the average gas mileage for a fleet of a particular model of car. Previous data collection of gas mileage gives her confidence that the standard deviation of the gas mileages that she collects will have a standard deviation of 3.3 miles per gallon. How many cars should she sample in order to be at least 95% certain that the difference between the population mean and sample mean is less than 1 mile per gallon?

The goal here is to find a sample size n such that

$$P(|\bar{X} - \mu| < 1) = 0.95$$

where $X_1, X_2, \ldots, X_n$ are the values collected from a population with a population mean miles per gallon μ and population standard deviation $\sigma = 3.3$. Using the central limit theorem, the probability can be approximated by

$$
\begin{aligned}
P(|\bar{X} - \mu| < 1) &= P(-1 < \bar{X} - \mu < 1) \\
&= P\left(-\frac{1}{3.3/\sqrt{n}} < \frac{\bar{X} - \mu}{3.3/\sqrt{n}} < \frac{1}{3.3/\sqrt{n}}\right) \\
&\cong P\left(-\frac{\sqrt{n}}{3.3} < Z < \frac{\sqrt{n}}{3.3}\right).
\end{aligned}
$$

Since this probability must be at least 0.95, and the 97.5th percentile of the standard normal distribution is approximately 1.96, the appropriate sample size n satisfies

$$\frac{\sqrt{n}}{3.3} = 1.96.$$

Solving for n yields $n = 41.8$. In order to achieve the prescribed precision, Cassie should collect the gas mileage on a sample of at least $n = 42$ cars.

The previous four examples have considered sample sizes of $n = 12$, $n = 100$, $n = 16$, and $n = 42$. In all cases, the central limit theorem has been invoked, and it is assumed that the accuracy is improved for larger n because the central limit theorem is an asymptotic result. Is there a general way to determine what n value is appropriate to invoke the central limit theorem? Although some elementary statistics textbooks state that a sample of at least $n = 30$ is necessary, the answer to the question is a bit more nuanced.

Figure 8.9 contains a 3×3 display of graphs of probability density functions of the *sample means* associated with the observations $X_1, X_2, \ldots, X_n$ drawn from three different continuous populations. Axis labels have been suppressed; the horizontal axis is x and the vertical axis is $f(x)$ for all graphs. The horizontal axes all range from 0 to 2, and the vertical axes range from 0 to the highest value that the probability density function achieves on $[0, 2]$. The three rows correspond to the probability distributions from which the observations were drawn: normal, uniform, and exponential. More specifically, the sampling distributions are $N(1, 1/4)$, $U(0, 2)$, and exponential(1). The parameters

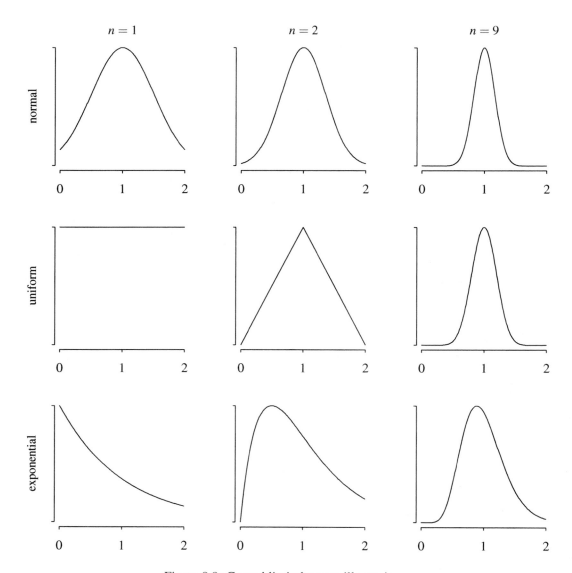

Figure 8.9: Central limit theorem illustrations.

for these distributions were chosen so that each has population mean 1. The population variances for the three population distributions are $1/4$, $1/3$, and 1, respectively. The three columns correspond to sample sizes: $n = 1$, $n = 2$, and $n = 9$. The first column of probability density functions corresponds to the population distribution.

The first row of probability density functions shows that the asymptotic result from the central limit theorem is satisfied exactly for all n when sampling from a normal population. The effect of increasing n on the sample mean is to simply reduce the variance because $\bar{X}_n \sim N(\mu, \sigma^2/n)$.

The second row of probability density functions shows that sampling from the symmetric, but not bell-shaped uniform distribution also has reduced variance with increasing n, and seems to be bell shaped by $n = 9$. The probability distribution of $\bar{X}_n$ when $n = 2$ is a triangular(0, 1, 2) distribution. The APPL code to plot the probability density function of the sample mean when $n = 9$ follows.

```
X := UniformRV(0, 2);
Y := ConvolutionIID(X, 9);
```

```
g := [[x -> x / 9], [0, 18]];
Xbar := Transform(Y, g);
PlotDist(Xbar, 0, 2);
```

The third row of probability density functions shows that sampling from the non-symmetric, and non-bell-shaped exponential distribution also has reduced variance with increasing n, but seems to have a slower convergence to the normal distribution than the uniform distribution. Using the moment generating function technique and the transformation technique in series, it can be determined that for an exponential(1) population, $\bar{X}_n \sim$ Erlang(n, n). Even for $n = 9$, the residual effect of the non-symmetry of the sampling distribution is still apparent.

The conclusions that can be drawn from Figure 8.9 are as follows.

- The distribution of $\bar{X}_n$ is exactly normal for all n when the population distribution is normally distributed.

- When applying the central limit theorem, a symmetric population is preferred to a non-symmetric population.

- When applying the central limit theorem, a nearly bell-shaped population is preferred to a non-bell-shaped population.

8.4   Exercises

8.1   There are $3n$ balls placed at random into n identical urns, that is, for each ball an urn is selected at random and a ball is placed in that urn. Assume that n is a positive integer greater than 1. Let the random variable X_n be the number of balls in the first urn.

(a) Find the probability mass function of X_n for a finite integer n.

(b) Find the limiting distribution of X_n.

8.2   Let $X_1, X_2, \ldots, X_n$ be mutually independent and identically distributed $U(0, 1)$ random variables. The radius of a circle is a continuous random variable having the same distribution as $n \cdot \min\{X_1, X_2, \ldots, X_n\}$.

(a) Find the probability density function of the area of the circle for any positive integer n.

(b) Find the limiting distribution of the area of the circle.

8.3   Let $X_1, X_2, \ldots, X_n$ be a random sample from an exponential population with mean θ.

(a) Find the moment generating function of the sample mean $\bar{X}_n$.

(b) Find the limiting moment generating function of the sample mean $\bar{X}_n$. What type of distribution is this?

(c) Find the moment generating function of the standardized random variable

$$\frac{\bar{X}_n - \theta}{\theta/\sqrt{n}}.$$

(d) Find the limiting moment generating function of

$$\frac{\bar{X}_n - \theta}{\theta/\sqrt{n}}.$$

What type of distribution is this?

8.4 Let X_n be the minimum value of a random sample $Y_1, Y_2, \ldots, Y_n$ taken from a continuous population with cumulative distribution function $F_Y(y)$. Let $X_n = \min\{Y_1, Y_2, \ldots, Y_n\}$. Find the limiting distribution of $nF_Y(X_n)$.

8.5 Let X_1, X_2 and X_3 be independent and identically distributed $U(0, 1)$ random variables. Find the 90th percentile of their sum:

   (a) approximately, using the central limit theorem,

   (b) approximately, using Monte Carlo simulation,

   (c) exactly, using the cumulative distribution function technique,

   (d) exactly, using the transformation technique,

   (e) exactly, using APPL.

8.6 A computer program rounds 12 real numbers, then adds them. If the error due to rounding can be assumed to be $U(-0.5, 0.5)$, find the probability that the sum of the real numbers differs from the sum of the integers created by rounding by less than 1

   (a) approximately, using the central limit theorem,

   (b) exactly, using APPL.

8.7 Let $X_n \sim \text{Poisson}(n)$, for $n = 1, 2, \ldots$. Find the limiting distribution of a standardized Poisson random variable

$$\frac{X_n - n}{\sqrt{n}}.$$

8.8 Show that $X_n \sim N\left(0, 1/n^2\right)$ converges in distribution to a degenerate distribution.

8.9 For the random variable X_n defined in Example 8.5, plot the population mean, variance, skewness and kurtosis of X_n for $n = 1, 2, \ldots, 20$ on four separate sets of axes. Comment on the asymptotic values of these four quantities.

8.10 Consider a large batch of light bulbs whose lifetimes are known to have exponential(1) lifetimes. Shanika knows that the population distribution is exponential, but she does not know the value of the population mean. She estimates the population mean lifetime of the light bulbs by averaging n sample lifetimes from bulbs chosen at random from the batch. Find the smallest value of n that assures, with probability of at least 0.95, that the sample mean is within 0.2 of the population mean

   (a) exactly,

   (b) approximately, using the central limit theorem.

8.11 Let $X_1, X_2, \ldots, X_{10}$ be mutually independent observations drawn from a population with probability density function

$$f_X(x) = x/2 \qquad 0 < x < 2.$$

Find the probability that the sample mean associated with these ten values exceeds 1.6 by

   (a) using the central limit theorem,

   (b) using APPL,

   (c) using Monte Carlo simulation.

Index